Alkali Cation Transport Systems in Prokaryotes

Alkali Cation Transport Systems in Prokaryotes

Edited by

Evert P. Bakker, Ph.D.

Professor of Microbiology
University of Osnabrück
Osnabrück, Germany

CRC Press

Boca Raton   Ann Arbor   London   Tokyo

Library of Congress Cataloging-in-Publication Data

Alkali cation transport systems in prokaryotes / edited by Evert P.
  Bakker.
    p. cm.
    Includes bibliographical references and index.
    ISBN 0-8493-6982-7
    1. Biological transport. 2. Ion channels. 3. Cations-
-Physiological transport. 4. Prokaryotes--Physiology. I. Bakker,
Evert P.
  QH509.A44   1992
  589.9'08'75--dc20   92-22124
       74759   CIP

© 1993 by CRC Press, Inc.

International Standard Book Number 0-8493-6982-7

Library of Congress Card Number 92-22124

Printed in the United States of America 1 2 3 4 5 6 7 8 9 0

Printed on acid-free paper

PREFACE

During the past 20 years much membrane research in prokaryotes has focused on the description of the proton circulation across the cell membrane and its role in energy coupling, and on the characterization of the proteins involved in this process. The transmembrane transport of the alkali-cations Na^+ and K^+ has initially only been studied by a few groups. This situation has gradually changed, because it was recognized that prokaryotes are much more diverse than people had thought. In particular, Na^+ ions can replace protons in all aspects of energy coupling. The intention of this book is to bring all of the new data and concepts together at a moment that (1) from a cell physiological point of view the alkali cation transport processes have been described satisfactorily, (2) many of the proteins that mediate these processes have been identified, (3) the primary sequences of several of these proteins are known and hypotheses can be built about their mode of action at the molecular level, and (4) none of these proteins have yet been crystallized.

The book is divided into three sections: I. Na^+ transport systems, II. K^+ transport systems, and III. NH_4^+ transport systems. The last section is included, because several K^+ uptake systems may also translocate NH_4^+ ions. Every section starts with an overview of the different aspects discussed in it. The chapters that follow describe either a single topic or a single (group of) organism(s) and can therefore be read separately. Section I focuses on the role that Na^+ ions and Na^+ transport processes play in pH homeostasis of the cytoplasm of prokaryotes (Chapters IB, IE, IF, and IJ), in replacing protons in the process of energy coupling (Chapters ID, IF, IG, IH, and IJ), and on the tight coupling of transmembrane Na^+ movement to the central reaction pathway of methane formation from carbon dioxide and hydrogen in methanogenic bacteria (Chapters IH and IJ). Section II contains a detailed description of the role of K^+ ions and K^+ transport across the cell membrane in the maintenance of turgor pressure in prokaryotes. Both K^+ uptake and K^+ efflux as well as K^+ release via stretch-activated channels are involved in this process. Chapters IIB, IIC, IID, IIE, and IIH give overviews about the properties of these transport systems. Other aspects that are discussed are the roles of K^+ in pH homeostasis (Chapters IIE and IIG) and in gene expression during osmoadaptation (Chapter IIF). The section ends with a review about the transmembrane channels, which bacteriocins or bacteriophages form in the cytoplasmic membrane of prokaryotes (Chapter IIJ). The first two chapters of Section III contain a detailed description of which ammonium transport systems are present in prokaryotes and which role ammonium transport plays in the regulation of nitrogen source utilization in these organisms. The last chapter of the book explains how chemostats can be used to detect futile transmembrane transport cycles in prokaryotes and that in many microorganisms K^+ transport systems may be involved in futile NH_4^+/NH_3 cycling across the cell membrane.

Evert P. Bakker
March 1992

THE EDITOR

Evert P. Bakker, Ph.D. is Professor of Microbiology of the Faculty of Biology and Chemistry at the University of Osnabrück, Osnabrück, Germany.

Dr. Bakker studied Chemistry and Biochemistry at the University of Amsterdam, The Netherlands. He received his M.Sc. and Ph.D. degrees in 1970 and 1974, respectively. After doing postdoctoral work at the Weizmann Institute of Science, Rehovot, Israel and the National Jewish Hospital and Research Center, Denver, Colorado, he came to Germany as a University assistant in 1978 and was appointed a C2 Professor (approximately equivalent to Associate Professor) of Microbiology at the University of Osnabrück in 1985.

Dr. Bakker is a member of the American Society for Microbiology, The Society for General Microbiology (U.K.), The Dutch and German Societies of Biochemistry, and of two German Societies for Microbiology (VAAM and DGHM).

Dr. Bakker has been the recipient of a Bruno Mendel Traveling Fellowship (1974 to 1975), a Weizmann Institute of Science Postdoctoral Fellowship (1975 to 1976) and three short-term EMBO Traveling Fellowships (1975, 1981, and 1983). He has received grants from the Deutsche Forschungsgemeinschaft, the Fonds der Chemischen Industrie, the European Community, and NATO. Dr. Bakker has published over 50 research papers. He divides his scientific interests between physiological aspects of cation transport in prokaryotes, acidophilic microorganisms and proteins, and aminoglycoside antibiotics.

CONTRIBUTORS

Tjakko Abee, Ph.D.
Department of Microbiology
University of Groningen
Haren, The Netherlands

Karlheinz Altendorf, Ph.D.
Department of Microbiology
University of Osnabrück
Osnabrück, Germany

Yasuhiro Anraku, Ph.D.
Department of Biology, Faculty of
Science
University of Tokyo
Tokyo, Japan

Evert P. Bakker, Ph.D.
Department of Microbiology
University of Osnabrück
Osnabrück, Germany

Eugene M. Barnes, Jr., Ph.D.
Department of Biochemistry
Baylor College of Medicine
Houston, Texas

Ian R. Booth, Ph.D.
Department of Molecular and Cell
Biology
Marischal College
University of Aberdeen
Aberdeen, Scotland

Ed T. Buurman, Ph.D.
Department of Microbiology
University of Amsterdam
Amsterdam, Holland

Peter Dimroth, Ph.D.
Department of Microbiology
ETH Zurich
Zurich, Switzerland

Roseileen M. Douglas, Ph.D.
Department of Molecular and Cell
Biology
Marischal College
University of Aberdeen
Aberdeen, Scotland

Gail P. Ferguson
Department of Molecular and Cell
Biology
Marischal College
University of Aberdeen
Aberdeen, Scotland

Alexandre Ghazi, Ph.D.
Laboratory of Biomembranes
University of Paris-Sud
Orsay, France

Gerhard Gottschalk, Ph.D.
Institute for Microbiology
University of Göttingen
Göttingen, Germany

Arthur A. Guffanti, Ph.D.
Department of Biochemistry
Mount Sinai School of Medicine
New York, New York

D. Mack Ivey, Ph.D.
Department of Biological Sciences
University of Arkansas
Fayetteville, Arkansas

Arumugam Jayakumar, Ph.D.
Institute for Molecular Genetics
Baylor College of Medicine
Houston, Texas

Yoshimi Kakinuma, Ph.D.
Faculty of Pharmaceutical Sciences
Chiba University
Chiba, Japan

Diethelm Kleiner, Ph.D.
Department of Microbiology
University of Bayreuth
Bayreuth, Germany

Wil N. Konings, Ph.D.
Department of Microbiology
University of Groningen
Haren, The Netherlands

Terry Ann Krulwich, Ph.D.
Department of Biochemistry
Mount Sinai School of Medicine
New York, New York

Andrew J. Lamb, Ph.D.
Department of Molecular and Cell
  Biology
Marischal College
University of Aberdeen
Aberdeen, Scotland

Lucienne Letellier, Ph.D.
Laboratory of Biomembranes
University of Paris-Sud
Orsay, France

Volker Müller, Ph.D.
Institute for Microbiology
Univeristy of Göttingen
Göttingen, Germany

Andrew W. Munro, Ph.D.
Department of Molecular and Cell
  Biology
Marischal College
University of Aberdeen
Aberdeen, Scotland

Oense M. Neijssel, Ph.D.
Department of Microbiology
University of Amsterdam
Amsterdam, Holland

Etana Padan, Ph.D.
Department of Molecular Microbial
  Ecology
Institute of Life Sciences
Hebrew University
Jerusalem, Israel

Graeme Y. Ritchie, Ph.D.
Department of Molecular and Cell
  Biology
Marischal College
University of Aberdeen
Aberdeen, Scotland

Peter Schönheit, Ph.D.
Department of Biology
Free University of Berlin
Berlin, Germany

Shimon Schuldiner, Ph.D.
Department of Molecular Microbial
  Ecology
Institute of Life Sciences
Hebrew University
Jerusalem, Israel

Annette Siebers, Ph.D.
Departments of Biochemistry and
  Microbiology
Biotechnology Laboratory
University of British Columbia
Vancouver, Canada

**M. Joost Teixeira de Mattos,
  Ph.D.**
Department of Microbiology
University of Amsterdam
Amsterdam, Holland

Hajime Tokuda, Ph.D.
Institute of Applied Microbiology
University of Tokyo
Tokyo, Japan

Ichiro Yamato, Ph.D.
Department of Biological Science
  and Technology
Science University of Tokyo
Chiba, Japan

Mario Zoratti, Ph.D.
Centro Studi Fisiologia Mitocondri
Consiglio Naz. Ricerche
Padova, Italy

TABLE OF CONTENTS

Section II: K⁺ Transport Systems

Section III: NH_4^+ Transport Systems

Section I: Na⁺ Transport Systems

Chapter IA

Na$^+$ TRANSPORT SYSTEMS IN PROKARYOTES

Etana Padan and Shimon Schuldiner

TABLE OF CONTENTS

I. A Na^+ CYCLE EXISTS IN ALL LIVING CELLS

All living cells, whether eukaryotic or prokaryotic, maintain a Na^+ cycle across the cytoplasmic membrane. This cycle is driven by Na^+ extruding systems which excrete the ion and maintain a Na^+ concentration gradient ($\Delta\bar{\mu}_{Na^+}$) directed inward (reviews in References 1 through 14). Na^+ reenters the cells via Na^+ gradient consumers, Na^+-coupled cotransport systems being the most widespread route (reviews in References 8 and 15). Na^+-motive flagella[16,17] and Na^+ gradient-driven ATP synthase are known for certain bacteria.[12,18] The energy stored in the Na^+ gradient, in the first half of the Na^+ cycle, is transduced to osmotic, mechanical, or chemical work in its second part. Hence, the $\Delta\bar{\mu}_{Na^+}$, like $\Delta\bar{\mu}_{H^+}$, is a convertible energy currency. No ion, other than H^+ and Na^+, is known to play such a central role in bioenergetics.

In addition to its importance in bioenergetics the Na^+ cycle has additional crucial roles in the physiology of all cells; it maintains low intracellular Na^+ and plays essential roles in homeostasis of volume and cytoplasmic pH (reviews in References 19 through 30).

Many of the specific prokaryotic Na^+ transport systems contributing to the Na^+ cycle of bacteria are thoroughly reviewed in this book. (References 7 through 14 refer to Chapters IB through IJ, respectively). In the overview we focus on molecular properties which differentiate or unite the prokaryotic Na^+ cycle and on the role of Na^+ ions in the physiology of the bacterial cell.

II. Na^+ EXTRUSION SYSTEMS IN PROKARYOTES ARE MOST VERSATILE

In animal cells the Na^+/K^+-ATPase uses ATP to energize the cytoplasmic membrane. It pumps Na^+ out, in exchange for K^+, generating a $\Delta\bar{\mu}_{Na^+}$ which serves as the primary energy source at the membrane. This Na^+ cycle we designate primary since it is driven by a primary Na^+ pump.

The cytoplasmic membrane of plants, fungi, and many bacteria is energized by primary proton pumps: H^+-ATPase of the P-type in plants and fungi and the F-type in most bacteria, excluding the archaebacteria, which possess the V-type H^+-ATPase.[31] The prokaryotes also possess primary proton pumps linked to electron transport (mitochondrial, chloroplast type) or photoreaction (bacteriorhodopsin). All these systems pump protons out, generating a $\Delta\bar{\mu}_{H^+}$, which serves as the primary energy source at the cytoplasmic membrane. Utilizing the $\Delta\bar{\mu}_{H^+}$ as a driving force, Na^+ extrusion in many plants, fungi, and bacteria is conducted by Na^+/H^+ antiporters[7] (Chapter IB), initiating a Na^+ cycle which we designate secondary since it is dependent on the primary proton cycle.

The dependence of the Na^+/H^+ antiporters on $\Delta\bar{\mu}_{Na^+}$ for Na^+ extrusion suggests that this form of Na^+ export is not efficient when $\Delta\bar{\mu}_{H^+}$ is limiting and/or when the Na^+ concentration is high and imposes a heavy load on the $\Delta\bar{\mu}_{H^+}$. Accordingly, alternative systems initiating primary Na^+ cycles exist in anaerobic and marine bacteria. Decarboxylases utilizing energy from decarboxylation reactions serve as primary Na^+ pumps in many anaerobes[9] (Chapter ID). Anaerobes also possess Na^+-ATPase[12,32,33] (Chapters IG and IID). In methanogens several metabolic reactions are coupled to the formation and utilization of the Na^+ gradient[13,14] (Chapters IH and II). In marine bacteria[1,34,35] and other halotolerant and alkaline-tolerant bacteria (for classification of alkaliphiles and alkaline-tolerant bacteria see Reference 36) primary Na^+ pumps are linked to electron transport[3] (see Chapter IF). The latter organisms also may possess Na^+-ATPase. It has been suggested that stress caused by low $\Delta\bar{\mu}_{H^+}$ induces the primary Na^+ pump even in organisms such as *Escherichia coli.*[41]

In contrast to this general theme, primary Na^+ pumps have not yet been discovered in either extreme halophiles or in extreme alkaliphiles (nonmarine nonhalotolerant). The primary proton cycles of these organisms operate against the heaviest load of extracellular Na^+ (>3.5 M NaCl)[42,43] or with the lowest $\Delta\bar{\mu}_{H^+}$ (-25 to -50 mV;[36] see, however, Chapter IG also), respectively, yet to extrude Na^+ they use the Na^+/H^+ antiporter-dependent secondary Na^+ cycle.[10,42,44-46]

The extreme halophiles seem to tolerate a high cytoplasmic concentration of Na^+ (up to 1 M).[47] It appears, therefore, that they compromise with a Na^+ gradient lower than 10, which can be easily produced by the Na^+/H^+ antiporter in spite of the heavy Na^+ load imposed on cell energetics. The properties of this antiporter appear to be unique; it requires a gating potential of 100 mV and has properties that restrict Na^+ back flow even when $\Delta\bar{\mu}_{H^+}$ is reversed.[48]

In the extreme alkaliphiles the low $\Delta\bar{\mu}_{H^+}$ is due to the Na^+/H^+ antiporter, which generates a reversed ΔpH (three units). This secondary Na^+ cycle has a crucial role in homeostasis of cytoplasmic pH of the extreme alkaliphiles, facing values of extracellular pH up to 11[10] (Chapter IE).

It may be concluded that prokaryotes have most versatile Na^+ extrusion systems. This versatility can be related to the modes of energization across the cytoplasmic membranes and to the constraints imposed by the multifarious nature of the environments microorganisms face.

III. SYSTEMS DRIVEN BY Na⁺ GRADIENT IN BACTERIA

Na^+ cotransport systems use the energy of the Na^+ electrochemical gradient to drive substrate accumulation. These cotransporters are the rule in animal cells. In accordance with the demonstration of both primary and secondary Na^+ cycles in bacteria, Na^+ cotransport systems are now known to be common in bacteria that grow under a wide variety of conditions[8,15] (see

Chapter IC). In many bacteria living at moderate conditions of salinity (<1 M) and pH (neutral), transport is mainly coupled to H^+ rather than to Na^+. In *E. coli* only four Na^+ cotransport systems are known.[49-52] Most transport systems in obligate halophiles and extreme alkaliphiles are Na^+ symporters.[10,42] In the case of the latter an obvious selective advantage is conferred since the ΔpH of opposite polarity lowers the available $\Delta\bar{\mu}_{H^+}$[10,36] (Chapter IE).

With the change to a primary Na^+ cycle in the anaerobes,[9,12-14] marine bacteria[11] and similar halotolerant alkaline-tolerant organisms,[2,3] Na^+ becomes the main coupling ion for transport. In marine bacteria[11,17] and extreme alkaliphilic bacteria[16] flagellar motility is also Na^+ dependent. Interestingly, protein translocation across the membrane in *Vibrio alginolyticus* has also been suggested to be dependent on $\Delta\bar{\mu}_{Na^+}$[53] (see Chapter IF).

A major consumer of energy is the reaction of ATP synthesis, which in all bacteria growing at moderate Na^+ and H^+ concentrations is driven, in a chemiosmotic mechanism, by $\Delta\bar{\mu}_{H^+}$ via F-type H^+-ATPase. This is also the case in extreme halophiles[42] and in the extreme nonmarine nonhalotolerant alkaliphiles[10] (Chapters IE and IG). The very low available $\Delta\bar{\mu}_{H^+}$ of the latter led Krulwich[10] to propose a non-Mitchellian mechanism, still involving the F-type H^+-ATPase. Na^+-ATPases accompany the primary Na^+ pumps in various anaerobes[12,32,33] (Chapter IG). Their existence has also been suggested in marine bacteria.[38-40] However, recently the *unc* operon has been cloned from *V. alginolyticus* and expressed in *E. coli*.[54] Since the purified and reconstituted enzyme exhibits only H^+-pumping activity, the authors conclude that if *V. alginolyticus* retains Na^+-ATPase it is not likely to be the F-type enzyme (Chapter IF).

In allowing the reuptake of Na^+, all $\Delta\bar{\mu}_{Na^+}$ consumers complete the Na^+ cycle. If no other substantial Na^+ leaks exist, they may also determine the rate of the Na^+ cycle.

It is apparent that the various Na^+ cycles observed seem to be an adaptation to a unique environment. The complete primary Na^+ cycle allows the organism to cope with a limited amount of $\Delta\bar{\mu}_{H^+}$ or none, in anaerobic niches. The possibility of alternating between primary and secondary Na^+ cycles affords to marine bacteria a very wide spectrum of adaptation in terms of pH and salinity.[11,39,55] Whereas the secondary Na^+ cycle of extreme halophiles differs from that of ordinary bacteria in its accommodation to higher intracellular Na^+, that of the extreme alkaliphiles is unique and essential to the adaptation to an extremely alkaline pH (moderate salinity, nonmarine) niche.[10]

Disregarding the specific Na^+ cycles, all bacteria have Na^+/H^+ antiporter(s) (Chapter IB). Even organisms that possess a primary Na^+ cycle still possess Na^+/H^+ antiporter(s)[10,45,56] (Chapters IG, IH, and IJ). This ubiquitousness is in accord with the important roles ascribed to these antiporters, in addition to their creation of the secondary Na^+ cycle. Since they exchange

$\Delta\bar{\mu}_{H^+}$ for $\Delta\bar{\mu}_{Na^+}$ they serve as energy buffers.[57,58] In methanogens they have even been suggested to produce $\Delta\bar{\mu}_{H^+}$ at the expense of the $\Delta\bar{\mu}_{Na^+}$[13,14] (Chapters IH and IJ). The Na$^+$/H$^+$ antiporters also serve (nonbioenergetically) most central roles in control and homeostasis mechanisms of the cell (see below and Chapter IB).

IV. A MOLECULAR APPROACH TO THE Na$^+$ TRANSPORTERS

Many of the Na$^+$ transport genes of bacteria have already been cloned, allowing molecular genetic studies of their products.[7-9,12,59,60] The physiological role of a transporter is studied by deleting the structural gene from the chromosome and comparing the mutant phenotype to that of the wild type. The deletion mutation is also used to explore how many transporters of similar function exist. For example, deleting *nhaA* showed that it is indispensable in *E. coli* for tolerance to high Li$^+$, high Na$^+$, and alkaline pH in the presence of Na$^+$.[61] It also revealed a new Na$^+$/H$^+$ antiporter system, *nhaB*[7] (Chapter IB). Study of the regulation of expression of the transport systems has also become possible by constructing protein and operon fusions with reporter genes. Studies with a protein fusion between *nhaA* and *lacZ* showed that *nhaA* is induced by Na$^+$ and Li$^+$, and alkalinization increases the sensitivity of the system to the ions.[62] Molecular biological studies are also most important in the search for the regulatory proteins involved.[63]

Molecular genetics provides important tools for overexpression of the transport protein, a crucial step for purification. As yet, few of the various prokaryotic Na$^+$ transporters have been purified to homogeneity and reconstituted in liposomes in a functional form. Using a site-specifically cleavable fusion protein, the proline/Na$^+$ cotransporter of *E. coli* has been purified[64] (Chapter IC). Overexpressing *nhaA* allowed the purification of a Na$^+$/H$^+$ antiporter of *E. coli*[65] (Chapter IB). The redox-linked Na$^+$ pump of *V. alginolyticus*,[11,66,67] the decarboxylation-linked Na$^+$ pump,[9] and the Na$^+$-ATPase of *Propionigenium modestum*[12,68,69] have been purified by conventional biochemical methods (see Chapters ID, IF, and IG). Purification of a functional protein is the only way to prove the involvement of a gene product in the transport reaction. Furthermore, it is crucial for the study of molecular properties of the transport systems, specifically those which cannot easily and/or unequivocally be determined in membrane preparations or in the intact cell.[65]

High-resolution structural information is essential to study the relationship between structure and function in a transporter. Unfortunately, however, none of the transporters has as yet been crystallized. Furthermore, since the molecular weight of most transporters is above 30,000, nuclear magnetic resonance (NMR) spectroscopy cannot as yet be applied to resolve their structure. Notwithstanding this as yet unachieved goal, it has become apparent that a combination of genetic and molecular genetic methods can be used to delineate amino acid residues that affect active transport. It should be borne in mind, however, that without atomic structure it is impossible to determine whether these amino acids are directly involved in the mechanism or indirectly affect a conformation which changes activity but is irrelevant to the mechanism.

An approach used to identify amino acids that affect transport activity is random mutagenesis of the chromosomal gene and selection for phenotypes modified in transport activity.[70,71] Identification of the mutation involves cloning and sequencing of the mutated gene. The melibiose carrier (MelB), which cotransports the sugar with either Na^+ or H^+, has been studied with this approach.[72] Proline 122 has been shown to be essential for cation specificity since its replacement by serine produces a protein which loses its H^+ coupling ability and becomes dependent on either Na^+ or Li^+. With this approach most interesting results pertaining to ion and substrate specificity have also been obtained with the proline transporter (PutP)[8,15,73] (Chapter IC).

Random mutagenesis is now conducted directly on the cloned transporter genes. This allows an easy verification of the mutation site and its identification by direct sequencing of the mutated cloned gene.[74] This approach was applied in *E. coli* to isolate mutants with alterations in substrate spceificity of the melibiose[75] (see below) and proline[8,15,76] transporters.

Site-directed mutagenesis also makes it possible to test the involvement of certain amino acids in the transport reaction.[77,78] This approach is particularly rewarding if previous data, biochemical or genetic, suggest such involvement. Chemical modification is one approach used to specify amino acid targets for site-directed mutagenesis. With these combined approaches a histidine was shown to play a crucial functional role in the *lac* carrier[77-79] and possibly also in the tetracycline transporter.[80]

Another approach used to identify important amino acid residues is based on the suggestion advanced with many transporters that conserved sequences form conceptual candidates for important functional regions.[81,82] With this approach Anraku and his colleagues[8,76,83] (Figure 1) identified a conserved sequence, with SOB motif, in Glts (the glutamate/Na^+ cotransporter), PutP (both in *E. coli*), and Sglt1, the rabbit intestine glucose/Na^+ cotransporter[84] (see Chapter IC). The SOB motif was also found in PanF (the pantothenate/Na^+ cotransporter of *E. coli*)[85] and Na^+/nucleoside[130] cotransporters.

Although we have not found the entire SOB motif in many other Na^+-coupled transporters, all of them except one contain the first glycine. Furthermore, we have found conserved amino acids which are not included in the SOB motif; the third amino acid following the glycine is always leucine and the eighth one is alanine (Figure 1). Charged amino acids are very rarely found in this segment, which would suggest that this motif is located in the apolar part of the membrane. The most extended homology in this area is observed between the various Na^+/H^+ antiporters, specially NhaA and NhaB.

TRANSPORTER START

Sglt1	378	L	R	G	L	M	L	S	V	M	M	A	S	L	M	S	S
Gat1	295	S	Y	G	L	G	L	G	S	L	I	A	L	G	S	Y	N
PutP	326	I	A	G	I	L	L	S	A	I	L	A	A	V	M	S	T
PanF	321	A	A	G	I	F	L	A	A	P	M	A	A	I	M	S	T
MelB	233	L	L	G	M	A	L	A	Y	N	V	A	S	N	I	I	T
NapA	63	E	I	G	V	I	L	L	M	F	L	A	G	L	E	S	D
NhaC	83	G	I	G	T	A	L	G	I	D	P	A	I	A	A	G	A
NhaB	295	I	I	G	V	W	L	V	T	A	L	A	L	H	L	A	E
NhaA	300	P	L	G	I	S	L	F	C	W	L	A	L	R	L	K	L
Nhe1	102	P	F	E	I	S	L	W	I	L	L	A	C	L	M	K	I

FIGURE 1. Conserved amino acids in Na⁺-driven transporters. Sglt1[84] is the rabbit (or human) Na⁺/glucose symporter. Gat1[121] is the rat GABA symporter. PutP,[122] PanF,[85] and MelB[123] are the *E. coli* proline, panthetonate, and melibiose symporters, respectively. The Na⁺/H⁺ antiporters are NhaA[125] and NhaB[129] of *E. coli*, NhaC[124] of *Bacillus firmus* OF4, NapA of *Enterococcus hirae*,[127] and Nhe1[126] of human. The full SOB motif identified by Deguchi and co-workers[8,83] is G(35 - 38X) A(4X) L(3X) GR (see Chapter IC).

The significance of these motifs will become apparent only when the structure of the transporters is resolved. Moreover, it is equally clear that dynamic information at high resolution is also required to solve the mechanism. In this respect, site-directed mutagenesis can be very useful to engineer a protein for various biophysical studies. For example, all the tryptophans of the *lac* carrier have been exchanged with phenylalanine. Since none of these residues was found to be essential, tryptophan could now be reintroduced at specific sites and its fluorescence used to study static and dynamic aspects of permease structure and function.[86]

V. Na⁺ AND H⁺ ARE INTERCHANGEABLE IN VARIOUS TRANSPORTERS

One of the most intriguing characteristics of some transport systems is the interchangeability between Na⁺ and H⁺. The F_1F_0-ATPase of *P. modestum* can use gradients of Na⁺ or H⁺ ions[68,69] (Chapter IG). A similar behavior has been ascribed to the enzyme of *V. alginolyticus*,[38] although there exists no direct evidence that supports this notion[54] (Chapter IF). The H⁺/melibiose cotransporter of *E. coli* can be driven by Na⁺ or a proton, depending on the substrate, and is inhibited by Li⁺.[49] A single mutation (proline 122 to serine

122) is sufficient to convert the system from H^+ selective to Li^+ selective and even Li^+ dependent.[72,87] Under appropriate conditions the eukaryotic Na^+/K^+-ATPase can drive a slow H^+/K^+ exchange[88] or a slow Na^+/H^+ exchange,[89] and the gastric H^+/K^+-ATPase can drive a slow Na^+/K^+ exchange.[90] Boyer[91] has suggested that transport systems which can transport either Na^+ or H^+ may behave like neutral crown ethers in binding both Na^+ ions, largely dehydrated, and H^+, as hydronium ions. Therefore, they can transport both Na^+ and H_3O^+ via the same site and with a similar mechanism.

Biochemical evidence suggests that substrate-coupled Na^+ and H^+ translocation may occur by similar pathways; Na^+ is a competitive inhibitor of melibiose-coupled H^+ translocation via *melB* permease.[92]

Assuming that mutations affecting substrate specificity of the carrier change amino acids forming the substrate recognition site, Botfield and Wilson[75] have selected 70 mutants resistant to inhibition by a nonmetabolizable melibiose analog (methyl β-D-thiogalactoside). Four widely separated regions of the protein, where amino acid substitutions alter substrate specificity, have been identified. This strong clustering of mutations suggests that these areas form the substrate recognition site. Most interestingly, the changes in the sugar specificity are almost always accompanied by changes in cation (H^+ or Na^+) recognition properties, suggesting a direct interaction between the sugar and the ionic substrates at the same site.

VI. Na$^+$ CHANNELS AND PASSIVE Na$^+$ LEAKS IN PROKARYOTES

It has been implied that the rate of Na^+ entry into the cells determines the rate of operation of the Na^+ cycle.[29] Indeed, the presence of substrates of Na^+-coupled cotransport systems markedly enhances acidification of the cytoplasm via the Na^+/H^+ antiporter during an alkaline shift of extreme alkaliphile[93] (Chapter IE). Nevertheless, both alkaliphiles and other bacteria may possess a symport-independent, pH-regulated Na^+ entry route as suggested for *Exiguobacterium aurantiacum*[94] and *E. coli*.[29] This raises the questions of the passive pathways of Na^+ leaks and Na^+ channels in prokaryotes.

The outer membrane of Gram-negative bacteria presents a barrier to the passage of macromolecules, but is relatively permeable to hydrophilic solutes due to the presence of porins. Porins are abundant proteins that form nonspecific channels across the outer membrane of *E. coli*.[95] Using salt taxis as a sensitive *in vivo* assay for outer membrane permeability it has been shown that porins hardly affect salt permeability via the outer membrane.[96] Relative to amino acid taxis, salt taxis is surprisingly little impaired by the loss of porins. Like the amino acids tested, these salts are sensed by one or more of the methyl-accepting chemotaxis proteins and therefore must penetrate the outer membrane. These results imply that an additional mechanism allows salt penetration through the outer membrane. Accordingly, study of *E. coli* cell membrane with the patch-clamp technique reveals new ion channels located within the outer membrane[97] (see, however, Chapter IIH also).

Although there is evidence for K^+ channels in the cytoplasmic membrane of *E. coli*[98,99] (Chapter IIH), there is no such evidence for Na^+ channels. Using NMR spectroscopy, Castle et al.[28] showed that the downhill Na^+ movement from de-energized cells (under anaerobic conditions or in the presence of uncoupler) is very slow, in agreement with the results of other studies (measuring fluxes of $^{22}Na^+$).[100-103] Although this slow Na^+ movement is probably mediated by Na^+/H^+ antiporter, the contribution of Na^+ leaks or channels has not been ruled out.

It is obvious that as long as Na^+ symporters or antiporters operate at the membrane, they complicate the determination of passive Na^+ routes. Deleting these systems by molecular genetic tools will greatly simplify these studies. A case in point is the recently constructed *E. coli* mutant deleted of both Na^+/H^+ antiporters, *nhaA* and *nhaB*.[7]

VII. TOXICITY OF Na⁺

There is a universal asymmetry in the distribution of Na^+ and K^+ ions across the plasma membrane: intracellular K^+ is high whereas intracellular Na^+ is low.[104] The reason could be the more ubiquitous occurrence of Na^+ as compared to K^+ and, therefore, first using the former for ion coupling. Alternatively, it is possible that Na^+ is toxic in comparison to K^+.

Nevertheless, all organisms require salts, and there is an optimum concentration of salts for each; accordingly, too little and too much are both avoided. *E. coli* has been shown to exhibit salt taxis. Although the threshold is high (0.1 to 1 m*M*), this is a powerful attraction, in some cases nearly as strong as for the best *E. coli* attractants known, the amino acids L-aspartate and L-serine. The optimum concentration for NaCl taxis is between 10 and 100 m*M*.[105] Higher concentrations of salt (for example, 400 to 500 m*M*) act as a repellent.[106] Lower concentrations are also less attractive, and pure water is avoided. The nature of the cation is critical, the order of taxis being NH_4^+ > Li^+ > Na^+ > K^+ > Rb^+ > Cs^+. Mg^{++} is the best divalent cation. The nature of the anion has a lesser but definite effect; Cl^- is the best halide. It appears that both methyl-accepting chemotaxis proteins, MCPI and MCPII, participate in salt taxis, the former being more important. Salt taxis is different from osmotaxis, which gives less than one tenth of the response and does not operate via MCP.

The repellent effect of NaCl suggests that Na^+, above a certain concentration (400 to 500 m*M* at pH 7),[106] has a deleterious effect on *E. coli* cells. This effect is due neither to increased osmolarity nor to increased ionic strength since osmotaxis is different from salt taxis (see above) and various salts have different specific effects.

Determination of NaCl concentrations inhibiting growth of the wild type may reflect the concentrations toxic to cell components exposed to the environment. However, it is also possible that intracellular Na^+ may be responsible for the inhibition, suggesting toxic effects of the ion on cytoplasmic components. In this case, and given the existence of Na^+ extrusion systems in all bacteria, it is predicted that impairments of Na^+ excretion will increase the Na^+ sensitivity of the cells.

A molecular genetic approach was undertaken in *E. coli* to explore its tolerance to Na^+. A mutant, NM81, impaired in its Na^+ extrusion capacity was constructed by deleting *nhaA*, the Na^+/H^+ antiporter gene.[61] Comparison of the salt tolerance of this mutant to the wild type revealed a specific toxic effect of Na^+. Thus, at a pH of 6.8 the mutant does not grow in the presence of a NaCl concentration above 600 mM, whereas the wild type grows at concentrations up to 900 mM. This effect is independent of ionic strength or osmolarity since the mutant grows at 900 mM KCl like the wild type. Furthermore, the toxicity is pH dependent, increasing upon alkalinization. At pH 8.6, whereas 100 mM NaCl has no effect on the wild type, it completely inhibits the growth of NM81. Deleting both *nhaA* and *nhaB* in mutant EP432 further increased the Na^+ sensitivity of the cells.[131] These cells are already inhibited at pH 7.5 by 100 mM NaCl. Since mutant EP432 does not possess any measurable Na^+/H^+ antiporter activity under the conditions used,[7] it is suggested that it does not extrude Na^+ and its intracellular Na^+ reaches toxic concentrations at much lower extracellular concentrations than the wild type (Chapter IB). Therefore, it is tempting to conclude that intracellular Na^+ has a specific toxic effect on a cytoplasmic component(s) in *E. coli*. It will be most interesting to measure these toxic intracellular Na^+ concentrations directly and compare them to cytoplasmic concentrations causing the Na^+ repellent response of *E. coli* (see above).

The increased sensitivity of the mutants to Na^+ caused by alkaline pH is intriguing. It may be related to the interchangeability of Na^+ and H^+ (see above). At alkaline pH Na^+ will compete better with H^+ and therefore may inhibit proton-requiring processes. It is also plausible that with alkalinization intracellular Na^+ reaches toxic levels at lower extracellular Na^+ concentrations. Indeed, the Na^+ gradient that the cells can maintain decreases when the medium pH is increased.[107] Another alternative is that homeostasis of intracellular pH is impaired in the mutants and Na^+ aggravates this condition.

Which reaction is the most sensitive to Na^+? If this reaction is essential, a straightforward answer is difficult to obtain by inhibiting or deleting it. Using peptidyl-puromycin synthesis on polyribosomes as an *in vitro* model for protein synthesis in *E. coli*, Pestka[108] showed that 100 to 200 mM NaCl inhibited the reaction. Interestingly, this *in vitro* inhibiting concentration of Na^+ is similar to the concentration inhibiting growth of the $\Delta nhaA \Delta nhaB$ mutant (EP432)[7] (Chapter IB).

In marked contrast to Na^+, K^+ activates a number of cell enzymes from all living cells (reviewed in Reference 109). Some of these effects are specific to the ion and independent of osmolarity or ionic strength. Activation of protein synthesis by K^+ was demonstrated by Ennis and Lubin[110] in studies of a K^+ transport mutant of *E. coli*. As cell K^+ fell and Na^+ rose, protein synthesis was progressively inhibited while DNA and RNA synthesis continued. The concentration of K^+ needed to activate enzymes (about 10 mM) is far below the intracellular concentration (above 150 mM) found in most bacteria (see also Chapter IIA).

There is no clear answer as to why K^+ is preferred to Na^+. It appears that ions of similar size, such as NH_4^+ and Rb^+ usually can replace K^+, while smaller monovalent cations with larger hydration shells, such as Na^+ and Li^+, often antagonize activation by K^+.[109]

VIII. DO BACTERIAL CELLS HAVE HOMEOSTASIS OF CYTOPLASMIC Na⁺ CONCENTRATION, AND HOW DO THEY RESPOND TO Na⁺ STRESS?

The demonstration of Na^+ toxicity and repellent activity on the one hand and Na^+ taxis on the other hand raises the question of homeostasis of intracellular Na^+ concentrations. By using a perfusion system which stably maintained the external pH of dense cultures of endogenously respiring *E. coli*, Pan and Macnab[107] extended studies by ^{31}P- and ^{23}Na-NMR spectroscopy of Na^+ and H^+ bioenergetics[27,28] to a wide pH range (pH 6.4 to 8.4). These studies substantiated previous results[22-26] showing that intracellular pH in neutrophilic bacteria is regulated to 7.6 to 7.8 over the entire range of pH permitting growth.

In marked contrast to the extensive homeostasis of pH, intracellular Na^+ was not found to be constant.[107] Measurement of $[Na^+]_{in}$ as a function of $[Na^+]_{out}$ in the range from 4 to 285 mM, at 25°C, an extracellular pH of 6.6 to 6.7, and constant osmolarity (375 mosm), shows that $\Delta pNa \{ - \log ([Na^+]_{in}/[Na^+]_{out}) \}$ is constant (1.24) and intracellular Na^+ is not regulated under these conditions varying from 0.23 to 16.4 mM, respectively. Lack of regulation of intracellular concentration of Na^+ was observed at various pH values between 6.25 and 7.6;[28] ΔpNa changes with pH (see below), but at each pH value it is constant and independent of extracellular Na^+ concentrations (80 and 180 mM at a constant osmolarity of 375 mosm).

These results show that extracellular pH affects intracellular Na^+ concentration. At 80 mM $[Na^+]_{out}$, increasing extracellular pH up to 7.4 causes a parallel decrease in ΔpH and ΔpNa, although the latter is always somewhat larger. Above pH 7.4 ΔpNa increases with external pH, but beyond pH 8 it again slowly decreases. These variations with pH suggest different modes of Na^+ export.[107] It also shows that intracellular Na^+ increases with pH at a

TABLE 1

**Regulation of Expression of *nhaA'-'lacZ* by *nhaR*
is Dependent on Intracellular Na$^+$**

Strain	Gene on plasmid	β-Galactosidase activity (units)	
		$-$Na$^+$	$+$Na$^+$
RK33Z	*nhaR*	30	170
RK33Z/pGM42T	*nhaR*	35	430
RK33Z/pGM36	*nhaA*	33	35
RK33Z/pEL24	*nhaB*	35	34

Note: RK33Z is an *E. coli* mutant carrying *nhaA'-'lacZ* instead of
the chromosomal *nhaA*.[63] Plasmid pGM42T[63] bears active
nhaR and an inactive *nhaA*; pGM36[125] and pEL24[129] carry
active *nhaA* and active *nhaB*, respectively. Induction was
carried out in the presence of 100 mM NaCl, and the activity
of β-galactosidase was determined as described.[63]

given extracellular Na$^+$ concentration. In the presence of 80 mM Na$^+$, intracellular Na$^+$ at pH 6.7 is 4.6 mM whereas at pH 8.4 it is 25.3 mM (see
also Chapter IB).

On the basis of the published data[107] it can be estimated that at neutral
pH and at 400 to 500 mM extracellular Na$^+$ the intracellular Na$^+$ is 40 to
50 mM. Since this extracellular concentration is the repellent concentration
(see above), we suggest that 40 to 50 mM intracellular Na$^+$ becomes toxic
to the cytoplasm.

External concentrations in the same range induce β-galactosidase activity
of a *nhaA'-lacZ* fusion (*in vivo*), and alkaline pH increases the sensitivity of
the expression system to the ions.[62] Our recent results[63] (and Table 1) show
that *nhaR*, a gene localized downstream from *nhaA*, is a positive regulator
of *nhaA*. NhaR works in trans and its effect is Na$^+$ dependent. This Na$^+$
dependency is related to intracellular rather than extracellular Na$^+$. Thus,
plasmids bearing *nhaA*[63] or *nhaB* (Table 1), which encode for heterologous
proteins with common function, Na$^+$/H$^+$ antiport, inhibit the effect of *nhaR*
on *nhaA* expression. It remains to be tested whether intracellular Na$^+$ is the
immediate signal of the control exerted by NhaR. Most interestingly, on the
basis of protein similarity, NhaR has been shown to belong to the LysR family
of positive regulators,[111] which share homologous sequences in the N-terminal
region of the protein, where a helix-turn-helix domain is found. Some of
these proteins are involved in the response of bacteria to various environmental
stresses. An example is OxyR,[112] which is crucial for adaptation to oxidative
damage. We postulate that *nhaR* is involved in a novel signal transduction
pathway responding to Na$^+$ stress. Taken together it is suggested that *E. coli*
cells react to a high Na$^+$ load, when the intracellular concentration approaches

the stress concentrations, by swimming away from the toxic ion and at the same time by initiating the Na^+-dependent *nhaR* response which leads to induction of *nhaA* for efficient excretion of the ion (see also Chapter IB).

On the other hand, it appears that as long as toxic concentrations are not reached, the cells can accommodate quite high variations in intracellular Na^+. Furthermore, as long as $[K^+]_{in}$ is high, which is the usual case, there is no reason to maintain constant $[Na^+]_{in}$ either for osmotic balance or for ionic strength. Hence, there is a wide flexibility in intracellular Na^+ below its toxic concentration. This is consistent with the function of Na^+/H^+ antiporters in pH homeostasis, manipulating the concentration of Na^+ to keep intracellular H^+ concentration constant.

IX. PHYSIOLOGICAL ROLES OF THE Na⁺ CYCLE

As discussed above, the bioenergetic roles of the Na^+ cycle are cardinal. It maintains a Na^+ gradient directed inward to drive active transport of solutes into cells. Since $\Delta\bar{\mu}_{H^+}$ and $\Delta\bar{\mu}_{Na^+}$ are exchangeable via the Na^+/H^+ antiporters, the Na^+ cycle also acts as a device to buffer energy, storing energy as either $\Delta\bar{\mu}_{H^+}$ and/or $\Delta\bar{\mu}_{Na^+}$ ready to be utilized as needed.[23,57,58] In excreting Na^+, the Na^+ cycle also maintains the intracellular Na^+ concentrations below the toxic levels (see above).

The Na^+ cycle also has essential regulatory roles. It is involved in pH homeostasis of the cytoplasm, a property that is shared by all bacteria.[7,10,22-26,29,107] The set point of constant intracellular pH of the neutrophilic bacteria is 7.6 to 7.8; that of the acidophiles is around 6.5 and of the alkaliphiles around 8.2. Therefore, at any extracellular pH of growth the rate of proton extrusion, which is maintained by the primary proton pumps, must be compensated for by an appropriate rate of proton uptake so as to maintain the constant set point of intracellular pH. Na^+/H^+ antiporters whose activity is dependent on pH, have been suggested to play this role.[22,113] A model of pH homeostasis in *E. coli* based on two antiporters with different constant stoichiometries, whose contribution changes with pH, has been advanced.[114]

A genetic approach has been undertaken to explore the role of the Na^+/H^+ antiporters in pH homeostasis. Different mutants impaired in antiporter activity were isolated in the extreme alkaliphile (*Bacillus alcalophilus*[10] as well as in *E. coli*.[52,115,116] These mutants have been shown to be impaired in pH homeostasis also. However, in neither of these works has the mutation lesion been proven to directly affect a Na^+/H^+ antiporter.

The molecular genetic approach undertaken in *E. coli* yielded a mutant deleted of *nhaA*, NM81.[61] This mutant, which does not contain the NhaA antiporter but contains an additional Na^+/H^+ antiporter, *nhaB*, and the K^+/H^+ antiporter,[117] grows at all pHs as long as Na^+ is not added, suggesting that *nhaB* and/or the K^+/H^+ antiporter can fulfill the role of pH regulator in the absence of Na$^+$. A mutant devoid of both antiporters ($\Delta nhaA \Delta nhaB$), can also grow at pH values above 8.0 in the absence of added Na$^+$.[7]

In the presence of Na$^+$, NM81 becomes most sensitive to pH. Thus, whereas in the presence of 100 mM NaCl it can grow at pH 7.5, at pH 8.6 the mutant is completely inhibited by this Na$^+$ load.[61] These results imply that NhaA is indispensable for growth at alkaline pH in the presence of Na$^+$. Furthermore, NhaA has been purified and reconstituted in proteoliposomes in a functional form. Its study in this form shows that it is markedly sensitive to pH.[65] Based on the rates of Na$^+$ downhill transport catalyzed by NhaA in the presence of a membrane potential and on its abundance in the membrane,[65] we can estimate the rates of transport in the intact cell (Figure 2). This calculation shows that at pH 6.5 the activity of NhaA is very low, 3 nmol/ mg of cell protein per minute, increasing dramatically with pH so that at pH 8.6 its activity is 400 nmol/mg of cell protein per minute. These results, which corroborate results from previous studies conducted with membrane vesicles and intact cells,[101,118] show that NhaA can function as both pH sensor and pH titrator (see Chapter IB).

Most importantly, the induction of *nhaA* by Na$^+$ (10 mM) is markedly affected by pH, increasing the level of the antiporter about tenfold over a pH range between 7.5 to 8.5[62] (see Figure 2), yielding when fully expressed a rate of 4000 nmol/mg of cell protein per minute. The rate of electron transport in *E. coli* over this pH range and in the presence of similar Na$^+$ concentrations is constant[22] and may account for a proton extrusion rate of about 1000 nmol/ mg of cell protein per minute. Therefore, we conclude that NhaA can easily modulate the rate of proton extrusion and regulate intracellular pH in *E. coli* over a wide range of Na$^+$ concentrations.

Since the rates of transport catalyzed by NhaB are low (6 to 7 nmol/mg cell protein per minute,[61] see also Figure 2) and independent of pH, we suggest that it contributes mainly under conditions at which NhaA is less active, i.e., low Na$^+$ and low pH (see also Chapter IB).

As discussed above, the rate of the Na$^+$ leakage into the cell is most probably the rate-limiting step for the operation of the Na$^+$ cycle. Na$^+$-solute cotransport is apparently the main route of Na$^+$ entry in *Bacillus firmus* since without the solute pH homeostasis is impaired.[10] In *E. coli*, however, regulation of intracellular pH has been shown in bacteria respiring on solutes which are taken up without Na^{+22} or on endogenous substrates.[107] As discussed above and by Booth[29], the question of reentry of Na$^+$ into the cells, to keep the Na$^+$ cycle rolling in *E. coli*, is most important but is still unknown.

Since many biomolecules (including all proteins and nucleic acids) are acid-base species whose conformations are pH dependent, the regulation of cytoplasmic pH serves a centrally important physiological function by providing a suitable environment for maintenance of stable structure and enzymatic activity. Furthermore, it appears that intracellular pH itself can serve

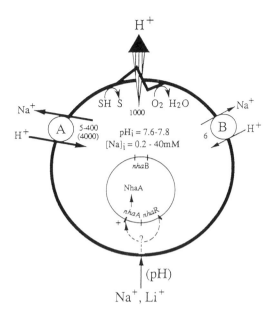

pH Activity Measured	6.6	7.6	8.5
	(nmole / min / mg cell protein)		
Total Downhill Transport	7	17	400
NhaA	<5	10	400
NhaB	6	7	7

FIGURE 2. The Na⁺/H⁺ antiporters of *E. coli* activity and regulation. The rates of downhill ²²Na⁺ transport catalyzed by NhaA and NhaB were measured in right-side out membrane vesicles prepared from wild type[128] and Δ*nhaA* or calculated from values obtained in proteoliposomes reconstituted with NhaA.[65] The rates of H⁺ transport coupled to respiration were calculated from the rate of electron transport[22] (see also Chapter IB).

as a specific signal for certain regulons.[25] The SOS regulon has been shown to be induced upon alkalinization of intracellular pH.[119] Remarkably, many aspects of the central regulatory roles played by the Na⁺ cycle in the pro-karyotes are shared by the eukaryotes. This includes the role of Na⁺/H⁺ antiporters in pH regulation[120] and in pH-related control mechanisms affecting cell proliferation.[20]

ACKNOWLEDGMENTS

We thank R. Karpel for critical reading of the manuscript. The research that is reported in this chapter was supported by a research grant from the United States - Israel Binational Science Foundation (BSF).

REFERENCES

1. **Dimroth, P.,** Sodium ion transport: decarboxylases and other aspects of sodium ion cycling in bacteria, *Microbiol. Rev.,* 51, 320, 1987.
2. **Skulachev, V. P.,** Bacterial sodium transport: bioenergetic functions of sodium ions, in *Ion Transport in Prokaryotes,* Rosen, B. P. and Silver, S., Eds., Academic Press, New York, 1987, 131.
3. **Skulachev, V. P.,** The sodium world, in *Membrane Bioenergetics,* Skulachev, V. P., Ed., Springer-Verlag, Berlin, 1988, 293.
4. **Rosen, B. P.,** Recent advances in bacterial ion transport, *Annu. Rev. Microbiol.,* 40, 263, 1986.
5. **Skou, J. C., Norby, J. G., Maunsbach, A. B., and Esmann, M.,** The Na^+, K^+ pump, Part A, in *Molecular Aspects,* Alan R. Liss, New York, 1988.
6. **Glynn, I. M. and Karlish, S. J. K.,** Occluded cations in active transport, *Annu. Rev. Biochem.,* 59, 171, 1990.
7. **Schuldiner, S. and Padan, E.,** Na^+/H^+ antiporters in *E. coli,* in *Alkali Cation Transport Systems in Prokaryotes,* Bakker, E. P., Ed., CRC Press, Boca Raton, FL, 1992. chap. IB.
8. **Yamato, I. and Anraku, Y.,** Na^+/substrate cotransport systems in prokaryotes, in *Alkali Cation Transport Systems in Prokaryotes,* Bakker, E. P., Ed., CRC Press, Boca Raton, FL, 1992. chap. IC.
9. **Dimroth, P.,** Na^+ extrusion coupled to decarboxylation reactions, in *Alkali Cation Transport Systems in Prokaryotes,* Bakker, E. P., Ed., CRC Press, Boca Raton, FL, 1992, chap. ID.
10. **Ivey, D. M., Guffanti, A. A., and Krulwich, T. A.,** The Na^+ cycle in alkaliphilic *Bacillus* species in *Alkali Cation Transport Systems in Prokaryotes,* Bakker, E. P., Ed., CRC Press, Boca Raton, FL, 1992, chap. IE.
11. **Tokuda, H.,** The Na^+ cycle in *Vibrio alginolyticus,* in *Alkali Cation Transport Systems in Prokaryotes,* Bakker, E. P., Ed., CRC Press, Boca Raton, FL, 1992, chap. IF.
12. **Dimroth, P.,** The Na^+-translocating ATP-synthetase from *Propionigenium modestrum,* in *Alkali Cation Transport Systems in Prokaryotes,* Bakker, E. P., Ed., CRC Press, Boca Raton, FL, 1992, chap. IG.
13. **Müller, V. and Gottschalk, G.,** Na^+ translocation in the course of methanogenesis from methanol or formaldehyde, in *Alkali Cation Transport Systems in Prokaryotes,* Bakker, E. P., Ed., CRC Press, Boca Raton, FL, 1992, chap. IH.
14. **Schönheit, P.,** The role of Na^+ in the first step of CO_2 reduction to methane in methanogenic bacteria, *Alkali Cation Transport Systems in Prokaryotes,* Bakker, E. P., Ed., CRC Press, Boca Raton, FL, 1992, chap. IJ.
15. **Maloy, S. R.,** Sodium coupled cotransport, in *Bacterial Energetics,* Krulwich, T. A., Ed., Academic Press, New York, 1990, 203.

16. **Hirota, N., Kitada, M., and Imae, Y.,** Flagellar motors of alkalophilic *Bacillus* are powered by an electrochemical potential gradient of Na⁺, *FEBS Lett.*, 132, 278, 1981.

17. **Dibrov, P. A., Kostyrko, V. A., Lazarova, R. L., Skulachev, V. P., and Smirnova, I. A.,** The sodium cycle. I. Na⁺ dependent motility and modes of membrane energization in marine alkalotolerant bacterium *Vibrio alginolyticus, Biochim. Biophys. Acta,* 850, 449, 1986.

18. **Hoffmann, A., Laubinger, W., and Dimroth, P.,** Na⁺-coupled ATP synthesis in *Propionigenium modestum*: is it a unique system?, *Biochim. Biophys. Acta,* 1018, 206, 1990.

19. **Boron, W. F.,** Transporting of H⁺ and of ionic weak acids and bases, *J. Membr. Biol.,* 72, 1, 1983.

20. **Pouyssegur, J., Franchi, A., Lagarde, A., and Sardet, C.,** Molecular genetics of the mammalian Na⁺/H⁺ antiporter, in *Na⁺/H⁺ Exchange,* Grinstein, S., Ed., CRC Press, Boca Raton, FL, 1988, 337.

21. **Haussinger, D.,** *pH Homeostasis — Mechanism and Control,* Academic Press, San Diego, 1988.

22. **Padan, E., Zilberstein, D., and Rottenberg, H.,** The proton electrochemical gradient in *E. Coli, Eur. J. Biochem.,* 63, 533, 1976.

23. **Padan, E., Zilberstein, D., and Schuldiner, S.,** pH homeostasis in bacteria, *Biochim. Biophys. Acta,* 650, 151, 1981.

24. **Padan, E. and Schuldiner, S.,** Intracellular pH regulation in bacterial cells, *Methods Enzymol.,* 125, 337, 1986.

25. **Padan, E. and Schuldiner, S.,** Intracellular pH and membrane potentials as regulators in the prokaryotic cell, *J. Membr. Biol.,* 95, 189, 1987.

26. **Slonczewski, J. L., Rosen, B. P., Alger, S. R., and Macnab, R. M.,** pH homeostasis in *Escherichia coli*: measurement by ³¹P nuclear magnetic resonance of methyl phosphonate and phosphate, *Proc. Natl. Acad. Sci. USA.,* 78, 6271, 1981.

27. **Castle, A. M., Macnab, R. M., and Schulman, R. G.,** Measurements of intracellular sodium concentration and sodium transport in *E. coli* by ²³Na⁺ NMR, *J. Biol. Chem.,* 261, 3288, 1986a.

28. **Castle, A. M., Macnab, R. M., and Schulman, R. G.,** Coupling between the sodium and proton gradients in respiring *Escherichia coli* cells measured by ²³Na⁺ NMR, *J. Biol. Chem.,* 261, 7797, 1986b.

29. **Booth, I. R.,** Regulation of cytoplasmic pH in bacteria, *Microbiol. Rev.,* 49, 395, 1985.

30. **Bakker, E. P.,** The role of alkali-cation transport in energy coupling of neutrophilic and acidophilic bacteria: an assessment of methods and concepts, *FEMS Microbiol. Rev.,* 75, 319, 1990.

31. **Nelson, N. and Taiz, L.,** The evolution of H⁺-ATPase, *Trends Biochem. Sci.,* 14, 113, 1989.

32. **Heefner, D. L. and Harold, F. M.,** ATP-linked sodium transport in *Streptococcus faecalis.* I. The sodium circulation, *J. Biol. Chem.,* 255, 11396, 1980.

33. **Heefner, D. L. and Harold, F. M.,** ATP-driven sodium pump in *Streptococcus faecalis, Proc. Natl. Acad. Sci. U.S.A.,* 79, 2798, 1982.

34. **Unemoto, T., Tokuda, H., and Hayashi, M.,** Primary sodium pumps and their significance in bacterial energetics, in *Bacterial Energetics,* Krulwich, T. A., Ed., Academic Press, New York, 1990, 33.

35. **Ken-Dror, S., Lanyi, J. K., Schobert, B., Silver, B., and Avi-Dor, Y.,** An NADH:quinone oxidoreductase of the halotolerant bacterium Ba1 is specifically dependent on sodium ions, *Arch. Biochem. Biophys.,* 244, 766, 1986.

36. **Krulwich, T. A. and Guffanti, A. A.,** Alkalophilic bacteria, *Annu. Rev. Microbiol.,* 43, 435, 1989.

37. **Chernyak, B. V., Dibrov, P. A., Glagolev, A. N., Sherman, M., and Skulachev, V. P.,** A novel type of energetics in a marine alkali-tolerant bacterium, *FEBS Lett.,* 164, 38, 1983.

38. **Dibrov, P. A., Lazarova, R. L., Skulachev, V. P., and Verkhovskaya, M. L.,** The sodium cycle. II. Na^+ coupled oxidative phosphorylation in *Vibrio alginolyticus* cells, *Biochem. Biophys. Acta,* 850, 458, 1986.

39. **Sakai, Y., Moritani, C., Tsuda, M., and Tsuchiya, T.,** A respiration-driven and artificially driven ATP synthesis in mutants of *Vibrio parahaemolyticus*, *Biochim. Biophys. Acta,* 973, 450, 1989.

40. **Sakai-Tomita, Y., Tsuda, M., and Tsuchiya, T.,** Na^+-coupled ATP synthesis in a mutant of *Vibrio parahaemolyticus* lacking H^+-translocating ATPase activity, *Biochem. Biophys. Res. Commun.,* in press.

41. **Dibrov, P. A.,** The role of sodium ion transport in *Escherichia coli* energetics, *Biochim. Biophys. Acta,* 1056, 209, 1991.

42. **Lanyi, J. K.,** The role of Na^+ in transport processes in bacterial membranes, *Biochim. Biophys. Acta,* 559, 377, 1979.

43. **Reed, R. H.,** Halotolerant and halophilic microorganisms, in *Microbes in Extreme Environments,* Herbert, R. A. and Codd, G. A., Eds., Academic Press, London, 1986, 55.

44. **Krulwich, T. A.,** Bioenergetics of alkalophilic bacteria, *J. Membr. Biol.,* 89, 113, 1986.

45. **Krulwich, T. A. and Ivey, D. M.,** Bioenergetics in extreme environments, in *Bacterial Energetics,* Krulwich, T. A., Ed., Academic Press, New York, 1990, 417.

46. **Krulwich, T. A., Guffanti, A. A., and Seto-Young, D.,** pH homeostasis and bioenergetic work in alkalophiles, *FEMS Microbiol. Rev.,* 75, 271, 1990.

47. **Kushner, D. J. and Kamekura, M.,** Physiology of halophilic eubacteria, in *Halophilic Bacteria,* Rodriguez-Valira, F., Ed., CRC Press, Boca Raton, FL, 1988, 109.

48. **Murakami, N. and Konishi, T.,** Cooperative regulation of the Na^+/H^+-antiporter in *Halobacterium halobium* by ΔpH and $\Delta\psi$, *Arch. Biochem. Biophys.,* 281, 13, 1990.

49. **Tsuchiya, T., Raven, J., and Wilson, T. H.,** Co-transport of Na^+ and methyl-β-D-thiogalactopyranoside mediated by the melibiose transport system of *Escherichia coli,* *Biochem. Biophys. Res. Commun.,* 76, 26, 1977.

50. **Stewart, L. M. W. and Booth, I. R.,** Na^+ involvement in proline transport in *E. coli,* *FEMS Microbiol. Lett.,* 19, 161, 1983.

51. **MacDonald, R. E., Lanyi, J. K., and Greene, R. V.,** Sodium-stiumlated glutamate uptake in membrane vesicles of *Escherichia coli*: the role of ion gradients, *Proc. Natl. Acad. Sci. U.S.A.,* 74, 3167, 1977.

52. **Ishikawa, T., Hama, H., Tsuda, M., and Tsuchiya, T.,** Isolation and properties of a mutant of *E. coli* possessing defective Na^+/H^+ antiporter, *J. Biol. Chem.,* 262, 7443, 1987.

53. **Tokuda, H., Kim, Y. J., and Mizushima, S.,** *In vitro* protein translocation into inverted membrane vesicles prepared from *Vibrio alginolyticus* is stimulated by the electrochemical potential of Na^+ in the presence of *Escherichia coli* SecA, *FEBS Lett.,* 264, 10, 1990.

54. **Krumholz, L. R., Esser, U., and Simoni, R. D.,** Characterization of the H^+ pumping F_1F_0 ATPase of *Vibrio alginolyticus,* *J. Bacteriol.,* 172, 6809, 1990.

55. **Tokuda, H., Asano, M., Shimamura, Y., Unemoto, T., Sugiyama, S., and Imae, Y.,** Roles of the respiratory Na^+ pump in bioenergetics of *Vibrio alginolyticus,* *J. Biochem. (Tokyo),* 103, 650, 1988.

56. **Krulwich, T. A.,** Na^+/H^+ antiporters, *Biochim. Biophys. Acta,* 726, 245, 1983.

57. **Oesterhelt, D., Hartmann, R., Michel, H., and Wagner, G.,** Light driven proton translocation and energy conservation by halobacteria, in *Energy Conservation in Biological Membranes,* Schager, G. and Klingenberg, M., Eds., Springer-Verlag, Heidelberg, 1978, 140.

58. **Schuldiner, S. and Fishkes, H.,** Sodium-proton antiport in isolated membrane vesicles of *E. coli,* *Biochemistry,* 17, 706, 1978.

59. **Kakinuma, Y., Igarashi, K., Konishi, K., and Yamato, I.,** Primary structure of the α-subunit of vacuolar-type Na^+-ATPase in *Enterococcus hirae, FEBS Lett.,* 292, 64, 1991.

60. **MacLeod, P. R. and MacLeod, R. A.,** Cloning in *Escherichia coli* K-12 of a Na+-dependent transport system from a marine bacterium, *J. Bacteriol.*, 165, 825, 1986.

61. **Padan, E., Maisler, N., Taglicht, D., Karpel, R., and Schuldiner, S.,** Deletion of *ant* in *E. coli* reveals its function in adaptation to high salinity and an alternative Na+/H+ antiporter system(s), *J. Biol. Chem.*, 264, 20,297, 1989.

62. **Karpel, R., Alon, T., Glaser, G., Schuldiner, S., and Padan, E.,** Expression of a sodium proton antiporter (NhaA) in *Escherichia coli* is induced by Na+ and Li+ ions, *J. Biol. Chem.*, 266, 21753, 1991.

63. **Rahav-Manor, O., Carmel, O., Karpel, R., Taglicht, D., Glaser, G., Schuldiner, S., and Padan, E.,** NhaR, a protein homologous to a family of bacterial regulatory proteins (LysR) regulates *nhaA*, the sodium proton antiporter gene in *Escherichia coli*, *J. Biol. Chem.*, 267, 10,433, 1992.

64. **Hanada, K., Yamato, I., and Anraku, Y.,** Purification and reconstitution of *Escherichia coli* proline carrier using a site specifically cleavable fusion protein, *J. Biol. Chem.*, 263, 7181, 1988.

65. **Taglicht, D., Padan, E., and Schuldiner, S.,** Overproduction and purification of a functional Na+/H+ antiporter coded by *nhaA* (*ant*) from *Escherichia coli*, *J. Biol. Chem.*, 266, 11,289, 1991.

66. **Hayashi, M. and Unemoto, T.,** FAD and FMN flavoproteins participate in the sodium transport respiratory chain NADH-guinone reductase of a marine bacterium *Vibrio alginolyticus*, *FEBS Lett.*, 202, 327, 1986.

67. **Tokuda, H.,** Respiratory Na+ pump and Na+-dependent energetics in *Vibrio alginolyticus*, *J. Bioenerg. Biomembr.*, 21, 693, 1989.

68. **Laubinger, W. and Dimroth, P.,** The sodium ion translocating adenosine triphosphatase of *Propiogenium modestrum* pumps protons at low sodium ion concentrations, *Biochemistry*, 28, 7194, 1988.

69. **Laubinger, W. and Dimroth, P.,** Characterization of the ATP synthase of *Propionigenium modestum* as a primary sodium pump, *Biochemistry*, 27, 7531, 1988.

70. **Mieschendahl, M., Buchel, D., Bocklage, H., and Muller-Hill, B.,** Mutations in the *lacY* gene of *Escherichia coli* define functional organization of lactose permease, *Proc. Natl. Acad. Sci. U.S.A.*, 78, 7652, 1981.

71. **Shuman, H. A. and Beckwith, J.,** *Escherichia coli* mutants that allow transport of maltose via the galactoside transport system, *J. Bacteriol.*, 137, 365, 1979.

72. **Yazyu, H., Shiota, S., Futai, M., and Tsuchiya, T.,** Alteration in cation specificity of the melibiose transport carrier of *Escherichia coli* due to replacement of proline 122 with serine, *J. Bacteriol.*, 162, 933, 1985.

73. **Ohsawa, M., Mogi, T., Yamamoto, H., Yamato, I., and Anraku, Y.,** Proline carrier mutant of *Escherichia coli* K-12 with altered cation sensitivity of substrate-binding activity: cloning, biochemical characterization, and identification of the mutation, *J. Bacteriol.*, 170, 5185, 1988.

74. **Brooker, R. J. and Wilson, T. H.,** Isolation and nucleotide sequencing of lactose carrier mutants that transport maltose, *Proc. Natl. Acad. Sci. U.S.A.*, 82, 3959, 1985.

75. **Botfield, M. C. and Wilson, T. H.,** Mutations that simultaneously alter both sugar and cation specificity in the melibiose carrier of *Escherichia coli*, *J. Biol. Chem.*, 263, 12,909, 1988.

76. **Yamato, T., Ohsawa, M., and Anraku, Y.,** Defective cation-coupling mutants of *Escherichia coli* Na+/proline symport carrier — characterization and localization of mutations, *J. Biol. Chem.*, 265, 2450, 1990.

77. **Kaback, H. R.,** Site-directed mutagenesis and ion-gradient driven active transport: on the path of the proton, *Annu. Rev. Physiol.*, 50, 243, 1988.

78. **Kaback, H. R., Bibi, E., and Roepe, P. D.,** β-Galactoside transport in *E. coli*: a functional dissection of *lac* permease, *Trends Biochem. Sci.*, 15, 309, 1990.

79. **Padan, E., Sarkar, H. K., Viitanen, P. V., Poonian, M. S., and Kaback, H. R.,** Site-specific mutagenesis of histidine residues in the *lac* permease of *Escherichia coli,* *Proc. Natl. Acad. Sci. U.S.A.,* 82, 6765, 1985.

80. **Yamaguchi, A., Adachi, K., Akasaka, T., Ono, N., and Sawai, T.,** Metal-tetracycline/ H^+ antiporter of *Escherichia coli* encoded by transposon Tn10, *J. Biol. Chem.,* 266, 6045, 1991.

81. **Maiden, M. C. J., David, E. O., Baldwin, S. A., Moore, D. C. M., and Henderson, P. J. F.,** Mammalian and bacterial sugar transport proteins are homologous, *Nature (London),* 325, 641, 1986.

82. **Ames, G. F. L., Mimura, C. S., and Shyamala, V.,** Bacterial periplasmic permeases belong to a family of transport proteins operating from *Escherichia coli* to human: traffic ATPases, *FEMS Microbiol. Rev.,* 75, 429, 1990.

83. **Deguchi, Y., Yamato, I., and Anraku, Y.,** Nucleotide sequence of *gltS,* the Na^+/ glutamate symport carrier gene of *Escherichia coli B, J. Biol. Chem.,* 265, 21,704, 1990.

84. **Hediger, M. A., Coady, M. J., Ikeda, T. S., and Wright, E. M.,** Expression cloning and cDNA sequencing of the Na^+/glucose cotransporter, *Nature (London),* 330, 379, 1987.

85. **Reizer, J., Reizer, A., and Saier, M. H., Jr.,** The Na^+/pantothenate symporter (PanF) of *Escherichia coli* is homologous to the Na^+/proline symporter (putP) of *E. coli* and the Na^+/glucose symporters of mammals, *Res. Microbiol.,* 141, 1069, 1990.

86. **Menezes, M. E., Roepe, P. D., and Kaback, H. R.,** Design of a membrane transport protein for fluorescence spectroscopy, *Proc. Natl. Acad. Sci. U.S.A.* 87, 1638, 1990.

87. **Shiota, S., Yamane, Y., Futai, M., and Tsuchiya, T.,** *Escherichia coli* mutants possessing a Li^+-resistant melibiose carrier, *J. Bacteriol.,* 162, 106, 1985.

88. **Hara, Y. and Nakao, M.,** ATP-dependent proton uptake by proteoliposomes reconstituted with purified Na^+, K^+-ATPase, *J. Biol. Chem.,* 261, 12,655, 1986.

89. **Polvani, C. and Blostein, R.,** Protons as substitutes for sodium and potassium in the sodium pump reaction, *J. Biol. Chem.,* 263, 16,757, 1988.

90. **Polvani, C., Sachs, G., and Blostein, R.,** Sodium ions as substitutes for protons in the gastric H,K-ATPase, *J. Biol. Chem.,* 264, 17854, 1989.

91. **Boyer, P. D.,** Bioenergetic coupling to protonmotive force: should we be considering hydronium ion coordination and not group protonation?, *Trends Biochem. Sci.,* 13, 5, 1988.

92. **Damiano-Forano, E., Bassilana, M., and Leblanc, G.,** Sugar binding properties of the melibiose permease in *Escherichia coli* membrane vesicles. Effect of Na^+ and H^+ concentrations, *J. Biol. Chem.,* 261, 6893, 1986.

93. **Krulwich, T. A., Guffanti, A. A., Bornstein, R. F., and Hoffstein, T.,** A sodium requirement for growth, solute transport and pH homeostasis in *Bacillus firmus* RAB, *J. Biol. Chem.,* 57, 1885, 1982.

94. **McLaggan, D., Selwyn, M. Y., and Dawson, A. P.,** Dependence on Na^+ of control of cytoplasmic pH in a facultative alkalophile, *FEBS Lett.,* 165, 254, 1984.

95. **Rosenbusch, J. P.,** Characterization of the major envelope protein from *Escherichia coli, J. Biol. Chem.,* 249, 8019, 1974.

96. **Ingham, C., Buechner, M., and Adler, J.,** Effect of outer membrane permeability on chemotaxis in *Escherichia coli, J. Bacteriol.,* 172, 3577, 1990.

97. **Buechner, M., Delcour, A. H., Martinac, B., Adler, J., and Kung, C.,** Ion channel activities in the *Escherichia coli* outer membrane, *Biochim. Biophys. Acta,* 1024, 111, 1990.

98. **Booth, I., Douglas, R. M., Ferguson, G. P., Lamb, A. J., Munro, A.W., and Ritchie, G. Y.,** Potassium-efflux systems, in *Alkali Cation Transport Systems in Prokaryotes,* Bakker, E. P., Ed., CRC Press, Boca Raton, FL, 1992, chap. IIE.

99. **Zoratti, M. and Ghazi, A.,** Stretch-activated channels in prokaryotes, in *Alkali Cation Transport Systems in Prokaryotes,* Bakker, E. P., Ed., CRC Press, Boca Raton, FL, 1992, chap. IIH.

100. **Tsuchiya, T. and Takeda, K.,** Extrusion of sodium ions energized by respiration and glycolysis in *Escherichia coli, J. Biochem.,* 86, 225, 1979.

101. **Borbolla, M. G. and Rosen, B. P.,** Energetics of sodium efflux from *Escherichia coli, Arch. Biochem. Biophys.,* 22, 98, 1984.

102. **Bassilana, M., Damiano, E., and Leblanc, G.,** Kinetic properties of Na⁺ antiport in *Escherichia coli* membrane vesicles. Effects of imposed electrical potential, proton gradient and internal pH, *Biochemistry,* 23, 5288, 1984.

103. **Bassilana, M., Damiano, E., and Leblanc, G.,** Relationship between the Na⁺ antiport activity and the components of the electrochemical proton gradient in *Escherichia coli* membrane vesicles, *Biochemistry,* 23, 1015, 1984.

104. **Harold, F. M.,** *The Vital Force: A Study of Bioenergetics,* W. H. Freeman, New York, 1986.

105. **Qi, Y. and Adler, J.,** Salt taxis in *Escherichia coli* bacteria and its lack in mutants, *Proc. Natl. Acad. Sci. U.S.A.,* 86, 8358, 1989.

106. **Li, C., Boileau, A. J., Kung, C., and Adler, J.,** Osmotaxis in *Escherichia coli, Proc. Natl. Acad. Sci. U.S.A.,* 85, 9451, 1988.

107. **Pan, J. W. and Macnab, R. M.,** Steady-state measurements of *Escherichia coli* sodium and proton potentials at alkaline pH support the hypothesis of electrogenic antiport, *J. Biol. Chem.,* 265, 9247, 1990.

108. **Pestka, S.,** Peptidyl-puromycin synthesis on polyribosomes from *Escherichia coli, Proc. Natl. Acad. Sci. U.S.A.,* 69, 624, 1972.

109. **Walderhaug, M. O., Dosch, D. C., and Epstein, W.,** Potassium transport in bacteria in *Ion Transport in Prokaryotes,* Rosen, B. P. and Silver, S., Eds., Academic Press, New York, 1987, 85.

110. **Ennis, H. L. and Lubin, M.,** Dissociation of ribonucleic acid and protein synthesis in potassium deprived bacteria, *Biochim. Biophys. Acta,* 50, 399, 1961.

111. **Henikoff, S., Haughn, G. W., Calvo, J. M., and Wallace, J. C.,** A large family of bacterial activator proteins, *Proc. Natl. Acad. Sci. U.S.A.,* 85, 6602, 1988.

112. **Storz, G., Tartaglia, L. A., and Ames, B. N.,** Transcriptional regulator of oxidative stress-inducible genes: direct activation by oxidation, *Science,* 248, 189, 1990.

113. **Mitchell, P.,** Chemiosmotic coupling in oxidative and photosynthetic phosphorylation, *Biol. Rev. Cambridge Philos. Soc.,* 41, 445, 1966.

114. **Macnab, R. M. and Castle, A. M.,** A variable stoichiometry model for pH homeostasis in bacteria, *Biophys. J.,* 52, 637, 1987.

115. **Zilberstein, D., Padan, E., and Schuldiner, S.,** A single locus in *E. coli* governs growth in alkaline pH and growth on carbon sources whose transport is sodium dependent, *FEBS Lett.,* 116, 177, 1980.

116. **McMorrow, I., Shuman, H. A., Sze, D., Wilson, D. M., and Wilson, T. H.,** Sodium/proton antiport is required for growth of *Escherichia coli* at alkaline pH, *Biochim. Biophys. Acta,* 981, 21, 1989.

117. **Brey, R. N., Beck, J. C., and Rosen, B. P.,** Cation/proton antiporter systems in *Escherichia coli, Biochem. Biophys. Res. Commun.,* 83, 1588, 1978.

118. **Leblanc, G., Bassilana, M., and Damiano, E.,** Na⁺/H⁺ exchange in bacteria and organelles, in *Na⁺/H⁺ Exchange,* Grinstein, S., Ed., CRC Press, Boca Raton, FL, 1988, 103.

119. **Schuldiner, S., Agmon, V., Brandsma, J., Cohen, A., Freidman, E., and Padan, E.,** Induction of SOS functions by alkaline intracellular pH in *Escherichia coli, J. Bacteriol.,* 168, 936, 1986.

120. **Grinstein, S., Rotin, D., and Mason, M. J.,** Na⁺/H⁺ exchange and growth factor-induced cytosolic pH changes. Role in cellular proliferation, *Biochim. Biophys. Acta,* 988, 73, 1989.

121. **Guastella, J., Nelson, N., Nelson, H., Czyzyk, L., Keynan, S., Miedel, M. C., Davidson, N., Lester, H. A., and Kanner, B. I.,** Cloning and expression of a rat brain GABA transporter, *Science,* 249, 1303, 1990.

122. **Nakao, T., Yamato, I., and Anraku, Y.,** Nucleotide sequence of *put*P, the proline carrier gene of *Escherichia coli* K12, *Mol. Gen. Genet.,* 208, 70, 1987.

123. **Yazyu, H., Shiota-Niiya, S., Shimamoto, T., Kanazawa, H., Futai, M., and Tsuchiya, T.,** Nucleotide sequence of the *mel*B gene and characteristics of deduced amino acid sequence of the melibiose carrier in *Escherichia coli, J. Biol. Chem.,* 259, 4320, 1984.

124. **Ivey, D. M., Guffanti, A. A., Bossewitch, J. S., Padan, E., and Krulwich, T. A.,** Molecular cloning and sequencing of a gene from alkaliphilic *Bacillus firmus* OF4 that functionally complements an *Escherichia coli* strain carrying a deletion in the *nha*A Na$^+$/H$^+$ antiporter gene, *J. Biol. Chem.,* 266, 23,483, 1991.

125. **Karpel, R., Olami, Y., Taglicht, D., Schuldiner, S., and Padan, E.,** Sequencing of the gene *ant* which affects the Na$^+$/H$^+$ antiporter activity in *E. coli, J. Biol. Chem.,* 263, 10,408, 1988.

126. **Sardet, C., Franchi, A., and Pouyssegur, J.,** Molecular cloning, primary structure and expression of the human growth factor-activatable Na$^+$/H$^+$ antiporter, *Cell,* 56, 271, 1989.

127. **Waser, M., Hess-Bienz, D., Davies, K., and Solioz, M.,** Cloning and disruption of a putative Na/H-antiporter gene of *Enterococcus hirae, J. Biol. Chem.,* 267, 5396, 1992.

128. **Karpel, R.,** Cloning Control and Physiological Role of the Gene *nha*A Encoding for Sodium Proton Antiporter in *E. coli,* Ph.D. thesis, Hebrew University, Jerusalem, 1990.

129. **Pinner, E., Padan, E., and Schuldiner, S.,** Cloning sequencing and expression of *nha*B gene, encoding a Na$^+$/H$^+$ antiporter in *Escherichia coli, J. Biol. Chem.,* 267, 11,064, 1992.

130. **Pajor, A. and Wright, E. M.,** personal communication.

131. **Pinner, E., Padan, E., and Schuldiner, S.,** unpublished results.

Chapter IB

Na$^+$/H$^+$ ANTIPORTERS IN *ESCHERICHIA COLI*

Shimon Schuldiner and Etana Padan

TABLE OF CONTENTS

I. INTRODUCTION

Sodium proton antiporters are ubiquituous membrane proteins which have been found in the cytoplasmic and subcellular membranes of cells of many different origins, including plants and animals, and in microorganisms such as bacteria, algae, and fungi.[1-5] They exchange Na^+ and H^+ ions and have been assigned central roles in signal transduction and cell homeostasis such as regulation of intracellular pH (pH_{in}), cell Na^+ content, and cell volume.[6-9]

The existence of cation/H^+ antiporters was first postulated by Mitchell[10] and demonstrated in mitochondria by Mitchell and Moyle.[11] In bacteria, antiporter activity was first reported in *Streptococcus faecalis* (now *Enterococcus hirae*).[12] Since then many studies have been done which point to the fact that while in animal cells the antiporter located in the cytoplasmic membrane is the main acid extruding system, in most bacterial cells the antiporter is the major Na^+-extruding mechanism driven by the inwardly directed H^+-electrochemical gradient. The list of bacteria possessing antiporter activity is continually growing and already includes alkaliphiles,[13] halophiles,[14] thermophiles,[15] photosynthetic bacteria[16] nitrogen-fixing bacteria[17] marine bacteria,[18] methanogens,[19] and Clostridia.[20] Several bacteria have been shown to possess primary Na^+ pumps and even a complete Na^+ cycle including synthesis of ATP driven by $\Delta\bar{\mu}_{Na^+}$[21,22] (see Chapter IG). There is also some evidence suggesting the operation of a primary Na^+ pump in *Escherichia coli* grown under low $\Delta\bar{\mu}_{H^+}$.[23] In the organisms in which the antiporter was shown to coexist with primary Na^+ pumps it has been suggested that, depending on the conditions, it converts $\Delta\bar{\mu}_{Na^+}$ to $\Delta\bar{\mu}_{H^+}$ (like eukaryotic proteins) or vice versa (like prokaryotic ones).[24]

In *E. coli* the antiporter activity has been studied using a wide variety of techniques at multiple levels from the intact cell[25-27] to the pure protein reconstituted in proteoliposomes[28] and including a large body of work in isolated membrane vesicles (see below for references on specific topics). In intact cells, coupling between the movement of Na^+ and H^+ was first studied by West and Mitchell,[25] who observed proton extrusion following addition of Na^+ to an anaerobic cell suspension. Coupling between $\Delta\bar{\mu}_{Na^+}$ and $\Delta\bar{\mu}_{H^+}$ was described in a series of studies by Macnab and collaborators.[27,29-31] Borbolla and Rosen[26] also demonstrated in intact cells that extrusion of Na^+ against its electrochemical gradient is coupled to $\Delta\bar{\mu}_{H^+}$ and that ATP is not required for this process. Coupling between the ions in subcellular preparations has been demonstrated in a variety of modes: (1) an imposed Na^+ gradient drives the generation of pH gradients in right-side-out membrane vesicles[32] and of $\Delta\bar{\mu}_{H^+}$ in proteoliposomes reconstituted with pure NhaA;[28] (2) $\Delta\bar{\mu}_{H^+}$ generated by respiration[33] or ΔpH generated by an ammonium gradient[28,34-36] drives Na^+ uptake against its concentration gradient in inverted membrane vesicles and in reconstituted systems with native or pure antiporter; (3) addition of Na^+ ions to inverted membrane vesicles in which a pH gradient was generated by respiration causes a decrease in the magnitude of the preexisting pH gradient[32,37,38] (see also Reference 33); and (4) the rate of Na^+ extrusion from right-side-out membrane vesicles[32,39,40] and from proteoliposomes reconstituted with pure NhaA[28] is dependent on the $\Delta \bar{\mu}_{H^+}$ across the membrane.

It was only during recent years[41] that it became evident that in wild-type *E. coli* cells there are two distinct systems that catalyze Na^+/H^+ exchange, namely NhaA and NhaB. Since these findings are central for our understanding of the activities, we will discuss the genetic evidence that proves this contention before we discuss other properties of the antiporters.

II. MOLECULAR GENETIC APPROACH

A genetic approach has been undertaken in several laboratories to study the Na^+/H^+ antiporter system in *E. coli*. Different mutants have been isolated which have lost the Na^+/H^+ antiporter activity as well as the capacity to grow at alkaline pH, implying the role of the antiporter in pH homeostasis.[42-44] The selections were based on the inability of the mutants to grow on carbon sources that require a sodium gradient for their uptake: a mutation in locus *phs* was obtained based on an inability to grow on both glutamate and melibiose,[42] Hit-1 could not grow on serine,[43] and HS3051 was isolated as a strain that could not grow on melibiose in the presence of 100 mM NaCl.[44] Thus far only the *phs* mutation[42] has been characterized in some detail. Mapping of the *phs* mutation showed it to be an allele of *rpoA*, the gene encoding the α-subunit of the RNA polymerase.[45] The mutation causes a relatively selective transcription defect which affects several genes that seem to have a common regulatory mechanism.[46] In conclusion, even though the classical genetic approach has yielded several mutants, none of them has yet been shown to be defective in a structural gene coding for the antiporter.

A different type of mutant, with increased rather than decreased antiporter activity, has been isolated by Tsuchiya, Wilson, and collaborators[47] based on its resistance to Li^+. Li^+ ions are toxic to *E. coli* cells mostly due to their effect on the cell pyruvate kinase.[48] When grown on melibiose as a carbon source the toxicity is augmented due to their effect on the melibiose transporter, which is inhibited by Li^+ ions. Tsuchiya, Wilson, and colleagues isolated a mutant in *E. coli* that tolerates Li^+ concentrations that are otherwise toxic to the wild-type cells,[47] This mutant harbors at least two mutations responsible for the acquirement of the resistance. One is in the *melB* allele (*melBLiD*), in which replacement of proline at position 122 with serine brings about a modification in the melibiose transporter such that it can now cotransport the sugar with Li^+.[47,49] The second mutation is in an additional locus, which brings about an enhanced antiporter activity capable of an increased excretion of the toxic ion (which is also a substrate of the antiporter). We have separated the two mutations and have shown that the one, which increased antiporter activity (*nhaA*up, previously called *ant*up), is necessary to confer resistance to toxic levels of Li[+] ions.[35] The finding of this phenotype allowed us to map this mutation to about 0.3 min on the *E. coli* chromosome.[35]

We have taken advantage of the toxicity of Li[+] ions and the resistance associated with high activity of the antiporter in order to clone the wild-type *nhaA* gene.[35,36] We assumed that when in high copy number (plasmidic + chromosomal) the wild-type *nhaA* would increase Na[+]/H[+] antiporter activity and thereby confer Li[+] resistance to cells, i.e., an NhaA[up] phenotype. We surveyed a library of plasmids containing inserts covering 15 kilobase pairs (kbp) from *car* to *dnaJ*,[50,51] which should include the wild-type locus affecting the antiporter, and succeeded in cloning in an insert bearing *nhaA*. This achievement initiated the molecular biology of the Na[+]/H[+] exchange.

To study the role of the antiporter, the strategy was to inactivate the chromosomal gene by replacing most or all of it with a selectable marker. The appropriate constructs were engineered in plasmids, taking care to include enough flanking sequences to allow for homologous recombination to take place. After selection and transduction into our isogenic strains, recombination in the desired locus was tested by Southern hybridizations.[41] We constructed Δ*nhaA* strains in which either two thirds of the gene[41] or the whole gene[53] was replaced with *kan* without disrupting any of the neighboring genes. The Δ*nhaA* strains obtained grow normally in low-sodium medium, indicating that, at least under these conditions, *nhaA* is not an essential gene. Δ*nhaA* cannot adapt to high sodium concentrations which do not affect the wild type (0.7 M NaCl at pH 6.8). The Na[+] sensitivity of Δ*nhaA* is pH dependent, increasing at alkaline pH (0.1 M NaCl at pH 8.5). The Δ*nhaA* strains also cannot challenge the toxic effects of Li[+] ions (0.01 M), a substrate of the Na[+]/H[+] antiporter system. It is concluded that *nhaA* is indispensable for adaptation to high salinity, for challenging Li[+] toxicity, and for growth at alkaline pH (in the presence of Na[+]). Although it confers resistance to Li[+], *nhaA* does not increase the limits of pH or salt that *E. coli* can cope with, suggesting that factors other than *nhaA* are limiting in setting the upper limits of tolerance. In addition, the fact that the Δ*nhaA* strain can grow at alkaline pH at Na[+] concentrations below 0.1 M suggests the presence of additional or alternative mechanisms for pH homeostasis.

Indeed, growth of the Δ*nhaA* strain is normal on carbon sources which require Na[+] ions for transport, suggesting that it can generate a sodium gradient large enough to support growth even in the absence of NhaA activity. Moreover, antiporter activity, as measured in everted membrane vesicles, decreases only to 50% of the wild-type level. A detailed analysis of the remaining antiporter activity in the Δ*nhaA* strain reveals that the kinetic properties of the residual activity differ from those displayed by the NhaA protein; (1) the K_m for transport of Li[+] ions is about 15 times higher, and (2) the activity is practically independent of intracellular pH. Hence, our results demonstrate the presence of a novel alternative Na[+]/H[+] antiporter(s) in *E. coli* (designated *nhaB*) additional to *nhaA* system.[41]

In addition to learning about the role of *nhaA* in the cell physiology and to the unveiling of the alternative system, construction of the Δ*nhaA* strain provided us with a system for cloning of genes coding for other antiporters by functional complementation. The Δ*nhaA* strain is sensitive to Na⁺, and its sensitivity to Li⁺ ions is highly increased. Transformation of a Δ*nhaA* strain by a multicopy plasmid carrying *nhaA* renders the transformants resistant to the ions. It was therefore anticipated that homologous and even heterologous antiporter genes would be able to complement the deletion. Using this paradigm we have succeeded in cloning three other genes: one from *Salmonella enteritidis* homologous to *nhaA*,[52] the *nhaB* gene from *E. coli*,[54] and a novel gene from the alkaliphilic *Bacillus firmus* OF4[55] (see Chapter IE) that codes for a putative antiporter. For the cloning of *nhaB* it was also strictly necessary to prepare the library from the Δ*nhaA* strain in order to prevent recloning of *nhaA*.[54]

Cloning of *nhaB* made possible the generation of Δ*nhaB* and Δ*nhaA*Δ*nhaB* strains which already supplied further invaluable information about the components involved in the metabolism of Na⁺ and H⁺ in *E. coli*. The Δ*nhaB* strain showed no impairment in its ability to adapt to high salt or alkaline pH or in its resistance to Li⁺.[56] These findings suggest that NhaA alone can cope with the salt and pH stress and that it has a capacity high enough for these functions. Also, as will be discussed below, expression of *nhaA* is highly regulated and increases significantly under the conditions in which it is essential: high salt, alkaline pH (in the presence of Na⁺ ions), and the presence of toxic Li⁺ ions.[53]

The double mutant, Δ*nhaA*Δ*nhaB*, grows very poorly in the presence of Na⁺ concentrations as low as 10 mM, which are the contaminating levels of L broth. At concentrations of 100 mM Na⁺ (pH 7.5), growth is completely arrested. However at pH 8.4 it grows in the absence of added Na⁺ (contaminating levels of 10 mM). Analysis of the antiporter activity in membranes prepared from the Δ*nhaA*Δ*nhaB* strain shows no residual activity of Na⁺/H⁺ antiporter.[56]

Based on these results we can tentatively conclude that, in relation to *nhaB*, *nhaA* is more flexible and capable of handling a higher load of Na⁺ within the entire pH range. Thus, it can support growth in the absence of *nhaB* even at high salinity and pH. *nhaB*, on the other hand, can support growth only at relatively low sodium concentrations. In addition, *nhaB* shows a higher affinity for Na⁺ than NhaA (see below), suggesting that the recurrent theme described for many other transport systems possibly holds also for the systems handling Na⁺ and H⁺: a low-affinity, high-capacity system (in our case *nhaA*) and another high-affinity, low-capacity system (*nhaB*) are required to cope with adaptation to a wide range of concentrations. Interestingly, and unlike most other chromosomally encoded mineral transport systems,[57] in the case of the Na⁺/H⁺ antiporters the high-capacity one is regulated.[53] Very little is known thus far about the regulation of *nhaB* or the kinetic properties of the system. Also, we still do not know what is the actual contribution of each protein in the wild-type strain under various conditions.

III. PROPERTIES OF *nhaA* AND *nhaB*

The nucleotide sequences of the *nhaA* and *nhaB* genes have been determined. They encode membrane proteins of molecular weight (M_r) 41,316[28,36] and 55,543,[54] respectively. A hydrophathic evaluation of the amino acid sequence of both proteins revealed the presence, respectively, of 11 and 12 putative transmembrane-spanning segments linked by hydrophilic segments of variable length. The proteins were specifically labeled using the T7 polymerase system described by Tabor and Richardson,[58] a system designed to label, in the intact cell, solely the product of genes cloned downstream of T7 promoters. In both cases a protein is labeled which displays an apparent M_r in sodium dodecylsulfate polyacrylamide gel electrophoresis lower than that expected from the analysis of the sequence (33 and 45 kDa, respectively[28,54]), a phenomenon which has been described for many hydrophobic proteins. The label is associated with the membrane fraction even after washes with 5 *M* urea, a finding which corroborates that both proteins are intrinsic membrane components. NhaA has been purified in a functional state and shown to catalyze Na^+/H^+ exchange (see also below and Reference 28).

Sequencing of the genes allowed for an accurate mapping in the *E. coli* chromosome (Figure 1): *nhaA* is located at 0.35 min distal to *dnaJ* and between two insertion sequences detected in various strains: IS186 and IS1. Between IS186 and *nhaA* we find *gef,* a member of a gene family encoding small, toxic proteins of approximately 50 amino acids.[59] Downstream of *nhaA* there is an additional open reading frame that had already been determined[51] and was recently found to be involved in regulation of expression of *nhaA*, designated *nhaR* (see also below and Reference 60). *rps20* is located further downstream of Is1.[51] *nhaB* is located at 25.5 min, between *fadR* and the *umuCD* operon. Downstream of *nhaB* there is a small potential open reading frame whose function is yet unknown.

A mutant, Hit1, which cannot grow on serine as a carbon source in a Na^+-dependent fashion was isolated.[43] Since this mutant showed impaired Na^+ extrusion capacity and lack of growth at alkaline pH (with glycerol as a carbon source), it was concluded that the mutation affects the Na^+/H^+ antiporter activity. Recently[95] the mutation was mapped at 25.6 min on the *E. coli* chromosome, suggesting that *hit1* resides in the *nhaB* locus. However, we could not complement Hit1 by the cloned *nhaB* either for growth on serine at neutral pH or for growth on glycerol at alkaline pH. The phenotype of Hit1 also differs from Δ*nhaB* (see above). Molecular characterization of the *hit1* mutation is required.

The sequences of the two antiporters, NhaA and NhaB, show very little similarity. The overall identity as determined with the algorithm of Smith and Waterman (as described in Reference 61) is 20%.

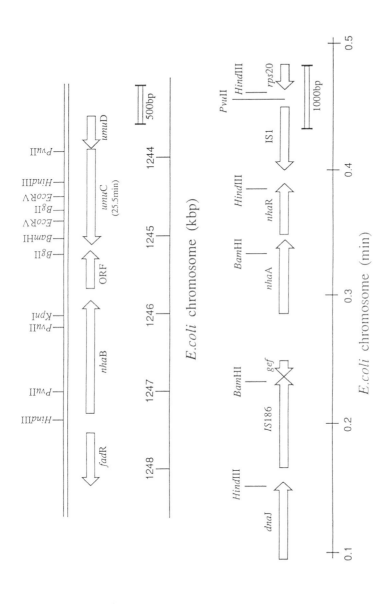

FIGURE 1. Chromosomal mapping of the genes involved in the Na⁺ cycle. The mapping of *nhaA*, *nhaR*, and *nhaB* is based on comparison of flanking sequences with published ones and comparison of restriction patterns with the restriction map of Kohara et al.[93,96]

A sizable number of genes coding for antiporters have already been cloned from various sources; in addition to *nhaA* and *nhaB*[54] from *E. coli*, these include *nhaA* from *S. enteritidis*[52] and the antiporter gene from *B. firmus*.[55] An additional gene has also been cloned from *E. hirae*.[62] From eukaryotic sources three sequences are already available: human,[63] rabbit heart,[64] and rabbit ileal villus cell from the basolateral membrane.[65] There seems to be very little conservation of sequences, not only between the eukaryotic and prokaryotic sequences, but also within the prokaryotic sequences themselves. Except for the two *nhaA* sequences from *E. coli* and *S. entritidis*, overall identities between each one of the prokaryotic sequences are in all the cases not higher than 20%, an identity level observed between almost any pair of unrelated membrane proteins. Moreover, we have not been successful at identifying recurrent motifs other than the one possibly shared by all the Na$^+$ transporters[66,67] (see Chapter IA).

IV. OVERPRODUCTION AND PURIFICATION OF NhaA

The only way to prove that a certain gene is coding for a protein that catalyzes a given activity is to purify the protein and then show that it is indeed responsible for the function and that the amino acid sequence of the isolated protein is at least in part identical with that deduced from the nucleotide sequence of the gene. With membrane proteins that catalyze only vectorial reactions this is not a trivial task, and it has been performed in very few instances. There are at least three reasons for the scarcity of membrane proteins that have been purified in a functional state: (1) the biochemistry of membrane proteins is more difficult than that of soluble proteins and necessitates the choice of proper detergents; (2) techniques for reconstitution are not always trivial and may be somewhat cumbersome and laborious for quick assessment of purifications; and (3) the proteins of interest are not always abundant, so large amounts of material must be processed.

In the case of the Na$^+$/H$^+$ antiporter, successful reconstitution has been reported already.[34,68] In both reports octyl glucoside was used as a detergent, the driving force being a ΔpH generated either by the reconstituted respiratory chain[68] or by an ammonium gradient.[34] Since the transporter could be labeled specifically using the T7 polymerase system,[36] we were able to follow its fate rapidly and efficiently during the various purification steps.[28] Probably the most important aid in the purification was the 200-fold overproduction of the antiporter. In order to overproduce the *nhaA* gene product, a plasmid was constructed (pEP3T) in which the promoterless gene was cloned downstream of the strong inducible *tac* promoter. When cells carrying pEP3T are induced with isopropylthiogalactoside (IPTG) a diffuse protein band of about 33 kDa becomes evident after 30 min, and induction is maximal after 90 to 120 min. The band is undetectable before induction, as expected from our estimate that under these conditions it represents only 0.1 to 0.2% of the membrane protein. At the end of the induction period it is the most abundant protein in the membrane. Essentially all NhaA appears to be membrane associated. Cell growth declines with kinetics similar to NhaA induction, demonstrating the detrimental effect of NhaA overproduction on growth. In dodecyl maltoside extracts of membranes prepared from induced cells the antiporter is about 20% of the total protein, about 100 to 200-fold more abundant than in extracts from wild-type cells. When membranes produced from such induced cells are solubilized and reconstituted, the Na⁺/H⁺ antiport activity observed is about 100-fold higher than the activity measured in proteoliposomes prepared from membranes derived from wild-type cells, suggesting that most, if not all, of the overproduced protein is catalytically active. An additional purification of only fourfold on 2-diethylaminoethanol (DEAE) and hydroxylapatite columns was sufficient to yield a highly purified fraction.[28]

When the activity of the purified protein (600 nmol/min/mg protein) is compared to the activity of reconstituted total membrane extract (120 nmol/min/mg protein), the rise in specific activity (fivefold) is comparable to the purification fold calculated from the ³⁵S labeling of the protein. Hence, the overall purification obtained, including the overproduction step, is about 400- to 500-fold.

This protein (which is the only antiporter purified thus far), when reconstituted into liposomes, catalyzes practically all the modes of action that were previously documented in intact cells and in membrane preparations (Figure 2). Thus, proteoliposomes reconstituted with NhaA accumulate Na⁺ ions against their concentration gradient upon imposition of a pH gradient generated by ammonium diffusion (Figure 2A and Reference 28; see Chapter IF for an explanation of the method). The activity displays an apparent K_m of 110 μM (pH 8.6).

Another mode of catalysis which has been measured in the NhaA proteoliposomes is downhill transport of ²²Na⁺ in the presence or in the absence of imposed $\Delta\bar{\mu}_{H^+}$ (Figure 2B). The most striking property of this reaction is that the antiporter is virtually turned off when the reconstituted proteoliposomes are loaded with potassium acetate and ²²Na⁺ at pH 6.5 and diluted into a medium of identical composition containing valinomycin and devoid of Na⁺. The rate of efflux increases upon imposition of either a proton gradient or an electrical potential across the membrane. Thus, in proteoliposomes diluted into media containing choline acetate and valinomycin, the membrane potential, generated by the outwardly directed K⁺ gradient, accelerates sodium efflux. Upon dilution of the proteoliposomes into media containing potassium gluconate, the proton gradient formed by the outwardly directed acetate gradient accelerates efflux many fold.[28]

The downhill movement of Na⁺ ions is coupled to the movement of H⁺ ions against their concentration gradient. This was tested in a series of

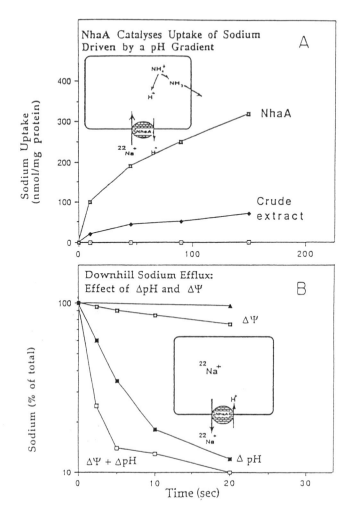

FIGURE 2. Modes of catalysis of NhaA. Proteoliposomes reconstituted with pure NhaA catalyze uphill uptake of ^{22}Na driven by a pH gradient (A) or downhill efflux of ^{22}Na$^+$ (B), which is influenced by ΔpH, by $\Delta\Psi$, and by the absolute value of the pH (C). The latter experiment is done under conditions in which generation of $\Delta\Psi$ is prevented by valinomycin (5 μm). ΔpH is short-circuited with a weak acid ([acetic acid]$_{in}$ = [acetic acid]$_{out}$ = 100 mM). For more details see the article by Taglicht et al.[28]

experiments in which proteoliposomes reconstituted with NhaA were loaded with 10 mM NaCl and with the pH indicator pyranine. Upon dilution of the proteoliposomes into a medium devoid of Na$^+$ a rapid acidification of the internal milieu is observed, as indicated by the changes in the fluorescence of the trapped pyranine (Figure 3 and Reference 28; see Section VI for the explanation of the assay). As expected, this acidification is prevented by, and reversed upon addition of, ammonium salts or nigericin.

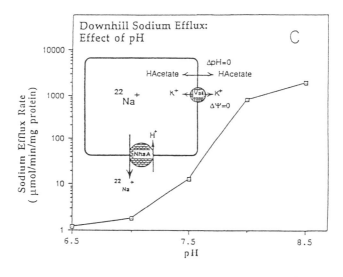

FIGURE 2C.

V. KINETIC PROPERTIES OF THE ANTIPORTERS

Most of the kinetic studies on the antiporter published thus far have been done in either whole cells or membrane vesicles from wild-type strains. As we now know, these membranes contain at least three proteins which exchange Na$^+$ and H$^+$ ions: NhaA, NhaB,[41,54] and the nonspecific Kha system.[37] Since Kha has a higher affinity for K$^+$ than for Na$^+$, in a high-K$^+$ medium its contribution is usually minimal. Still, we have two systems whose relative contributions under most conditions are not yet well characterized. However, since there is a strong dependence of the antiporter activity on pH,[28,39,40] and since NhaB shows practically no dependence on pH,[41,54] we can assume, until further information becomes available, that the dominant activity measured at pH 8.0 is that of NhaA.

Another complication of many of the kinetic measurements is the use in many of the experiments of a very indirect and qualitative measurement of the antiporter activity — namely, the ability of Na$^+$ or Li$^+$ ions to collapse a pH gradient as measured by a fluorescent acridine derivative.[32] The ability of various concentrations of the ions to change the steady-state level of fluorescence has been measured,[32,38,41,47] a parameter which is not necessarily linearly related to the rate of the antiporter activity. The range of apparent K$_m$ values obtained with this technique in wild-type strains (pH 8.0) is 3.0 to 16 mM (Table 1);[32,38,47] in NhaA^{up} strains or strains containing a multiple dose of *nhaA* the range is between 0.68 and 4.5 mM.[41,47] ΔnhaA strains yield an apparent K$_m$ value of 0.25 mM for NhaB,[41] and strains containing a multiple dose of *nhaB* display an even lower K$_m$ of 40 to 70 μM.[54] Therefore, from the data existent in the literature, taken as a whole it would seem that the

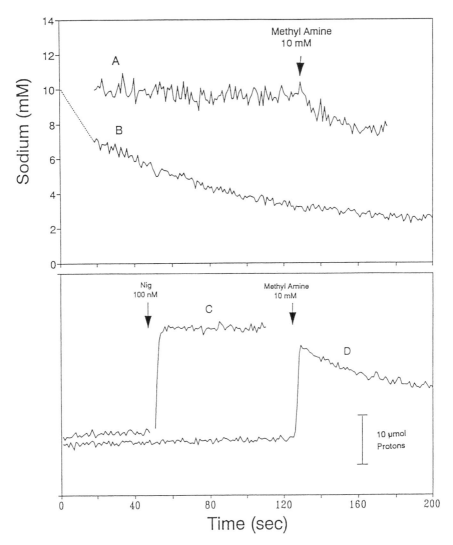

FIGURE 3. NhaA catalyzes large Na$^+$ movements only when generation of $\Delta\tilde{\mu}_{H^+}$ is prevented. Proteolipsomes were reconstituted with purified NhaA[28](4 µg protein per milliliter) in the presence of either NaCl (15 m*M*) and SBFI (0.5 m*M*, traces A and B) or pyranine (0.05 m*M*, traces C and D). The buffers used for loading the liposomes also contained 10 m*M* potassium phosphate, pH 7.5 (10 m*M*), and either 140 m*M* KCl (traces A, C, and D) or 100 m*M* potassium acetate (trace B). The external probe was removed by gel filtration through a 2-ml Sephadex® G-50 column preswollen in the same buffer in which the proteoliposomes were loaded. The proteoliposomes were diluted 50-fold into the same buffer containing valinomycin (20 n*M*) but devoid of NaCl. Where indicated by arrows, nigericin (100 n*M*) or methyl amine (10 m*M*) was added. SBFI (sodium-binding benzofuran isophtalate) fluorescence was measured at 510 nm (with a 12-nm slit) using an excitation light at 335 nm (with a 3-nm slip). Pyranine fluorescence was measured at 510 nm (3-nm slit) using an excitation light at 465 nm (3-nm slit). Additions of aliquots of sodium or acid in the presence of nigericin were used to calibrate the fluorescent response.

TABLE 1
Activities Catalyzed by the Na$^+$/H$^+$ Antiporter in Various Experimental Systems

Activity	Strain	Preparation	pH$_{out}$	Km (mM)	Vm (nmol/min/mg protein)	Ref.
ΔpH-driven ²²Na$^+$ uptake	TA15	ISO vesicles	8.8	0.5	30	35
	W3133-2S	ISO vesicles	8.8	0.4	114	35
	Pure NhaA	Proteoliposomes	8.6	0.1	3250	28
	W3133-2S	Proteoliposomes	8.0	2.4	670	34
	ML308-225	ISO vesicles	7.3	>25	nd	33
			pH$_{in}$			
Downhill Na$^+$ efflux from RSO vesicles	RA11	RSO vesicles	6.8	40	380	39
	RA11	RSO vesicles	7.0	20	380	39
	RA11	RSO vesicles	7.5	7	380	39
	RA11	RSO vesicles	7.7	3.5	380	39
		Genotype	**pH$_{out}$**			
Activity measured with acridine orange in ISO vesicles	ML308-225	nhaA$^+$ nhaB$^+$	8	3-5		32, 38
	W3133-2	nhaA$^+$ nhaB	8	16		47
	W3133-2S	nhaA^{up} nhaB$^+$	8	4.5		41, 47
	NM81	ΔnhaA nhaB$^+$	8	0.25		41
	TA15/pGM12	nhaA$^+$ nhaB$^+$/p-nhaA	8	0.68		41
	NM81/PEL24	ΔnhaA nhaB$^+$/p-nhaB	8	0.05		56

Note: ISO vesicles = inside-out membrane vesicles; RSO vesicles = right-side-out membrane vesicles; nd = not determined.

range of measurements of apparent K_m of NhaA fall, at pH 8, in the low millimolar range, while NhaB is in the submillimolar range, suggesting that NhaA is the low-affinity system while NhaB has a higher affinity for Na^+ ions.

The apparent K_m values obtained when measuring $\Delta\bar{\mu}_{H^+}$-driven $^{22}Na^+$ uptake in the vesicles from equivalent strains cannot be compared directly since the pH of the reaction is different. At pH 7.3 the system did not seem to saturate even at 50 mM,[33] at pH 8.0 it was 2.4 mM,[34] and at pH 8.8 it was 0.5 mM.[35] The apparent K_m of the pure NhaA at pH 8.6 is 0.1 mM.[28] The published data, taken at face, would suggest a dependence of the apparent K_m on pH, similar to what has been claimed for the efflux reaction (see below). However, we should stress that the three K_m values compared were obtained in three different preparations (inverted membrane vesicles, or proteoliposomes reconstituted with crude extract or with NhaA).

In experiments in which $^{22}Na^+$ efflux from right-side-out vesicles from wild-type strains was measured, the apparent K_m was dependent on the internal pH of the vesicle: 40 mM at pH 6.8 and 3.5 mM at pH 7.7, with intermediate values of 20 and 7 mM when the pH was 7 and 7.5, respectively.[39,40] The authors interpreted this effect as competition between internal H^+ and Na^+ ions.

An apparent gating of $^{22}Na^+$ efflux from right-side-out vesicles by ΔpH was reported by Bassilana et al.[39,40] The authors considered the possibility that this is not an intrinsic gating phenomenon, as suggested for the antiporter of *Habbacterium halobium*,[14] but rather an effect of low pH_{in} on the protein. Similarly, an apparent gating for uptake of $^{22}Na^+$ was reported in inverted membrane vesicles by Nakamura et al.[34] Interestingly, in inverted membrane vesicles the antiporter activity, as measured by the rate of $^{22}Na^+$ uptake driven by an ammonium gradient[34] or by the acridine orange technique,[41] seems to show a dependence on the external pH (which corresponds to the internal pH of cells or right-side-out membranes). The existence of a gating phenomenon or a need for energy even for downhill transport has also been claimed in intact cells: when $^{22}Na^+$-loaded cells were diluted into sodium-free medium, most of the radioactivity remained in the cells unless an energy source was added.[26] It would seem, however, that the lack of activity in the absence of $\Delta\bar{\mu}_{H^+}$ is not necessarily an intrinsic energetic requirement, but is most likely due to a kinetic block. In proteoliposomes reconstituted with NhaA, very rapid downhill efflux could be detected in the absence of $\Delta\bar{\mu}_{H^+}$ provided the reaction was carried out at the appropriate pH.[28]

Homologous sodium/sodium exchange or counterflow had not been observed until now in any of the systems used. External sodium has no effect on efflux of $^{22}Na^+$ from either intact cells,[26] right-side-out membrane vesicles,[39] or proteoliposomes reconstituted with NhaA,[28] a phenomenon that suggests that the translocation of sodium across the membrane may be the rate-limiting step in the overall reaction. Another possibility to explain the lack of effect of sodium in *trans* is that the antiporter is very assymetric and it does not recognize the sodium ions in the external face of the cell or the membrane. Although it is very appealing to postulate an assymetry of this type, since it would indeed facilitate the function of a protein whose function is to extrude Na$^+$, there is not yet solid experimental data to support it.

VI. THE Na$^+$/H$^+$ ANTIPORT ACTIVITY OF NhaA IS ELECTROGENIC

The issue of whether the antiporter is electrogenic has been a matter of some controversy over the years. Although it was first suggested that the exchange is electroneutral,[25] further evidence has indicated an electrogenic exchange (H$^+$:Na$^+$ > 1).[27,29,31,32,38,39] It has also been proposed that below a certain pH$_{out,}$ the antiporter is electroneutral, while above it it is electrogenic.[32] This question was addressed in an extensive and careful study by Macnab and colleagues.[27,29-31] In this study, steady-state values of $\Delta\bar{\mu}_{H^+}$ and $\Delta\bar{\mu}_{Na^+}$ were measured under various conditions in endogenously respiring *E. coli* using [23] Na$^+$- and ^{31}P-NMR spectroscopy. Na$^+$ extrusion and maintenance of a low intracellular Na$^+$ concentration were found to correlate with the development and maintenance of $\Delta\bar{\mu}_{H^+}$. At pH 6.7 a concentration ratio ([Na$^+$]$_{out}$/[Na$^+$]$_{in}$) of about 25 was observed; this was independent of extracellular Na$^+$ concentrations over the measured range of 4 to 285 m*M*, indicating that intracellular Na$^+$ concentration is not regulated (Chapter IA). When the gradients were measured at various pH values it was found that in the acidic to neutral pH range the Na$^+$ chemical potential followed the proton chemical potential quite closely, always exceeding it slightly. Above pH 7.4 there was a progressive divergence between the two values. Thus, whereas the ΔpH continued to decrease, reached zero at pH 7.5, and changed signs (pH$_{in}$ becoming more acidic than pH$_{out}$), ΔpNa [ΔpNa = ($-$log [Na$^+$]$_{in}$/[Na$^+$]$_{out}$)] practically leveled off at a value of 25 to 40 mV, corresponding to a Na$^+$ concentration gradient of 2.5- to 5-fold at the alkaline pH values. As a consequence, the apparent overall stoichiometry changes from 1.1 at pH$_{out}$ = 6.5 to 1.4 at pH$_{out}$ = 8.5.[31]

It was suggested that this change in apparent overall stoichiometry might reflect a change in the relative rates of two antiporters with different stoichiometries rather than a change in stoichiometry of a single protein.[30] Our studies with purified NhaA support the notion that it is electrogenic: as described above, downhill movement of ^{22}Na is stimulated by the imposition of a membrane potential (negative inside) even at pH values as low as 6.5 (Figure 2). The downhill movement of Na$^+$ via NhaA generates a membrane potential, as suggested by the fact that in the absence of valinomycin the efflux rate of ^{22}Na$^+$ is severalfold lower than in its presence.[28] (Figure 2). Direct measurement of a membrane potential with Oxonol VI further supports the rheogenic nature of the antiporter[69] (see following paragraph).

Direct measurement of the stoichiometry, as evaluated from rate measurements, is now approachable with the pure NhaA, and efforts in this direction are now being made in our laboratory. Na^+ movements are being followed by monitoring either $^{22}Na^+$ or changes in fluorescence of a novel sodium indicator, SBFI[69] (sodium-binding benzofuran isophtalate). Some of the problems which we face in these experiments beautifully illustrate important properties of the antiporter: massive (measurable) movements of Na^+ can be detected only under conditions in which the formation of $\Delta\bar{\mu}_{H^+}$ by the antiporter is prevented or when preformed gradients are collapsed. Representative results are shown in Figure 3: proteoliposomes loaded with Na^+ lose very small amounts of the ion (A), while a pH gradient already reaches its maximal value as soon as it can be measured (C). Addition of 10 mM methylamine transiently alkalinizes the internal milieu and allows for exit of some of the Na^+ ions. Addition of three identical aliquots of methylamine are necessary to release most of the internal Na^+. If the generation of $\Delta\bar{\mu}_{H^+}$ is prevented by doing the experiment in the presence of potassium acetate and valinomycin, half of the internal Na^+ is lost after about 60 s (B). In conclusion, it is enough to move a very small amount of Na^+ ions to allow for the rapid generation of $\Delta\bar{\mu}_{H^+}$ so that perturbations in the intracellular pH can be corrected quickly and efficiently without massive movements of Na^+. In these experiments a rough estimate of the stoichiometry of H^+/Na^+ can be obtained from the rate of buildup of the pH gradient and the rate of release of Na^+ after the addition of methylamine: the values are slightly higher than $2H^+/Na^+$.

A stoichiometry of $2H^+/Na^+$ has been estimated also in similar experiments using a thermodynamic rather than a kinetic approach. The size of the $\Delta\bar{\mu}_{H^+}$ generated by a Na^+ gradient can be predicted from the fact that at equilibrium $\Delta\bar{\mu}_{Na^+} = n\Delta\bar{\mu}_{H^+}$, which means that $\Delta pNa^+ = n\Delta pH + (n - 1)\Delta\psi$. We have measured with Oxonol VI the size of the $\Delta\psi$ generated at various ΔpNa^+, in the presence of nigericin, which allows for an electroneutral exchange of K^+ and H^+ and thereby discharges the ΔpH. The magnitude of the $\Delta\psi$ generated at various pH values (7.1 to 8.2) was consistent with a stoichiometry of two.[69]

Our results indicate that the apparent changes in stoichiometry measured in the intact cell at alkaline pH are not due to a change in stoichiometry of NhaA, but rather to its relative contribution to the Na^+ cycle.

VII. THE pH SENSOR IN NhaA

The proposed role of the antiporter in pH_{in} regulation, i.e., acidification of the cytoplasm at alkaline extracellular pH, implied that the activity should be dependent on and/or regulated by pH so that the higher the pH the higher the activity.[70] Studies of Na^+ effects on lactose-dependent H^+ circulation suggested that the Na^+/H^+ antiporter was more active at alkaline than at neutral pH.[71] Studies in right-side-out membrane vesicles by Leblanc and collaborators[39,40] showed that the activity of the antiporter is extremely dependent on pH$_{in}$. In these studies, right-side-out membrane vesicles were loaded with ^{22}Na$^+$ at various pH values and the rate of downhill efflux (V$_{Na+}$) was monitored. The authors found that imposition of $\Delta\bar{\mu}_{H^+}$ stimulated Na$^+$ efflux at all external pH values between 5.5 and 7.5. The contributions of the electrical ($\Delta\Psi$) and chemical (ΔpH) potential to the acceleration mechanism were studied by their selective dissipation with valinomycin and nigericin in respiring vesicles and by imposition of artificial gradients. $\Delta\Psi$ stimulated efflux even at acid pH (5.5), provided the internal pH was increased by imposition of a pH gradient, suggesting that at this pH the antiporter is also electrogenic. The effect of ΔpH on V$_{Na+}$ is dependent on the external pH value: at low pH the relation is nonlinear and indicates the existence of an apparent threshold. This threshold progressively decreases as the pH rises and at pH 7 and above disappears completely. These variations in behavior can be accounted for by variations in pH$_{in}$. The authors propose that the high internal H$^+$ concentration inhibits Na$^+$ efflux by competition, as suggested by the change in the apparent K$_m$ for Na$^+$ from 40 mM at pH$_{in}$ 6.8 to 3.5 mM at pH$_{in}$ 7.7. These important conclusions are complicated somewhat by several factors which could not be controlled at the time: (1) "nonstimulated" passive leaks also increased with pH and, above pH 7.5, were too high to ignore or to correct for; (2) the existence of more than one transport system for Na$^+$ was not known, and the relative contributions of each therefore could not be analyzed; and (3) changes in pH$_{in}$ always necessitated changes in ΔpH as well, and therefore the two factors could not be isolated from one another. The first two problems do not exist in proteoliposomes reconstituted with pure NhaA.

As described above, when efflux of ^{22}Na$^+$ from proteoliposomes is monitored, it is stimulated upon imposition of ΔpH, a stimulation which can be ascribed, at least in part, to the change in pH$_{in}$. Efflux has also been measured under conditions where the only driving force is the gradient of Na$^+$ generated upon dilution. In these experiments, generation of $\Delta\Psi$ or ΔpH by the action of the antiporter is prevented by the presence of valinomycin and high concentrations of K$^+$ inside and outside the proteoliposomes and with the penetrating weak acid acetate which rapidly short-circuits any pH gradient generated. Under these conditions we have measured a stimulation of up to 2000-fold in the efflux rate from proteoliposomes reconstituted with NhaA upon increase of the pH from 6.5 to 8.5. This stimulation of the purified NhaA is solely a kinetic effect since no H$^+$ ion gradient is generated under the conditions tested. Moreover, the stimulation reflects the effect solely on NhaA, without background activity of the other Na$^+$/H$^+$ antiporter(s) (NhaB). In the experiments described above, both internal and external pH were modified, and we do not know yet whether a change in the *cis* side alone is sufficient to bring about the stimulation. Also, we still have to determine how the kinetic parameters are affected by pH. The downhill efflux rates are very high, increasing up to 2.2 mmol/min/mg at pH 8.5 (Na^+_{in} = 15 mM), a value which corresponds to a turnover number of 10^3 sec^{-1}. This turnover number is one of the highest reported thus far for an ion-coupled transport system and is only ten times lower than the turnover number of the erythrocyte anion exchanger[72] (see also Chapter IIC for turnover rates of K^+ transport).

We suggest that the steep pH dependence of NhaA defines a "set point" for the activity such that NhaA is practically inactive at pH values below the intracellular homeostatic one (7.6 to 7.8). When the pH increases, the antiporter is activated so that it can acidify the cytoplasm back to the "resting pH_{in}" in a self-regulated mechanism. This idea of a molecular pH meter and titrator in the same molecule seems to be quite a successful one since it was chosen also by completely different molecules: the animal Na^+/H^+ antiporter[73] and the nonerythroid Cl/HCO_3 exchanger.[74,75] The set point of the human protein seems to be regulated by various hormones through phosphorylation of the cytoplasmic domain of the protein.[76] We do not know yet whether the set point of NhaA or the Cl/HCO_3 exchanger is regulated or modulated by physiological factors. Also, it remains to be tested whether the H^+ sensing and the H^+ transporting sites are identical or overlap somehow.

VIII. PHARMACOLOGY OF THE ANTIPORTER

Amiloride has been reported as an inhibitor of most but not all antiporters[77] in eukaryotes[78] and in some bacteria as well.[55] There is some controversy in the literature as to the effect of amiloride-like compounds on *E. coli* antiporter activity. Amiloride was claimed to inhibit Na^+-dependent changes in pH gradients in everted vesicles (K_i = 40 μM).[79] These data were interpreted as an effect on the antiporter. However, results both in our laboratory[97] and in others[80] detected a potent uncoupler activity of amiloride. We were unable to detect any significant inhibition of downhill sodium efflux catalyzed by purified NhaA, with either amiloride or the MK-685 and benzamil derivatives.[28]

Of a long list of chemical modifiers tested (reviewed in Reference 80) the only one that was found to inhibit the antiporter was the histidyl reagent diethylpyrocarbonate (DEPC). DEPC was found to inhibit Na^+ efflux in a specific way since hydroxylamine reverses the inactivation. DEPC also inhibits Na^+ efflux by purified NhaA.[69]

IX. REGULATION OF EXPRESSION OF *nhaA*

The existence of a multicomponent system responsible for homeostatis of Na^+ and H^+ ions requires the cell to carefully regulate each individual transporter. Thus far, we only have information on the regulation of expression of *nhaA*. In the *nhaA* gene we have mapped two promoters by primer extension in the 5' upstream region.[53] In addition, a quite extensive putative secondary structure in the RNA can be predicted in the 5' end of the gene,[36] and the first codon is GTG rather than ATG.[28] GTG has been found to mediate initiation of translation in about 8% of the documented *E. coli* proteins,[81] and it has been suggested that it may be used in mRNAs that are poorly translated. Also, the codon usage in *nhaA* is classical of poorly expressed proteins.[52] We estimate that under the growth conditions which are standard in our laboratory (L broth adjusted to pH 7.5 in which the Na⁺ is replaced with K⁺ and the contamination levels of Na⁺ are around 10 mM, or minimal salt medium to which sodium is not added) NhaA is a minor component of the membrane (less than 0.2%, or the equivalent of less than 500 copies per cell[28]). Expression with an exogenous promoter (*tac*) is much higher when the regulatory sequences of *nhaA* are deleted,[28] implying that, at least under some conditions, the upstream region has an inhibitory effect on expression of *nhaA*.

We have constructed a chromosomal translation fusion between *nhaA* and *lacZ* (*nhaA'-'lacZ*) and have found that the levels of expression are very low unless Na⁺ or Li⁺ is added. Na⁺ and Li⁺ ions increase expression in a time- and concentration-dependent manner:[53] maximal increase is detected when the cells are exposed to 50 to 100 mM of either ion for a period of 2 h. The effect is specific to the nature of the cation and is not related to a change in osmolarity. When either a functional NhaA or NhaB is coexpressed to high levels from multicopy plasmids, the effect of 100 mM Na⁺ is undetectable.[60] Since the antiporter genes do not share significant homology we conclude that it is their activity, rather than their DNA or proteins, which is responsible for eliminating the induction by Na⁺. Since the antiporter activity reduces intracellular Na⁺, the findings suggest that intracellular (rather than extracellular) Na⁺ is the signal for this process. In the same vein, alkaline pH potentiates the effect of the ions: while 10 mM Na⁺ has no effect at pH 7.5, its effect is maximal at pH 8.5. This synergistic effect with pH could reflect, at least in part, the fact that the Na⁺ gradient which the cell can maintain decreases, and therefore [Na]$_{in}$ increases with an increase in pH of the medium.[29,31]

The pattern of regulation of *nhaA* reflects, therefore, its role in adaptation to high salinity and alkaline pH in *E. coli*, as revealed by the analysis of the Δ*nhaA* strains. This pattern also suggests the possibility of involvement of novel regulatory genes in addition to *nhaA* and *nhaB*.

A gene downstream of *nhaA*, *nhaR* (previously known as *antO*),[82] or 28-kDa protein,[51] is proposed to play a role in the regulation of *nhaA*. In addition to its proximity to *nhaA*, and due to the fact that there are no conspicuous consensus sequences of either terminators or promoters between the two genes, expression with foreign promoters cloned upstream of *nhaA* brings about expression of *nhaR* as well.[36] A multiple dose of *nhaR* enhances the Na⁺-dependent induction of the *nhaA'-'lacZ* fusion (see Table 1 of Chapter IA). The fact that the dose level affects the induction by Na⁺, but not the basic level of expression, suggests that the Na^+ induction involves *nhaR* either directly or indirectly. NhaR exerts its effect in *trans,* as shown in the latter experiments. Furthermore, extracts derived from cells overexpressing *nhaR* exhibit DNA binding capacity specific to the upstream sequences of *nhaA*, as observed by gel retardation assays.[60] Inactivation of chromosomal *nhaR* by insertion unveils a phenotype of sensitivity to Li^+ higher than that displayed by the wild type.[60] A change of tolerance toward Na^+ in these cells becomes apparent only at pH 8.5, under conditions which the load seems to be more pronounced, as suggested by the phenotype of the Δ*nhaA* strain and by the pattern of regulation of *nhaA*. Both phenotypes are corrected by *nhaR* in *trans.*[60]

On the basis of the latter results, it is proposed that NhaR is a positive regulator of *nhaA*. This suggestion is in accordance with the fact that NhaR belongs to the OxyR-LysR family of positive regulators first described by Henikoff et al.[82] and also studied by Christman et al.[83] All the proteins in this group have in their N-terminus a conserved helix-turn-helix domain which is supposed to bind to DNA. Several of these proteins are involved in the response of the organism to stress, such as, for example, OxyR, which is essential for the resistance of the organism to oxidative stress.[84] NhaR may represent a component of another type of stress response essential for tolerance to Na^+ and Li^+.

The regulation of gene expression of anion and cation transport systems is gaining more and more attention. Some of them, such as, for example, the Pst, Kdp, Fur, magnesium, and sulfate transport systems, are quite well understood (reviewed recently by Silver and Walderhaug[57]). Regulation of these systems seems to fall into two categories: two component pairs of regulatory proteins or a single one that both senses and regulates. Also, a common theme of the chromosomally encoded mineral transport systems studied thus far is the presence of multiple transport systems for each ion: a low-affinity, high-capacity transport system which is expressed constitutively, and a high-affinity, high-capacity transport system which is carefully regulated in response to ion starvation. In iron-regulated systems, the sensor and effector activities are carried out by the same protein, Fur. The protein responds to the intracellular, rather than extracellular, iron levels as opposed to the systems that sense inorganic phosphate (Pi) and Mg^{++}, which respond to the extracellular levels. One possible reason is that intracellular Fe^{++} levels are generally low, whereas intracellular Pi and Mg^{++} always remain high (above millimolar levels) to support metabolism even when the extracellular concentration drops toward zero.[57] Hence, regulation of iron transport can employ an intracellular signal. The regulation of the Na^+ transport systems seem to respond to intracellular Na^+ sharing some of the properties displayed by the Fur-regulated systems rather than with the other ions.

However, Fur is a repressor of transcription, whereas PhoB and NhaR belong to the family of positive activators of transcription, similar to the case of plasmid-based toxic ion resistance systems,[57] in which ion binding seems to lead to the stimulation of mRNA synthesis instead of its repression.

X. THE ROLE OF Na$^+$/H$^+$ ANTIPORTERS IN REGULATION OF pH

One of the central roles assigned to the antiporter is in the regulation of intracellular pH (pH$_{in}$), mainly at alkaline extracellular pH. In *E. coli*, pH$_{in}$ has been shown to be clamped at around 7.8 despite huge changes in the extracellular medium pH.[6,7,70] Since then, many bacteria as well as eukaryotic cells have been shown to strictly maintain a constant cytoplasmic pH at around neutrality.[6-9,13,31,70,94] Relatively small increases in pH$_{in}$ stop cell division and activate expression of specific genes[85] and of regulons.[86,87] It is therefore not surprising that both eukaryotic and prokaryotic cells have evolved several pH$_{in}$ regulative mechanisms to eliminate metabolically induced changes in pH$_{in}$ or to counter extreme environmental conditions.[2,7,9,63,76]

We have proposed that Na$^+$/H$^+$ antiporters in conjunction with the primary H$^+$ pumps are responsible for homeostasis of intracellular pH in *E. coli*[6,32,70] (see also Chapter IA). This suggestion had its most dramatic experimental validation in alkaliphiles, in which it was shown that Na$^+$ ions are required for acidification of the cytoplasm and for growth.[88-90] In neutrophiles such as *E. coli* there is no direct evidence that supports this contention since it is not clearly established that Na$^+$ is required for growth at alkaline pH. In some alkaliphiles the requirement for Na$^+$ is not easy to demonstrate either, presumably due to a very high affinity for Na$^+$ (as low as 0.5 m*M*), such that the contamination present in most media suffices to support growth.[91] McMorrow et al.[44] have taken special precautions to reduce Na$^+$ to very low levels (5 to 15 μ*M*) and have reported a strict requirement for Na$^+$ (saturable at 100 μ*M*) for growth of *E. coli* at pH 8.5. This range of concentrations of Na$^+$ required for growth is well within the range of the K$_m$ of the NhaB system (40 to 70 μ*M*).[54]

The Na$^+$/H$^+$ antiporter mutants generated in our laboratory[41,56] show that, in the absence of added Na$^+$, cells can grow at alkaline pH even without the antiporter genes. When Na$^+$ is added (>0.1 *M*), *nhaA,* but not *nhaB* alone, can cope with the load. As described above, *nhaA* seems to be more flexible and capable of handling higher loads, as suggested not only by the phenotypes obtained but also because this system is highly regulated, both at the level of the protein[28] and at the level of gene expression.[53,60] In all of the above documented cases we are assuming that growing cells regulate their pH, but no measurements of internal pH are available yet with any of the mutants described.

Although the finding described strongly support the involvement of the antiporters in regulation of pH, at least in the presence of sodium ions, there are still many unknown components. The primary means of H$^+$ extrusion under aerobic conditions is the electron transport chain, which in the absence of permeable ions would develop a negligible ΔpH and a large $\Delta\Psi$. There is now evidence suggesting that Na^+ and K^+ are involved in the modulation of $\Delta\bar{\mu}_{H^+}$ components: electrogenic uptake of K^+, with or without H^+, would compensate for charge extrusion and thus permit the development of ΔpH; an electrogenic antiport of Na^+ and H^+ would be a means of generating an inverted ΔpH. The reported change in apparent overall stoichiometry may reflect the change in the relative rates of NhaA and NhaB, two antiporters which may have different stoichiometries. We know that NhaA is electrogenic, with a stoichiometry of close to two.[69] We do not know whether NhaB is electrogenic. However, we have measured rates of downhill Na^+ transport in membrane vesicles of $\Delta nhaA$ and wild-type strains at various pH values assuming that they reflect the rate of NhaB and NhaB + NhaA, respectively. From the maximal rate of NhaA activity in its purified state and from its estimated abundance, we have calculated its contribution to the overall activity of the cell under various conditions. It seems that NhaB is the main activity at acid pH while NhaA is dominant at alkaline pH.

NhaA is a highly active system: the maximal value of downhill Na^+ efflux at pH 8.5 is around 400 nmol/min/mg cell protein, and it can increase up to 4000 when *nhaA* is fully induced. Since the rate of H^+ extrusion through the respiratory chain is around 1000 nmol/min/mg cell protein and the rate of K^+ transport through the Trk systems is aroung 500[92] (Chapter IIC), our measurements show that indeed NhaA can quantitatively account for a rapid response to changes in ion content (Chapter IA). Quantitative measurement of NhaB activity and study of *nhaB* regulation will provide a more detailed understanding of the Na^+ and H^+ cycles in *E. coli*. Understanding of these cycles will provide important clues for our comprehension of the process of adaptation to extreme pH and salt environments.

ACKNOWLEDGMENTS

We thank D. Taglicht for critical reading of the manuscript and for suggestions. The research from the authors' laboratory that is reported in this chapter was supported by a research grant from the United States-Israel Binational Science Foundation (BSF).

REFERENCES

1. **Krulwich, T. A.,** Na^+/H^+ antiporters, *Biochim. Biophys. Acta,* 726, 245, 1983.
2. **Grinstein, S., Ed.,** Na^+/H^+ Exchange, CRC Press, Boca Raton, FL, 1988.
3. **Haigh, J. R. and Phillips, J. H.,** A sodium/proton antiporter in chromaffin granule membranes, *Biochem. J.,* 257, 499, 1989.

4. **Blumwald, W. and Poole, R. J.**, Na$^+$/H$^+$ antiporter in isolated tonoplast vesicles from storage tissue of *Beta vulgaris*, *Plant Physiol.*, 78, 163, 1985.

5. **Katz, A., Kaback, R., and Avron, M.**, Na$^+$/H$^+$ antiport in isolated plasma membrane vesicles from the halotolerant alga *Dunaliella salina*, *FEBS Lett.*, 202, 141, 1986.

6. **Padan, E., Zilberstein, D., and Schuldiner, S.**, pH homeostasis in bacteria, *Biochim. Biophys. Acta*, 650, 151, 1981.

7. **Booth, I. R.**, Regulation of cytoplasmic pH in bacteria, *Microbiol. Rev.*, 49, 395, 1985.

8. **Pouyssegur, J., Sardet, C., Franchi, A., L'Allemain, G., and Paris, S.**, A specific mutation abolishing Na$^+$/H$^+$ antiport activity in hamster fibroblasts precludes growth at neutral and acidic pH, *Proc. Natl. Acad. Sci. U.S.A.*, 81, 4833, 1984.

9. **Grinstein, S., Rotin, D., and Mason, M. J.**, Na$^+$/H$^+$ exchange and growth factor-induced cytosolic pH changes. Role in cellular proliferation, *Biochim. Biophys. Acta*, 988, 73, 1989.

10. **Mitchell, P.**, Coupling of phosphorylation to electron and hydrogen transfer by a chemiosmotic type of mechanism, *Nature (London)*, 191, 144, 1961.

11. **Mitchell, P. and Moyle, J.**, Respiration-driven proton translocation in rat liver mitochondria, *Biochem. J.*, 105, 1147, 1967.

12. **Harold, F. M. and Papineau, D.**, Cation transport and electrogenesis by *Streptococcus faecalis*, *J. Membr. Biol.*, 8, 45, 1972.

13. **Krulwich, T. A.**, Bioenergetics of alkalophilic bacteria, *J. Membr. Biol.*, 89, 113, 1986.

14. **Lanyi, J. K.**, The role of sodium ions in bacterial membrane transport, *Biochim. Biophys. Acta*, 559, 377, 1979.

15. **Goto, K., Hirata, H., and Kagawa, Y.**, A stable Na$^+$/H$^+$ antiporter of thermophilic bacterium PS3, *J. Bioenerg. Biomembr.*, 12, 297, 1980.

16. **Padan, E. and Vitterbo, A.**, Cation transport in cyanobacteria, *Methods Enzymol.*, 167, 561, 1987.

17. **Bhattacharyya, P. and Barnes, E. M. J. R.**, Proton-coupled sodium uptake by membrane vesicles from *Azotobacter vinelandii*, *J. Biol. Chem.*, 253, 3848, 1978.

18. **Niven D. F. and MacLeod, B. A.**, Sodium ion-proton antiport in a marine bacterium, *J. Bacteriol.*, 134, 782, 1978.

19. **Muller, V., Blaut, M., and Gottschalk, G.**, Generation of a transmembrane gradient of Na$^+$ in *Methanosarcina barkeri*, *Eur. J. Biochem.*, 162, 461, 1987.

20. **Terracciano, J. S., Schreurs, W. J .A., and Kashket, E.R.**, Membrane H$^+$ conductance of *Clostridium thermoaceticum* and *Clostridium acetobutilicum*: evidence for electrogenic sodium proton antiport in *Clostridium thermoaceticum*, *Appl. Env. Microbiol.*, 53, 782, 1987.

21. **Skulachev, V. P.**, Bacterial sodium transport: bioenergetic functions of sodium ions, in *Ion Transport in Prokaryotes*, Rosen, B. P. and Silver, S., Eds., Academic Press, New York, 1987, 131.

22. **Dimroth, P.**, Sodium ion transport: decarboxylases and other aspects of sodium ion cycling in bacteria, *Microbiol. Rev.*, 51, 320, 1987.

23. **Dibrov, P.**, The role of sodium ion transport in *Escherichia coli* energetics, *Biochim. Biophys. Acta*, 1056, 209, 1991.

24. **Kaesler, B. and Schonheit, P.**, The sodium cycle in methanogenesis CO$_2$ reduction to the formaldehyde level in methanogenic bacteria is driven by a primary electrochemical potential of Na$^+$ generated by formaldehyde reduction to CH$_4$, *Eur. J. Biochem.*, 186, 309, 1989.

25. **West, I. C. and Mitchell, P.**, Proton/sodium antiport in *E. coli*, *Biochem. J.*, 144, 87, 1974.

26. **Borbolla, M. G. and Rosen, B. P.**, Energetics of sodium efflux from *Escherichia coli*, *Arch. Biochem. Biophys.*, 22, 98, 1984.

27. **Castle, A. M., Macnab, R.M., and Schulman, R.G.**, Measurements of intracellular sodium concentration and sodium transport in *E. coli* by ^{23}Na$^+$ NMR, *J. Biol. Chem.*, 261, 3288, 1986.

28. **Taglicht, D., Padan, E., and Schuldiner, S.,** Overproduction and purification of a functional Na^+/H^+ antiporter coded by *nha*A (*ant*) from *Escherichia coli*, *J. Biol. Chem.*, 266, 11289, 1991.

29. **Castle, A. M., Macnab, R. M., and Shulman, R. G.,** Coupling between the sodium and proton gradients in respiring *Escherichia coli* cells measured by ^{23}Na NMR, *J. Biol. Chem.*, 261, 7797, 1986.

30. **Macnab, R. M. and Castle, A. M.,** A variable stoichiometry model for pH homeostasis in bacteria, *Biophys. J.*, 52, 637, 1987.

31. **Pan, J. W. and Macnab, R. M.,** Steady-state measurements of *Escherichia coli* sodium and proton potentials at alkaline pH support the hypothesis of electrogenic antiport, *J. Biol. Chem.*, 265, 9247, 1990.

32. **Schuldiner, S. and Fishkes, H.,** Sodium-proton antiport in isolated membrane vesicles of *E. coli*, *Biochemistry*, 17, 706, 1978.

33. **Reenstra, W. W., Patel, L., Rottenberg, H., and Kaback, H. R.,** Electro-chemical proton gradient in inverted membrane vesicles from *Escherichia coli*, *Biochemistry*, 19, 1, 1980.

34. **Nakamura, T., Hsu, C., and Rosen, B. P.,** Cation/proton antiport systems in *Escherichia coli* — solubilization and reconstitution of ΔpH driven sodium/proton and calcium/proton antiporters, *J. Biol. Chem.*, 261, 678, 1986.

35. **Goldberg, B. G., Arbel, T., Chen, J., Karpel, R., Mackie, G. A., and Schuldiner, S.,** Characterization of Na^+/H^+ antiporter gene of *E. coli*, *Proc. Natl. Acad. Sci. U.S.A.*, 84, 2615, 1987.

36. **Karpel, R., Olami, Y., Taglicht, D., Schuldiner, S., and Padan, E.,** Sequencing of the gene *ant* which effects the Na^+/H^+ antiporter activity in *E. coli*, *J. Biol. Chem.*, 263, 10408, 1988.

37. **Brey, R. N., Beck, J. C., and Rosen, B. P.,** Cation/proton antiporter systems in *Escherichia coli*, *Biochem. Biophys. Res. Commun.*, 83, 1588, 1978.

38. **Beck, J. C. and Rosen, B. P.,** Cation/proton antiport systems in *Escherichia coli*: properties of the sodium/proton antiporter, *Arch. Biochem. Biophys.*, 194, 208, 1979.

39. **Bassilana, M., Damiano, E., and Leblanc, G.,** Kinetic properties of Na^+-H^+ antiport in *Escherichia coli* membrane vesicles: Effects of imposed electrical potential, proton gradient and internal pH, *Biochemistry*, 23, 5288, 1984.

40. **Bassilana, M., Damiano, E., and Leblanc, G.,** Relationship between the Na^+-H^+ antiport activity and the components of the electrochemical proton gradient in *Escherichia coli* membrane vesicles, *Biochemistry*, 23, 1015, 1984.

41. **Padan, E., Maisler, N., Taglicht, D., Karpel, R., and Schuldiner, S.,** Deletion of *ant* in *E. coli* reveals its function in adaptation to high salinity and an alternative Na^+/H^+ antiporter system(s), *J. Biol. Chem.*, 264, 20297, 1989.

42. **Zilberstein, D., Padan, E., and Schuldiner, S.** A single locus in *E. coli* governs growth in alkaline pH and on carbon sources whose transport is sodium dependent. *FEBS Lett.*, 116, 177, 1980.

43. **Ishikawa, T., Hama, H., Tsuda, M., and Tsuchiya, T.,** Isolation and properties of a mutant of *E. coli* possessing defective Na^+/H^+ antiporter, *J. Biol. Chem.*, 262, 7443, 1987.

44. **McMorrow, I., Shuman, H. A., Sze, D., Wilson, D. M., and Wilson, T. H.,** Sodium/proton antiport is required for growth of *Escherichia coli* at alkaline pH, *Biochim. Biophys. Acta*, 981, 21, 1989.

45. **Rowland, G. C., Giffard, P. M., and Booth, I. R.,** Genetic studies of the *phs* locus of *Escherichia coli*, a mutation causing pleiotropic lesions in metabolism and pH homeostasis, *FEBS Lett.*, 173, 295, 1984.

46. **Giffard, P. M., Rowland, G. C., Kroll, R. G., Stewart, L. M. D., Bakker, E. P., and Booth, I. R.,** Phenotypic properties of a unique *rpo*A mutation (*phs*) of *Escherichia coli*, *J. Bacteriol.*, 164, 904, 1985.

47. **Niiya, S., Yamasaki, K., Wilson, T. H., and Tsuchiya, T.,** Altered cation coupling to melibiose transport in mutants of *Escherichia coli, J. Biol. Chem.,* 257, 8902, 1982.

48. **Umeda, K., Shiota, S., Futai, M., and Tsuchiya, T.,** Inhibitory effect of Li⁺ on cell growth and pyruvate kinase activity of *Escherichia coli, J. Bacteriol.,* 160, 812, 1984.

49. **Yazyu, H., Shiota, S., Futai, M., and Tsuchiya, T.,** Alteration in cation specificity of the melibiose transport carrier of *Escherichia coli* due to replacement of proline 122 with serine, *J. Bacteriol.,* 162, 933, 1985.

50. **Mackie, G. A.,** Cloning of fragments of *dap*B2 DNA and identification of the *dap*B gene product, *J. Biol. Chem.,* 255, 8928, 1980.

51. **Mackie, G. A.,** Structure of the DNA distal to the gene for ribosomal protein S20 in *E. coli* K12: presence of a strong terminator and an IS1 element, *Nucleic Acids Res.,* 14, 6965, 1986.

52. **Pinner, E., Carmel, O., Bercovier, H., Sela, S., Padan, E., and Schuldiner, S.,** Cloning, sequencing and expression of the *nha*A and *nha*R genes from *Salmonella enteritidis, Arch. Microbiol.,* 157, 323, 1992.

53. **Karpel, R., Alon, T., Glaser, G., Schuldiner, S., and Padan, E.,** Expression of a sodium proton antiporter (NhaA) in *Escherichia coli* is induced by Na⁺ and Li⁺ ions, *J. Biol. Chem.,* 266, 21753, 1991.

54. **Pinner, E., Padan, E., and Schuldiner, S.,** Cloning, sequencing and expression of the *nha*B gene, encoding a Na⁺/H⁺ antiporter in *E. coli, J. Biol. Chem.,* 267, 11,064, 1992.

55. **Ivey, D. M., Guffanti, A. A., Bossewitch, J. S., Padan, E., and Krulwich, T. A.,** Molecular cloning and sequencing of a gene from alkaliphilic *Bacillus firmus* OF4 that functionally complements an *Escherichia coli* strain carrying a deletion in the *nha*A Na⁺/H⁺ antiporter gene, *J. Biol. Chem.,* 266, 23483, 1991.

56. **Pinner, E., Kotler, Y., Padan, E., and Schuldiner, S.,** in preparation.

57. **Silver, S. and Walderhaug, M.,** Gene regulation of plasmid and chromosomal-determined inorganic ion transport in bacteria, *Microbiol. Rev.,* in press.

58. **Tabor, S. and Richardson, C. C.,** A bacteriophage T7 RNA polymerase/promoter system for controlled exclusive expression of specific genes, *Proc. Natl. Acad. Sci. U.S.A.,* 82, 1074, 1985.

59. **Poulsen, L. K., Larsen, N. W., Molin, S., and Andersson, P.,** A family of genes encoding a cell-killing function may be conserved in all gram-negative bacteria, *Mol. Microbiol.,* 3, 1463, 1989.

60. **Rahav-Manor, O., Carmel, O., Karpel, R., Taglicht, D., Glaser, G., Schuldiner, S., and Padan, E.,** NhaR, a protein homologous to a family of bacterial regulatory proteins (LysR) regulates *nha*A, the sodium proton antiporter gene in *E. coli, J. Biol. Chem.,* 267, 10,433, 1992.

61. **Devereux, J., Haeberli, P., and Smithies, O.,** A comprehensive set of sequence analysis programs for the VAX, *Nucleic Acids Res.,* 12, 387, 1984.

62. **Waser, M., Hess-Bienz, D., Davies, K., and Solioz, M.,** Cloning and disruption of a putative NaH-antiporter gene of *Enterococcus hirae, J. Biol. Chem.,* 267, 5396, 1992.

63. **Sardet, C., Franchi, A., and Pouyssegur, J.,** Molecular cloning, primary structure and expression of the human growth factor-activatable Na⁺/H⁺ antiporter, *Cell,* 56, 271, 1989.

64. **Fliegel, L., Sardet, C., Pouyssegur, J., and Barr A.,** Identification of the protein and cDNA of the cardiac Na⁺/H⁺ exchanger, *FEBS Lett.,* 279, 25, 1991.

65. **Tse, C. M., Ma, A. I., Yang, V. W., Watson, A. J. M., Levine, S., Montrose, M. H., Potter, J., Sardet, C., Pouyssegur, J., and Donowitz, M.,** Molecular cloning and expression of a cDNA encoding the rabbit ileal villus cell basolateral membrane Na⁺/H⁺ exchanger, *EMBO J.,* 10, 1957, 1991.

66. **Deguchi, Y., Yamato, I., and Anraku, Y.,** Nucleotide sequence of *glt*S, the Na⁺/glutamate symport carrier gene of *Escherichia coli, J. Biol. Chem.,* 265, 21704, 1990.

67. **Padan, E. and Schuldiner, S.**, Na$^+$-transport systems in prokaryotes, in *Alakli Cation Transport Systems in Prokaryotes,* Bakker, E. P., Ed., CRC Press, Boca Raton, FL, 1992, chap. IA.

68. **Tsuchyia, T., Misawa, A., Miyaka, Y., Yamasaki, K., and Niiya, S.**, Solubilization and reconstitution of membrane energy-transducing systems in *E. coli, FEBS Lett.,* 142, 231, 1982.

69. **Taglicht, D.**, Ph.D. thesis, *The sodium proton antiporter from* Escherichia coli: *A molecular and biochemical approach,* Hebrew University of Jerusalem, 1992.

70. **Padan, E., Zilberstein, D., and Rottenberg, H.**, The proton electrochemical gradient in *E. coli, Eur. J. Biochem.,* 63, 533, 1976.

71. **Zilberstein, D., Schuldiner, S., and Padan, E.**, The electrochemical gradient of protons and active transport of lactose in *E. coli* cells, *Biochemistry,* 18, 669, 1979.

72. **Kabantchik, J.**, The anion transport system of red blood cell membranes, in *Blood Cell Biochemistry,* Vol. 1, Harris, J. R., Ed., Plenum Press, New York, 1990, 337.

73. **Aronson, P. S.**, Kinetic properties of the plasma membrane Na$^+$/H$^+$ exchanger, *Annu. Rev. Physiol.,* 47, 545, 1985.

74. **Olsnes, S., Tonnessen, T. I., and Sandvig, K.**, pH-regulated anion antiport in nucleated mammalian cells, *J. Cell Biol.,* 102, 967, 1986.

75. **Olsnes, S., Tonnessen, T. I., Ludt, J., and Sandvig, K.**, Effect of intracellular pH on the rate of chloride uptake and efflux in different mammalian cell lines, *Biochemistry,* 26, 2778, 1987.

76. **Sardet, C., Counillon, L., Franchi, A., and Pouyssegur, J.**, Growth factors induce phosphorylation of the Na$^+$/H$^+$ antiporter, a glycoprotein of 110 kD, *Science,* 247, 723, 1990.

77. **Raley-Susman, K. M., Cragoe, E. J., Jr., Sapolsky, R. M., and Kopito, R. R.**, Regulation of intracellular pH in cultured hippocampal neurones by an amiloride-insensitive Na$^+$/H$^+$ exchanger, *J. Biol. Chem.,* 266, 2739, 1991.

78. **Benos, D. J.**, Amiloride: chemistry, kinetics and structure activity relationships, in *Na$^+$/H$^+$ Exchange,* Grinstein, S., Ed., CRC Press, Boca Raton, FL, 1988, 121.

79. **Mochizuki-Oda, N. and Oosawa, F.**, Amiloride sensitive Na$^+$/H$^+$ antiporter in *Escherichia coli, J. Bacteriol.,* 163, 395, 1985.

80. **Leblanc, G., Bassilana, M., and Damiano, E.**, Na$^+$/H$^+$ exchange in bacteria and organelles, in *Na$^+$/H$^+$ Exchange* Grinstein, S., Ed., CRC Press, Boca Raton, FL, 1988, 103.

81. **Gold, L. and Stromo, G.**, Translational initiation in *E. coli* and *S. typhimurium* in, *Cellular and Molecular Biology,* Neidhardt, F. C., Ed., American Society for Microbiology, Washington, D.C., 1987, 1302.

82. **Henikoff, S., Haughn, G. W., Calvo, J. M., and Wallace, J. C.**, A large family of bacterial activator proteins, *Proc. Natl. Acad. Sci. U.S.A.,* 85, 6602, 1988.

83. **Christman, M. F., Storz, G., and Ames, B. N.**, OxyR, a positive regulator of hydrogen peroxide-inducible genes in *Escherchia coli* and *Salmonella typhimurium,* is homologous to a family of bacterial regulatory proteins, *Proc. Natl. Acad. Sci. U.S.A.,* 86, 3484, 1989.

84. **Storz, G., Tartaglia, L. A., and Ames, B. N.**, Transcriptional regulator of oxidative stress-inducible genes: direct activation by oxidation, *Science,* 248, 189, 1990.

85. **Bingham, R. J., Hall, K. S., and Slonczewski, J. L.**, Alkaline induction of a novel gene locus, *alx,* in *Escherichia coli, J. Bacteriol.,* 172, 2184, 1989.

86. **Schuldiner, S., Agmon, V., Brandsma, J., Cohen, A., Friedman, E., and Padan, E.**, Induction of SOS functions by alkaline intracellular pH in *Escherichia coli, J. Bacteriol.,* 168, 936, 1986.

87. **Padan, E. and Schuldiner, S.**, Intracellular pH and membrane potential as regulators in the prokaryotic cell, *J. Membr. Biol.,* 95, 189, 1987.

88. **Krulwich, T. A., Guffanti, A. A., Bornstein, R. F., and Hoffstein, T.,** A sodium requirement for growth, solute transport and pH homeostasis in *Bacillus firmus* RAB, *J. Biol. Chem.*, 257, 1885, 1982.

89. **Krulwich, T. A., Federbush, J. G., and Guffanti, A. A.,** Presence of a non-metabolizable solute that is translocated with Na+ enhances Na+-dependent pH homeostasis in an alkalophilic *Bacillus*, *J. Biol. Chem.*, 260, 4055, 1985.

90. **McLaggan, D. M., Selwyn, J., and Dawson, A. P.,** Dependence on Na+ of control of cytoplasmic pH in a facultative alkalophile, *FEBS Lett.*, 165, 254, 1984.

91. **Sugiyama, S. H., Matsukura, H., and Imae, Y.,** Relationship between Na+-dependent flagellar rotation and amino acid transport in alkalophilic *Bacillus*, *FEBS Lett.*, 182, 265, 1985.

92. **Walderhaug, M. O., Dosch, D. C., and Epstein, W.,** Potassium transport in bacteria, in *Ion Transport in Prokaryotes,* Rosen, B. P. and Silver, S., Eds., Academic Press, San Diego, 1987, 85.

93. **Kohara, Y., Akiyama, K., and Isono, K.,** The physical map of the whole *E. coli* chromosome: applications of a new strategy for rapid analysis and sorting of a large genomic library, *Cell,* 50, 495, 1987.

94. **Slonczewski, J. L., Rosen, B. P., Alger, S. R., and Macnab, R. M.,** pH homeostasis in *Escherichia coli*; measurement by ^{31}P nuclear magnetic resonance of methyl phosphonate and phosphate, *Proc. Natl. Acad. Sci. U.S.A.,* 78, 6271, 1981.

95. **Thelen, P., Tsuchiya, T., and Goldberg, E. B.,** Characterization and mapping of a major Na+/H+ antiporter gene of *Escherichia coli, J. Bacteriol.*, 173, 6553, 1991.

96. **Medigue, C., Viari, A., Henaut, A., and Danchin, A.,** *Escherichia coli* molecular genetic map (1500 kbp): update II, *Mol. Microbiol.,* 5, 2629, 1991.

97. **Schuldiner, S. and Padan, E.,** unpublished results.

Chapter IC

Na+/SUBSTRATE SYMPORT IN PROKARYOTES

Ichiro Yamato and Yasuhiro Anraku

TABLE OF CONTENTS

0-8493-6982-7/93/$0.00 + $.50
© 1993 by CRC Press, Inc.

I. INTRODUCTION

Highlighted by Mitchell's chemiosmotic theory,[1] it is widely accepted that the ion circuit (movement) including H^+, Na^+, and K^+ across biological membranes plays important roles in biological energy transduction, such as the respiratory proton pump, H^+- or Na^+-coupled ATP synthesis, active transport, and flagellar motors. Comprehensive studies of cellular chemiosmotic work in general and of primary energy-transducing machineries in particular during the last decade have strengthened the concept and demonstrated that the ion circuits under functional cooperation indeed regulate ionic and solute homeostasis in the cytoplasm. The mechanisms by which a downhill proton (sodium ion) flux is coupled to an uphill transport of solute (cations and organic solutes) across the membrane have been studied extensively, and a wide variety of H^+ (Na^+)/substrate symporters and antiporters have been characterized. In prokaryotic cells, the symporters and antiporters, which also are called secondary active transport systems, are almost exclusively driven by a proton (sodium ion)-motive force generated from redox-reaction-coupled primary H^+ (Na^+) pumps and H^+ (Na^+)-translocating ATPases.

This chapter describes the molecular biological and bioenergetic aspects of Na^+/substrate symport in prokaryotes. Current studies dealing with the structure-function relationship of Na^+ symport carriers are discussed in detail, with special reference to the mechanism of the H^+/lactose symport in *Escherichia coli*.[2-5] Several comprehensive reviews for the Na^+ circuit and physiological aspects of Na^+ symport have been published.[6-15] For updating current information, readers should consult the other chapters of Section I of this book.

II. Na^+/SUBSTRATE SYMPORT

Phenomenologically, the Na^+/substrate symport, driven by a sodium ion-motive force across the membrane, establishes a large concentration gradient of substrate in the cytoplasm against the medium and provides cells with nutritional requirements for growth. In most prokaryotes, sodium ions influxed in this manner are pumped out via a Na^+/H^+ antiporter to regenerate the $\Delta\bar{\mu}_{Na^+}$[16] (see Chapters IA and IB of this book). Thus, thermodynamic coupling between the Na^+/substrate symport and the Na^+/H^+ antiport ensures that the cells turnover a Na^+ circuit and accumulate nutrients homeostatically in the cytoplasmic pool.

Recent advances in gene technology and protein engineering have potentiated molecular biological studies of elementary steps in Na^+ symport over the thermodynamic description. Therefore, such fundamental questions can be addressed as why prokaryotes have acquired a Na^+/substrate symport system, how those symport machineries are built up, how a symport carrier senses the sodium ion-motive force, by which mechanism the syn-coupled fluxes of Na$^+$ and substrates in the symport carrier in the cytoplasmic membrane take place, and how the carrier regulates those elementary steps in coupled translocation.

To answer these questions, biochemical investigations of a variety of Na$^+$ symport carriers utilizing advanced gene and protein technology are required. For this purpose, bacterial systems, especially those of *E. coli*, are useful. Amplification of a carrier is feasible and facilitates quantitative analyses of binding and transport activities in membranes[17-21] and purification on a large scale.[22,23]

III. DISTRIBUTION OF Na$^+$/SUBSTRATE SYMPORT IN PROKARYOTES

All living organisms have acquired chemiosmotic systems to maintain a low cytoplasmic Na$^+$ pool irrespective of the presence or absence of a high Na$^+$ concentration in the surrounding medium. Thus, *Klebsiella aerogenes* has a primary Na$^+$ pump, oxaloacetate decarboxylase, that functions in the anaerobic degradation of citrate[24] (see also Chapter ID), whereas many marine bacteria possess a respiration-coupled Na$^+$ pump[25] (see also Chapter IF). Potent Na$^+$/H$^+$ antiport systems are present in most heterotrophic eubacteria (see also Chapter IB), and Na$^+$/K$^+$-ATPases are commonly distributed in most eukaryotic cells and tissues. Such chemiosmotic work generates an electrochemical gradient of Na$^+$ across the membrane, and the $\Delta\bar{\mu}_{Na^+}$ thus created can be utilized as a driving force for Na$^+$/substrate symport.[6,10-12,26-29]

Table 1 shows examples of Na$^+$ symport systems that have been described recently. In some cases, the dependence of transport activity on [Na$^+$]$_{out}$ is not clear enough to indicate Na$^+$ symport, but those that are listed here show that Na$^+$-coupled systems are present in a wide range of organisms. The older aspects of Na$^+$-dependent amino acid transport systems have been described.[26] It should be noted that molecular biological studies on the Na$^+$ circuit and related processes in bacteria living in extreme environments, including archaebacteria such as alkalophiles,[8] thermophiles,[12] halophiles,[28] and acidophiles,[30] are rapidly developing, due mainly to an increasing interest in microbial ecology in extreme environments and to increased feasibility of gene and protein technologies.

The thermodynamic aspect for cation-coupled symport was first formulated by West and Mitchell,[31] who measured a stoichiometric change in the medium pH upon addition of lactose in starved *E. coli* cells. Similar experiments proving concomitant movement of Na$^+$ or Li$^+$ with a fixed ratio of proline,[32] melibiose,[33,34] or serine/threonine[35] followed which demonstrated Na$^+$ symport in *E. coli* K-12. Later, using intact cells or cytoplasmic membrane vesicles of the right-side-out orientation, transport activities of substrates such as glutamate in *E. coli* B[36,37] and amino acids and sugars[28,29] in marine pseudomonads were found to depend quantitatively on the Na$^+$ in the medium,

TABLE 1
Distribution of Na$^+$ Symport Systems in Prokaryotes

Bacterium	Substrate	Miscellaneous	Ref.
Alteromonas halo-planktis	Arg, Lys, Glu, Leu, Asp, Ala, Ser, Gly, fucose, succinate, citrate, galactose	Marine bacterium	97
Bacillus	Ser, methylamine	Alkalophilic	8, 98
B. firmus Rab	Asp, Met, malate	Alkalophilic	99
B. stearothermophilus	Glu	Thermophilic	100
B. subtilis	Glu		101
Chromatium vinosum	Ala, α-aminoisobutyrate		102
Clostridium fervidus	Amino acids	Thermophilic	103
Deleya strains	Met	Marine bacteria	29
Desulfococcus multivorans	SO$_4^{-2}$	Marine bacterium	104
Escherichia coli	Pro		18, 22, 32, 39, 40, 44
	Melibiose		23, 33, 74, 105, 106
	Glu		36, 37, 43, 73
	Ser, Thr		35
Extreme halophilic archaebacteria	Mono- and tri- carboxylic acids	Alkaliphilic	107
Fusobacterium nucleatum	Glu		108
Halobacterium halobium	Most amino acids	Halophilic	109
Klebsiella pneumoniae	Citrate		110
Methanococcus voltae	Ile, Val, Leu, Ser, Pro, Gly, Phe	Marine bacterium	111, 112
Pseudomonas aeruginosa	Leu, Ileu, Val		41, 42, 113
P. doudoroffi	Glu, acetate, succinate	Marine bacterium	28
Rhizobium meliloti	Glycine betaine		114
Ruminal bacterium	Glu, Gln, Ser, His		115
Peptostreptococcus	Leu, Ile, Val		116
Streptococcus bovis	Ser, Thr, Ala		117
Salmonella typhimurium	Pro		47, 118
Staphylococcus aureus	Taurine		119
V. alginolyticus	Sucrose	Marine bacterium	120
V. costicola	α-Aminoisobutyrate		121, 122
V. fischeri	Ala, Arg, Gly, succinate		97

$\Delta \bar{\mu}_{Na^+}$ (not $\Delta \bar{\mu}_{H^+}$) or ΔpNa, and thus to be catalyzed by the mechanism of Na$^+$ symport.

IV. GENETICS AND MOLECULAR BIOLOGY OF Na$^+$ SYMPORT CARRIERS

A. GENETIC STUDIES

Genetic and biochemical studies of lactose transport in *E. coli* have established that the transport activity is catalyzed by the *lacY* gene product,[38] or lactose carrier (permease). It mediates the H$^+$/lactose symport reaction in a monomeric state.[2] In a similar manner, genetic studies on Na$^+$-dependent transport systems have been performed, and several transport mutants with different phenotypes showing resistance to substrate analogues[39-42] and defects of transport activities[43-47] have been isolated. In some cases, Li$^+$ toxicity to cells was utilized for the selection of transport mutants.[48,49] More recently, several researchers have succeeded in isolating and characterizing mutants that are altered in substrate specificities and in cation coupling.[48-52] The mutants involved in the last groups are of particular interest in analyzing the structure-function relationship of the Na$^+$/substrate symport carriers.

Table 2 summarizes typical examples of mutants defective in Na$^+$/substrate symport, derived from *E. coli*, *Salmonella typhimurium*, and *Pseudomonas aeruginosa*.

B. CLONING AND NUCLEOTIDE SEQUENCING OF CARRIER GENES

Usually, carrier genes are cloned and expressed in homologous organisms using complementation activities, as was the case of the *putP* gene[53] of *E. coli*. Now, advanced cloning techniques are available which allow gene cloning by complementation in heterologous organisms[54] and by DNA probes synthesized according to amino acid sequencing data obtained from purified carriers — for example, the alanine carrier in *Bacillus stearothermophilus* PS3.[55]

The amino acid sequences of several carrier proteins were deduced from nucleotide sequencing data, and their hydropathy profiles revealed 10 to 12 membrane-spanning regions in the molecules.[9,56] A certain hydropathy profile does not necessarily imply that all hydrophobic regions are membrane-spanning domains and that they are in an α-helical conformation, but the few examples of bacterial inner membrane proteins examined by X-ray crystallography or electron microscopy[57,58] have shown that this is the case.

Other important information obtained from DNA sequencing results from similarity searches such as for a family of H$^+$/sugar symporters[56] and of transport systems that contain a periplasmic binding protein.[59] At present, carrier protein similarity has not been found in general.

TABLE 2
Typical Mutations and Clones of Na$^+$ Symport Systems in Prokaryotes

Bacterium	Gene (substrate)	Selection	Cloning	Ref.
E. coli	putP (proline)	Substrate analogue resistant		39, 40, 50
		Transport defective		44, 52
			Complement	53, 96, 123, 124
	gltS (glutamate)	Transport defective		43
			Complement or gene dose	82, 125
	melB (melibiose)	Transport defective		45, 46
		Li$^+$ toxicity		48
			Complement	126–128
Salmonella typhimurium	putP (proline)	Transport defective		47
		Substrate analogue resistant		51
		Li$^+$ toxicity		49
			Complement	129, 130
Pseudomonas aeruginosa	braB (branched-chain amino acids)	Substrate analogue resistant		41, 42
			Complement	131, 132

C. REGULATION OF GENE EXPRESSION

We examined the regulation of gene expression for the Na^+/proline carrier in *E. coli*.[60] The *putP* gene is located in a proline utilization system, the *put* regulon. Proline is utilized as the sole source of nitrogen and carbon. Consistent with this nutritional importance, the expression of *putP* is under catabolite repression and nitrogen control and is repressed by proline. The repressor is thought to be the *putA* gene product (proline oxidase). However, this gene expression does not seem to be regulated significantly by the $[Na^+]$ inside or outside of the cell.

The regulation of gene expression of carrier proteins relevant to the Na^+ cycle has not yet been studied extensively. However, this very example of the *put* regulon suggests that the gene expression of Na^+ symporting carriers is regulated by the nutritional demand of cells, not by $[Na^+]$ or by a gradient of $[Na^+]$. This also seems true for the *mel* operon[61] and the *gltS* gene[43] in *E. coli*.

V. BIOCHEMISTRY OF Na⁺ SYMPORT CARRIERS

A. SOLUBILIZATION, PURIFICATION, AND RECONSTITUTION

Biochemical studies, including purification and reconstitution of Na^+ symport carriers, are essential for elucidation of the Na^+ symport mechanism. The use of a stable and reliable reconstituted proteoliposome system is of particular importance because it allows the investigation of the elementary steps of symport reaction.

Solubilization, purification, and reconstitution of a carrier protein was first reported by Kasahara and Hinkle[62] for a glucose transporter of erythrocytes. Since then, several conventional methods for purification of a number of carrier proteins have been accomplished by means of DEAE ion-exchange chromatography. Highly reliable purification protocols were established for the H^+/lactose symporter of *E. coli*.[63-65] The *lac* carrier seems to be very hydrophobic and unstable even in the presence of long alkyl chain detergents such as dodecylmaltoside. The protein is basic in nature as deduced from the DNA sequence. However, this property was only utilized in one purification protocol;[65] otherwise, carriers were mostly purified or enriched as the flow-through fraction from a DEAE column. The stable Na^+/alanine symport carrier from thermophilic bacteria seemed to be susceptible to an extensive cycle of column chromatography and has been purified by conventional chromatographic techniques similar to those for soluble proteins.[66]

Currently, if a gene of interest has once been cloned and characterized, it can be manipulated to produce a novel product for a specific use. One example is a gene fusion. In particular, the protein fusion designed for site-specific cleavage[22,67] has been extremely useful for the purification of the Na^+/proline carrier from *E. coli* membranes (Figure 1). A fusion gene of the *putP* gene connected with *lacZ* (the β-galactosidase gene) through a linker

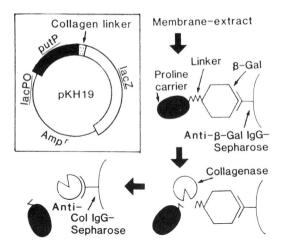

FIGURE 1.   Scheme of the purification procedure for the proline carrier protein of *E. coli.*[22] The proline carrier was produced as a fusion protein with β-galactosidase with a collagen linker in between. The fusion protein was solubilized with dodecylmaltoside, affinity purified with the antibody-Sepharose® column, and specifically cleaved by collagenase. The insert shows the structure of pKH19 used for the production of the fusion protein. *lacOP* indicates the promotor and operator region of the lactose operon of *E. coli*, *putP* is the gene for the proline carrier, and *lacZ* is the gene for the β-galactosidase. For details see text. (From Hanada, K., Yamato, I., and Anraku, Y., *J. Biol. Chem.*, 263, 7181, 1988. With permission.)

oligonucleotide which corresponds to the amino acid sequence of collagen was constructed.[67] The fusion gene product was bifunctionally active and supported cell growth of the *putP* mutant. The cytoplasmic membranes containing a large amount of the fusion protein were solubilized with dodecylmaltoside in the presence of *E. coli* phospholipid, and the solubilized fraction was applied to an anti-β-galactosidase IgG affinity column. After washing the column with a buffer containing dodecylmaltoside and phospholipid, the adsorbed fusion protein was treated with collagenase and the carrier moiety alone specifically eluted from the column. The eluent containing the proline carrier and a small amount of collagenase was then applied to an anticollagenase IgG column, and the flow-through fraction was collected as the fraction of purified carrier protein. This purified fraction was active in a reconstitution system driven by a Na^+ electrochemical gradient.[22] Several useful fusion vectors are now commercially available, such as pRIT (protein A fusion vector, Pharmacia LKB Biotechnology, Inc., Tokyo, Japan), pMal (maltose binding protein fusion vector, New England Biolabs, Inc., Massachusetts), and so on.

A second example is the use of the high efficiency of the bacteriophage T7 promoter. The high level of expression causes a transformed cell to form inclusion bodies, which can be solubilized by urea, and the target product can be purified by conventional techniques.[23] This procedure appears simple and useful, but we believe that, in principle, the use of a site-specifically cleavable fusion protein is more rational.

Several procedures have been reported for reconstitution of carrier proteins into liposomes by dilution,[63] dialysis,[65] or freeze-thaw sonication.[22,62,64,68] Octylglucoside and Triton® X-100 are widely used detergents in any experimental protocol. For the proline carrier of *E. coli,* however, we had to use dodecylmaltoside to maintain the carrier protein in stable soluble form. Since this detergent has a low critical micelle concentration, only the freeze-thaw sonication method can be used for the reconstitution of the proline carrier.[22,68]

Transport assays in reconstituted proteoliposomes provide information regarding the basic properties of the carrier function. For example, the $H^+/$lactose carrier of *E. coli* is functional as a monomer because even where the carrier/liposome ratio is reduced to less than one the carrier still retains normal $\Delta\Psi$-dependent uptake activity.[69] This kind of study has not been done for Na^+ symport carriers. There seems to be phospholipid specificity for the optimal carrier activity of the lactose carrier,[70] but for Na^+ symport carriers information of this kind is not yet available.

B. MECHANISM OF TRANSPORT ACTIVITY

The chemiosmotic theory suggests that from a thermodynamic point of view the $\Delta\bar{\mu}_{Na^+}$ forms the driving force for the uphill transport of substrate via the Na^+ symport carrier.[4] Experimentally this has been verified by the Na^+ dependence of the transport activities, concomitant movement of Na^+ with substrate transport, and substrate accumulation depending on $\Delta\bar{\mu}_{Na^+}$. Na^+ symport by the proline or melibiose carriers alone has been demonstrated unequivocally[22,23] in the reconstituted proteoliposome system, where interfering fluxes of Na^+ due to the Na^+/H^+ antiporter, for example, are absent.

The stoichiometries and coupling mechanisms were inferred from the kinetic experiments of the transport reactions[71] in intact cells and membrane vesicles. A one to one stoichiometry was observed for the flux ratio of H^+ or Na^+ to the substrate using a pH or Na^+ electrode in $H^+/$lactose,[31] $Na^+/$proline,[32] and $Na^+/$melibiose[33] symport systems. However, measurements of flux ratios and the kinetic parameters of a transport reaction are not sufficient to answer the question of the mechanism of Na^+ transport via the carrier, or how Na^+ (cation) movement drives substrate movement.

Mechanisms of secondary active transport systems have been discussed by Wright[71] and Sanders et al.[72] The general mechanistic model of secondary active transport is shown in Figure 2A. This general model typically can be divided into (1) ordered,[3,18,20,73] (2) random,[4] and (3) selective binding models.[5] Most investigators usually use information from transport activity measurement to formulate mechanisms. However, caution is required for modeling a certain mechanism from the transport activities alone because the transport assay itself may not be complete or accurate enough to distinguish each possibility.

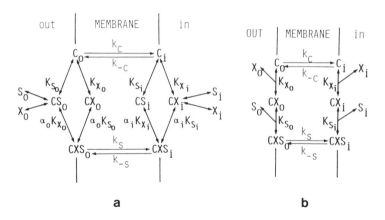

FIGURE 2. General model[4,71] and the ordered binding model (affinity model)[3,18,20,73] of the secondary active transport system. C, S, and X are the carrier, substrate, and symported coupling cation (cosubstrate), respectively. k_c, k_{-c}, k_s and k_{-s} are the rate constants of translocation as shown in the figure. K_X and K_S are the dissociation constants of substrate and cosubstrate, respectively, to the carrier, and α_i and α_o are constants. Subscripts i and o represent the inside and outside of the membrane. a, general model; b, the ordered binding or affinity model.

We have studied the binding properties of carrier proteins for proline and glutamate, especially the Na^+-dependent binding activity of substrates. From these studies it is concluded that the binding of Na^+ to the carrier causes a conformational change in the protein that results in an increase in its affinity for the substrate, as shown in Figure 2B (ordered binding or affinity model).[18,20,73] Thus, by assuming that the Na^+ affecting the binding of the carrier is the coupling ion itself in the transport reaction, the affinity model is proposed as a general mechanism for the symport cycle. This assumption that the ion affecting the binding is the coupling ion itself has not been confirmed experimentally in a reconstituted proteoliposome system. However, since the ions that affect solute binding are the same as the ones that drive transport and since mutants defective in ion coupling in the transport reaction also show the same defect in the ion-dependent substrate binding,[3,18,39,52] the assumption seems correct. This affinity model logically predicts the conformational change of the carrier by the binding of the coupling ions to acquire affinity for the substrate. A similar model has also been proposed by Bassilana et al.[21,74] based on their extensive studies of the coupling ion-dependent binding of substrate to the melibiose carrier in *E. coli*.

Since the affinity model explains the transport characteristics of glutamate,[73] lactose,[3] and proline[18] well, alternative models (the random binding or selective binding model) can be ruled out. Furthermore, the charge-relay (or proton-relay)[2,75-77] mechanism which Kaback et al.[2,75] proposed as the mechanism for the movement of the coupling ion in the H^+/lactose symport

TABLE 3
Kinetic Parameters of Binding and Transport Reactions of the Proline and Glutamate Carriers in *E. coli*

Substrate	Kd_{Na^+}	Kt_{Na^+}	Kd_{Na^+}/Kt_{Na^+}	Kd_S	Kt_S	Kd_S/Kt_S	Ref.
Glutamate	10 mM	20 mM	1	1 μM	2 μM	1	19, 73
Proline	10 mM	30 μM	100	8 μM	2 μM	1	17, 18, 32

Note: Kd and Kt are the dissociation constant in the binding reaction from the carrier and the Michaelis constant of the transport reaction, respectively. The subscripts Na⁺ and S indicate that those values are the Kd or Kt for Na⁺ and the substrate, respectively.

carrier also remains to be tested from the point of view of the affinity model of the transport cycle. The reason why most symport cycles occur according to the affinity model is not well understood. However, this mechanism may guarantee the stoichiometric and efficient energy transduction of the system without leakage.[134]

The kinetic parameters obtained for the transport and binding reactions of the glutamate and proline carriers are shown in Table 3. The catalytic mechanisms of the melibiose and glutamate carriers are understandable because the affinities for Na⁺ (Li⁺) and substrates of the carriers in binding reactions are of the same order of magnitude as those in the transport reactions.[73,74] However, the affinity for Na⁺ of the proline carrier in the binding reaction was about 100 times lower than that in the transport reaction. This was explained by the functional asymmetry of the carrier[18] (that is, the binding site for the coupling cation orienting to the cytoplasmic side has a higher affinity for the cation than the site orienting to the periplasmic side). It should be mentioned that the mechanism of Na⁺ symport has been predicted using the affinity model studied with the best characterized carrier systems. The binding assay may be helpful in elucidating the mechanism of Na⁺ symport in other solute transport systems among various prokaryotes.

VI. STRUCTURE AND FUNCTION OF THE Na⁺ SYMPORT CARRIERS

In order to understand symport function, knowledge of the molecular structure of carriers or the structural basis of the affinity model is essential. In general, symport carriers are postulated to have 10 to 12 membrane-spanning segments.[9] However, detailed X-ray crystallographic or electron microscopic analyses are not yet available. Thus, we will discuss this important problem by referring to currently available data on mutations, similarity search, chemical modifications, immunochemical investigations, topology study, and so on.

A. MUTATIONS

Mutagenesis *in vivo* and *in vitro* has been widely used to obtain transport mutants with different phenotypes. They have been selected and classified as transport-defective, altered substrate specificity, and altered coupling cation specificity mutants (Table 2). Special selection was used to identify coupling mutants of the melibiose carrier of *E. coli*. The transport of melibiose is coupled to either Li^+ or Na^+. Therefore, the cells in which the carrier is amplified properly should accumulate much Li^+ simultaneously with melibiose from the medium and eventually suffer the toxic effects of Li^+. Thus, mutants that are resistant to Li^+ in the presence of melibiose may have low coupling efficiency of melibiose transport with Li^+, but normal efficiency with Na^+.[48] Using Li^+ toxicity, Myers and Maloy[49] obtained many proline carrier mutants from *E. coli* and *S. typhimurium*. They showed that both the amino-terminal (N-terminal) and carboxyl-terminal (C-terminal) parts of the proline carriers are involved in Na^+ symport coupling, whereas the middle part is involved in substrate recognition.[49,51] In contrast to their notion, Ohsawa et al.[52] have isolated and characterized a coupling mutant that has a point mutation in its mid-region. Several other mutants defective in coupling and/or substrate recognition in the proline carrier have also been reported.[50]

Melibiose and proline carriers have been studied by site-directed mutagenesis. His-94 is shown to be indispensable for transport of the melibiose carrier, and it is postulated to be related to the coupling mechanism,[74,78] presumably as in the proton-relay mechanism of the H^+/lactose carrier.[79,80] In the Na^+/proline carrier, Cys-344 or Cys-281, which is located close to the substrate binding site, binds *N*-ethylmaleimide.[81] However, there is still insufficient information from these mutants to elucidate the structure-function relationship of the carriers, and these kinds of studies should be pursued further together with physicochemical determination of the tertiary structures of the symporter molecules.

B. SIMILARITY SEARCH

Similarities of protein primary structures are informative in finding several essential and functional domains, such as the ATP binding site in the *hisP* gene product for the periplasmic binding protein-dependent histidine transport system.[59] However, Na^+-coupled symport carriers do not show extensive similarities. Only the Na^+/glutamate and Na^+/proline carriers of *E. coli* and the Na^+/glucose carrier of intestinal brush border membranes contain a sequence of five amino acids which has been proposed as the Na^+-binding motif[82] (**G**—about 30—**AXXXXLXXXGR**). The importance of this motif remains to be confirmed by site-directed mutagenesis. At present, the nucleotide sequences of only a few other Na^+ symporter genes have been determined, and the deduced amino acid sequences seem to contain similar regions for the Na^+-binding motif (see Figure 3).

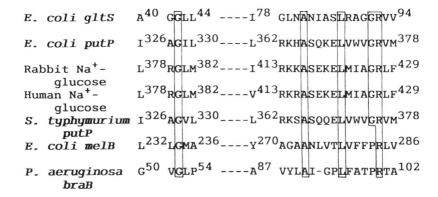

FIGURE 3. Conserved amino acids in the sequences of several Na⁺ symport carrier proteins. Identical amino acids in the seven amino acid sequences are shown in boxes. Shown from the top to the bottom lines: the glutamate carrier in *E. coli*;[82] the proline carrier in *E. coli*;[96] the Na⁺/glucose cotransporter of the rabbit intestine;[133] the Na⁺/glucose cotransporter of the human intestine;[83] the proline carrier in *S. typhimurium*;[130] the melibiose carrier in *E. coli*;[126] the branched-chain amino acid carrier in *Pseudomonas aeruginosa* PAO.[131]

Similarity studies also can be used for the analysis of the evolutional relationship between proteins of various species. The nucleotide and amino acid sequences of the proline carrier of *E. coli* are quite similar to those of the intestinal Na⁺/glucose transporter of mammalian cells[83] since they show 28% identity and 53% similarity in the amino acid sequence, as well as 44% identity in the DNA sequence. The same applies to the similarity between the glucose transporters in erythrocytes and a family of bacterial sugar transport proteins.[56,84] The construction of a phylogenetic tree for these groups of transport genes may bring about a new aspect of the development of eukaryotes.

C. CHEMICAL MODIFICATION

Chemical modification is the complementary biochemical approach to genetics to investigate the structure-function relationships among proteins. The proline carrier of *E. coli* was modified using *N*-ethylmaleimide,[85] dicyclohexylcarbodiimide, and diethyl pyrocarbonate. *N*-Ethylmaleimide modifies the Cys residues in the proline carrier, thus inactivating binding and transport activities. The inactivation is protected by proline in a Na⁺- or Li⁺-dependent manner.[85] This suggests that the reactive Cys residue(s), the modification of which by *N*-ethylmaleimide causes the inhibition of proline binding and transport activities, is located close to the binding sites for proline and the coupling cation. Dicyclohexylcarbodiimide and diethyl pyrocarbonate modify the carboxyl group of Glu and Asp and the imidazole group of His, respectively. These reagents also inactivate the binding and transport activities of the proline carrier, and Na⁺ plus proline protects against their inactivation.[135] This observation suggests that active Glu/Asp and His residues are also located near the ligand binding sites.

These results, combined with those of a molecular biological study,[81] suggest that either Cys-281 or Cys-344 is the important cysteine close to the substrate binding site.[136]

D. IMMUNOCHEMICAL STUDY

Kaback and colleagues[86] raised monoclonal antibodies against the $H^+/$ lactose symport carrier in *E. coli,* which can discriminate substrate binding and translocation reactions. In their studies, however, the epitopes that reacted with the antibodies were not identified. Specific monoclonal antibodies against Na^+-symport carriers have not yet been obtained.

Polyclonal antibodies against synthesized peptides have been obtained and used to study the structure of carriers. Antibodies against the C-terminal peptides of melibiose carrier[87] and proline carrier[88] in *E. coli* have been used to demonstrate that the C-terminal part of these carriers is on the cytoplasmic side of the membrane. The same is true for the lactose carrier.[89,90] Thus, most carrier proteins may orient their C-termini toward the cytosol. An antibody against the hydrophilic portion (amino acids 33 to 45) of the proline carrier protein was raised, but it did not react with native or SDS-denatured carrier, suggesting that this portion is buried within the molecule and is not accessible from outside.[137]

E. TOPOLOGY STUDY

Tn*phoA* or Tn*bla* is a useful tool for the investigation of the topology of membrane proteins.[91,92] Several carrier proteins have been studied using Tn*phoA* and shown to have a configuration consistent with the model postulated from hydropathy analysis.[93,94] With this approach, the lactose carrier protein[95] of *E. coli* showed the same configuration as that predicted from the hydropathy analysis. This kind of information also may be useful in considering the tertiary structure of Na^+-symport carrier proteins.

F. SECONDARY STRUCTURE MODEL OF THE Na^+/PROLINE SYMPORT CARRIER OF *E. COLI*

Genetic and biochemical studies concerning the structure-function relationship of the proline carrier described in this section are summarized in Figure 4. The hydropathy profile of the deduced amino acid sequence suggests 12 membrane-spanning segments.[96] The C-terminal portion is thought to be on the cytoplasmic side, judging from immunochemical studies.[88] Thus, the orientation of the carrier protein is postulated to be that shown in Figure 4. The hydrophilic segment between membrane-spanning segments I and II is thought to be folded inside the membrane because the antibody raised against the synthetic amino acid sequence in this region does not react with the carrier molecule.[137] The hydrophilic segments between segments VI and VII and

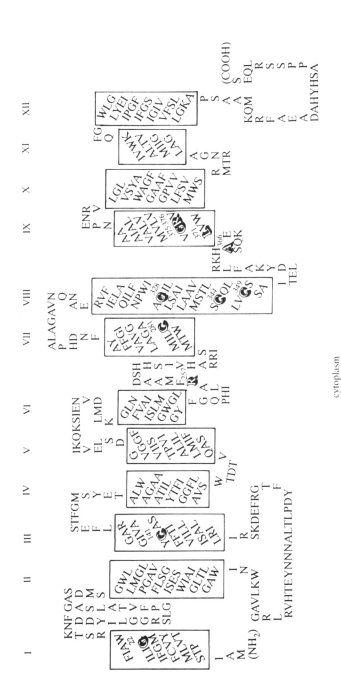

FIGURE 4. Model of the secondary structure of the proline carrier from *E. coli*. Hydrophobic segments, shown in *boxes* as transmembrane domains, are connected by hydrophilic loops. Amino acid residues that have been shown to be related in cation coupling or substrate recognition sites are indicated with their residue numbers as a *superscript*. For details, see text.

between VIII and IX are postulated to be folded inside the molecule since they have the relevant amino acid residues with the substrate- or Na^+-binding activities.[52,82] Furthermore, an amino acid change from Gly-22 to Glu-22 or Cys-141 to Tyr-141 causes a remarkable decrease in affinity for Na^+, suggesting that these residues are related to the binding of the coupling cation.[50] Therefore, the above hydrophilic segments with membrane-spanning segments I, III, VII, VIII, and IX may constitute the Na^+ and proline binding sites in the folded configuration. As Henderson and Maiden have discussed,[56] extensive alignment and computer graphics may help build a working model of tertiary structure for carriers.

VII. PERSPECTIVE

Information is accumulating about the properties and function of secondary active transport systems in prokaryotes and eukaryotes. Their importance in cell physiology, especially in bioenergetic aspects of ion and solute homeostasis, is now well recognized. More information from a wider range of organisms is needed for a deeper understanding of nature. Furthermore, the evolution and physiological diversification of transport carriers are attractive targets for future research. In this context, the authors are especially interested in the evolutionary branching of H^+- and Na^+-coupled symport processes and the origin of symporters and antiporters in prokaryotes to adapt and overcome the surrounding environment.

As presented here, the Na^+/amino acid symport cycle catalyzing proline or glutamate transport in *E. coli* can be described by the ordered binding model.[18,73] It should be tested whether other symport systems function according to the same mechanism. The mechanism predicts certain constraints on the structure of carrier proteins in the membrane, which should be elucidated by further studies of the structure-function relationship and especially by X-ray crystallography and three-dimensional modeling.

During the verification of mechanism of the transport reaction, the proline carrier was shown to have unique features of binding and transport activities.[18] At present, we explain this property by assuming that the carrier functions asymmetrically in the membrane.[18] This unique property of the carrier may be important for cell physiology, to which our future investigations should be addressed.

During the similarity study it was found that many carrier proteins are composed of a two-domain structure, consisting of N- and C-terminal halves, which suggests that the present-day genes for secondary active transport carriers have evolved by tandem duplication of homo- or heterogenes of an ancestral type having five or six membrane-spanning segments.[9] This idea is of interest when applied to the origin and evolution of carrier proteins. A soluble binding protein or even an enzyme has one binding site for a substrate, which approaches the binding site in the protein molecule from the outside.

In general, the substrate and the product enter and exit from that binding site via the same pathway. However, a carrier protein must have one binding site for the substrate and two pathways for the substrate to reach the binding site. As already discussed, symport carriers possess two independent binding sites, one for the substrate and the other for the coupling cation. This suggests that two ancestral genes underwent tandem fusion or insertion into the other to form a fused or mosaic protein. This kind of a complex protein may have two independent binding sites, which the substrates can access by two independent paths, thus creating a coupling device in the molecule that is obligatory for functioning as a member of the sym- and antiporter protein family.

VIII. SUMMARY

In this chapter we describe current evidence showing the spreading and characteristics of Na^+/substrate symport systems among prokaryotic organisms. Of these systems, the proline and melibiose symport systems in *E. coli* are the best characterized. We discuss and evaluate these systems as examples of how secondary carrier proteins may function. On the basis of data from studies of the Na^+/proline and Na^+/glutamate carriers, the symport cycle is postulated to occur via the ordered binding of the substrates to the carrier protein. Biochemical and gene manipulation techniques used for the purification and reconstitution of the proline and melibiose carriers are also described. Finally, recent and future aspects of structure-function relationship studies are discussed in some detail, emphasize that the present level of knowledge remains insufficient to construct a tertiary structure of the membrane protein consistent with the molecular model of how secondary active transport carriers function.

REFERENCES

1. **Mitchell, P.,** Coupling of phosphorylation to electron and hydrogen transfer by a chemiosmotic type of mechanism, *Nature, (London)*, 191, 144, 1961.
2. **Kaback, H. R., Bibi, E., and Roepe, P.D.,** β-Galactoside transport in *E. coli*: a functional dissection of *lac* permease, *Trends Biochem. Sci.*, 15, 309, 1990.
3. **Yamato, I. and Anraku, Y.,** Dependence on pH of substrate binding to a mutant lactose carrier, *lacY*ᵘⁿ in *Escherichia coli*. A model for H⁺/lactose symport, *Biochem. J.*, 258, 389, 1989.
4. **Wright, J. K., Seckler, R., and Overath, P.,** Molecular aspects of sugar:ion cotransport, *Annu. Rev. Biochem.*, 55, 225, 1986.
5. **Page, M. G. P.,** The role of protons in the mechanism of galactoside transport via the lactose permease of *Escherichia coli*, *Biochim. Biophys. Acta*, 897, 112, 1987.
6. **Maloy, S. R.,** Sodium-coupled cotransport, *Bacteria*, 12, 203, 1990.

7. **Dimroth, P.**, Mechanisms of sodium transport in bacteria, *Philos. Trans. R. Soc. London Ser. B,* 326, 465, 1990.

8. **Krulwich, T. A. and Guffanti, A. A.**, The sodium cycle of extreme alkalophiles: a secondary sodium/hydrogen antiporter and sodium/solute symporter, *J. Bioenerg. Biomembr.,* 21, 663, 1989.

9. **Maloney, P. C.**, Microbes and membrane biology, *FEMS Microbiol. Rev.,* 87, 91, 1990.

10. **Lewis, S. A. and Donaldson, P. J.**, Sodium dependence of cation permeabilities and transport, *Curr. Top. Membr. Transp.,* 34, 83, 1989.

11. **Leblanc, G., Pourcher, T., and Bassilana, M.**, Molecular biology and bacterial secondary transporters, *Biochimie,* 71, 969, 1989.

12. **De Vrij, W., Speelmans, G., Heyne, R. I. R., and Konings, W. N.**, Energy transduction and amino acid transport in thermophilic aerobic and fermentative bacteria, *FEMS Microbiol. Rev.,* 75, 183, 1990.

13. **Gottschalk, G. and Blaut, M.**, Generation of proton and sodium motive forces in methanogenic bacteria, *Biochim. Biophys. Acta,* 1018, 263, 1990.

14. **Lancaster, J. R., Jr.**, Sodium, protons, and energy coupling in the methanogenic bacteria, *Biochim. Biophys. Acta,* 21, 717, 1989.

15. **Skulachev, V. P.**, Bacterial sodium energetics, *FEBS Lett.,* 250, 106, 1989.

16. **Pan, J. W. and Macnab, R. M.**, Steady-state measurements of *Escherichia coli* sodium and proton potentials at alkaline pH support the hypothesis of electrogenic antiport, *J. Biol. Chem.,* 265, 9247, 1990.

17. **Mogi, T. and Anraku, Y.**, Mechanism of proline transport in *Escherichia coli* K12.II. Effect of alkaline cations on binding of proline to a H^+/proline symport carrier in cytoplasmic membrane vesicles, *J. Biol. Chem.,* 259, 7797, 1984.

18. **Yamato, I. and Anraku, Y.**, Mechanism of Na^+/proline symport in *Escherichia coli*: reappraisal of the effect of cation binding to the Na^+/proline symport carrier, *J. Membr. Biol.,* 114, 143, 1990.

19. **Fujimura, T., Yamato, I., and Anraku, Y.**, Mechanism of glutamate transport in *E. coli* B. I. Proton-dependent and sodium ion-dependent binding of glutamate to a glutamate carrier in the cytoplasmic membrane, *Biochemistry,* 22, 1954, 1983.

20. **Yamato, I. and Rosenbusch, J. P.**, Dependence on pH of substrate binding to lactose carrier in *Escherichia coli* cytoplasmic membrane, *FEBS Lett.,* 151, 102, 1983.

21. **Damiano-Forano, E., Bassilana, M., and Leblanc, G.**, Sugar binding properties of the melibiose permease in *Escherichia coli* membrane vesicles. Effect of Na^+ and H^+ concentrations, *J. Biol. Chem.,* 261, 6893, 1986.

22. **Hanada, K., Yamato, I., and Anraku, Y.**, Purification and reconstitution of *Escherichia coli* proline carrier using a site specifically cleavable fusion protein, *J. Biol. Chem.,* 263, 7181, 1988.

23. **Roepe, P. D. and Kaback, H. R.**, Isolation and functional reconstitution of soluble melibiose permease from *Escherichia coli, Biochemistry,* 29, 2572, 1990.

24. **Dimroth, P.**, Sodium ion transport decarboxylases and other aspects of sodium ion cycling in bacteria, *Microbiol. Rev.,* 51, 320, 1987.

25. **Unemoto, T., Tokuda, H., and Hayashi, M.**, Primary sodium pumps and their significance in bacterial energetics, *Bacteria,* 12, 33, 1990.

26. **Anraku, Y.**, Transport and utilization of amino acids by bacteria, in *Microorganisms and Nitrogen Sources,* Payne, J. W., Ed., John Wiley & Sons, New York, 1980, 9.

27. **MacDonald, R. E., Greene, R. V., and Lanyi, J. K.**, Light-activated amino acid transport systems in *Halobacterium halobium* envelope vesicles: role of chemical and electrical gradients, *Biochemistry,* 16, 3227, 1977.

28. **Wisse, G. A. and MacLeod, R. A.**, Role of sodium in growth, respiration and membrane transport in the marine bacterium *Pseudomonas doudoroffii* 70, *Arch. Microbiol.,* 153, 64, 1989.

29. **Berthelet, M. and MacLeod, R. A.,** Effect of sodium concentration and nutritional factors on the lag phase and exponential growth rates of the marine bacterium *Deleya aesta* and of other marine species, *Appl. Environ. Microbiol.,* 55, 1754, 1989.

30. **Matin, A.,** Bioenergetics parameters and transport in obligate acidophiles, *Biochim. Biophys. Acta,* 1018, 267, 1990.

31. **West, I. C. and Mitchell, P.,** Stoichiometry of lactose-H⁺ symport across the plasma membrane of *Escherichia coli, Biochem. J.,* 132, 87, 1973.

32. **Chen, C.-C., Tsuchiya, T., Yamane, Y., Wood, J. M., and Wilson, T. H.,** Na⁺(Li⁺)-proline cotransport in *Escherichia coli, J. Membr. Biol.,* 84, 157, 1985.

33. **Tsuchiya, T. and Wilson, T. H.,** Cation-sugar cotransport in the melibiose transport system of *Escherichia coli, Membr. Biochem.,* 2, 63, 1978.

34. **Tsuchiya, T., Oho, M., and Shiota-Niiya, S.,** Lithium ion-sugar cotransport via the melibiose transport system in *Escherichia coli*: measurement of Li⁺ transport and specificity, *J. Biol. Chem.,* 258, 12765, 1983.

35. **Hama, H., Shimamoto, T., Tsuda, M., and Tsuchiya, T.,** Properties of a sodium-coupled serine-threonine transport in *Escherichia coli, Biochim. Biophys. Acta,* 905, 231, 1987.

36. **MacDonald, R. E., Lanyi, J. K., and Greene, R. V.,** Sodium-stimulated glutamate uptake in membrane vesicles of *Escherichia coli*: the role of ion gradients, *Proc. Natl. Acad. Sci. U.S.A.,* 74, 3167, 1977.

37. **Tsuchiya, T., Hasan, S. M., and Raven, J.,** Glutamate transport driven by an electrochemical gradient of sodium ions in *Escherichia coli, J. Bacteriol.,* 131, 848, 1977.

38. **Rickenberg, H. V., Cohen, G. N., Buttin, G., and Monod, J.,** La galactoside-permease d'*Escherichia coli, Ann. Inst. Pasteur (Paris),* 91, 829, 1956.

39. **Tristram, H. and Nealse, S.,** The activity and specificity of the proline permease in wild-type and analogue-resistant strains of *Escherichia coli, J. Gen. Microbiol.,* 50, 121, 1968.

40. **Wood, J. M. and Zadworny, D.,** Characterization of an inducible porter required for L-proline catabolism by *Escherichia coli* K12, *Can. J. Biochem.,* 57, 1191, 1979.

41. **Hoshino, T. and Kageyama, M.,** Mutational separation of transport systems for branched-chain amino acids in *Pseudomonas aeruginosa, J. Bacteriol.,* 151, 620, 1982.

42. **Hoshino, T., Tsuda, M., Iino, T., Nishio, K., and Kageyama, M.,** Genetic mapping of *bra* genes affecting branched-chain amino acid transport in *Pseudomonas aeruginosa, J. Bacteriol.,* 153, 1272, 1983.

43. **Marcus, M. and Halpern, Y. S.,** Genetic analysis of the glutamate permease in *Escherichia coli* K-12, *J. Bacteriol.,* 97, 1118, 1969.

44. **Motojima, K., Yamato, I., and Anraku, Y.,** Proline transport carrier-defective mutants of *Escherichia coli* K12: properties and mapping, *J. Bacteriol.,* 136, 5, 1978.

45. **Schmitt, R.,** Analysis of melibiose mutants deficient in α-galactosidase and thiomethylgalactoside permease II in *Escherichia coli* K-12, *J. Bacteriol.,* 96, 462, 1968.

46. **Lopilato, J., Tsuchiya, T., and Wilson, T. H.,** Role of Na⁺ and Li⁺ in thiomethylgalactoside transport by the melibiose transport system of *Escherichia coli, J. Bacteriol.,* 134, 147, 1978.

47. **Dendinger, S. and Brill, W. J.,** Regulation of proline degradation in *Salmonella typhimurium, J. Bacteriol.,* 103, 144, 1970.

48. **Kawakami, T., Akizawa, Y., Ishikawa, T., Shimamoto, T., Tsuda, M., and Tsuchiya, T.,** Amino acid substitutions and alteration in cation specificity in the melibiose carrier of *Escherichia coli, J. Biol. Chem.,* 263, 14276, 1988.

49. **Myers, R. S. and Maloy, S. R.,** Mutations of *putP* that alter the lithium sensitivity of *Salmonella typhimurium, Mol. Microbiol.,* 2, 749, 1988.

50. **Yamato, I., Ohsawa, M., and Anraku, Y.,** Defective cation-coupling mutants of *Escherichia coli* Na⁺/proline symport carrier: characterization and localization of mutations, *J. Biol. Chem.,* 265, 2450, 1990.

51. **Dila, D. and Maloy, S.,** Proline transport in *Salmonella typhimurium: putP* permease mutants with altered substrate specificity, *J. Bacteriol.,* 168, 590, 1986.

52. **Ohsawa, M., Mogi, T., Yamamoto, H., Yamato, I., and Anraku, Y.,** Proline carrier mutant of *Escherichia coli* K-12 with altered cation sensitivity of substrate-binding activity: cloning, biochemical characterization, and identification of the mutation, *J. Bacteriol,* 170, 5185, 1988.

53. **Motojima, K., Yamato, I., Anraku, Y., Nishimura, A., and Hirota, Y.,** Amplification and characterization of the proline transport carrier of *Escherichia coli* K-12 by using *ProT+* hybrid plasmids, *Proc. Natl. Acad. Sci. U.S.A.,* 76, 6255, 1979.

54. **MacLeod, P. and MacLeod, R.,** Cloning in *Escherichia coli* K-12 of a Na$^+$-dependent transport system from a marine bacterium, *J. Bacteriol.,* 165, 825, 1986.

55. **Morosawa, H., Akiyama, S., Kamata, H., Hamamoto, T., Ohta, T., Kagawa, Y., and Hirata, H.,** Cloning of the gene for alanine carrier protein from *Bacillus stearothermophilus* PS3, *Seikagaku,* 62, 754, 1990.

56. **Henderson, P. J. F. and Maiden, M. C. J.,** Homologous sugar transport proteins in *Escherichia coli* and their relatives in both prokaryotes and eukaryotes, *Philos. Trans. R. Soc. London Ser. B,* 326, 391, 1990.

57. **Deisenhofer, J., Epp, O., Miki, K., Huber, R., and Michel, H.,** Structure of the protein subunits in the photosynthetic reaction centre of *Rhodopseudomonas viridis* at 3A resolution, *Nature (London),* 318, 618, 1985.

58. **Henderson, R. and Unwin, P. N. T.,** Three-dimensional model of purple membrane obtained by electron microscopy, *Nature (London),* 257, 28, 1975.

59. **Ames, G. F.-L.,** Energetics of periplasmic transport systems, *Bacteria,* 12, 225, 1990.

60. **Nakao, T., Yamato, I., and Anraku, Y.,** Mapping of the multiple regulatory sites for *putP* and *putA* expression in the *putC* region of *Escherichia coli, Mol. Gen. Genet.,* 214, 379, 1988.

61. **Okada, T., Uenuma, K., Niiya, S., Kanazawa, H., Futai, M., and Tsuchiya, T.,** Role of inducer exclusion in preferential utilization of glucose over melibiose in diauxic growth of *Escherichia coli, J. Bacteriol.,* 146, 1030, 1981.

62. **Kasahara, M. and Hinkle, P. C.,** Reconstitution and purification of the D-glucose transporter from human erythrocytes, *J. Biol. Chem.,* 252, 7384, 1977.

63. **Newman, M. J., Foster, D. L., Wilson, T. H., and Kaback, H. R.,** Purification and reconstitution of functional lactose carrier from *Escherichia coli, J. Biol. Chem.,* 256, 11804, 1981.

64. **Wright, J. K. and Overath, P.,** Purification of the lactose:H$^+$ carrier of *Escherichia coli* and characterization of galactoside binding and transport, *Eur. J. Biochem.,* 138, 497, 1984.

65. **Page, M. G. P., Rosenbusch, J. P., and Yamato, I.,** The effect of pH on proton sugar symport activity of the lactose permease purified from *Escherichia coli, J. Biol. Chem.,* 263, 15897, 1988.

66. **Hirata, H., Kambe, T., and Kagawa, Y.,** A purified alanine carrier composed of a single polypeptide from thermophilic bacterium PS3 driven by either proton or sodium ion gradient, *J. Biol. Chem.,* 259, 10653, 1984.

67. **Hanada, K., Yamato, I., and Anraku, Y.,** Construction and properties of bifunctionally active membrane-bound fusion proteins: *Escherichia coli* proline carrier linked with β-galactosidase, *J. Biol. Chem.,* 262, 14100, 1987.

68. **Hanada, K., Yamato, I., and Anraku, Y.,** Solubilization and reconstitution of proline carrier in *Escherichia coli*; quantitative analysis and optimal conditions, *Biochim. Biophys. Acta,* 939, 282, 1988.

69. **Costello, M. J., Escaig, J., Matsushita, K., Viitanen, P. V., Menick, D. R., and Kaback, H. R.,** Purified *lac* permease and cytochrome *o* oxidase are functional as monomers, *J. Biol. Chem.,* 262, 17072, 1987.

70. **Chen, C.-C. and Wilson, T. H.,** The phospholipid requirement for activity of the lactose carrier of *Escherichia coli, J. Biol. Chem.,* 259, 10150, 1984.

71. **Wright, J. K.,** The kinetic mechanism of galactoside/H⁺ cotransport in *Escherichia coli, Biochim. Biophys. Acta,* 855, 391, 1986.

72. **Sanders, D., Hansen, U.-P., Gradmann, D., and Slayman, C. L.,** Generalized kinetic analysis of ion-driven cotransport systems: a unified interpretation of selective ionic effects on Michaelis parameters, *J. Membr. Biol.,* 77, 123, 1984.

73. **Fujimura, T., Yamato, I., and Anraku, Y.,** Mechanism of glutamate transport in *E. coli* B. II. Kinetics of glutamate transport driven by artificially imposed proton and sodium ion gradients across the cytoplasmic membrane, *Biochemistry,* 22, 1959, 1983.

74. **Pourcher, T., Bassilana, M., Sarkar, H. K., Kaback, H. R., and Leblanc, G.,** The melibiose/sodium symporter of *Escherichia coli*: kinetic and molecular properties, *Philos. Trans. R. Soc. London Ser. B,* 136, 411, 1990.

75. **Kaback, H. R.,** Use of site-directed mutagenesis to study the mechanism of a membrane transport protein, *Biochemistry,* 26, 2071, 1987.

76. **Kaback, H. R.,** Site-directed mutagenesis and ion-gradient driven active transport: on the path of the proton, *Annu. Rev. Physiol.,* 50, 243, 1988.

77. **Yamaguchi, A., Adachi, K., Akasaka, T., Ono, N., and Sawai, T.,** Metal-tetracycline/ H⁺ antiporter of *Escherichia coli* encoded by a transposon *Tn*10. Histidine 257 plays an essential role in H⁺ translocation, *J. Biol. Chem.,* 266, 6045, 1991.

78. **Pourcher, T., Sarkar, H. K., Bassilana, M., Kaback, H. R., and Leblanc, G.,** Histidine-94 is the only important histidine residue in the melibiose permease of *Escherichia coli, Proc. Natl. Acad. Sci. U.S.A.,* 87, 468, 1990.

79. **Puttner, I. B. and Kaback, H. R.,** *lac* permease of *Escherichia coli* containing a single histidine residue is fully functional, *Proc. Natl. Acad. Sci. U.S.A.,* 85, 1467, 1988.

80. **Puttner, I. B., Sarkar, H. K., Padan, E., Lolkema, J. S., and Kaback, H. R.** Characterization of site-directed mutants in the *lac* permease of *Escherichia coli* . I. Replacement of histidine residues, *Biochemistry,* 28, 2525, 1989.

81. **Yamato, I. and Anraku, Y.,** Site-specific alteration of cysteine 281, cysteine 344, and cysteine 349, in the proline carrier of *Escherichia coli, J. Biol. Chem.,* 263, 16055, 1988.

82. **Deguchi, Y., Yamato, I., and Anraku, Y.,** Nucleotide sequence of *gltS,* the Na⁺/ glutamate symport carrier gene of *Escherichia coli* B, *J. Biol. Chem.,* 265, 21704, 1990.

83. **Hediger, M. A., Turk, E., and Wright, E. M.,** Homology of the human intestinal Na⁺/glucose and *Escherichia coli* Na⁺/proline cotransporters, *Proc. Natl. Acad. Sci. U.S.A.,* 86, 5748, 1989.

84. **Maiden, M. C. J., Davis, E. O., Baldwin, S. A., Moore, D. C. M., and Henderson, P. J. F.,** Mammalian and bacterial sugar transport proteins are homologous, *Nature (London),* 325, 641, 1987.

85. **Hanada, K., Yamato, I., and Anraku, Y.,** Identification of proline carrier in *Escherichia coli* K-12, *FEBS Lett.,* 191, 278, 1985.

86. **Carrasco, N., Viitanen, P., Herzlinger, D., and Kaback, H. R.,** Monoclonal antibodies against the *lac* carrier protein from *Escherichia coli*. I. Functional studies, *Biochemistry,* 23, 3681, 1984.

87. **Botfield, M. C. and Wilson, T. H.,** Peptide-specific antibody for the melibiose carrier of *Escherichia coli* localizes the carboxyl terminus to the cytoplasmic face of the membrane, *J. Biol. Chem.,* 264, 11649, 1989.

88. **Komeiji, Y., Hanada, K., Yamato, I., and Anraku, Y.,** Orientation of the carboxyl terminus of the Na⁺/proline symport carrier in *Escherichia coli, FEBS Lett.,* 256, 135, 1989.

89. **Seckler, R. and Wright, J. K.,** Sidedness of native membrane vesicles of *Escherichia coli* and orientation of the reconstituted lactose:H⁺ carrier, *Eur. J. Biochem.,* 142, 269, 1984.

90. **Carrasco, N., Herzlinger, D., Mitchell, R., De Chiara, S., Danho, W., Gabriel, T. F., and Kaback, H. R.**, Intramolecular dislocation of the COOH terminus of the *lac* carrier protein in reconstituted proteoliposomes, *Proc. Natl. Acad. Sci. U.S.A.*, 81, 4672, 1984.

91. **Manoil, C. and Beckwith, J.**, A genetic approach to analyzing membrane protein topology, *Science*, 233, 1403, 1986.

92. **Broome-Smith, J. K. and Spratt, B. G.**, A vector for the construction of translational fusions to TEM β-lactamase and the analysis of protein export signals and membrane protein topology, *Gene*, 49, 341, 1986.

93. **Gott, P. and Boos, W.**, The transmembrane topology of the *sn*-glycerol 3-phosphate permease of *Escherichia coli* analyzed by *phoA* and *lacZ* protein fusions, *Mol. Microbiol.*, 2, 655, 1988.

94. **Lloyd, A. D. and Kadner, R. J.**, Topology of the *Escherichia coli uhpT* sugar-phosphate transporter analyzed by using Tn*phoA* fusions, *J. Bacteriol.*, 172, 1688, 1990.

95. **Calamia, J. and Manoil, C.**, *lac* permease of *Escherichia coli*: Topology and sequence elements promoting membrane insertion, *Proc. Natl. Acad. Sci. U.S.A.*, 87, 4937, 1990.

96. **Nakao, T., Yamato, I., and Anraku, Y.**, Nucleotide sequence of *put*P, the proline carrier gene of *Escherichia coli* K12, *Mol. Gen. Genet.*, 208, 70, 1987.

97. **Droniuk, R., Wong, P., Wisse, G., and MacLeod, R.**, Variation in quantitative requirements for Na$^+$ for transport of metabolizable compounds by the marine bacteria *Alteromonas haloplanktis* 214 and *Vibrio fischeri*, *Appl. Environ. Microbiol.*, 53, 1487, 1987.

98. **Kitada, M. and Horikoshi, K.**, Bioenergetic properties of alkalophilic *Bacillus* sp. strain C-59 on an alkaline medium containing potassium carbonate, *J. Bacteriol.*, 169, 5761, 1987.

99. **Guffanti, A., Cohn, D., Kaback, H., and Krulwich, T.**, Relationship between the Na$^+$/H$^+$ antiporter and Na$^+$/substrate symport in *Bacillus alcalophilus*, *Proc. Natl. Acad. Sci. U.S.A.*, 78, 1481, 1981.

100. **De Vrij, W., Bulthuis, R. A., Van Iwaarden, P. R., and Konings, W. N.**, Mechanism of L-glutamate transport in membrane vesicles from *Bacillus stearothermophilus*, *J. Bacteriol.*, 171, 1118, 1989.

101. **Kusaka, I. and Kanai, K.**, Characterization of glutamate transport system in hydrophobic protein (H-protein) of *Bacillus subtilis*, *Biochim. Biophys. Acta*, 552, 492, 1979.

102. **Pettitt, C., Davidson, V., Cobb, A., and Knaff, D.**, Sodium-dependent α-aminoisobutyrate transport by the photosynthetic purple sulfur bacterium *Chromatium vinosum*, *Arch. Biochem. Biophys.*, 216, 306, 1982.

103. **Speelmans, G., De Vrij, W., and Konings, W. N.**, Characterization of amino acid transport in membrane vesicles from the thermophilic fermentative bacterium *Clostridium fervidus*, *J. Bacteriol.*, 171, 3788, 1989.

104. **Warthmann, R. and Cypionka, H.**, Sulfate transport in *Desulfobulbus propionicus* and *Desulfococcus multivorans*, *Arch. Microbiol.*, 154, 144, 1990.

105. **Prestidge, L. S. and Pardee, A. B.**, Second permease for methylthio-β-D-galactoside in *Escherichia coli*, *Biochim. Biophys. Acta*, 100, 591, 1965.

106. **Botfield, M. C., Wilson, D. M., and Wilson, T. H.**, The melibiose carrier of *Escherichia coli*, *Res. Microbiol.*, 141, 328, 1990.

107. **Kevbrina, M. V., Zoyagintseva, I. S., and Plakunov, V. K.**, The uptake of monocarboxylic and tricarboxylic acids by extreme halophilic archaebacteria, *Mikrobiologiya*, 58, 703, 1989.

108. **Robrish, S. A. and Thompson, J.**, Sodium requirement for glutamate-dependent sugar transport by *Fusobacterium nucleatum* ATCC 10953, *Curr. Microbiol.*, 19, 329, 1989.

109. **Lanyi, J.**, The role of Na$^+$ in transport processes in bacterial membranes, *Biochim. Biophys. Acta*, 559, 377, 1979.

110. **Dimroth, P. and Thomer, A.,** Solubilization and reconstitution of the sodium-dependent citrate carrier of *Klebsiella pneumoniae, J. Biol. Chem.,* 265, 7721, 1990.
111. **Jarrell, K., Bird, S., and Sprott, G.,** Sodium-dependent isoleucine transport in the methanogenic archaebacterium *Methanococcus voltae, FEBS Lett.,* 166, 357, 1984.
112. **Ekiel, I., Jarrell, K., and Sprott, G.,** Amino acid biosynthesis and sodium-dependent transport in *Methanococcus voltae* as revealed by ^{13}C NMR, *Eur. J. Biochem.,* 23, 1683, 1985.
113. **Hoshino, T. and Kageyama, M.,** Sodium-dependent transport of L-leucine in membrane vesicles prepared from *Pseudomonas aeruginosa, J. Bacteriol.,* 237, 73, 1979.
114. **Fougere, F. and Le Rudulier, D.,** Uptake of glycine betaine and its analogs by bacteroids of *Rhizobium meliloti, J. Gen. Microbiol.,* 136, 157, 1990.
115. **Chen, G. and Russell, J. B.,** Transport and deamination of amino acids by a Gram-positive, monensin-sensitive ruminal bacterium, *Appl. Environ. Microbiol.,* 56, 2186, 1990.
116. **Chen, C. and Russell, J. B.,** Sodium-dependent transport of branched-chain amino acids by a monensin-sensitive ruminal *Peptostreptococcus, Appl. Environ. Microbiol.,* 55, 2658, 1989.
117. **Russell, J. B., Strobel, H. J., Driessen, A. J. M., and Konings, W. N.,** Sodium-dependent transport of neutral amino acids by whole cells and membrane vesicles of *Streptococcus bovis,* a ruminal bacterium, *J. Bacteriol.,* 170, 3531, 1988.
118. **Cairney, J., Higgins, C. F., and Booth, I. R.,** Proline uptake through the major transport system of *Salmonella typhimurium* is coupled to sodium ions, *J. Bacteriol.,* 160, 22, 1984.
119. **Bieber, E. and Wilkinson, B.,** Sodium-dependent uptake of taurine in encapsulated *Staphylococcus aureus* strain M, *Biochim. Biophys. Acta,* 770, 127, 1984.
120. **Kakinuma, Y. and Unemoto, T.,** Sucrose uptake is driven by the sodium electrochemical potential in the marine bacterium *Vibrio alginolyticus, J. Bacteriol.,* 163, 1293, 1985.
121. **Hamaide, F., Sprott, G., and Kushner, D.,** Energetics of sodium dependent α-aminoisobutyric acid transport in the moderate halophile *Vibrio costicola, Biochim. Biophys. Acta,* 766, 77, 1984.
122. **MacLeod, R.,** Salt requirements for membrane transport and solute retention in some moderate halophiles, *FEMS Microbiol. Rev.,* 39, 109, 1986.
123. **Wood, J. M., Zadworny, D., Lohmeiner, E., and Weiner, J. H.,** Hybrid plasmids complement a *putP* mutation in *Escherichia coli* K12, *Can. J. Biochem.,* 57, 1328, 1979.
124. **Mogi, T., Yamamoto, H., Nakao, T., Yamato, I., and Anraku, Y.,** Genetic and physical characterization of *putP,* the proline carrier gene of *Escherichia coli* K12, *Mol. Gen. Genet.,* 202, 35, 1986.
125. **Deguchi, Y., Yamato, I., and Anraku, Y.,** Molecular cloning of *gltS* and *gltP,* which encode glutamate carriers of *Escherichia coli* B, *J. Bacteriol.,* 171, 1314, 1989.
126. **Yazyu, H., Shiota-Niiya, S., Shimamoto, T., Kanazawa, H., Futai, M., and Tsuchiya, T.,** Nucleotide sequence of the *melB* gene and characteristics of deduced amino acid sequence of the melibiose carrier in *Escherichia coli, J. Biol. Chem.,* 259, 4320, 1984.
127. **Tsuchiya, T., Ottina, K., Moriyama, Y., Newman, M. J., and Wilson, T. H.,** Solubilization and reconstitution of the melibiose carrier from a plasmid-carrying strain of *Escherichia coli, J. Biol. Chem.,* 257, 5125, 1983.
128. **Hanatani, M., Yazyu, H., Shiota-Niiya, S., Moriyama, Y., Kanazawa, H., Futai, M., and Tsuchiya, T.,** Physical and genetic characterization of the melibiose operon and identification of the gene products in *Escherichia coli, J. Biol. Chem.,* 259, 1807, 1984.
129. **Hahn, D. R., Myers, R. S., Kent, C. R., and Maloy, S. R.,** Regulation of proline utilization in *Salmonella typhimurium*: molecular characterization of the *put* operon, and DNA sequence of the *put* control region, *Mol. Gen. Genet.,* 213, 125, 1988.

130. **Miller, K. and Maloy, S.**, DNA sequence of the *putP* gene from *Salmonella typhimurium* and predicted structure of proline permease, *Nucleic Acids Res.*, 18, 3057, 1990.

131. **Hoshino, T., Kose, K., and Uratani, T.**, Cloning and nucleotide sequence of the gene *braB* coding for the sodium-coupled branched-chain amino acid carrier in *Pseudomonas aeruginosa* PAO, *Mol. Gen. Genet.*, 220, 461, 1990.

132. **Uratani, Y. and Hoshino, T.**, Difference in sodium requirement of branched chain amino acid carrier between *Pseudomonas aerurinosa* PAO and PML strains is due to substitution of an amino acid at position 292, *J. Biol. Chem.*, 264, 18944, 1989.

133. **Hediger, M. A., Coady, M. J., Ikeda, T. S., and Wright, E. M.**, Expression cloning and cDNA sequencing of the Na^+/glucose co-transporter, *Nature (London)*, 330, 379, 1987.

134. **Yamato, I.**, Ordered binding model as a general mechanistic mechanism for secondary active transport systems, *FEBS Lett.*, 298, 1, 1992.

135. **Hanada, K., Yoshida, T., Yamato, I., and Anraku, Y.**, unpublished results, 1990.

136. **Hanada, K., Yoshida, T., Yamato, I., and Anraku, Y.**, Sodium ion and proline binding sites in the Na^+/proline symport carrier of *Escherichia coli*, *Biochim. Biophys. Acta*, 1105, 61, 1992.

137. **Komeiji, Y., Hanada, K., and Yamato, I., and Anraku, Y.**, unpublished observation, 1989.

Chapter ID

Na$^+$ EXTRUSION COUPLED TO DECARBOXYLATION REACTIONS

Peter Dimroth

TABLE OF CONTENTS

0-8493-6982-7/93/$0.00 + $.50

I. INTRODUCTION

The chemiosmotic theory introduced by Mitchell describes coupling of exergonic and endergonic membrane reactions via a proton circuit.[1-3] Bacterial cells contain primary pumps within their membrane that pump protons from the cytoplasm to the environment at the expense of light or chemical energy. Examples of these primary energy converters are the photochemical reaction centers, bacteriorhodopsin, and the diverse respiratory chains. The electrochemical gradient of protons thus established provides the driving force for a number of energy-requiring processes catalyzed by protein complexes located within the cell membrane. Examples are the numerous H$^+$-coupled transport systems, the synthesis of ATP by F$_1$F$_0$-ATPase, reversed electron transfer, NADPH synthesis from NADH by transhydrogenase, and the movement of the flagellar motor.

Considerable evidence has accumulated during the last decade that chemiosmotic energy transduction is not restricted to H$^+$ as the coupling ion, but likewise applied to Na$^+$-coupled energy transformations.[4-8] Secondary Na$^+$/solute symport systems are well established in many bacterial species[9-11] (see Chapter IC). Organisms living at high Na$^+$ concentrations (halophilic and marine bacteria) or at high pH values (alkaliphilic bacteria) use these Na$^+$-coupled transport systems primarily (see Chapters IE and IF), while the Na$^+$/solute symporters are less frequent in neutrophilic freshwater bacteria.

In marine vibrios[12] and alkaliphilic bacteria[13] the $\Delta\bar{\mu}_{Na^+}$ also provides the driving force for the flagellar motor. The energetic link between $\Delta\bar{\mu}_{H^+}$ and $\Delta\bar{\mu}_{Na^+}$ usually is the Na$^+$/H$^+$ antiporter, which extrudes Na$^+$ ions from the cells coupled to the $\Delta\bar{\mu}_{H^+}$-driven uptake of protons[11] (see Chapter IB). In addition, a few bacterial species dispose of a primary Na$^+$ pump to form $\Delta\bar{\mu}_{Na^+}$.[4-8] The first primary Na$^+$ pump discovered in a bacterium was oxaloacetate decarboxylase of *Klebsiella pneumoniae* fermenting citrate.[14-16] Later, methylmalonyl-CoA decarboxylase[17,18] and glutaconyl-CoA decarboxylase[19] were found to be other members of the Na^+-transport decarboxylase family. In *Vibrio alginolyticus*, a respiratory Na^+ translocation was discovered and shown to be located at the NADH:ubiquinone oxidoreductase segment of the respiratory chain[20,21] (Chapter IF). The most prominent example of a bacterium using Na^+ bioenergetics is *Propionigenium modestum*.[22] All ATP synthesis in this bacterium depends on a Na^+ circuit between the Na^+-pumping methylmalonyl-CoA decarboxylase and the Na^+-translocating F_1F_0-ATPase[23] (Chapter IG).

These and other examples of bacteria employing Na^+ bioenergetics are summarized in Table 1. They clearly indicate the occurrence of primary Na^+ transport systems in bacteria. Like proton gradients the generated Na^+ electrochemical gradients provide a source of stored energy for the cell that can be used to drive endergonic membrane reactions such as ATP synthesis, active transport, or the movement of the flagellar motor.

II. PHYSIOLOGICAL FUNCTION OF SODIUM ION TRANSPORT DECARBOXYLASES

A. CITRATE FERMENTATION IN *KLEBSIELLA PNEUMONIAE*

The fermentation pathways performed by a number of anaerobic bacteria involve the decarboxylation of a carboxylic acid. The energy liberated in this reaction ($\Delta G_0' \approx -28$ kJ/mol) is used by certain decarboxylases located in the bacterial membrane to pump Na^+ ions out of the cells.[4-7] This mechanism of energy conservation was discovered for oxaloacetate decarboxylase of *K. pneumoniae* fermenting citrate. The enzymes of the citrate fermentation pathway are induced in *Klebsiella* by citrate under anaerobic conditions and are employed for citrate degradation.[30,31] Citrate is transported into the cell in symport with Na^+ ions and is subsequently cleaved by citrate lyase to acetate and oxaloacetate. The latter is decarboxylated by the membrane-bound oxaloacetate decarboxylase to pyruvate. The decarboxylation is coupled to the electrogenic extrusion of Na^+ ions form the inside of the bacterial cell to the outside. Pyruvate formate lyase splits pyruvate to acetyl-CoA and formate. Phosphotransacetylase converts acetyl-CoA to acetyl phosphate, and acetate kinase catalyzes the synthesis of ATP by substrate-level phosphorylation from ADP and acetyl phosphate. Thus, 1 mol of ATP is synthesized in these cells per mole of citrate fermented. The reactions of the entire pathway are shown schematically in Figure 1.[4]

B. SODIUM ION CIRCUIT BETWEEN OXALOACETATE DECARBOXYLASE AND CITRATE CARRIER

Of special interest from the point of view of membrane bioenergetics are oxaloacetate decarboxylase and the citrate carrier. The membrane-bound oxaloacetate decarboxylase is specifically activated by Na^+ ions and functions as an electrogenic Na^+ pump, transporting 2 mol Na^+ ions out of the cell

TABLE 1
A Selection of Primary Na$^+$ Transport Systems in Bacteria

Bacteria	Growth Condition	System	Ref.
Klebsiella pneumoniae	Fermentation of citrate	Oxaloacetate decarboxylase	16
		NADH:ubiquinone oxidoreductase	24
Salmonella typhimurium	Fermentation of citrate	Oxaloacetate decarboxylase	25
Veillonella alcalescens	Fermentation of lactate	Methylmalonyl-CoA decarboxylase	18
Propionigenium modestum	Fermentation of succinate	Methylmalonyl-CoA decarboxylase	22
		F_1F_0 ATPase	23
Acidaminococcus fermentans	Fermentation of glutamate	Glutaconyl-CoA decarboxylase	19
Peptostreptococcus asaccharolyticus	Fermentation of glutamate	Glutaconyl-CoA decarboxylase	26
Clostridium symbiosum	Fermentation of glutamate	Glutaconyl-CoA decarboxylase	26
Fusobacterium nucleatum	Fermentation of glutamate	Glutaconyl-CoA decarboxylase	26
Methanosarcina barkeri	Fermentation of CO_2 to CH_4	A step between methylene-tetrahy-dromethanopterin and methyl-coenzyme M	27
Vibrio alginolyticus	Aerobic	NADH:ubiquinone oxidoreductase	21
Halotolerant bacterium Ba1	Aerobic	NADH:ubiquinone oxidoreductase	21
Vitreoscilla	Aerobic	Cytochrome *o*	29

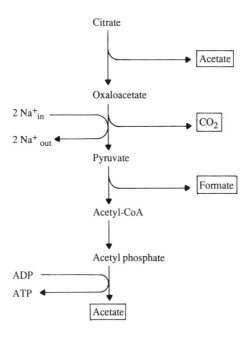

FIGURE 1.   Citrate fermentation pathway of *Klebsiella pneumoniae*.

per mole oxaloacetate decarboxylated.[32] The thus-generated Na⁺ ion gradient is required to drive the uptake of the growth substrate citrate into the cells.[33,34] In order not to be repelled by the membrane potential (inside negative), the uptake of the citrate trianion requires a cotransport with cations carrying at least three positive charges. Genetic evidence inferred the existence of three different citrate carriers in *K. pneumoniae*.[35] Aerobic cells express a H⁺/citrate symporter, and anaerobic cells express a Na⁺/citrate symporter.[33] The genes for these two different citrate carriers have been cloned and their nucleotide sequences have been determined.[36,37]

The anaerobically expressed citrate carrier was extracted from the bacterial membrane with Triton® X-100 and reconstituted into proteoliposomes. These proteoliposomes catalyzed Na⁺-dependent citrate counterflow, or citrate uptake, upon the application of a Na⁺ concentration gradient, indicating that the anaerobically induced citrate carrier is a Na⁺/citrate symporter.[34] A Na⁺ cycle is thus operating in *K. pneumoniae*: Na⁺ ions taken up in symport with citrate are exported again by the action of oxaloacetate decarboxylase. While the stoichiometry of the decarboxylase is two Na⁺ per oxaloacetate,[32] that of the citrate carrier is not yet clear. As mentioned above at minimum three monovalent cations must be symported with the citrate. These could be three Na⁺ ions, two Na⁺ ions and a proton, or other combinations of these ions. Proton extrusion is expected to proceed without ATP hydrolysis by secretion of the metabolic end products acetate, formate, and bicarbonate in symport with protons. If the excretion of these end products together with protons had to proceed against a large $\Delta\bar{\mu}_{H^+}$, the end-product concentrations inside the cell would increase. This could result in a self-poisoning of the cells.[33,34] In contrast, a $\Delta\bar{\mu}_{Na^+}$ does not interfere with a proton-coupled secretion of the end products as long as this is an electroneutral event. The interference with $\Delta\bar{\mu}_{H^+}$ by the extrusion of the acidic end products is probably the reason why the oxaloacetate decarboxylase of *K. pneumoniae* pumps Na^+ ions rather than protons.

C. OTHER SODIUM ION TRANSPORT DECARBOXYLASES

A sodium ion-pumping oxaloacetate decarboxylase is also present in citrate-fermenting *Salmonella typhimurium* cells, where the physiological role of the Na^+ gradient established by this enzyme may be the same as in *K. pneumoniae*.[25] Other enzymes belonging to the group of Na^+ ion-translocating decarboxylases are methylmalonyl-CoA decarboxylase[17,18] and glutaconyl-CoA decarboxylase.[19] Methylmalonyl-CoA decarboxylase is involved in propionate fermentation from succinate in *Veillonella alcalescens*[18] and *P. modestum*.[22] *P. modestum* can grow on succinate. The entire metabolic energy of such cells derives from the $\Delta\bar{\mu}_{Na^+}$ established upon decarboxylation of methylmalonyl-CoA. *P. modestum*, therefore, contains a Na^+-translocating F_1F_0-ATPase for ATP synthesis at the expense of the electrochemical Na^+ ion gradient.[23] In contrast, *V. alcalescens* lacks a Na^+-translocating ATPase and is therefore unable to grow with succinate as the only carbon and energy source. These cells grow on the free energy obtained from the conversion of lactate to acetate, propionate, CO_2, and H_2. The branch of the pathway leading to acetate yields ATP by substrate-level phosphorylation, but in addition produces NADH in the initial oxidation of lactate to pyruvate, which must be reoxidized. For this purpose, part of the pyruvate is carboxylated to oxaloacetate under consumption of ATP. Subsequently, the oxaloacetate is reduced to succinate, thereby consuming NADH and regenerating part of the ATP via $\Delta\bar{\mu}_{H^+}$ generated by fumarate reductase.[38] Additional energy in the form of a $\Delta\bar{\mu}_{Na^+}$ is conserved during the subsequent conversion of succinate to propionate and CO_2 by the methylmalonyl-CoA decarboxylase Na^+ pump.[17,18] The Na^+ gradient in *V. alcalescens* may be used to drive transport processes, e.g., the uptake of the growth substrate lactate. The glutaconyl-CoA decarboxylase-derived $\Delta\bar{\mu}_{Na^+}$ may have a similar role as a driving force for the uptake of the growth substrate glutamate in glutamate-fermenting bacteria that contain this enzyme.

III. DISCOVERY OF THE OXALOACETATE DECARBOXYLASE Na^+ PUMP

Anaerobic growth of *K. pneumoniae* is Na^+ dependent. This dependence is based on the Na^+ requirement of oxaloacetate decarboxylase[39] and, in

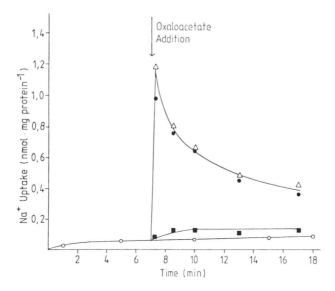

FIGURE 2. Kinetics of Na^+ uptake by French press vesicles from *K. pneumoniae*. The transport was initiated by oxaloacetate addition as indicated (●). Symbols: vesicles preincubated with avidin (■); vesicles preincubated with avidin-biotin complex (△); control without oxaloacetate addition (○). (From Dimroth, P., *Ann. N.Y. Acad. Sci.*, 447, 72, 1985. With permission.)

addition, on the Na^+ requirement for citrate uptake.[33] Remarkable properties of the crude oxaloacetate decarboxylase were reported by Stern.[39] The enzyme was associated with the particulate material of the cell, required Na^+ (but not divalent cation) for catalytic activity, and was inhibited by avidin, indicating that biotin is involved in the catalysis. These properties would be quite unusual if the only function of the decarboxylase was catalysis of a step in the pathway of citrate degradation. The membrane-linked character of the enzyme and its activation by Na^+ were reminiscent, however, of transport enzymes such as the Na^+/K^+-ATPase and suggested to us that oxaloacetate decarboxylase might act as a Na^+ pump.[14-16] A further indication that this hypothesis was feasible came from the negative value of the free energy change of the reaction of about 28 kJ/mol, which would be sufficient to drive uphill transport of Na^+ ions.

Experiments performed with inverted vesicles from *K. pneumoniae* revealed a rapid accumulation of Na^+ inside these vesicles upon oxaloacetate decarboxylation, followed by Na^+ efflux after the substrate had been consumed (Figure 2).[16,32] In contrast, no appreciable Na^+ uptake was observed in the absence of oxaloacetate or after inhibition of the enzyme with avidin. These results demonstrated the function of oxaloacetate decarboxylase as a Na^+ pump. The free energy of the decarboxylation reaction is used to drive the active transport of Na^+ ions. This was a completely unexpected and new type of conversion of chemical energy which exists in addition to the well-known energy conservation mechanisms of electron transport and substrate-level phorphorylation.

Subsequently, oxaloacetate decarboxylase of *S. typhimurium*,[25] methylmalonyl-CoA decarboxylase of *V. alcalescens*,[17,18] and glutaconyl-CoA decarboxylase of *Acidaminococcus fermentans*[19] and of *Fusobacterium nucleatum*[26] were demonstrated to function as similar Na^+ pumps.

IV. PURIFICATION AND SUBUNIT COMPOSITION OF THE SODIUM ION-TRANSLOCATING DECARBOXYLASES

All sodium ion transport decarboxylases share a number of properties: (1) they are membrane-bound enzyme complexes, (2) they are specifically activated by Na^+ ions, (3) they catalyze an electrogenic Na^+ transport across the membrane to the side opposite to that of substrate decarboxylation, (4) they contain covalently bound biotin, (5) they have a similar subunit composition, and (6) analogous subunits have similar functions.[4-7] The presence of covalently bound biotin has greatly facilitated the purification of these enzymes by avidin-Sepharose® affinity chromatography.[40-42] After disruption of the bacterial cells by passage through a French press, the membranes were isolated by differential centrifugation and the enzymes were extracted with Triton® X-100. The solubilized decarboxylases were passed over a monomeric avidin-Sepharose® affinity column, from which they were eluted specifically with biotin. An analysis of the enzyme samples thus obtained by SDS-polyacrylamide gel electrophoresis showed that they were of high purity.

A comparison of the subunit compositions of several sodium ion transport decarboxylases is shown in Table 2. The oxaloacetate decarboxylases of *K. penumoniae*[43] and *S. typhimurium*[25] have an almost identical pattern of three subunits (α, β γ). Subunits corresponding in molecular mass to the α-, β-, and γ-chains are also present in methylmalonyl-CoA decarboxylase,[18] and subunits corresponding to the α- and β-chains are present in glutaconyl-CoA decarboxylase.[26] A characteristic of the decarboxylases acting on thioester substrates is the presence of a separate biotin carrier protein subunit, whereas in oxaloacetate decarboxylase the biotin is bound to the α-chain.

It is remarkable that the sodium ion transport decarboxylases, which have three different substrate specificities and are found in bacteria that are not closely related phylogenetically, share a similar composition of the α- and β-subunits. It will be shown below that these subunits perform analogous functions within each enzyme complex. The sodium ion transport decarboxylases are therefore a closely related family of enzymes.

TABLE 2
Subunit Composition of the Sodium Ion Transport Decarboxylases

Substrate	Organism	Apparent Molecular Mass (in kilodaltons)				Ref.
		α-Chain	β-Chain	Biotin carrier	γ- (or δ-) Chain	
Oxaloacetate	*Klebsiella pneumoniae*	64	34	α-chain	9 (γ-chain)	43,52,54,64
	Salmonella typhimurium	63	33	α-chain	9 (γ-chain)	25,64
Methylmalonyl-CoA	*Veillonella alcalescens*	60	33	18.5 (γ-chain)	14 (δ-chain)	18
Glutaconyl-CoA	*Acidaminococcus fermentans*	65	33	24 (γ-chain)	14 (δ-chain)	26
	Fusobacterium nucleatum	65	33	19 (γ-chain)	16 (δ-chain)	26
	Clostridium symbiosum	65	33	22–24 (γ-chain)	14 (δ-chain)	26

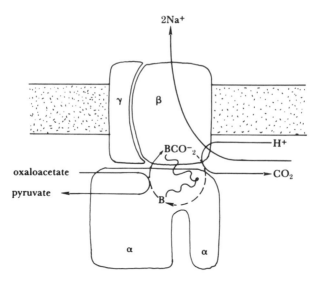

FIGURE 3. A model linking structure and function of the oxaloacetate decarboxylase subunits. B = biotin. (From Dimroth, P., *Philos. Trans. R. Soc. London Ser. B*, 326, 465, 1990. With permission.)

V. DECARBOXYLATION MECHANISM

A. SUBSTRATE STEREOCHEMISTRY AND ACTIVATION BY METAL IONS

During the decarboxylation of oxaloacetate,[44] glutaconyl-CoA,[45] or methylmalonyl-CoA[46] the stereochemical configuration is retained. In this respect these reactions are the same as all other investigated reactions catalyzed by biotin enzymes.[47] As biotin enzymes the decarboxylases are completely inhibited by avidin. Sodium ions specifically activate these enzymes with a K_m of about 1 mM. A less efficient activation of oxaloacetate decarboxylase (K_m $\approx$ 25 mM) and glutaconyl-CoA decarboxylase (K $\approx$ 100 mM) has also been observed with Li$^+$ ions.[40-42]

B. CATALYSIS OF PARTIAL REACTIONS BY INDIVIDUAL SUBUNITS

The decarboxylation mechanism of all sodium ion transport decarboxylases involves two distinct steps, each of which is catalyzed by a subunit of similar molecular weight of these enzyme complexes. An essential feature of the reaction mechanism is the intermediate formation of protein-bound carboxybiotin. In oxaloacetate decarboxylase, the prosthetic biotin group is attached to a specific domain (α_2) of the α-subunit (Figure 3, Table 2) and in the CoA-ester decarboxylases to a separate protein subunit with M_r 18,000 to 24,000 (γ-chain, Table 2). The reaction cycle is initiated by a transfer of the carboxyl group from the substrate to the prosthetic biotin group on the enzyme. This reaction is catalyzed by the α-subunit of each enzyme complex, which functions as a carboxyltransferase (Equation 1).[43,48,49]

$$R-COO^- + \text{biotin-protein} \rightleftarrows RH + {}^-OOC\text{-biotin-protein} \qquad (1)$$

The carboxyltransferases of the three different decarboxylases are functional in the complete absence of Na$^+$ ions. They are peripheral membrane proteins that, when dissociated from the other subunits, are soluble in the absence of detergent.[43,48,49]

The carboxyltransferase reaction is succeeded by the Na$^+$-dependent decarboxylation of the carboxybiotin. It has been shown that the lyase uses H$^+$ as a substrate and releases CO$_2$ as a product rather than consuming H$_2$O and generating HCO$_3^-$.[18,43]

$$^-OOC\text{-biotin-protein} + H^+ + 2Na^+_{in} \rightleftarrows CO_2 + \text{biotin-protein} + 2Na^+_{out} \qquad (2)$$

In oxaloacetate decarboxylase this reaction is catalyzed by the β- (or $\beta + \gamma$) subunits and in the CoA-ester decarboxylases by the β- or ($\beta + \delta$)-subunits, respectively. These polypeptides have been shown to be firmly embedded within the lipid bilayers.

C. DISSOCIATION AND RECONSTITUTION OF ENZYME COMPLEXES

1. Oxaloacetate Decarboxylase

To elucidate the above-mentioned functions of the subunits of the sodium ion transport decarboxylases, the enzyme complexes had to be dissociated. Oxaloacetate decarboxylase dissociates upon freezing and thawing in the presence of chaotropic salts, e.g., LiCl or LiClO$_4$, and reassembles upon dilution, as shown by the disappearance and recovery, respectively, of catalytic activity.[50] The pH is of central importance for the dissociation and reconstitution of the enzyme complex. Whereas an acidic pH promotes dissociation, a more alkaline pH promotes reconstitution. Plots of the initial rate of either dissociation or reconstitution against pH gave sigmoidal curves with inflection points around pH 6.5.[5,50] The results suggest that a single ionizable group with a pK of about 6.5 (possibly contributed by a histidine residue) is involved in these processes. If this group is protonated, the enzyme complex is destabilized and dissociation takes place; the uncharged form of the residue appears to be required for the reconstitution of the decarboxylase from its subunits.

After dissociation, the biotin-containing α-chain could be separated from the β- and γ-subunits by chromatography on avidin-Sepharose® in the denaturing solvent.[50] Subsequently, the enzyme complex was reconstituted from the isolated α- and ($\beta + \gamma$)-subunits. The reconstituted enzyme recovered all functions of the native decarboxylase, including that of Na$^+$ transport.

The Na$^+$ transport function was also recovered by a sequential reconstitution initiated with the incorporation of the (β + γ)-subunits into proteoliposomes and completed by adding the α-subunit.[50] Interestingly, the Na$^+$ conductance of the proteoliposomes was low and was not significantly increased if only the β- and γ-subunits were incorporated. A freely open Na$^+$ channel is therefore apparently not formed by the β- and γ-subunits.[63]

2. Methylmalonyl-CoA Decarboxylase

Dissociation of the β-subunit from the methylmalonyl-CoA decarboxylase enzyme complex was achieved by prolonged washing of an avidin-Sepharose® column with the enzyme bound via its prosthetic biotin group. The β-chain-depleted enzyme was inactive as methylmalonyl-CoA decarboxylase, but retained the carboxyltransferase activity (Equation 1).[49] The carboxyltransferase was not completely specific for methylmalonyl-CoA, but also accepted malonyl-CoA as a substrate. The enzyme could thus catalyze the carboxyl transfer reaction from methylmalonyl-CoA to acetyl-CoA, yielding propionyl-CoA and malonyl-CoA, and vice versa.[49]

3. Glutaconyl-CoA Decarboxylase

A partial dissociation was also described for glutaconyl-CoA decarboxylase.[48] It was observed that small concentrations of a primary alcohol, e.g., 2% 1-butanol, completely inactivated the decarboxylase activity, but left part of the carboxyltransferase activity intact. The butanol treatment specifically precipitated the β-chain, the presumed catalyst of the lyase reaction (Equation 2). Interestingly, inactivation of the β-subunit by the organic solvent was specifically inhibited by 50 mM Na$^+$ salts. Previously, it had been shown that similar Na$^+$ concentrations protected the β-subunits of all three decarboxylases from tryptic hydrolysis.[18,19,43] Thus, this subunit has a binding site for Na$^+$ which, when occupied, forces the protein into a more resistant conformation. This notion is in accord with the catalysis of the Na$^+$-dependent reaction step of the catalytic cycle by the β-subunit and further suggests that this protein is responsible for the pumping of Na$^+$ ions.

D. MODEL OF OXALOACETATE DECARBOXYLASE STRUCTURE AND FUNCTION

A model of the location of the individuals subunits of oxaloacetate decarboxylase in the membrane and their function in catalysis, based on the data discussed above, is shown in Figure 3.[5] It appears from this model that the gross structure of oxaloacetate decarboxylase may be similar to that of the F$_1$F$_0$-ATPases. The α-chain of oxaloacetate decarboxylase corresponds to the F$_1$ moiety, and the β- and γ-subunits correspond to the F$_0$ part. The arrangement of the subunits of the other decarboxylases in the membrane is probably analogous.

From a kinetic study of the reaction mechanism of oxaloacetate decarboxylase it was concluded that the overall decarboxylation involves catalysis at two different and independent sites.[51] Site 1 catalyzes the carboxyl transfer from oxaloacetate to enzyme-bound biotin. The carboxybiotin moves to site 2, where a proton is added and CO_2 is released, if this site is occupied with Na^+. The model derived from enzyme kinetics is thus in complete harmony with the results on subunit structure and function described above.

VI. STRUCTURAL STUDIES

A. DOMAIN STRUCTURE OF THE α-SUBUNIT

The model of structure and function of oxaloacetate decarboxylase, shown in Figure 3, indicates two different domains in the peripheral α-subunit: the carboxyltransferase domain (α_1) and the biotin domain (α_2). This model is based in part on the results of limited proteolytic digestion of the α-chain with trypsin, chymotrypsin, and thermolysin, which degrade the α-chain rapidly into a large C-terminal and a smaller biotin-containing N-terminal fragment, accompanied by the loss of catalytic activity. The results of Figure 4 show the effect of trypsin on the digestion of the oxaloacetate decarboxylase subunits and on its enzymic activity. The α-subunit is completely degraded within 8 min yielding two main fragments: a C-terminal fragment of M_r 51,000 (α_1) and an N-terminal biotin-containing fragment of M_r 10,200 (α_2). Fragment α_2 contained biotin, as indicated by Western blot analysis and visualization with avidin conjugated to alkaline phosphatase.[52] The proteolysis led to the complete loss of oxaloacetate decarboxylase activity (Figure 4A). The kinetics of the proteolysis indicate that the digestion of the α-chain to the α_1-fragment probably proceeds through a fragment of slightly higher molecular weight. In addition, part of the α-chain is degraded to a fragment with a molecular weight between that of α_1 and the β-chain. The β- and the γ-subunits appear to be more resistant to tryptic hydrolysis. Investigations by electron microscopy revealed a cleft in the α-subunit with the prosthetic biotin group located at its bottom in close proximity to the β- and γ-subunits.[53] This location is in accord with the catalytic mechanism, which predicts the flip-flop movement of the biotin between the carboxyltransferase site on the α-subunit and the lyase catalytic site on the β- (or β + γ) subunit(s).

B. PRIMARY STRUCTURE OF OXALOACETATE DECARBOXYLASE

In order to elucidate the primary structure, the genes encoding oxaloacetate decarboxylase of *K. pneumoniae* have been cloned in *E. coli* on a cosmid vector. Three clones which are able to grow on citrate were isolated; two could utilize this carbon source only aerobically, while the third also performed the fermentation of citrate.[35] The cosmid of the latter strain contained the genes encoding oxaloacetate decarboxylase and the Na^+-dependent citrate

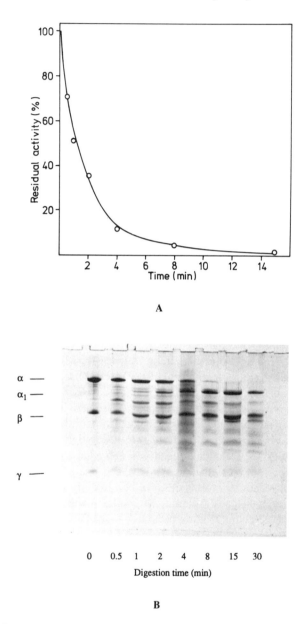

A

B

FIGURE 4. Limited proteolysis of oxaloacetate decarboxylase with trypsin. The incubation mixture (25°C) contained in 0.5 ml 20 mM Tris/HCl buffer, pH 7.5, 100 μg of oxaloacetate decarboxylase and 0.5 μg of trypsin. For residual oxaloacetate decarboxylase activity determinations, samples (50 μl) were taken at the times indicated in A and added to 10 μg trypsin inhibitor. Subsequently, the oxaloacetate decarboxylase activities were determined. In a parallel experiment (B), the proteolysis of samples (50 μl) was terminated by addition to 50 μl 10% formic acid. After evaporation in a Speed/Vac apparatus, the residues were dissolved in sample buffer and subjected to SDS-gel electrophoresis in a 10% polyacrylamide gel. The samples were taken after 0, 0.5, 1, 2, 4, 8, 15, and 30 min incubation with trypsin.

carrier in an operon-like structure. The genes encoding the decarboxylase were subcloned on plasmids and their nucleotide sequences were determined.[52,54] The arrangement of the oxaloacetate decarboxylase genes on the genome was γ-α-β. The amino acid sequence deduced from the nucleotide sequence was secured by protein sequencing of about one third of the entire molecule.

A remarkable feature of the sequence is the striking homology of the carboxyltransferase domain of oxaloacetate decarboxylase with the 5S subunit of the transcarboxylase enzyme complex of *Propionibacterium shermanii* that catalyzes exactly the same carboxyltransfer reaction as the α-chain of oxaloacetate decarboxylase.[52] Further homology was found between the biotin-containing C-terminal domain of the α-subunit and other biotin-containing proteins. Like in most biotin proteins, the prosthetic group of oxaloacetate decarboxylase was bound to a lysine, 35 residues upstream of the C-terminus. No apparent homology of the α-subunit with other biotin enzymes was detectable in the region between the N-terminal and the C-terminal domains. This region contained an extended alanine- and proline-rich sequence, which could function as a hinge allowing the flip-flop movement of the biotin prosthetic group between the two catalytic sites on the α- and (β + γ)-subunits, respectively.[52] An alanine-proline hinge a distinct distance from the prosthetic group appears to be a general motif for enzyme complexes in which this group must move between different subunits. Another example is the dihydrolipoamide acetyltransferase subunit of the pyruvate dehydrogenase multienzyme complex of *E. coli*,[55] for which it has been demonstrated by nuclear magnetic resonance (NMR) spectroscopic techniques that the alanine-proline hinge is highly mobile.[56]

The structural relationship between the α-subunit of oxaloacetate decarboxylase and the 5S and 1.2S subunits of transcarboxylase is interesting from an evolutionary point of view. It suggests that enzyme complexes with very distinct overall functions, such as the Na⁺-transporting oxaloacetate decarboxylase and the transcarboxylase, have developed by an assembly of subunits that derived in part from a common ancestral gene.

The sequences of the β- and γ-subunits are not related to any other known sequence.[54] Please note that the published sequence of the β-subunit[54] is in error in its C-terminal part due to a cloning artifact. The correct sequence of the *K. pneumoniae* β-subunit, together with the sequence of all three oxaloacetate decarboxylase subunits from *S. typhimurium,* has now been determined.[64] The sequences of the decarboxylases from these two organisms are highly homologous.[64] The β-subunit contains extended stretches of hydrophobic amino acid residues, in accord with the location of this subunit, firmly embedded within the membrane. While the β-subunit certainly traverses the membrane several times, the γ-subunit probably contains only one transmembrane helix in the amino-terminal part and a hydrophilic carboxyl-terminal part that is certainly not embedded within the lipid bilayer.[54]

VII. MECHANISMS OF Na$^+$ TRANSLOCATION

A. BIOENERGETICS AND STOICHIOMETRY

To investigate the Na$^+$ transport activity of the decarboxylases, the purified enzymes were incorporated into liposomes.[19,57,58] Most experiments were performed with proteoliposomes reconstituted by the detergent dilution method, with octylglucoside as the detergent. Inspection by electron microscopy of proteoliposomes containing methylmalonyl-CoA decarboxylase revealed that each proteoliposome, with a mean diameter of 62 nm, contained on average nine to ten enzyme molecules incorporated into the phospholipid bilayer and one to two enzyme molecules in the proper orientation to perform inwardly directed Na$^+$ pumping.[59]

Upon the addition of substrate to proteoliposomes containing oxaloacetate decarboxylase or methlymalonyl-CoA decarboxylase, a sevenfold Na$^+$ concentration gradient (50 mV) was established.[16,32,60] Simultaneously, a membrane potential was generated which was about 60 mV at its maximum. Accordingly, the total sodium-motive force was about 110 mV. To pump 1 mol of Na$^+$ ions against an electrochemical gradient of 110 mV requires 10.5 kJ. Since the standard free energy change of the decarboxylation reaction ($\Delta G_0'$) is approximately -28 kJ/mol, it is high enough to pump 2 to 3 mol of Na$^+$ ions per mole of substrate decarboxylated.

A direct measurement of the stoichiometry was performed by determining Na$^+$ uptake and substrate decarboxylation in parallel. With all three decarboxylases the stoichiometry was about two Na$^+$ ions per decarboxylation event, if determined during the initial stage of the transport process.[19,32,61] The stoichiometry changed when steeper Na$^+$ concentration gradients developed because decarboxylation continued at the same rate even after a constant internal Na$^+$ concentration was reached. Nevertheless, transport and chemical reaction were partially coupled because the decarboxylation rate increased about twofold after monensin was added, which completely abolished the Na$^+$ gradient.[19,32,60] An increase in the decarboxylation rate was also observed after dissipation of the membrane potential with valinomycin and K$^+$.[60] In the presence of this ionophore, the initial rates of Na$^+$ uptake and malonyl-CoA decarboxylation increased in parallel over the valinomycin-free controls so that the coupling ratio was about the same, i.e., two Na$^+$ per malonyl-CoA decarboxylated.[60]

B. MODE OF COUPLING BETWEEN DECARBOXYLATION AND Na$^+$ TRANSPORT

As mentioned above, the rate of malonyl-CoA decarboxylation was not independent of the development of $\Delta\Psi$ and ΔpNa^+, although the decarboxylation continued even after the Na$^+$ gradient was fully established. A type of coupling could therefore be anticipated that was independent from net Na$^+$ accumulation. In an investigation of Na$^+$ transport by methylmalonyl-CoA decarboxylase-containing proteoliposomes it was found that the enzyme not only catalyzed net Na^+ uptake, but also an exchange of external and internal Na^+ ions consisting of Na^+ influx plus Na^+ efflux.[60] It was shown that Na^+ influx rather than net uptake is coupled to malonyl-CoA decarboxylation with a stoichiometry of 2:1 during all stages of the development of the electrochemical gradient of Na^+ ions. This means that the efflux rate increases during development of the Na^+ gradient and becomes equal to the rate of Na^+ influx in the steady state where the Na^+ concentration gradient is at its maximum level.

These results provide important insights into the coupling mechanism between malonyl-CoA decarboxylation and Na^+ transport and may be of significance for other primary converters of chemical energy into ion gradients, as well. A hypothetical scheme of the Na^+ translocation mechanism is shown in Figure 5.[60] In the first step (1), the prosthetic biotin group on the γ-subunit of methylmalonyl-CoA decarboxylase (Table 2) is carboxylated by carboxyltransfer from malonyl-CoA, yielding acetyl-CoA. This step is catalyzed by the α-subunit. The carboxybiotin then moves into the decarboxylase catalytic site on the β-subunit (2). With the carboxybiotin bound, the β-subunit assumes a conformation where it binds Na^+ ions with high affinity from the external surface of the proteoliposomes (3). The binding of Na^+ ions triggers the chemical decarboxylation of carboxybiotin that is associated with a conformational change of the β-subunit, by which the Na^+ ions become exposed to the inner surface of the proteoliposomes (4). The Na^+ ions then dissociate from the β-subunit, which in this conformation has a low Na^+ binding affinity (5). The system is now ready to initiate a new round of the Na^+ pumping cycle.

From the strict coupling of Na^+ influx to malonyl-CoA decarboxylation it seems that steps 1, 3, 4, and 5 should operate precisely as indicated in the scheme. The efflux of Na^+ observed at high $\Delta\tilde{\mu}_{Na^+}$ is most likely catalyzed by reaction 2. This step involves a conformational transition of the β-subunit, by which the Na^+ binding site accessible from the internal surface of the proteoliposomes becomes exposed to the external surface. Depending on the magnitude of the $\Delta\tilde{\mu}_{Na^+}$, this conformational transition of subunit β could proceed without or with Na^+ translocation from the inside to the outside. If the $\Delta\tilde{\mu}_{Na^+}$ is fully developed, all β-subunits may be saturated with Na^+ ions and the maximum number of Na^+ ions (two per reaction) may be translocated from the inside to the outside.[60]

The above-mentioned details of the coupling mechanism of the methylmalonyl-CoA decarboxylase Na^+ pump were elucidated by following the flux of Na^+ ions using the isotope $^{22}Na^+$. This methodology cannot be applied to H^+-coupled transport systems, and the mechanistic details of these systems are therefore more difficult to decipher. We do not know yet whether the features of the coupling mechanism determined for methylmalonyl-CoA decarboxylase are unique to this enzyme and perhaps the family of

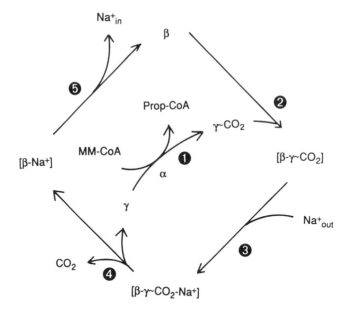

FIGURE 5.   Model of coupling between methylmalonyl-CoA decarboxylation and Na^+ transport into reconstituted proteoliposomes. (1) Carboxyl transfer from methylmalonyl-CoA to the prosthetic biotin group in the γ-subunit, catalyzed by the α-subunit; (2) movement of carboxybiotin on γ to the decarboxylase binding site on the free β-subunit; (3) binding of Na^+ ions from the outside to the β-γ~CO_2 complex; (4) decarboxylation of carboxybiotin with a conformational change in β exposing the Na^+ ions to the inner surface of the membrane; (5) dissociation of Na^+ to the inside of the proteoliposomes. MM-CoA: methylmalonyl-CoA, Prop-CoA: propionyl-CoA. (From Hilpert, W. and Dimroth, P., *Eur. J. Biochem.*, 195, 79, 1991. With permission.)

Na^+-translocating decarboxylases or may more generally apply to primary transport systems which convert chemical energy into an electrochemical ion gradient.

VIII. REVERSIBILITY

A fundamental feature of coupled vectorial transport systems is reversibility. The reactions catalyzed by the Na^+-translocating decarboxylases are completely irreversible if catalyzed by the soluble enzymes. To imitate the physiological situation in which the decarboxylations are coupled to Na^+ transport, the reversibility has been studied using oxaloacetate decarboxylase and methylmalonyl-CoA decarboxylase reconstituted into proteoliposomes.[61] According to Equation 3, the reversibility could be measured by the isotopic exchange between $^{14}CO_2$ and the carboxylated substrated (R–COO^-)

$$R\text{–}COO^- + 2Na^+_{out} + H^+ \rightleftarrows RH + CO_2 + 2Na^+_{in} \qquad (3)$$

Proteoliposomes containing oxaloacetate decarboxylase catalyzed a $^{14}CO_2$-oxaloacetate exchange, and methylmalonyl-CoA decarboxylase-containing proteoliposomes catalyzed a $^{14}CO_2$-malonyl-CoA exchange. These exchange reactions are absolutely dependent on the Na⁺ ion gradient established during decarboxylation of part of the substrate since no exchange takes place in the presence of monensin, which abolishes the gradient. Net carboxylation of acetyl-CoA to malonyl-CoA was observed when a large inwardly directed Na⁺ concentration gradient was applied to methylmalonyl-CoA decarboxylase-containing proteoliposomes; none was observed in the absence of a Na⁺ gradient.[61]

A transcarboxylase system was constructed by reconstituting oxaloacetate decarboxylase and methylmalonyl-CoA decarboxylase into the same liposomes. The Na⁺ gradient developed by one of the enzymes functioning as a decarboxylase could be used by the other enzyme to drive the carboxylation of its decarboxylated substrate (Figure 6).[61] This system catalyzed the carboxylation of acetyl-CoA to malonyl-CoA by decarboxylation of oxaloacetate to pyruvate and vice versa. The Na⁺ gradient provides the energetic coupling between the exergonic decarboxylation and the endergonic carboxylation reaction. No carboxylation, therefore, occurred if the Na⁺ gradient was dissipated with monensin. These are the first demonstrated examples of the energetically unfavorable carboxylation reactions being driven by a Na⁺ gradient rather than by ATP hydrolysis.

The classical transcarboxylase of *Propionibacterium shermanii*[62] catalyzes the same overall reaction as the membrane-linked transcarboxylase system constructed with the two decarboxylases. Moreover, carboxyltransferase and biotin carboxyl carrier protein components participate in both processes, but the coupling mechanisms are completely different. In soluble transcarboxylase, during the decarboxylation of the first substrate, the carboxyl group is linked to the biotin prosthetic group, from which it is directly transferred to the second substrate in the carboxylation reaction. Decarboxylation and carboxylation are thus coupled by an "energy-rich" chemical bond, the carboxyl group attached to biotin on the enzyme. The mechanism is distinct from the vectorial energy coupling in the membrane-linked transcarboxylase system consisting of oxaloacetate decarboxylase and methylmalonyl-CoA decarboxylase. These two carboxylation mechanisms parallel the two mechanisms of ATP synthesis: substrate-level phosphorylation and phosphorylation coupled to vectorial H⁺ movement.

IX. OUTLOOK

The sodium ion transport decarboxylases provide unique systems to study the mode of ion translocation through biological membranes and the coupling between the vectorial and the chemical reaction. While much has been learned about the basic features of these and other primary transport systems, such

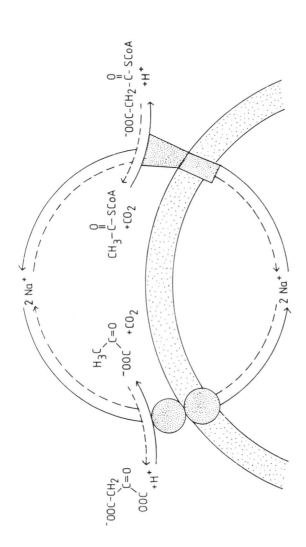

FIGURE 6. Na$^+$ circuit mediating the transcarboxylation from oxaloacetate and acetyl-CoA to pyruvate and malonyl-CoA and vice versa. (Reprinted with permission from Dimroth, P. and Hilpert, W., *Biochemistry*, 23, 5360, 1984. Copyright 1984 American Chemical Society.)

as ATPases and respiratory chain components, we are far from understanding the mechanism of any one of these systems in molecular detail. Advantages of the decarboxylases in elucidating these features may be their relatively small size and simple structure, with a composition of only three to four subunits, and the ease of purifying sizable quantities in a short time. Another distinct advantage is the use of Na$^+$ as a coupling ion because, in contrast to H$^+$, there is practically no limitation in varying the Na$^+$ concentration and because Na$^+$ fluxes can be quantitated more readily than H$^+$ fluxes. Furthermore, the development of methods to dissociate and reassociate the enzyme complexes and reconstitute their transport function by incorporation into proteoliposomes will help to elucidate details of the mechanism of these catalysts. The determination of the primary structure of oxaloacetate decarboxylase is the first step in combining functional with structural data. If other sequences of the related decarboxylases become available, one may be able to identify the conserved amino acid residues within the protein and determine whether these are of essence for one of the functions of these enzymes using site-directed mutagenesis or chemical modification techniques. These data will be a valuable contribution to our goal of understanding the mechanism of a vectorial catalyst in molecular detail. Eventually, however, a high-resolution structure will be required to put all the data together into an atomic model and, hopefully, to elucidate the details of the function of these interesting catalysts.

REFERENCES

1. **Mitchell, P.,** *Chemiosmotic Coupling and Energy Transduction,* Glynn Research, Bodmin, Cornwall, England, 1968.
2. **Nicholls, D. G.,** *Bioenergetics. An Introduction to the Chemiosmotic Theory,* Academic Press, London, 1982.
3. **Harold, F. M.,** *A Study of Bioenergetics,* W. H. Freeman, New York, 1986.
4. **Dimroth, P.,** Sodium ion transport decarboxylases and other aspects of sodium cycling in bacteria, *Microbiol. Rev.,* 51, 320, 1987.
5. **Dimroth, P.,** Mechanisms of sodium transport in bacteria, *Philos. Trans. R. Soc. London Ser. B,* 326, 465, 1990.
6. **Dimroth, P.,** Energy transductions by an electrochemical gradient of sodium ions, in *41 Colloquium Mosbach: The Molecular Basis of Bacterial Metabolism,* Hauska, G. and Thauer, R., Eds., Springer-Verlag, Berlin, 1990, 114.
7. **Dimroth, P.,** Na$^+$-coupled alternate to H$^+$-coupled primary transport systems in bacteria, *BioEssays,* 13, 463, 1991.
8. **Unemoto, T., Tokuda, H., and Hayashi, M.,** Primary sodium pumps and their significance in bacterial energetics, *Bacteria,* 12, 33, 1990.
9. **Lanyi, J.,** The role of Na$^+$ in transport processes of bacterial membranes, *Biochim. Biophys. Acta,* 559, 377, 1979.
10. **Maloy, S. R.,** Sodium-coupled cotransport, *Bacteria,* 12, 203, 1990.
11. **Krulwich, T. A.,** Na$^+$/H$^+$ antiporters, *Biochim. Biophys. Acta,* 726, 245, 1983.

12. **Dibrov, P. A., Kostyrko, V. A., Lazarova, R. L., Skulachev, V. P., and Smirnova, I. A.,** The sodium cycle. I. Na$^+$-dependent motility and modes of membrane energization in the marine alkalotolerant *Vibrio alginolyticus, Biochim. Biophys. Acta,* 850, 449, 1986.

13. **Hirota, N. and Imae, Y.,** Na$^+$-driven flagellar motors of an alkalophilic *Bacillus* strain YN-1, *J. Biol. Chem.,* 258, 10577, 1983.

14. **Dimroth, P.,** A new sodium transport system energized by the decarboxylation of oxaloacetate, *FEBS Lett.,* 122, 234, 1980.

15. **Dimroth, P.,** The role of biotin and sodium in the decarboxylation of oxaloacetate by the membrane-bound oxaloacetate decarboxylase from *Klebsiella aerogenes, Eur. J. Biochem.,* 121, 435, 1982.

16. **Dimroth, P.,** The generation of an electrochemical gradient of sodium ions upon decarboxylation of oxaloacetate by the membrane-bound and Na$^+$-activated oxaloacetate decarboxylase from *Klebsiella aerogenes, Eur. J. Biochem.,* 121, 443, 1982.

17. **Hilpert, W. and Dimroth, P.,** Conversion of the chemical energy of methylmalonyl-CoA decarboxylation into a Na$^+$ gradient, *Nature (London),* 296, 584, 1982.

18. **Hilpert, W. and Dimroth, P.,** Purification and characterization of a new sodium transport decarboxylase. Methylmalonyl-CoA decarboxylase from *Veillonella alcalescens, Eur. J. Biochem.,* 132, 579, 1983.

19. **Buckel, W. and Semmler, R.,** Purification, characterization and reconstitution of glutaconyl-CoA decarboxylase, a biotin-dependent sodium pump from anaerobic bacteria, *Eur. J. Biochem.,* 136, 427, 1983.

20. **Tokuda, H. and Unemoto, T.,** Characterization of the respiration-dependent Na$^+$ pump in the marine bacterium *Vibrio alginolyticus, J. Biol. Chem.,* 257, 10007, 1982.

21. **Tokuda, H. and Unemoto, T.,** Na$^+$ is translocated at NADH:quinone oxidoreductase segment in the respiratory chain of *Vibrio alginolyticus, J. Biol. Chem.,* 259, 7785, 1984.

22. **Hilpert, W., Schink, B., and Dimroth, P.,** Life by a new decarboxylation-dependent energy conservation mechanism with Na$^+$ as coupling ion, *EMBO J.,* 3, 1665, 1984.

23. **Laubinger, W. and Dimroth, P.,** Characterization of the Na$^+$-stimulated ATPase of *Propionigenium modestum* as an enzyme of the F_1F_0 type, *Eur. J. Biochem.,* 168, 475, 1987.

24. **Dimroth, P. and Thomer, A.,** A primary respiratory Na$^+$ pump of an anaerobic bacterium: the Na$^+$-dependent NADH:quinone oxidoreductase of *Klebsiella pneumoniae, Arch. Microbiol.,* 151, 439, 1989.

25. **Wifling, K. and Dimroth, P.,** Isolation and characterization of oxaloacetate decarboxylase of *Salmonella typhimurium,* a sodium ion pump, *Arch. Microbiol.,* 152, 584, 1989.

26. **Beatrix, B., Bendrat, K., Rospert, S., and Buckel, W.,** The biotin-dependent sodium ion pump glutaconyl-CoA decarboxylase from *Fusobacterium nucleatum, Arch. Microbiol.,* 154, 362, 1990.

27. **Müller, V., Winner, C., and Gottschalk, G.,** Electron transport-driven sodium extrusion during methanogenesis from formaldehyde + H$_2$ by *Methanosarcina barkeri, Eur. J. Biochem.,* 178, 519, 1988.

28. **Ken-Dror, S., Lanyi, J. K., Schobert, B., Silver, B., and Avi-Dor, Y.,** An NADH:quinone oxidoreductase of the halotolerant bacterium Ba$_1$ is specifically dependent on sodium ions, *Arch. Biochem. Biophys.,* 244, 766, 1986.

29. **Efiok, B. J. S. and Webster, D. A.,** A cytochrome that can pump sodium ion, *Biochem. Biophys. Res. Commun.,* 173, 370, 1990.

30. **Antranikian, G. and Giffhorn, F.,** Citrate metabolism in anaerobic bacteria, *FEMS Microbiol. Rev.,* 46, 175, 1987.

31. **Dimroth, P.,** The role of vitamins and their carrier proteins in citrate fermentation, in *The Roots of Modern Biochemistry,* Kleinkauf, H., Döhren, H., and Jaenicke, L., Eds., Walter de Gruyter, Berlin, 1988, 191.

32. **Dimroth, P.**, Biotin-dependent decarboxylases as energy transducing systems, *Ann. N.Y. Acad. Sci.*, 447, 72, 1985.
33. **Dimroth, P. and Thomer, A.**, Citrate transport in *Klebsiella pneumoniae*, *Biol. Chem. Hoppe-Seyler*, 367, 813, 1986.
34. **Dimroth, P. and Thomer, A.**, Solubilization and reconstitution of the Na⁺-dependent citrate carrier of *Klebsiella pneumoniae*, *J. Biol. Chem.*, 265, 7721, 1990.
35. **Schwarz, E. and Oesterhelt, D.**, Cloning and expression of *Klebsiella pneumoniae* genes coding for citrate transport and fermentation, *EMBO J.*, 4, 1599, 1985.
36. **van der Rest, M. E., Schwarz, E., Oesterhelt, D., and Konings, W. N.**, DNA sequence of a citrate carrier of *Klebsiella pneumoniae*, *Eur. J. Biochem.*, 189, 401, 1990.
37. **Siewe, R.**, Molekularbiologische Charakterisierung der Citrat Transportsysteme von *Klebsiella pneumoniae*, M.Sc. thesis, University of Osnabrück, Osnabrück, Germany, 1991.
38. **De Vries, W., Rietveld-Struijk, T. R. M., and Stouthamer, A. T.**, ATP formation associated with fumarate and nitrate reduction in growing cultures of *Veillonella alcalescens*, *Antonie van Leeuwenhoek J. Microbiol. Serol.*, 43, 153, 1977.
39. **Stern, J. R.**, Oxaloacetate decarboxylase of *Aerobacter aerogenes*. Inhibition by avidin and requirement for sodium ion, *Biochemistry*, 6, 3545, 1967.
40. **Dimroth, P.**, Preparation, characterization, and reconstitution of oxaloacetate decarboxylase from *Klebsiella aerogenes*, a sodium pump, *Methods Enzymol.*, 125, 530, 1986.
41. **Hilpert, W. and Dimroth, P.**, Sodium pump methylmalonyl-CoA decarboxylase from *Veillonella alcalescens*, *Methods Enzymol.*, 125, 540, 1986.
42. **Buckel, W.**, Biotin-dependent decarboxylases as bacterial sodium pumps: purification and reconstitution of glutaconyl-CoA decarboxylase from *Acidaminococcus fermentans*, *Methods Enzymol.*, 125, 547, 1986.
43. **Dimroth, P. and Thomer, A.**, Subunit composition of oxaloacetate decarboxylase and charterization of the a-chain as carboxyltransferase, *Eur. J. Biochem.*, 137, 107, 1983.
44. **Dimroth, P.**, Characterization of a membrane-bound biotin enzyme: oxalocetate decarboxylase from *Klebsiella aerogenes*, *Eur. J. Biochem.*, 115, 353, 1981.
45. **Buckel, W.**, Substrate stereochemistry of the biotin-dependent sodium pump glutaconyl-CoA decarboxylase and the mechanism of glutaconate CoA-transferase, *Eur. J. Biochem.*, 156, 259, 1986.
46. **Hoffmann, A. and Dimroth, P.**, Stereochemistry of the methylmalonyl-CoA decarboxylation reaction, *FEBS Lett.*, 220, 121, 1987.
47. **Rétey, J. and Robinson, J. A.**, *Stereospecificity in Organic Chemistry and Enzymology*, Verlag Chemie, Weinheim, Germany, 1982, 129.
48. **Buckel, W. and Liedtke, H.**, Specific cleavage of the sodium pump glutaconyl-CoA decarboxylase by primary alcohols, *Eur. J. Biochem.*, 156, 251, 1986.
49. **Hoffmann, A., Hilpert, W., and Dimroth, P.**, The carboxyltransferase activity of the sodium-ion-translocating methylmalonyl-CoA decarboxylase of *Veillonella alcalescens*, *Eur. J. Biochem.*, 179, 645, 1989.
50. **Dimroth, P. and Thomer, A.**, Dissociation of the sodium-ion-translocating oxaloacetate decarboxylase of *Klebsiella pneumoniae* and reconstitution of the active complex from the isolated subunits, *Eur. J. Biochem.*, 175, 175, 1988.
51. **Dimroth, P. and Thomer, A.**, Kinetic analysis of the reaction mechanism of oxaloacetate decarboxylase from *Klebsiella aerogenes*, *Eur. J. Biochem.*, 156, 157, 1986.
52. **Schwarz, E., Oesterhelt, D., Reinke, H., Beyreuther, K., and Dimroth, P.**, The sodium ion translocating oxaloacetate decarboxylase of *Klebsiella pneumoniae*. Sequence of the biotin-containing α-subunit and relationship to other biotin-containing enzymes, *J. Biol. Chem.*, 263, 9640, 1988.
53. **Däkena, P., Rohde, M., Dimroth, P., and Mayer, F.**, Oxaloacetate decarboxylase from *Klebsiella pneumoniae*: size and shape of the enzyme, and localization of its prothetic biotin group by electron microscopic affinity labeling, *FEMS Microbiol. Lett.*, 55, 35, 1988.

54. **Laussermair, E., Schwarz, E., Oesterhelt, D., Reinke, H., Beyreuther, K., and Dimroth, P.,** The sodium ion translocating oxaloacetate decarboxylase of *Klebsiella pneumoniae*. Sequence of the integral membrane-bound subunits β and γ, *J. Biol. Chem.,* 264, 14710, 1989.

55. **Stephens, P. E., Darlison, M. G., Lewis, H. M., and Guest, J. R.,** The pyruvate dehydrogenase complex of *Escherichia coli, Eur. J. Biochem.,* 133, 481, 1983.

56. **Radford, S. E., Lane, E. D., Perham, R. N., Miles, J. S., and Guest, J. R.,** Domains in pyruvate dehydrogenase complex, *Biochem. J.,* 247, 641, 1987.

57. **Dimroth, P.,** Reconstitution of sodium transport from purified oxaloacetate decarboxylase and phospholipid vesicles, *J. Biol. Chem.,* 256, 11974, 1981.

58. **Hilpert, W. and Dimroth, P.,** Reconstitution of Na$^+$ transport from purified methylmalonyl-CoA decarboxylase and phospholipid vesicles, *Eur. J. Biochem.,* 138, 579, 1984.

59. **Rohde, M., Däkena, P., Mayer, F., and Dimroth, P.,** Morphological properties of proteoliposomes reconstituted with the Na$^+$ pump methylmalonyl-CoA decarboxylase from *Veillonella alcalescens, FEBS Lett.,* 195, 280, 1986.

60. **Hilpert, W. and Dimroth, P.,** On the mechanism of sodium ion translocation by methylmalonyl-CoA decarboxylase from *Veillonella alcalescens, Eur. J. Biochem.,* 195, 79, 1991.

61. **Dimroth, P. and Hilpert, W.,** Carboxylation of pyruvate and acetyl coenzyme A by reversal of the Na$^+$ pumps oxaloacetate decarboxylase and methylmalonyl-CoA decarboxylase, *Biochemistry,* 23, 5360, 1984.

62. **Wood, H. G. and Barden, R. E.,** Biotin enzymes, *Annu. Rev. Biochem.,* 46, 385, 1977.

63. **Dimroth, P. and Thomer, A.,** *Biochemistry,* in press, 1992.

64. **Woehlke, G., Wifling, K., and Dimroth, P.,** Sequence of the sodium ion pump oxaloacetate decarboxylase from *Salmonella typhimurium, J. Biol. Chem.,* in press, 1992.

Chapter IE

THE Na⁺ CYCLE IN ALKALIPHILIC *BACILLUS* SPECIES

D. Mack Ivey, Arthur A. Guffanti, and Terry Ann Krulwich

TABLE OF CONTENTS

I. INTRODUCTION

A Na$^+$ cycle has been characterized physiologically in extremely alkaliphilic *Bacillus* species and has been shown to play a crucial role in bacteria for which pH homeostasis is a central challenge. The challenge arises from the need to maintain a cytoplasmic pH that is much more acidic than the external medium when the latter is well above pH 10. In our laboratory we have shown that several different alkaliphilic *Bacillus* species grow under continuous culture conditions, at a constant pH of 10.5 and higher on malate-containing media, with growth yields that are comparable to those of non-alkaliphilic *Bacillus* species growing on comparable media at pH 7.[1] So far, the highest pH that has been examined under such conditions has been pH 10.7, so the upper limit of pH has not yet been defined rigorously. In our laboratory we have not experienced profound difficulty in measuring pH values in this range, although certain precautions are taken. We employ a combination electrode with a glass bulb that is calibrated with a standard buffer provided at pH 10.01. Using this standardization, we then check the linearity of the response to pH by testing the pH of a set of additional standards that we prepare to known theoretical pH values in the desired, very alkaline range, e.g., between pH 9.2 and 10.7. During growth experiments with batch cultures or chemostats, pH measurements are made on samples that are removed from the culture since the standardization of a continually submerged electrode might be unreliable in this range of pH.

Some alkaliphilic *Bacillus* species are obligately alkaliphilic, growing well only at pH values above 9 on nonfermentable carbon sources.[2] These species, e.g., *Bacillus alcalophilus* and *Bacillus firmus* RAB, may be precluded from growing at near-neutral pH by properties of their membrane lipids that cause membrane leakiness in that range of pH.[3,4] The physical basis for this particular susceptibility is not yet known. However, the limitation that it places on the pH range for growth sets parallel limits on the dimensions of the swings in pH that the organisms need to handle. By contrast, other alkaliphilic *Bacillus* species, e.g., *Bacillus firmus* OF4 and *Bacillus* YN2000, are facultative strains that can grow well at pH values between 7.5 and at least 10.5.[1,5,6] It is notable that the obligately alkaliphilic strains grow slightly better at the upper edge of the pH range than related facultative strains when assayed in a chemostat competition experiment.[4] The lipids that preclude growth at near-neutral pH probably confer an advantage in the highly alkaline range. Importantly, both the facultative and obligate alkaliphiles are extremophiles that grow well at pH values of 10 and higher. The major difference between the two types is in their ability to grow at near-neutral pH. In the alkaliphile field, the designation "tolerant", i.e., alkaline-tolerant, has a distinct meaning, having generally been used to refer to bacteria that can grow well at pH values of about pH 9. The energetics of this latter group is qualitatively different from that of the extreme alkaliphiles of both types.

The Na^+ cycle of extreme alkaliphiles is mediated by several electrogenic Na^+/H^+ antiporters, possibly assisted by cytoplasmic Na^+-binding proteins that interact with them, and a group of Na^+/solute symporters. These antiporters and symporters are all secondary active transporters. The antiporters are energized by the electrochemical proton gradient that is established by respiration or that can be imposed artificially under experimental conditions.[7,8] The antiporters achieve the rapid, electrogenic exchange of extracellular protons for intracellular Na^+ so that respiring cells maintain an intracellular pH (pH_{in}) of about 8.3 during growth at pH 10.5.[1,2] The fluxes must be rapid, i.e., at least as rapid as primary proton extrusion, and the porters must be largely constitutive in order for the alkaliphile to withstand sudden changes in external pH. The alkaliphiles are remarkable with respect to this capacity when contrasted, for example, with a conventional neutrophilic bacterium such as *Escherichia coli*. This was shown in experiments in which cells of *B. firmus* RAB that were first equilibrated at pH 8.5 were subjected to a rapid alkalinization of the medium to pH 10.5.[9] In the absence of Na^+, the pH_{in} immediately rose to 10.5. By contrast, as long as Na^+ was present in sufficient concentration, the bacteria exhibited almost no change in pH_{in} right after the shift. Even more strikingly, if a solute that allowed Na^+ reentry was also included in the medium, then the alkalinization of the exterior was first followed by a slight *decline* in pH_{in}, i.e., an overshoot type of phenomenon, followed by maintenance of a steady pH_{in} at the preshift value. A different alkaliphile, *Exiguobacterium aurantiacum*, also exhibits Na^+-dependent pH homeostasis that spares the organism any major change in pH_{in} during a large upward change in the external pH.[10,11] By contrast, when *E. coli* cells are subjected, in a similar protocol, to a more modest upshift in the external pH, there is first a transient failure of pH homeostasis even in the presence of Na^+, followed by the reestablishment of a cytoplasmic pH that is below the pH of the medium, but somewhat higher than the preshift pH_{in}.[12]

In spite of the effectiveness of the alkaliphile's mechanism for pH homeostasis, this physiological function may be the limiting one with respect to the upper limit of pH for growth. This is suggested by the observation that *B. firmus* RAB generates genetic variants with high frequency when plated on malate-containing media, at pH 10.5, that contain suboptimal (2 to 3 mM) concentrations of Na^+. The variants thus isolated are capable of growing better than the parent strain at these low Na^+ concentrations and concomitantly grow better than the parent strain at extremely high initial growth pH values; these changes are associated with an even more rapid total cellular Na^+/H^+ antiport activity than is observed in the parent strain.[13] As in the wild-type parent, no primary Na^+ extrusion was observed in the variants.

In addition to the striking role of the Na^+ cycle in pH homeostasis in alkaliphilic bacteria, the Na^+ cycle also accounts for the active transport of numerous solutes and for the maintenance of a cytoplasmic Na^+ concentration that is lower than the outside concentration. With respect to solute transport, the use of Na^+ as the coupling ion for Na^+/solute symport, rather than protons, overcomes the problem of the low bulk proton-motive force. That is, the combined effects of primary proton extrusion and secondary Na^+/H^+ antiport result in a low electrochemical proton gradient and a substantial electrochemical Na^+ gradient which then is utilized for the accumulation of many solutes.[2,14] For the nonmarine alkaliphilic *Bacillus* species, the external Na^+ concentrations range from about 1 mM to 50–150 mM under the usual growth conditions.[15] However, we have found that these bacteria will grow in the presence of NaCl concentrations up to 1.5 M. When confronted with a NaCl concentration in the upper end of the range, the alkaliphile may require more Na^+ extrusion capacity than is provided by the constitutive antiport level that so adequately accommodates pH homeostasis. Perhaps, as in *E. coli*,[16,17] regulation of antiporter synthesis by the external Na^+ concentration will be a feature of the Na^+ cycle.

We will review the physiological properties and recent molecular information about the Na^+/H^+ antiporters and Na^+/solute symporters that are the known components of the alkaliphile Na^+ cycle. We note, however, that other alkaliphile solute transport systems, which are dependent on ATP rather than Na^+, are responsible for the uptake of several carbohydrate substrates.[2] In addition, as we focus on secondary porters coupled to Na^+, we do not mean to preclude the possible involvement, in the pH homeostasis of alkaliphiles, of ion fluxes other than Na^+ or of certain additional pathways of Na^+ that have yet to be demonstrated directly; at least one of the latter possibilities, a pH-regulated Na^+ channel, has been proposed,[10,11,18] as will be discussed further below.

II. Na^+/H^+ ANTIPORTERS

A. PHYSIOLOGY
1. Evidence for a Role in pH Homeostasis

Physiological studies of Na^+/H^+ antiporters in whole alkaliphile cells and membranes thus far represent determinations of the entire antiporter complement of the membrane taken together, complicated further by the presence of numerous other ion-translocating elements. The most compelling evidence for the role of Na^+/H^+ antiporters in pH homeostasis at external pH values in the alkaline range comes from pH shift experiments of the type conducted with the alkaliphiles *E. aurantiacum*[10,11] and *B. firmus* RAB.[9] As noted in Section I, in the absence of Na^+, equilibrated cells that are subjected to a sudden upward shift in the external pH respond with an immediate cytoplasmic alkalinization to the new external pH. This indicates the absence of either passive or energy-dependent but Na^+-independent mechanisms that can maintain a relatively acidified cytoplasm in the face of such an alkaline shift. By contrast, several facultative alkaliphiles have been shown to possess a K^+/ H^+ antiporter activity that appears to function in the maintenance of a slightly alkaline cytoplasmic pH relative to the external pH during growth in the near-neutral range.[19,20] One would predict that all extreme alkaliphiles of this type would require Na$^+$ for growth at high pH. Indeed, most alkaliphilic *Bacillus* species can be easily shown to require Na$^+$ for growth.[21] In some other instances, the Na$^+$ requirement for growth can only be demonstrated if special care is taken to reduce the inevitable contaminating Na$^+$, e.g., by using defined media and by using plasticware exclusively for media preparation and growth.[22]

The other strong evidence for a role of the alkaliphile Na$^+$ cycle in pH homeostasis is genetic. This includes the variants mentioned in Section I, in which enhanced ability to grow at remarkably high initial pH values as well as suboptimal Na$^+$ concentrations correlates with an increase in Na$^+$/H$^+$ antiport activity.[13] It also includes several mutants, isolated from different wild-type alkaliphiles in different laboratories, in which the loss of the ability to grow at pH values above about pH 9 was accompanied by the loss of cell and/or membrane Na$^+$/H$^+$ antiport activity.[23-25] In view of emerging indications that alkaliphiles have more than one Na$^+$/H$^+$ antiporter (see Section II.B.1), it may be that mutational loss of one antiporter is insufficient to give a phenotype. It will be especially helpful when strategies for transposon mutagenesis have been refined in the alkaliphile and mutants that are negative for growth at high pH can be isolated from transposition libraries that are screened at different pH values and different Na$^+$ concentrations to sort out the possible roles of different antiporter species. As such genetic approaches do become available, it will also be of interest to reinvestigate early observations of pleiotropy among at least some nonalkaliphilic mutant isolates. These poor-growing strains were compromised in several Na$^+$-dependent functions.[25] At first this pleiotropy was thought to represent the sharing of a common Na$^+$-translocating element by different porters,[26] but this possibility having been eliminated,[2] there remains the possibility of a mutation in some common regulatory element. The pleiotropic nonalkaliphilic mutants grow extremely poorly. The consequent ease of contamination by other bacteria has made these strains difficult to study, but when marked with a transposon carrying a selectable marker their genetic basis might become accessible to investigation.

2. Assay and Characterization

The properties of total cellular and membrane Na$^+$/H$^+$ antiport activity have been well characterized in alkaliphiles at a physiological level. Some assays are based upon measurements of ^{22}Na$^+$ fluxes induced by an imposed pH gradient[27] or by an imposed diffusion potential.[8] Alternatively, an electron donor or ATP may be added (e.g., to everted membrane preparations) to develop an electrochemical proton gradient which then drives the Na$^+$ flux.[28,29] Other antiport assays follow Na$^+$-dependent proton fluxes, as monitored either by the distribution of radiolabeled weak acids or bases,[28,29] or by fluorescent probes that distribute with the pH gradient or exhibit pH-dependent fluorescence changes within a vesicle.[24] Using such assays, different alkaliphilic *Bacillus* species have been found to catalyze very active membrane Na^+/H^+ antiport, exhibiting a broad range of affinities for Na^+. Usually the affinity, described as an apparent K_m for Na^+, is in the mid-millimolar range. However, some alkaliphilic *Bacillus* species exhibit higher affinities, correlating roughly with the concentration range of the Na^+ requirement for growth.[21]

All the alkaliphile Na^+/H^+ antiport activities described to date can apparently substitute Li^+ for Na^+. Li^+ appears to be a competitive inhibitor of Na^+ flux via the antiporter.[8] In fact, it was the observation that the Na^+/H^+ antiporter could utilize Li^+ as substrate while the Na^+/solute symporters of the same alkaliphiles could not[30] that first ruled out our hypothesis that the pleiotropy of some nonalkaliphilic mutant strains resulted from the mutational alteration of a Na^+-translocating element that was common to the Na^+-coupled antiporters and symporters. In contrast to findings with some mammalian and archaebacterial Na^+/H^+ antiporters[31] (see also Chapter IJ), no specific inhibitor for the bacterial antiporters has been identified. A modest inhibitory effect of an amiloride analog has been noted for the *nha*C gene product-dependent increase in Na^+/H^+ antiport (see Section II.B.1). Generally, however, amiloride does not have a pronounced effect on pH homeostasis in alkaliphilic *Bacillus* species, although it has inhibited the Na^+-dependent motility of *B. firmus* RAB.[32]

When the Na^+/H^+ antiport of *B. alcalophilus* was energized in starved cells by the imposition of a valinomycin-mediated potassium diffusion potential, the rate of $^{22}Na^+$ efflux showed a linear relationship to the magnitude of the potential that was imposed.[8] When efflux was driven by respiration, dissipation of the transmembrane electrical potential component of the electrochemical proton gradient was inhibitory to Na^+/H^+ antiport in various assays of alkaliphile cell and membrane preparations.[7,24] These observations are consistent with an overall electrogenic antiport. In fact, only an electrogenic antiport with a $H^+:Na^+$ stoichiometry greater that unity can explain the pattern of electrochemical proton and sodium ion gradients that are formed through the combined action of respiration and secondary ion transport at alkaline pH. It is likely, however, that one or more of the antiporters of facultatively alkaliphilic species may be electroneutral, including at least one that is active at near-neutral pH.[7] There is also a report of a facultatively anaerobic alkaliphile whose Na^+/H^+ antiporter and electrochemical gradient profiles are unusual and thus far incompletely understood.[33]

With respect to pH effects upon alkaliphile Na^+/H^+ antiport, some general physiological characteristics have been noted. First, as predicted from the capacity for pH homeostasis, the overall Na^+/H^+ antiport activity of pH 7.5- and pH 10.5-grown alkaliphile cells was comparable.[34] Nonetheless, as the controls of gene expression are characterized for specific antiporter-encoding genes, we anticipate that distinct differences in response to pH and high Na$^+$ will emerge between genes, as well as in the activity of the porters, as in *E. coli*.[16,35] The relationship between the whole cell or membrane Na$^+$/H$^+$ antiport activity and the assay pH is also of interest. Clearly, as in *E. coli*,[36] a cytoplasmic pH near neutral is inhibitory to antiport.[8] By contrast, cells that were loaded with ^{22}Na$^+$ and equilibrated so that the pH$_{in}$ was 9, and then energized by a diffusion potential at various external pH values, exhibited little difference in the rate of efflux over the range between pH 7 and 9.[8] At external pH values approaching 10, a decline in the rate of antiport was observed using this protocol,[8] even though the rate of Na$^+$/H$^+$ antiport of comparable respiring cells at pH 10.5 is so fast that it is difficult to measure without lowering the assay temperature or imposing an inwardly directed Na$^+$ gradient.[13] This difference in antiport rate as a function of energization mode has also been demonstrated very clearly by Kitada et al.[24] in alkaliphilic *Bacillus* sp. N-6.[24] It may reflect an adverse effect of the decreasing substrate proton concentration at extremely high pH values. Perhaps this effect is somehow mitigated when energization involves an actual proton pumping event rather than an imposed potential that is based on another cation. However, even at the highest pH values at which growth has been observed, alkaliphilic *B. firmus* strains exhibit diffusion potential-dependent antiport activity that is significant, albeit lower than the rates observed in respiring cells at the same pH. ATP synthesis, the only other alkaliphile process known to require the inward translocation of a proton in the alkaliphilic *Bacillus* species, becomes completely unresponsive to an imposed diffusion potential at much lower pH values (near pH 9), differing markedly from antiport in this respect.[37]

B. MOLECULAR CHARACTERIZATION

In order for the characterization of the Na$^+$/H$^+$ antiporters in alkaliphilic *Bacillus* species to move beyond an aggregate physiological characterization, identification and study of the genes encoding the antiporters are required. In *E. coli*, for example, the identification and characterization of the first prokaryotic antiporter gene, *nhaA*, facilitated both the purification of the gene product and detailed studies of the regulation at various levels.[35,38] Several different strategies have been initiated to develop these powerful approaches in alkaliphiles. Horikoshi and colleagues[39] have isolated two independent mutants of facultatively alkaliphilic *Bacillus* C-125 that no longer are able to grow at pH values above 9; the mutants were shown to have lost the capacity for acidifying their cytoplasm relative to the external medium at high pH. These strains were transformed with plasmid libraries of DNA from the parent strain. Two transformants, one from each mutant strain, had restored growth at pH 10.3. The plasmids from the transformants contained apparently overlapping 2-kilobase (kb) inserts that together encompassed approximately 2.5 kb of alkaliphile DNA. Since each plasmid could complement only the mutant strain from which it was isolated, Kudo et al.[39] suggested that two independent but linked loci required for growth at high pH are contained on these plasmids. Sequence data have not been presented. Until the genetics of alkaliphilic *Bacillus* species are more highly developed, this approach, preferably achieved by use of transposon mutagenesis, will at least allow correlations to be made between the restoration of a wild-type phenotype to specific mutants and cloned genes that can be characterized.

We have recently taken advantage of the fact that a strain (NM81) of *E. coli* that carries a deletion in the *nhaA* Na^+/H^+ antiporter gene will not grow in the presence of 0.5 to 0.7 M NaCl or in the presence of 0.1 M LiCl in melibiose-containing media.[35] This phenotype is reversed by a plasmid containing the *nhaA* gene. We transformed *E. coli* NM81 with plasmid libraries of *B. firmus* OF4 DNA and characterized a number of transformants that were no longer sensitive to 0.6 M NaCl.[40] The transformants, which we will briefly describe, fell into three categories.

1. Transformants Conferring Stable and Enhanced Na^+/H^+ Antiporter Activity

We have reported the characterization of two plasmids that are strong candidates for containing structural genes encoding Na^+/H^+ antiporters.[40] One plasmid, designated pJB10 and containing an incomplete open reading frame, allowed us to isolate pM4.1, which contains the intact gene (*nhaC*) for a hydrophobic, 42-kDa protein. A model based on the sequence and hydropathy analysis is shown in Figure 1, which also indicates the residues that show a region of modest sequence similarity to the human Na^+/H^+ exchanger characterized by Sardet et al.[41] The antiporter activity conferred upon the *E. coli* NM81 strain by pJB10 was stimulated by relatively high pH (pH 8.5), consistent with a role in pH homeostasis in the alkaline range. The gene was expressed from the T7 promoter in *E. coli*, and the gene product was shown to be membrane bound. The published sequence of this gene included the putative ribosome binding site, but no additional upstream sequence. We have recently used inverse polymerase chain reaction (PCR) to clone the region upstream of *nhaC* and have sequenced an additional 140 bp of the 5' noncoding sequence (Figure 2). This sequence has two regions that are candidates for the promoter of the *nhaC* gene; no other regulatory sites are obvious. The nucleotide sequence downstream of the *nhaC* gene contains an open reading frame that is predicted to encode a low-M_r basic protein that will be discussed in Section II.B.2.

A second transformant with stable Na^+/H^+ antiporter activity in membranes of *E. coli* NM81 was shown to contain a plasmid (pEM271) that had lost the ability to complement the deletion strain upon retransformation. The original transformant, designated NM8191, was cured of pEM271 and shown still to contain markedly enhanced Na^+/H^+ exchange.[40] In collaboration with Dr. Etana Padan and colleagues, we are investigating the possibility that an antiporter gene from the alkaliphile has recombined into the NM81

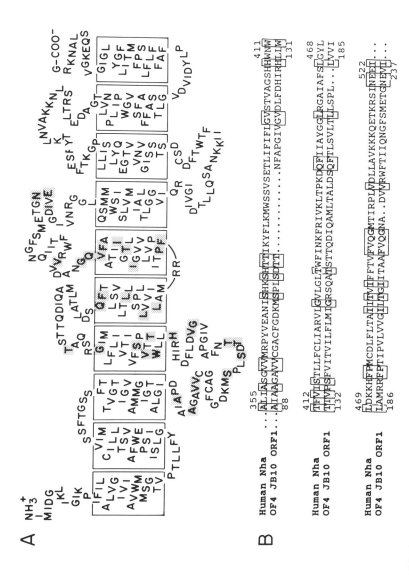

FIGURE 1. A model of the secondary structure of the product of the *nhaC* gene of alkaliphilic *Bacillus firmus* OF4.[40] The topological model is based on hydropathic analysis, with boxed regions representing membrane-spanning regions. Shaded residues, which correspond to the boxed residues of the sequence alignment, are identical to corresponding residues in the human Na⁺/H⁺ antiporter.[41] (From Ivey, D. M. et al., *J. Biol. Chem.*, 266, 23482, 1991. With permission.)

```
  1    TCATGATGACCATTATTATTGCCGAGTATGTTTGTGTTAGAAGGTTGAACCACATATTCCG    60
                                                           -10
                                  -35

 61    CTAGGTGTATGTCTTCGAATTGCGGCTGTTTTTTGGGCGCTATTTAAGAGATGGTTGGAA   120

                                                        -10

121    TGATCTAGAGAGTGCGATGATTGATGGATTGAAAATAGGGATCAAGCCTATTTTTATTTT   180
             M   I   D   G   L   K   I   G   I   K   P   I   F   I   L
             NhaC
```

FIGURE 2. Nucleotide sequence of the region upstream of the *nhaC* gene of *Bacillus firmus* OF4. Possible promoter sequences are underlined, and the putative ribosome binding site is shaded. The sequence of the *nhaC* coding region is given in Ivey et al.,[40] and the entire sequence can be obtained from GenBank/EMBL (Heidelberg, Germany) with the accession number M73530.

chromosome. Phage lysates from the pEM271-cured strain *E. coli* NM8191 have been used to transduce the Na^+-resistance determinant back into NM81, consistent with the notion that a single transferable determinant, perhaps originating from the alkaliphile, restores a wild-type phenotype to the *nhaA* deletion strain. We are currently attempting to reclone the determinant from NM8191.

2. Transformants Conferring Enhanced Na^+/H^+ Antiport Activity that Is Unstable in an Everted Vesicle Assay System

A series of clones from *B. firmus* OF4 DNA libraries is depicted in Figure 3. These clones all reverse the ion-sensitive growth phenotype of *E. coli* NM81, but are not likely to be the structural genes for Na^+/H^+ antiporters since deduced amino acid sequences do not predict integral membrane proteins of the general size and type so far found for this type of porter and other ion-coupled porters. The plasmid pRVH is a subclone of pM4.1 containing the open reading frame downstream of the *nhaC* gene that was described above. Membranes prepared from *E. coli* strain NM81 harboring pRVH show slightly enhanced Na^+/H^+ antiport activity over the residual activity found in control NM81 membranes, and the enhancement shows a decay during the first hour after membrane preparation. Similar results have been obtained with NM81 transformants harboring two other clones, pJB20 and pJB22, that had been isolated from a screen of the library prepared in pGEM3Zf(+) that had also yielded the *nhaC*-containing pJB10.[40] The results of Na^+/H^+ antiport assays of membranes from the pJB22/NM81 transformant are particularly striking with respect to cold sensitivity and time-dependent decay of the significantly enhanced membrane antiport activity.

As noted above, sequence analysis of each of these clones reveals no open reading frame predicted to encode a hydrophobic protein with multiple membrane-spanning regions indicative of a porter. What does emerge from the analyses is that each clone contains one or more open reading frames predicted to encode a low-M_r protein (Table 1). In two instances there is sufficient sequence similarity with genes in the data base to suggest the possibility that these alkaliphile gene products may turn out to be cation-binding proteins. The putative product of the pRVH open reading frame shares sequence similarity with the a subunit of the Na^+/K^+-ATPase[42] in a region where similarity to the Na^+/Ca^{2+} exchanger of cardiac sarcolemma has also been noted.[43] The small protein encoded by pJB22 shares extensive sequence similarity with CadC of *Staphylococcus aureus* cadmium resistance plasmid pI258.[44] CadC has been proposed to function as an accessory protein to CadA, the cadmium efflux ATPase.[45] We are intrigued by the possibility that these cloned alkaliphile products enhance Na^+/H^+ antiport in the NM81 strain by acting as cytoplasmic Na^+-binding proteins that can deliver substrate to (enhance the local cation concentration near?) the remaining Na^+/H^+ antiporter in that strain. The *in vitro* membrane assay of such enhancement would depend

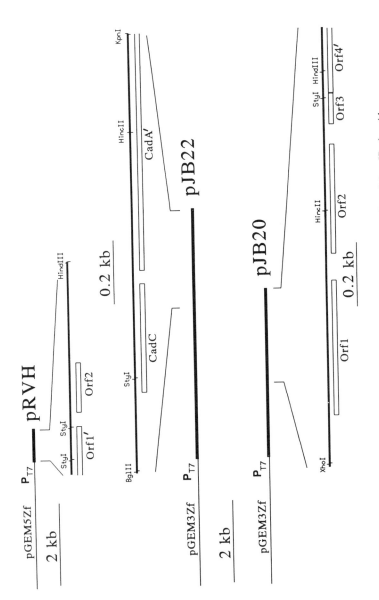

FIGURE 3. Organization of genes encoding the low-M_r proteins of the pJB plasmids.

TABLE 1
Properties of Deduced Products of
the Short Open Reading Frames of
the pJB Plasmids

Open reading frame		M_r	pI
pRVH	orf2	7,130	12.1
pJB20	orf1	20,800	10.4
	orf2	10,400	4.7
	orf3	3,780	10.3
	orf4	13,107[a]	7.7
pJB22	cadC	14,000	6.3

[a] Incomplete open reading frame.

on the amount of binding protein retained in the preparations and the stability of the protein-protein and protein-lipid interactions involved in that retention and in functional association. It would not be surprising, then, to observe variability between preparations, instability, and cold lability. Perhaps a component of the alkaliphile's remarkable capacity for Na⁺/H⁺ antiport is a set of cytoplasmic ion-binding proteins that function with respect to substrate extrusion in a manner analogous with the function of periplasmic binding proteins for certain solute uptake systems.[46]

3. Transformants that Do Not Enhance Na⁺/H⁺ Antiport Activity in Membrane Preparations

Two interesting and very incompletely understood clones fit into this category. The first, pJB14, was isolated during the same screen that yielded the other pJB transformants, on the basis of enhanced ability of pJB14/NM81 transformants to grow in the presence of 0.6 M NaCl at pH 7.5. This pGEM3Zf(+)-derived plasmid contains a very short (138-bp) insert that is predicted to encode a 45-amino acid-residue peptide with strong sequence similarity to a region of the family of ATP-binding proteins of the periplasmic permeases.[47] The domain encoded by pJB14 would lie just distal to the Walker B ATP-binding site[48] at the C-terminal end of the putative ATP-binding protein. We do not currently understand the basis for this clone's enhancement of growth of NM81 transformants in the presence of Na⁺.

The second clone, pM3.1, contains a 6-kb insert in the vector pSPT19 (Boehringer-Mannheim, Indianapolis, Indiana) and was originally isolated by hybridization to the pJB10 insert. Although plate assays indicated that this clone slightly enhanced the resistance of NM81 to 0.6 M NaCl at pH 7.5, no enhancement of membrane antiport was found, and subsequent sequence analysis of the clone did not clearly show the basis for the original hybridization to the pJB10 probe. With those caveats, we present the characteristics of a pM3.1 subclone because of the origin of the gene and the likelihood of its deduced product(s) being an interesting integral membrane protein whose relationship to the Na^+ cycle may be clarified by further work.

A 2-kb region of pM3.1 was subcloned and sequenced, and it was shown to contain two open reading frames. The first is predicted to encode a 14.6-kDa, hydrophobic protein with at least four membrane-spanning regions (Figure 4A). This protein does not strongly match any sequence in the protein data bases. It does show a slight similarity throughout its entire length to the galactoside permease of *Klebsiella pneumoniae*,[49] but assays of suitable transformants do not support the view that pM3.1 encodes either a galactoside or galactose permease. The second open reading frame overlaps with the first, a feature fairly common in operons of *Bacillus* species that may indicate translational coupling of the two genes.[50] The putative product, a 23-kDa, largely hydrophobic protein, again has no match in the protein data bases, except for the N-terminal 30 residues. This region has a consensus fatty acid acylation sequence (NH_3–MSRKKITALCIAGLLTLFSAGC...), which strongly suggests that this protein is tethered to the membrane via a lipoamide anchor (Figure 4B).[51] The low isoelectric point (pI = 4.5) of the entire protein suggests an extracellular location.[52] This is the second report of a lipoamide-anchored extracellular protein in an alkaliphilic *Bacillus*, the other being a lipoprotein penicillinase of *Bacillus* sp. strain 170,[53] which is quite similar to the lipoprotein penicillinases of *B. licheniformis*[54] and *B. cereus*.[55] Of interest in the present context of transport proteins is that the only other known lipoproteins in Gram-positive bacteria are the high-affinity binding proteins MalX and AmiA of *Streptococcus pneumoniae*.[56]

III. Na^+/SOLUTE SYMPORTERS

As noted in Section I, Na^+-dependent transport of solutes including amino acids,[6,57,58] organic acids,[57] carbohydrates,[57] and polyamines[59] is the norm in alkaliphiles. Na^+ symport with amino acids has been demonstrated in facultative alkaliphiles,[6] obligate alkaliphiles,[21,58] and more recently in a thermophilic alkaliphile.[60] Use of Na^+ bypasses the problem of the very low electrochemical proton gradient and provides an entry route for Na^+. During imposition of sudden alkaline shifts in the external pH of cells equilibrated at pH 8.5, not only does the presence of a nonmetabolizable substrate for a Na^+-coupled symporter enhance the steadiness of pH_{in} in the presence of a substantial concentration of Na^+ (50 mM), but it also renders a suboptimal concentration of Na^+ (2 mM) effective for pH homeostasis in *B. firmus* RAB.[9] There is, however, the cogent suggestion that alkaliphiles must have a discrete pH-regulated entry route for Na^+. This suggestion arises from work on *Exiguobacterium aurantiacum* in which McLaggan et al.[11] observed that Na^+-dependent acidification of the cytoplasm occurs in buffer containing ammonium sulfate, magnesium sulfate, chloride ion, and glucose, none of which is thought to be symported with Na^+. Moreover, when glucose was

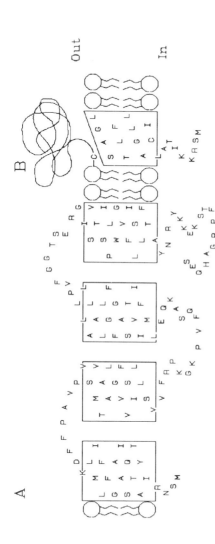

FIGURE 4. A model of the secondary structure of the putative products of the pSK2 open reading frames. The topology of the membrane-imbedded product (A) is based on hydropathy plots, and the topology of the lipoprotein (B) is inferred from localization studies of other lipoproteins[51] and from the low pI.[52] The nucleotide sequence of these open reading frames has been submitted to the GenBank/EMBL (Heidelberg, Germany) Data Bank with accession number M79460.

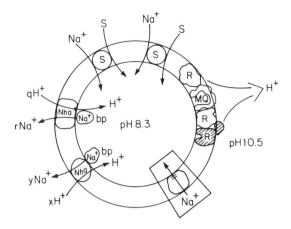

FIGURE 5. A model depicting the Na$^+$ cycle in the alkaliphilic *Bacillus* species. Distinct electrogenic Na$^+$/H$^+$ antiporters (nha) function secondarily to primary proton extrusion by the respiratory chain. The antiporters may capture Na$^+$ on the cytoplasmic side from one or more binding proteins (bp). Na$^+$-coupled solute symporters (S) complete the Na$^+$ cycle. In the rectangle is shown a possible pH-regulated Na$^+$ channel of the type proposed by McLaggan et al.[11]

added at an external pH of 9.7 in the presence of 10 mM NaCl, the pH$_{in}$ rose from about 8.3 to 8.5 while the internal concentration of Na$^+$ fell. The authors pointed out that if Na$^+$ was moving outward through a Na$^+$/H$^+$ antiporter, its efflux should have been accompanied by a decrease in pH$_{in}$. Precisely this latter sort of concomitant decrease in pH$_{in}$ was observed during experiments in which ^{22}Na$^+$ efflux from starved whole cells of alkaliphilic *B. firmus* OF4 was energized by the addition of malate.[7] Perhaps the *E. aurantiacum* system is more complex than the *Bacillus* system or the use of a fermentable carbon source is a complicating factor. The model depicted in Figure 5 illustrates the Na$^+$ cycle that we have elaborated in the alkaliphilic *Bacillus* species to date. Shown in the figure are two distinct Na$^+$/H$^+$ antiporters functioning electrogenically and secondarily to primary proton extrusion by the respiratory chain; of course, we do not know whether there are surely two, whether there are only two, and whether there also may be an electroneutral antiporter that functions at near-neutral pH. Each of the two antiporters shown is also depicted as capturing Na$^+$ on the cytoplasmic side from a binding protein. This speculation is based on the initial results from the functional complementation studies with *E. coli* that were discussed in an earlier section (II.B). A couple of Na$^+$-coupled solute symporters, representing a large number that exist in the membrane, are shown as completing the Na$^+$ cycle. In the rectangle, a possible pH-regulated Na$^+$ channel of the type proposed by McLaggan et al.[11] is also indicated.

The transport of α-aminoisobutyric acid (AIB) has been particularly well characterized in alkaliphiles and may be used as a paradigm for other systems. About 15 years ago Koyama et al.[61] and, soon after, Kitada and Horikoshi[62] demonstrated Na$^+$-dependent AIB uptake in alkaliphilic *Bacillus* 8–1. The Na$^+$ dramatically lowered the K_m for AIB uptake, but had no effect on the V_{max}. Other cations fail to substitute for Na$^+$, and amiloride has no inhibitory effect.[30,32,61,62] Transient uptake of AIB can be driven by an imposed ΔpNa^+ in membrane vesicles[63] and is also driven by an electron donor-generated $\Delta\Psi$ in the presence of Na$^+$.[64] Studies of passive solute efflux and exchange for the Na$^+$-coupled solute systems of *B. alcalophilus* have shown the Na$^+$-dependence and sensitivity to inhibition by the $\Delta\Psi$ of both processes.[65] The latter sensitivity is taken as suggestive that the tertiary complex between carrier, solute, and Na$^+$ is positively charged. In spite of long-standing and numerous physiological studies of Na$^+$/solute symport in alkaliphilic *Bacillus* species, there has been no characterization of a gene encoding one of these symporters or purification and study of one of these symporters in a reconstituted system.

IV. THE ABSENCE OF A PRIMARY Na$^+$ CYCLE IN EXTREME ALKALIPHILES

The evidence that extremely alkaliphilic *Bacillus* species lack primary Na$^+$ pumps of the type found in several of the other bacteria described in this volume has been reviewed elsewhere.[7,14,66] We will only briefly summarize that evidence here and then comment on the likely advantage of the presence of a primary proton cycle plus secondary Na$^+$ cycle pattern in the extreme alkaliphiles vs. the apparent presence of both primary proton and Na$^+$ cycles (probably together with full secondary Na$^+$ cycles) in certain other bacteria. The absence of primary, respiration-coupled Na$^+$ pumps in bacteria such as *B. firmus* RAB and OF4 is evidenced by

- The complete sensitivity of ^{22}Na$^+$ extrusion by respiring cells to abolition of the proton motive force under conditions in which possible inhibitory effects of the pH$_{in}$ on extrusion were eliminated[7]
- The absence of Na$^+$ stimulation of oxygen uptake by respiring cells[5]
- The absence of Na$^+$ stimulation of NADH oxidation by everted membrane vesicles[5]

The absence of a Na$^+$-coupled ATPase (synthase) has been shown by

- The failure of imposed chemical and electrochemical gradients of Na$^+$ that energize Na$^+$/solute symport to energize ATP synthesis by starved alkaliphile cells, whereas imposed proton gradients energize ATP synthesis[67]
- Lack of a Na$^+$ requirement or of stimulation thereby for ATP synthesis by alkaliphile cells or ADP + P$_i$-loaded right-side-out vesicles[68]
- Absence of a Na$^+$-stimulated ATPase in the membranes of alkaliphile cells[29,66]

- Failure of ATP to support $^{22}Na^+$ uptake by everted membrane vesicles unless an electrochemical proton gradient is allowed to develop as an intermediary[29]

- Inhibition of almost all of the membrane-associated ATPase activity by antibody raised against the F_1-ATPase,[69] and the exclusively proton-translocating (and not Na^+-translocating) nature of the F_1F_0-ATPase of two different alkaliphiles[66,70,71]

Strictly, speaking, it would still be possible that the alkaliphile possesses more than one type of F_0, but recent molecular biological studies in our lab indicate that there is only one F_1F_0-ATPase, encoded by one operon in the *B. firmus* OF4 chromosome.

For some years, as we developed most of the evidence cited above for the absence of a dominant, primary Na^+ cycle in the extreme alkaliphiles, other investigators maintained the view that these were the very organisms most likely to have such a cycle.[72,73] This expectation, which was also ours when we initiated work on the alkaliphiles,[58] arose from a focus on how oxidative phosphorylation occurred in the alkaliphile growing at pH 10.5, under conditions in which the bulk electrochemical proton gradient was very low. It is now clear both to us and to others that the alkaliphiles resolve this problem without the use of Na^+ as a coupling ion, without intracellular organelles or membrane-associated compartments,[74-76] and, possibly, by a mechanism other than manipulation of the H^+/ATP stoichiometry.[14,77] Hoffmann and Dimroth[78] have suggested that it would be possible to account for ATP synthesis in *B. alcalophilus* at pH 10.3 by a chemiosmotic mechanism, with the synthase using a H^+/ATP stoichiometry just a bit on the high side, i.e., 4.4. Their calculations depend on measurements of the $\Delta\Psi$ using probe concentrations (50 nM) that are in a range that is generally avoided. Earlier studies by others indicated that such probe concentrations were too low to saturate the external binding sites and also equilibrate with the transmembrane gradient in a *Bacillus* species,[79] and our own recent data reinforce that concern and do not support the report[78] of unusually high $\Delta\psi$ values in alkaliphilic *Bacillus* species.[77] If the values reported by Hoffmann and Dimroth[78] were nonetheless correct, a strictly chemiosmotic interpretation of ATP synthesis in their study is complicated by their finding that the ATP/ADP ratio and growth rate of *B. alcalophilus* were both highest when the magnitude of the putative driving force was not. In addition, since pH 10.3 is not the maximum pH for growth of the extreme alkaliphiles, the putative driving force would be expected to drop further (and the proposed H^+/ATP then increase further) as the growth pH increased. Finally, in our own consideration of how ATP synthesis is energized by the extreme alkaliphile at the most alkaline edge of its pH range, we have encompassed the striking observation of a difference between artificial and respiration-generated gradients with respect to their efficacy in energizing ATP synthesis at pH values above 9.5.[77] As discussed in detail elsewhere,[77] this phenomenon would seem to eliminate even a highly variable H$^+$/ATP stoichiometry as a resolution to the problem of ATP synthesis in the alkaliphile. If a high stoichiometry alone resolved the problem, it should work just as well with an imposed potential as a respiration-generated potential of the same magnitude. We have proposed that the mechanism may involve direct intramembrane transfers between proton-translocating residues on particular electron transport proteins and the F$_0$ portion of the ATP synthase[14,77] and that in the highly adapted alkaliphiles there may be special features of the F$_0$ that promote such transfers and prevent protons that arrive at the F$_0$ within the membrane from leaking outward rather than moving productively inward.[77,80] The bulk $\Delta\Psi$ would still be required because it is a major part of the total energetic driving force and would be expected to be necessary to move the protons inward, whatever their point of entry into the F$_0$. It should be noted, though, that because of the electrogenic nature of the Na$^+$/H$^+$ antiporters and their critical role in pH homeostasis, abolition of the bulk electrochemical proton gradient would certainly arrest growth of extreme alkaliphiles independently of its effect on ATP synthesis.

Since at least one other nonalkaliphilic bacterium does have a Na$^+$-coupled synthase,[81] it is notable that this was not the solution employed by the extreme alkaliphiles. Perhaps, the alkaliphiles' own mode of energy coupling offers some advantage that will become clear when the mechanism is better understood, and/or Na$^+$-coupled synthases do not work well enough to function as sole support of oxidative phosphorylation in an obligate aerobe. Since the alkaliphile apparently has an available alternative to the strict coupling of oxidative phosphorylation to a bulk electrochemical ion gradient with a fixed H$^+$/ATP stoichiometry, then the absence of a primary Na$^+$ cycle in the extreme alkaliphile may be advantageous. Na$^+$ extrusion routes can be restricted to electrogenic antiports at high pH, serving the vital function of pH homeostasis without the complicating factor of serving oxidative phosphorylation. It is a similar argument that encourages the positing of a pH-regulated channel for Na$^+$ reentry that would specifically serve pH homeostasis at high pH. If, nonetheless, the Na$^+$ symporters turn out to play the entire or even a major role in the Na$^+$ reentry part of the cycle, it will be of great interest to examine the multiple levels of control that allow pH homeostasis and solute uptake to be served by the same porters.

ADDENDUM IN PROOF

Additional studies of the pJB20 clone from *B. firmus* OF4 that enhances the Na$^+$ resistance of Na$^+$/H$^+$ antiporter-deficient *E. coli* strains have shown that the incomplete acidic orf4 (Figure 3) is required for the enhancement of the resistance, but neither orf1 nor orf2 is required and the clone does not result in an increase in the membrane Na$^+$/H$^+$ antiporter activity. The enhanced resistance may result from nonspecific binding of Na$^+$ by an overexpressed product of truncated orf4. The sequence for this entire clone has a GenBank accession number of Z14112. The sequence of the alkaliphile *cad*C gene, referred to in Table 1, has now been presented.[82]

ACKNOWLEDGMENT

This work was supported by grant DCB9018231 from the National Science Foundation.

REFERENCES

1. **Guffanti, A. A. and Hicks, D. B.,** Molar growth yields and bioenergetic parameters of extremely alkaliphilic *Bacillus* species in batch cultures, and growth in a chemostat at pH 10.5, *J. Gen. Microbiol.,* 137, 2375, 1991.
2. **Krulwich, T. A. and Guffanti, A. A.,** Alkalophilic bacteria, *Annu. Rev. Microbiol.,* 43, 435, 1989.
3. **Clejan, S. and Krulwich, T. A.,** Permeability studies of lipid vesicles from alkalophilic *Bacillus firmus* showing opposing effects of membrane isoprenoid and diacylglycerol fractions and suggesting a possible basis for obligate alkalophily, *Biochim. Biophys. Acta,* 946, 40, 1988.
4. **Dunkley, E. A., Jr., Guffanti, A. A., Clejan, S., and Krulwich, T. A.,** Facultative alkaliphiles lack fatty acid desaturase activity and lose the ability to grow at near-neutral pH when supplemented with an unsaturated fatty acid, *J. Bacteriol.,* 173, 1331, 1991.
5. **Guffanti, A. A., Finkelthal, O., Hicks, D. B., Falk, L. H., Sidhu, A., Garro, A., and Krulwich, T. A.,** Isolation and characterization of new facultatively alkalophilic strains of *Bacillus* species, *J. Bacteriol.,* 167, 766, 1986.
6. **Sugiyama, S., Matsukura, H., Koyama, N., Nosoh, Y., and Imae, Y.,** Requirement of Na$^+$ in flagellar rotation and amino-acid transport in a facultatively alkalophilic *Bacillus, Biochim. Biophys. Acta,* 852, 38, 1986.
7. **Krulwich, T. A. and Guffanti, A. A.,** The Na$^+$ cycle of extreme alkalophiles: a secondary Na$^+$/H$^+$ antiporter and Na$^+$/solute symporters, *J. Bioenerg. Biomembr.,* 21, 663, 1989.
8. **Garcia, M. L., Guffanti, A. A., and Krulwich, T. A.,** Characterization of the Na$^+$/H$^+$ antiporter of alkalophilic bacilli *in vivo*: $\Delta\Psi$-dependent ^{22}Na$^+$ efflux from whole cells, *J. Bacteriol.,* 156, 1151, 1983.
9. **Krulwich, T. A., Federbush, J. G., and Guffanti, A. A.,** Presence of a non-metabolizable solute that is translocated with Na$^+$ enhances Na$^+$-dependent pH homeostasis in an alkalophilic *Bacillus, J. Biol. Chem.,* 260, 4055, 1985.
10. **McLaggan, D., Selwyn, M. J., and Dawson, A. P.,** Dependence on Na$^+$ of control of cytoplasmic pH in a facultative alkalophile, *FEBS Lett.,* 165, 254, 1984.
11. **McLaggan, D., Selwyn, M. J., Dawson, A. P., and Booth, I. R.,** Role of Na$^+$ in pH homeostasis by the alkalophilic bacterium *Exiguobacterium auranticum, J. Gen. Microbiol.,* 137, 1709, 1991.
12. **Zilberstein, D., Ophir, I. J., Padan, E., and Schuldiner, S.,** Na$^+$ gradient-coupled porters of *Escherichia coli* share a common subunit, *J. Biol. Chem.,* 257, 3692, 1982.
13. **Krulwich, T. A., Guffanti, A. A., Fong, M. Y., Falk, L., and Hicks, D. B.,** Alkalophilic *Bacillus firmus* RAB generates variants which can grow at lower Na$^+$ concentrations than the parental strain, *J. Bacteriol.,* 165, 884, 1986.

14. **Ivey, D. M., Hicks, D. B., Guffanti, A. A., Sobel, G., and Krulwich, T. A.,** The problem of the electrochemical proton potential in alkaliphilic bacteria, *Mosbach Colloq.,* 41, 105, 1990.

15. **Krulwich, T. A. and Ivey, D. M.,** Bioenergetics in extreme environments, in *The Bacteria: A Treatise on Structure and Function,* Vol. 12, Krulwich, T. A., Ed., Academic Press, Orlando, FL, 1990, 417.

16. **Karpel, R., Alon, T., Glaser, G., Schuldiner, S., and Padan, E.,** Expression of a sodium proton antiporter (*Nha*A) in *Escherichia coli* is induced by Na+ and Li+ ions, *J. Biol. Chem.,* 266, 21753, 1991.

17. **Padan, E. and Schuldiner, S.,** Na+ transport systems in prokaryotes, in *Alkali Cation Transport Systems in Prokaryotes,* Bakker, E. P., Ed., CRC Press, Boca Raton, FL, 1992, chap. 1.

18. **Booth, I. R.,** Regulation of cytoplasmic pH in bacteria, *Microbiol. Rev.,* 49, 359, 1985.

19. **Koyama, N. and Nosoh, Y.,** Effect of potassium and sodium ions on the cytoplasmic pH of an alkalophilic *Bacillus, Biochim. Biophys. Acta,* 812, 206, 1985.

20. **Koyama, N., Wakabayashi, K., and Nosoh, Y.,** Effect of K+ on the membrane functions of an alkalophilic *Bacillus, Biochim. Biophys. Acta,* 898, 293, 1987.

21. **Krulwich, T. A., Guffanti, A. A., Bornstein, R. F., and Hoffstein, J.,** A sodium requirement for growth, solute transport, and pH homeostasis in *Bacillus firmus* RAB, *J. Biol. Chem.,* 257, 1885, 1982.

22. **Krulwich, T. A., Hicks, D. B., Seto-Young, D., and Guffanti, A. A.,** The bioenergetics of alkalophilic bacilli, *Crit. Rev. Microbiol.,* 16, 15, 1988.

23. **Koyama, N., Ishikawa, Y., and Nosoh, Y.,** Dependence of the growth of pH-sensitive mutants of a facultatively alkalophilic *Bacillus* on the regulation of cytoplasmic pH, *FEMS Microbiol. Lett.,* 34, 193, 1986.

24. **Kitada, M., Onda, K., and Horikoshi, K.,** The sodium/proton antiport system in a newly isolated alkalophilic *Bacillus* sp. *J. Bacteriol.,* 171, 1879, 1989.

25. **Krulwich, T. A., Mandel, K. G., Bornstein, R. F., and Guffanti, A. A.,** A non-alkalophilic mutant of *Bacillus alcalophilus* lacks the Na+/H+ antiporter, *Biochem. Biophys. Res. Commun.,* 91, 58, 1979.

26. **Guffanti, A. A., Cohn, D. E., Kaback, H. R., and Krulwich, T. A.,** A relationship between sodium-coupled antiporters and symporters in *Bacillus alcalophilus, Proc. Natl. Acad. Sci. U.S.A.,* 78, 1481, 1981.

27. **Seto-Young, D., Garcia, M. L., and Krulwich, T. A.,** Reconstitution of a bacterial Na+/H+ antiporter, *J. Biol. Chem.,* 260, 11393, 1985.

28. **Mandel, K. G., Guffanti, A. A., and Krulwich, T. A.,** Monovalent cation/proton antiporters in membrane vesicles from *Bacillus alcalophilus, J. Biol. Chem.,* 225, 7391, 1980.

29. **Guffanti, A. A.,** ATP dependent Na+/H+ antiport activity in *Bacillus alcalophilus* requires generation of an electrochemical gradient of protons, *FEMS Microbiol. Lett.,* 17, 307, 1983.

30. **Sugiyama, S., Matsukura, H., and Imae, Y.,** Relationship between Na+-dependent cytoplasmic pH homeostasis and Na+-dependent flagellar rotation and amino acid transport in alkalophilic *Bacillus, FEBS Lett.,* 182, 265, 1985.

31. **Benos, D. J.,** Amiloride: a molecular probe of sodium transport in tissues and cells, *Am. J. Physiol.,* 242, C131, 1982.

32. **Sugiyama, S., Cragoe, E. J., Jr., and Imae, Y.,** Amiloride, a specific inhibitor for the Na+-driven flagellar motors of alkalophilic *Bacillus, J. Biol. Chem.,* 263, 8215, 1988.

33. **Koyama, N., Niimura, Y., and Kozaki, M.,** Bioenergetic properties of a facultative anaerobic alkalophile, *FEMS Microbiol. Lett.,* 49, 123, 1988.

34. **Krulwich, T. A., Guffanti, A. A., and Seto-Young, D.,** pH homeostasis and bioenergetic work in alkalophiles, *FEMS Microbiol. Rev.,* 6, 271, 1990.

35. **Padan, E., Maisler, N., Taglicht, D., Karpel, R., and Schuldiner, S.,** Deletion of *ant* in *Escherichia coli* reveals its function in adaptation to high salinity and an alternative Na$^+$/H$^+$ antiporter system(s), *J. Biol. Chem.,* 264, 20297, 1989.

36. **Bassilana, M., Damiano, E., and Leblanc, G.,** Kinetic properties of Na$^+$-H$^+$ antiport activity in *Escherichia coli* membrane vesicles: effects of imposed electrical potential, proton gradient, and internal pH, *Biochemistry,* 23, 5288, 1984.

37. **Guffanti, A. A., Fuchs, R. T., Schneier, M., Chiu, E., and Krulwich, T. A.,** A transmembrane electrical potential generated by respiration is not equivalent to a diffusion potential of the same magnitude for ATP synthesis by *Bacillus firmus* RAB, *J. Biol. Chem.,* 259, 2971, 1984.

38. **Taglicht, D., Padan, E., and Schuldiner, S.,** Overproduction and purification of a functional Na$^+$/H$^+$ antiporter coded by *nha*A (*ant*) from *Escherichia coli, J. Biol. Chem.,* 266, 11289, 1991.

39. **Kudo, T., Hino, M., Kitada, M., and Horikoshi, K.,** DNA sequences required for the alkalophily of *Bacillus* sp. strain C-125 are located close together on its chromosomal DNA, *J. Bacteriol.,* 172, 7282, 1990.

40. **Ivey, D. M., Guffanti, A. A., Bossewitch, J. S., Padan, E., and Krulwich, T. A.,** Molecular cloning and sequencing of a gene from alkaliphilic *Bacillus firmus* OF4 that functionally complements an *Escherichia coli* strain carrying a deletion in the nhaA Na$^+$/H$^+$ antiporter gene, *J. Biol. Chem.,* 266, 23483, 1991.

41. **Sardet, C., Franchi, A., and Pouyssegur, J.,** Molecular cloning, primary structure, and expression of the human growth factor-activatable Na$^+$/H$^+$ antiporter, *Cell,* 56, 271, 1989.

42. **Lebovitz, R. M., Takeyasu, K., and Fambrough, D. M.,** Molecular characterization and expression of the (Na$^+$ + K$^+$)-ATPase alpha-subunit in *Drosophila melanogaster, EMBO J.,* 8, 193, 1989.

43. **Nicoll, D. A., Longoni, S., and Philipson, K. D.,** Molecular cloning and functional expression of the cardiac sarcolemmal Na$^+$-Ca^{2+} exchanger, *Science,* 250, 562, 1990.

44. **Nucifora, G., Chu, L., Misra, T. K., and Silver, S.,** Cadmium resistance from *Staphylococcus aureus* plasmid pI258 *cad*A gene results from a cadmium-efflux ATPase, *Proc. Natl. Acad. Sci. U.S.A.,* 86, 3544, 1989.

45. **Yoon, K. P. and Silver, S.,** A second gene in the *Staphylococcus aureus cad*A cadmium resistance determinant of plasmid pI258, *J. Bacteriol.,* 173, 7636, 1991.

46. **Ames, G. F., Mimura, C. S., and Shyamala, V.,** Bacterial periplasmic permeases belong to a family of transport proteins operating from *Escherichia coli* to human: traffic ATPases, *FEMS Microbiol. Rev.,* 6, 429, 1990.

47. **Mimura, C. S., Holbrook, S. R., and Ames, G. F.-L.,** Structural model of the nucleotide-binding conserved component of periplasmic permeases, *Proc. Natl. Acad. Sci. U.S.A.,* 88, 84, 1991.

48. **Walker, J. E., Saraste, M., Runswick, M. J., and Gay, N. J.,** Distantly related sequences in the alpha- and beta-subunits of ATP synthase, myosin, kinases and other ATP-requiring enzymes and a common nucleotide binding fold, *EMBO J.,* 1, 945, 1982.

49. **Buvinger, W. E. and Riley, M.,** Nucleotide sequence of *Klebsiella pneumoniae lac* genes, *J. Bacteriol.,* 163, 850, 1985.

50. **Zalkin, H. and Ebbole, D. J.,** Organization and regulation of genes encoding biosynthetic enzymes in *Bacillus subtilis, J. Biol. Chem.,* 263, 1595, 1988.

51. **Hayashi, S. and Wu, H. C.,** Lipoproteins in bacteria, *J. Bioenerg. Biomembr.,* 22, 451, 1990.

52. **van der Laan, J. C., Gerritse, G., Mulleners, L. J. S. M., van der Hoek, R. A. C., and Quax, W. J.,** Cloning, characterization, and multiple chromosomal integration of a *Bacillus* alkaline protease gene, *Appl. Environ. Microbiol.,* 57, 901, 1991.

53. **Kato, C., Nakano, Y., and Horikoshi, K.,** The nucleotide sequence of the lipo-penicillinase gene of alkalophilic *Bacillus* sp. strain 170, *Arch. Microbiol.,* 151, 91, 1989.

54. **Neugebauer, K., Sprengel, R., and Schaller, H.**, Penicillinase from *Bacillus licheni-formis*: nucleotide sequence of the gene and implications for the biosynthesis of a secretory protein in a Gram-positive bacterium *Nucleic Acids Res.*, 9, 2577, 1981.

55. **Hussain, M., Pastor, F. I., and Lampen, J. O.**, Cloning and sequencing of the *blaZ* gene encoding beta-lactamase III, a lipoprotein of *Bacillus cereus* 569/H, *J. Bacteriol.*, 169, 579, 1987.

56. **Martin, B., Alloing, G., Boucraut, C., and Claverys, J. P.**, The difficulty of cloning *Streptococcus pneumoniae mal* and *ami* loci in *Escherichia coli*: toxicity of *malX* and *amiA* gene products, *Gene*, 80, 227, 1989.

57. **Ando, A., Kusaka, I., and Fukui, S.**, Na⁺ dependent active transport systems in an alkalophilic *Bacillus*, *J. Gen. Microbiol.*, 128, 1057, 1982.

58. **Guffanti, A. A., Susman, P., Blanco, R., and Krulwich, T. A.**, The protonmotive force and α-aminoisobutyric acid transport in an obligately alkalophilic bacterium, *J. Biol. Chem.*, 253, 708, 1978.

59. **Chen, K. Y. and Cheng, S.**, Polyamine metabolism in an obligately alkalophilic *Bacillus alcalophilus* that grows at pH 11.0, *Biochem. Biophys. Res. Commun.*, 150, 185, 1988.

60. **Kitada, M., Wiijayanti, L., and Horikoshi, K.**, Biochemical properties of a thermophilic alkalophile, *Agric. Biol. Chem.*, 51, 2429, 1987.

61. **Koyama, N., Kiyomiya, A., and Nosoh, J.**, Na⁺-dependent uptake of amino acids by an alkalophilic *Bacillus*, *FEBS Lett.*, 72, 771, 1976.

62. **Kitada, M. and Horikoshi, K.**, Sodium-ion stimulated α-[1-C14]aminoisobutyric acid uptake in alkalophilic *Bacillus* species, *J. Bacteriol.*, 131, 784, 1977.

63. **Kitada, M. and Horikoshi, K.**, Further properties of sodium ion-stimulated α-[1-C14]aminoisobutyric acid uptake in alkalophilic *Bacillus* species, *J. Biochem. (Tokyo)*, 87, 1279, 1980.

64. **Kitada, M. and Horikoshi, K.**, Sodium-ion stimulated amino acid uptake in membrane vesicles of alkalophilic *Bacillus* No. 8–1, *J. Biochem. (Tokyo)*, 88, 1757, 1980.

65. **Bonner, S., Mann, M., Guffanti, A. A., and Krulwich, T. A.**, Na⁺/solute symport in membrane vesicles from *Bacillus alcalophilus*, *Biochim. Biophys. Acta*, 679, 315, 1982.

66. **Hicks, D. B. and Krulwich, T. A.**, Purification and reconstitution of the F₁F₀-ATP synthase from alkaliphilic *Bacillus firmus* OF4. Evidence that the enzyme translocates H⁺ but not Na⁺, *J. Biol. Chem.*, 265, 20547, 1990.

67. **Guffanti, A. A. and Krulwich, T. A.**, ATP synthesis is driven by an imposed ΔpH or $\Delta\mu_{H^+}$ but not by an imposed ΔpNa^+ or $\Delta\mu_{Na^+}$ in alkalophilic *Bacillus firmus* OF4 at high pH, *J. Biol. Chem.*, 263, 14748, 1988.

68. **Guffanti, A. A., Bornstein, R. F., and Krulwich, T. A.**, Oxidative phosphorylation by membrane vesicles from *Bacillus alcalophilus*, *Biochim. Biophys. Acta*, 635, 619, 1981.

69. **Hicks, D. B. and Krulwich, T. A.**, The membrane ATPase of alkalophilic *Bacillus firmus* RAB is an F₁-type ATPase, *J. Biol. Chem.*, 261, 12896, 1986.

70. **Hoffmann, A., Laubinger, W., and Dimroth, P.**, Na⁺-coupled ATP synthesis in *Propionigenium modestum*: is it a unique system?, *Biochim. Biophys. Acta*, 1018, 206, 1990.

71. **Hoffman, A. and Dimroth, P.**, The ATPase of *Bacillus alcalophilus* — purification and properties of the enzyme, *Eur. J. Biochem.*, 194, 423, 1990.

72. **Dibrov, P. A., Lazarova, R. L., Skulachev, V. P., and Verkhovskaya, M. L.**, The sodium cycle. II. Na⁺-coupled oxidative phosphorylation in *Vibrio alginolyticus* cells, *Biochim. Biophys. Acta*, 850, 458, 1986.

73. **Laubinger, W. and Dimroth, P.**, Characterization of the ATP synthase of *Propionigenium modestum* as a primary sodium pump, *Biochemistry*, 27, 7531, 1988.

74. **Krulwich, T. A.**, The fine structure of obligately alkalophilic bacilli, *FEMS Microbiol. Lett.*, 13, 299, 1982.

75. **Rohde, M., Mayer, F., Hicks, D. B., and Krulwich, T. A.,** Immunoelectron microscopic localization of the F_1F_0 ATPase (ATP synthase) on the cytoplasmic membrane of alkalophilic *Bacillus firmus* RAB, *Biochim. Biophys. Acta,* 985, 233, 1989.

76. **Khan, S., Ivey, D. M., and Krulwich, T. A.,** Membrane ultrastructure of alkaliphilic *Bacillus* species studied by rapid-freeze electron microscopy, *J. Bacteriol.,* 174, 5123, 1992.

77. **Guffanti, A. A. and Krulwich, T. A.,** Features of apparent nonchemiosmotic energization of oxidative phosphorylation by alkaliphilic *Bacillus firmus* OF4, *J. Biol. Chem.,* 267, 9580, 1992.

78. **Hoffmann, A. and Dimroth, P.,** The electrochemical proton potential of *Bacillus alcalophilus, Eur. J. Biochem.,* 201, 467, 1991.

79. **Zaritsky, A., Kihara, M., and Macnab, R. M.,** Measurement of membrane potential in *Bacillus subtilis*: a comparison of lipophilic cations, rubidium ion, and a cyanine dye as probes, *J. Membr. Biol.,* 63, 215, 1981.

80. **Ivey, D. M. and Krulwich, T. A.,** Organization and nucleotide sequence of the *atp* genes encoding the ATP synthase from alkaliphilic *Bacillus firmus* OF4, *Mol. Gen. Genet.,* 229, 292, 1991.

81. **Laubinger, W. and Dimroth, P.,** Characterization of the Na^+-stimulated ATPase of *Propionigenium modestum* as an enzyme of the F_1F_0 type, *Eur. J. Biochem.,* 168, 475, 1987.

82. **Ivey, D. M., Guffanti, A. A., Shen, Z., Kudyan, N., and Krulwich, T. A.,** The *cadC* gene product of alkaliphilic *Bacillus firmus* OF4 partially restores Na^+-resistance to an *Escherichia coli* strain lacking a Na^+/H^+ antiporter (NhaA), *J. Bacteriol.,* 174, 4878, 1992.

Chapter IF

The Na$^+$ CYCLE IN *VIBRIO ALGINOLYTICUS*

Hajime Tokuda

TABLE OF CONTENTS

I. INTRODUCTION

All living organisms contain systems that generate and transduce energy in membranes. The chemiosmotic theory[1] of Mitchell explains how energy coupling in membranes takes place. Electron transfer via the respiratory chain or photoredox chain causes the extrusion of H^+, leading to the generation of an electrochemical potential difference of H^+ (proton motive force) across the membranes. It is now established that the proton motive force can drive energy-dependent reactions such as the active transport of solutes, ATP synthesis, and flagellar rotation in bacteria. These processes are coupled to the backflow of H^+ through the systems that mediate these reactions. The circulation of H^+, therefore, plays a central role in these energy-transducing reactions. A certain number of solute transport systems in nonhalophilic bacteria are energized by the electrochemical potential of Na^+ (sodium motive force) instead of the proton motive force.[2,3] However, in these bacteria, the sodium motive force is generated via a Na^+/H^+ antiport system that is energized by the proton motive force.[4] Therefore, the circulation of H^+ is still essential for Na^+-dependent solute transport in nonhalophilic bacteria. On the other hand, various halophilic bacteria, including the marine bacterium *Vibrio alginolyticus*, have been found to possess primary Na^+ pumps coupled to chemical reactions such as respiration, decarboxylation, or ATP hydrolysis[5] (see Chapters IA, ID, and IG). These findings indicate that proton circulation is not the only mode of energy coupling in bacteria. In this chapter, the role of the Na^+ cycle in the bioenergetics of *V. alginolyticus* is discussed.

II. REPLACEMENT OF INTRACELLULAR K^+ BY EXTRACELLULAR Na^+ THROUGH K^+/H^+ AND Na^+/H^+ ANTIPORT SYSTEMS

Establishment of a simple method of replacing intracellular K^+ with extracellular Na^+ was essential for understanding the mechanism underlying Na^+ circulation in *V. alginolyticus*. Cells of *V. alginolyticus* growing on a synthetic medium maintain intracellular concentrations of K^+ and Na^+ of about 0.4 M and 80 mM, respectively.[6] Repeated washing of such cells with a 0.4 M NaCl solution does not change the internal concentrations of these cations, whereas treatment of the cells with a diethanolamine solution (pH 8.5) and subsequent dilution of the diethanolamine-treated cells in a 0.4 M NaCl solution cause the almost complete replacement of intracellular K^+ by extracellular Na^+.[7] This replacement takes place in two steps (Figure 1). The first step (K^+ depletion) involves a K^+/H^+ antiport system, and the second one (Na^+ loading) depends on a Na^+/H^+ antiport system. The release of intracellular K^+ down the concentration gradient occurs via the K^+/H^+ antiport system,[8] which becomes active at a cytoplasmic pH value higher than about 8.0. Membranes are permeable to the unprotonated form of diethano-

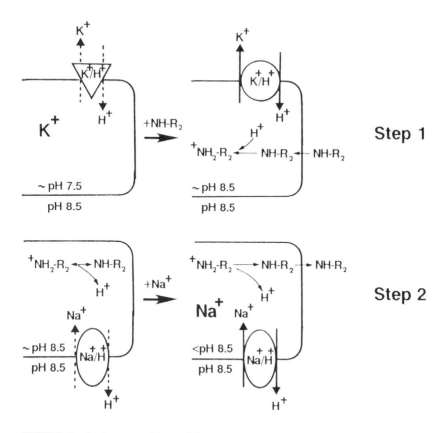

FIGURE 1. Replacement of intracellular K^+ by extracellular Na^+ in *Vibrio alginolyticus*. NH–R₂, diethanolamine.

lamine, whereas they are not to its protonated form. After unprotonated diethanolamine enters cells, its protonation takes place, which causes alkalization of the cytoplasm from about pH 7.5 to 8.5. Treatment of cells with diethanolamine thus induces the downhill efflux of K^+. The efflux of K^+ is coupled to the influx of H^+. The cytoplasmic pH, however, remains high because of the compensatory influx of diethanolamine. The release of K^+ therefore continues until almost all of it is replaced by diethanolamine. The dilution of diethanolamine-loaded cells in a 0.4 M NaCl solution causes the efflux of unprotonated diethanolamine. Depronotation of diethanolamine then takes place in the cells, leading to acidification of the cytoplasm. Extracellular Na^+ enters the cells in exchange for H^+ via the Na^+/H^+ antiport system. The successful replacement of intracellular K^+ with extracellular Na^+ by this method per se indicates that both K^+/H^+ and Na^+/H^+ antiport systems are present in *V. alginolyticus*.

III. RESPIRATORY Na$^+$ PUMP

A. *IN VIVO* STUDIES

The relationship between the external pH and the magnitude of the membrane potential ($\Delta\Psi$) was examined in detail in Na$^+$-loaded cells of *V. alginolyticus*. It was found that such cells are able to generate $\Delta\Psi$ (negative inside) at an alkaline pH even in the presence of a proton conductor, carbonyl cyanide *m*-chlorophenyl hydrazone (CCCP).[9] Although $\Delta\Psi$ generated at pH 6.5 collapses with 10 μM CCCP, $\Delta\Psi$ at pH 8.5 is unaffected by up to 50 μM CCCP. The generation of CCCP-resistant $\Delta\Psi$ is dependent on respiration and is always accompanied by acidification of the cytoplasm, leading to the generation of ΔpH (acidic inside) of a magnitude similar to that of $\Delta\Psi$.[10] These results indicate that the membranes of *V. alginolyticus* treated with CCCP at pH 8.5 are permeable to H$^+$ although $\Delta\Psi$ is generated. The generation of CCCP-resistant $\Delta\Psi$ at pH 8.5 is possible only if an electrogenic primary pump other than one that pumps protons operates in this organism.

Since the generation of CCCP-resistant $\Delta\Psi$ requires Na$^+$, the extrusion of Na$^+$ by Na$^+$-loaded cells was examined (Figure 2).[10] The active extrusion of Na$^+$ at pH 6.5 is sensitive to CCCP (Figure 2A), whereas that at pH 8.5 is resistant to CCCP, leading to the generation of a sodium concentration gradient across the membrane (Figure 2B). The CCCP-resistant Na$^+$ extrusion at pH 8.5 requires respiration (Figure 2B), but is independent of ATP. From these results it was concluded that *V. alginolyticus* possesses a unique pump which extrudes Na$^+$ as a direct result of respiration at alkaline pH.

When the respiratory Na$^+$ pump is functioning, the growth of *V. alginolyticus* is resistant to CCCP.[11] At pH 6.5, where the activity of the Na$^+$ pump is low, growth is completely inhibited by 5 μM CCCP. On the other hand, cells grow at pH 8.5 even in the presence of 50 μM CCCP. The pH profile of CCCP-resistant growth parallels that of the generation of CCCP-resistant $\Delta\Psi$. These results, taken together, suggest that the CCCP-resistant generation of the sodium motive force by the Na$^+$ pump makes the growth of *V. alginolyticus* resistant to CCCP. The isolation of mutants unable to grow in the presence of CCCP at pH 8.5 was then attempted. Two kinds of mutants, Nap1 and Nap2, were found to be defective in the Na$^+$ pump. Both mutants extrude Na$^+$ in the absence of CCCP, whereas in its presence the extrusion of Na$^+$ is completely inhibited even at pH 8.5 (Figure 2C).[12] $\Delta\Psi$ generated by the mutants collapses with CCCP under all conditions examined.[13] Furthermore, the CCCP-induced accumulation of H$^+$ inside cells is not observed with the mutants. It should be noted that the mutants still retain the Na$^+$/H$^+$ antiport system driven by the proton motive force and generate the sodium motive force at both pH 6.5 and 8.5 in the absence of CCCP.

B. Na$^+$-MOTIVE NADH-QUINONE REDUCTASE

It had been known that the NADH oxidase of *V. alginolyticus* requires Na$^+$ for the activity at the NADH-quinone reductase segment,[14] although the

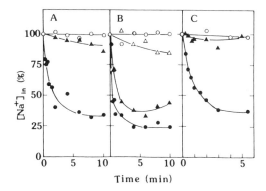

FIGURE 2. Active extrusion of Na^+ by the primary respiratory Na^+ pump and the secondary Na^+/H^+ antiport system in *Vibrio alginolyticus*.[5] Na^+-loaded cells of the wild type (A and B) and its Na^+ pump-defective mutant, Nap1 (C), were equilibrated with $^{22}Na^+$ in medium containing 0.4 M NaCl at pH 6.5 or 8.5. Na^+ extrusion at 25°C was assayed at either pH 6.5 (A) or 8.5 (B and C) in the presence of 20 mM glycerol. KCl (10 mM) was added to initiate the active extrusion of Na^+ (●). The internal level of $^{22}Na^+$ remained constant at about 0.4 M in the absence of KCl (○). Where specified, 10 μM CCCP (▲) or 10 μM CCCP plus 10 mM KCN (△) was added with KCl. (From Unemoto, T., Tokuda, H., and Hayashi, M., *The Bacteria, Vol. 12, Bacterial Energetics*, Krulwich, T. A., Ed., Academic Press, San Diego, 1990, 33. With permission.)

physiological meaning of this Na^+ requirement was unclear. Respiratory activities were examined in membranes isolated from the wild type, Nap1, Nap2, and Nap2R, a spontaneous revertant of Nap2.[15] The Na^+-dependent NADH oxidase activity observed in the wild type is completely absent in Nap1 and Nap2, whereas the NADH oxidase activity of Nap2R requires Na^+. In the presence of NADH, inside-out membrane vesicles prepared from wild-type cells require Na^+ for the generation of $\Delta\Psi$ (positive inside) and accumulate Na^+ against its electrochemical gradient.[16] The accumulation of Na^+ is resistant to CCCP, indicating that it is not driven by the proton motive force. In contrast, 2-heptyl-4-hydroxyquinoline *N*-oxide (HQNO), which has been shown to inhibit Na^+-dependent NADH-quinone reductase, also inhibits the accumulation of Na^+. Furthermore, essentially the same results were obtained from proteoliposomes reconstituted with the NADH oxidase.[17] These results, taken together, indicate that the Na^+-dependent NADH-quinone reductase segment of NADH oxidase is the Na^+ pump. Although neither Nap1 nor Nap2 possesses the Na^+-dependent NADH-quinone reductase, NADH-linked reduction of ubiquinone-1 (Q1) to ubiquinol-1 (QH$_2$) takes place in both mutants.[15] Detailed examinations have revealed that there are two kinds of NADH-quinone reductases in the wild-type cells; one is Na^+-dependent (NQR1) and the other is Na^+-independent (NQR2). The mutants specifically lack NQR1, which is the Na^+ pump.

TABLE 1
Properties of Subunits Composing the
Respiratory Na⁺ Pump of *Vibrio*
alginolyticus

Subunit	Mol wt	Cofactor	Function
α	52,000	FMN	Na$^+$ site (?)
β	46,000	FAD	NADH dehydrogenase
γ	32,000	—	Increases the affinity of β for Q1

C. PURIFICATION OF THE RESPIRATORY Na$^+$ PUMP

Purification of NQR1 revealed that it is composed of three subunits, α, β, and γ, with apparent molecular weights of 52,000, 46,000 and 32,000, respectively (Table 1).[18,19] NQR1 contains these subunits in equimolar quantities. The β-subunit exhibits NADH dehydrogenase activity and oxidizes both NADH and nicotinamide hypoxanthine dinucleotide (deamino-NADH) in the presence of menadione or Q1 as an electron acceptor. The isolated β-subunit reduces Q1 to ubisemiquinone (Q·$^-$), but not to QH$_2$. The formation of QH$_2$ requires both the α- and γ-subunits. The β-subunit does not require Na$^+$ for its activity and is insensitive to HQNO. On the other hand, the intact NQR1 requires Na$^+$ and is highly sensitive to HQNO. The β-subunit contains 1 mol of noncovalently bound FAD (flavin adenine dinucleotide), whereas the α-subunit contains FMN (flavin mononucleotide). The γ-subunit increases the affinity of the β-subunit for Q1. All three subunits are required for the reconstitution of the Na$^+$-dependent quinone reductase activity. A schematic model for the mode of electron transfer and extrusion of Na$^+$ at the NQR1 segment is shown in Figure 3.

The purified NQR1 complex has been reconstituted into liposomes prepared from soybean phospholipids.[19,20] The reconstituted proteoliposomes generate $\Delta\Psi$ upon the addition of both NADH and Q1 in the presence of Na$^+$. When assayed in the absence of Na$^+$, however, no $\Delta\Psi$ is generated. The generation of $\Delta\Psi$ is strongly inhibited by HQNO. The addition of menadione (which accepts electrons at the β-subunit) in place of Q1 fails to generate $\Delta\Psi$.

When the inactive NQR1 prepared from Nap2 is incubated with the β-subunit purified from the wild type, the active NQR1 complex is reconstituted,[21] indicating that Nap2 has an altered β-subunit. In contrast, membranes prepared from Nap1 lack all the subunits of NQR1.[22] When Nap1 is conjugated with the wild-type strain as a DNA donor, transconjugants that have recovered the NQR1 appear.[22] The wild type of *V. alginolyticus* retains two kinds of plasmids. These results, taken together, indicate that the Na$^+$ pump of *V. alginolyticus* may be encoded by one of these plasmids, which can be transferred from the wild type to the Nap1 strain through conjugation.

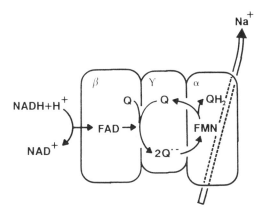

FIGURE 3. Subunit arrangement and electron transfer pathway in the Na^+-motive NADH-quinone reductase.

The mechanism whereby the NADH-quinone reductase segment extrudes Na^+ remains to be clarified. The corresponding enzyme system in mitochondria, known as complex I, contains FMN as a sole flavin. The translocation of H^+ across the membrane of mitochondria was proposed to be mediated by flavosemiquinone and/or ubiseminquinone radicals.[23] It seems unlikely, however, that these semiquinone radicals function as direct Na^+ carriers. Therefore, an indirect coupling mechanism is more plausible for the Na^+-motive NQR1 complex. Such a mechanism has been proposed by Wikström and Krab[24] for the H^+ pump activity of the bc_1 complex of the respiratory chain.

D. DISTRIBUTION AND COMMON PROPERTIES OF Na^+-MOTIVE NADH-QUINONE REDUCTASE

In addition to *V. alginolyticus*, a respiration-driven Na^+ pump has been reported to be present in *V. costicola*,[25] halotolerant Ba_1,[26,27] *V. parahaemolyticus*,[28,29] *Alteromonas* sp. strain 201,[29,30] and *Klebsiella pneumoniae*.[31] All these bacteria retain the Na^+-dependent NADH-quinone reductase segment. Furthermore, the enzymes of these bacteria show striking similarities in mode of electron transfer and enzymatic properties to NQR1 of *V. alginolyticus*. The enzymes are highly sensitive to HQNO, able to oxidize deamino-NADH, and form $Q^{\cdot-}$ as an intermediate of Q1 reduction. Inhibition of NADH dehydrogenase activity by Ag^+ is also a property that these enzymes have in common. Whether or not the respiratory Na^+ pumps of these bacteria exhibit any similarity at the molecular level would be an interesting subject for future investigations.

IV. Na$^+$:SOLUTE COTRANSPORT SYSTEMS

The active uptake of all amino acids in *V. alginolyticus* is dependent on Na$^+$.[32] Other cations such as Li$^+$, K$^+$, Rb$^+$, and Cs$^+$ cannot substitute for Na$^+$. The sodium motive force was found to be the driving force for the uptake of α-aminoisobutyric acid (AIB), a nonmetabolizable amino acid analogue. If Na$^+$-loaded cells are used, AIB uptake requires K$^+$ or Rb$^+$ in addition to Na$^+$, whereas only Na$^+$ is essential in control cells.[32] It was found that K$^+$ is required as a countercation for the active extrusion of Na$^+$ from Na$^+$-loaded cells. AIB uptake observed in the wild-type cells at pH 8.5 is resistant to CCCP, whereas it is inhibited at pH 6.5.[9] Furthermore, AIB uptake by Nap1 and Nap2 cells was sensitive to CCCP even at pH 8.5.[12] These results, taken together, indicate that the direct driving force for AIB uptake is the sodium motive force. The proton motive force is required only when the Na$^+$ pump activity is low, i. e., at acidic pH or in the mutants, Nap1 and Nap2. Under these conditions, the extrusion of Na$^+$ is carried out by the Na$^+$/H$^+$ antiport system, which is dependent on the proton motive force. Essentially the same conclusion was drawn for the driving force for sucrose uptake.[33]

It may be worth discussing the reason why Na$^+$-loaded cells require K$^+$ (or Rb$^+$) for the extrusion of Na$^+$. The Na$^+$-loaded cells retain about 0.4 *M* Na$^+$ inside them. The internal concentration of Na$^+$ remains at this level unless K$^+$ (or Rb$^+$) is added as a countercation.[32] The extrusion of Na$^+$ is accompanied by the stoichiometric uptake of K$^+$ (or Rb$^+$). So far, only K$^+$ and Rb$^+$ have been found to be effective countercations for Na$^+$ extrusion. These results indicate that the bulk extrusion of Na$^+$ is possible only when the osmotic and/or electrical imbalance caused by the Na$^+$ extrusion through the Na$^+$/H$^+$ antiport system or the Na$^+$ pump is compensated for by the influx of another cation. The influx of a large amount of cations is possible only via the K$^+$ transport system, which catalyzes the uptake of Rb$^+$ as well as K$^+$ in *V. alginolyticus*. Therefore, K$^+$ or Rb$^+$ is required for the extrusion of Na$^+$.

V. Na$^+$-DRIVEN FLAGELLAR ROTATION

Flagellar rotation is also driven by the sodium motive force in *V. alginolyticus*.[13,34] The motility of the wild-type cells is resistant to CCCP at alkaline pH, but not at acidic pH. The motility of Nap1 and Nap2 is completely inhibited by CCCP at any pH value examined.[13] Na$^+$-loaded cells require K$^+$ as well as Na$^+$ for motility. The effect of the intracellular concentration of Na$^+$ on motility was examined in detail with Na$^+$-loaded and control cells.[35] The motility of control cells, which contain about 50 m*M* Na$^+$ internally, increases with increasing Na$^+$ concentration in the medium. Significant motility is observed even at 10 m*M* Na$^+$, indicating that an inwardly directed

Na$^+$ concentration gradient is not required for motility. On the other hand, the Na$^+$-loaded cells show only slight motility even with 0.4 M Na$^+$ in the medium. An increase in Na$^+$ in the medium up to 0.5 M causes little improvement of their motility. The magnitude of the membrane potential is similar in Na$^+$-loaded cells and control cells. The poor motility of the Na$^+$-loaded cells, therefore, is not due to a decrease in the driving force. Extrusion of Na$^+$ initiated by the addition of K$^+$ causes significant stimulation of the motility of the Na$^+$-loaded cells. These results, taken together, indicate that intracellular Na$^+$ kinetically interferes with the rotation of flagellar motors. It is assumed that the force-generating unit of flagellar motors possesses binding sites for Na$^+$ on both sides of the membranes. The force for motor rotation may be generated via the sodium motive force during the step of Na$^+$ transfer from the external binding sites to the internal binding sites. If the affinity of Na$^+$ for the internal binding sites is not low, an increase in the intracellular Na$^+$ concentration results in the inhibition of Na$^+$ release from the internal binding sites, leading to poor rotation of the motor. Essentially the same inhibitory effect of internal Na$^+$ is observed in AIB uptake.[32,35] In contrast, sucrose uptake is affected little by Na$^+$ loading.[35] These results indicate that the inhibition of the Na$^+$-coupled system by the high intracellular Na$^+$ concentration is system specific.

VI. DOES Na$^+$-TRANSLOCATING ATPase EXIST IN *V. ALGINOLYTICUS?*

The presence of Na$^+$-translocating ATPase in *V. alginolyticus* was first proposed by Chernyak et al.[34] They reported that the CCCP-resistant motility of *V. alginolyticus* at pH 9 was not completely inhibited by KCN, whereas it was completely inhibited upon the combined addition of KCN and vanadate. Moreover, the motility in the absence of CCCP was found to be sensitive to the combined addition of KCN and arsenate. They assumed that, in addition to the respiratory Na$^+$ pump, a Na$^+$-motive ATPase which is sensitive to vanadate and arsenate exists in *V. alginolyticus*. Dibrov et al.[36] then reported that the imposition of a Na$^+$ pulse on energy-depleted cells of *V. alginolyticus* causes a transient increase in the ATP level, which is sensitive to an ionophore, monensin. It is not mentioned, however, whether Na$^+$-dependent ATP synthesis is sensitive to vanadate or not. Subcellular vesicles prepared from *V. alginolyticus* were shown to take up Na$^+$[37] as well as H$^+$[38] upon the addition of ATP. *N,N′*-dicyclohexylcarbodiimide, a specific inhibitor of F$_1$F$_0$-ATPase, was found to arrest the uptake of both Na$^+$ and H$^+$. From these observations, Skulachev[39] proposed the possibility that the F$_1$F$_0$-ATPase of *V. alginolyticus* extrudes not only H$^+$ but also Na$^+$. It has been shown that the F$_1$F$_0$-type ATPase of *Propionigenium modestum* extrudes Na$^+$,[40] whereas the extrusion of H$^+$ occurs in the absence of Na$^+$[41] (see Chapter IG).

The *unc* operon, which contains the structural genes encoding the F$_1$F$_0$-ATPase of *V. alginolyticus*, was cloned by Krumholz et al.[42] These genes were expressed in an *Escherichia coli unc* deletion strain, and the ATPase was purified and reconstituted into proteoliposomes. The reconstituted enzyme catalyzes the pumping of protons coupled to ATP hydrolysis, whereas no Na^+ pump activity is detected. The purified enzyme does not require Na^+ for the hydrolysis of ATP. Since the enzyme cloned is the only F_1F_0-type ATPase present in *V. algionolyticus*, Krumholz et al. concluded that if the organism contains a Na^+-translocating ATPase, it is not likely to be an F_1F_0-type ATPase.

Sakai et al.[43] isolated mutants lacking an F_1F_0-ATPase from *V. parahaemolyticus*, which is taxonomically similar to *V. alginolyticus* and possesses a respiratory Na^+ pump.[28,29] Both the wild-type and mutants cells were found to synthesize ATP upon the imposition of a Na^+ gradient, suggesting that a Na^+-dependent ATP synthase, which may be different from the F_1F_0-ATPase, is present in this organism.

It is important to point out that the active extrusion of Na^+ from Na^+-loaded cells of Nap1 and Nap2 is completely inhibited by CCCP,[12] suggesting that no primary Na^+ pump other than the respiratory one exists in *V. alginolyticus*. Furthermore, the motility of Nap1 and Nap2 is completely inhibited by CCCP even at alkaline pH.[13] These observations are therefore inconsistent with results reported by Chernyak et al.,[34] who examined motility and proposed the existence of another type of Na^+ pump. The increase in the level of ATP caused by the Na^+ pulse has been assumed to represent the direct energization of ATP synthase by the Na^+ gradient. It seems possible, however, that the Na^+ gradient energizes the active uptake of nutrients which have leaked out from the cells during energy starvation and are thus present in the medium. The uptake of nutrients induced by the Na^+ pulse may result in the transient increase in the ATP level.

Since the presence of Na^+-ATPase in *V. alginolyticus* remains to be confirmed, further studies are essential to obtain supporting evidence.

VII. CONCLUSION

It is now certain that *V. alginolyticus* generates and utilizes two kinds of energy, the proton motive force and the sodium motive force (Figure 4). The sodium motive force is generated by either the primary Na^+ pump or the coupling of the Na^+/H^+ antiport system with the proton motive force. Extrusion of Na^+ is specifically coupled to NADH-quinone reductase. In contrast, quinol oxidase functions to extrude H^+,[38,44] leading to the generation of the proton motive force. The active uptake of various solutes and the rotation of flagellar motors are driven directly by the sodium motive force. Protein translocation across the membrane is also suggested to be dependent on the sodium motive force.[45] Oxidative phosphorylation in *V. alginolyticus* is dependent on the proton motive force, although the sodium motive force also may be able to drive the synthesis of ATP under certain conditions.

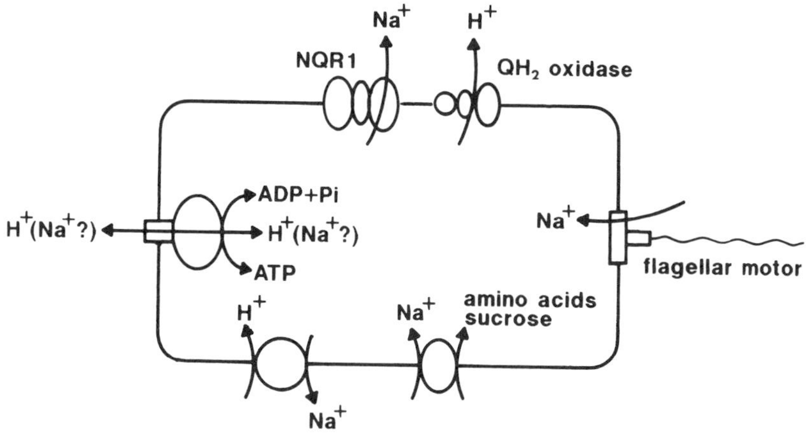

FIGURE 4. The Na⁺ cycle in *Vibrio alginolyticus*.

Since a similar respiratory Na⁺ pump has been found in various halophilic bacteria, it is plausible that the Na⁺ pump is advantageous for cells living under Na⁺-rich conditions. As discussed elsewhere,[46] the extrusion of Na⁺ by the primary Na⁺ pump seems to be more economical than that through the coupling of the Na⁺/H⁺ antiport system with the proton motive force. The concentration of Na⁺ in the environment at about neutral pH is much higher than that of H⁺. The concentration gradient generated across the membrane, therefore, may be more stable with Na⁺ than with H⁺. Generation of the proton motive force through H⁺ extrusion causes alkalization of the cytoplasm. To maintain the cytoplasmic pH at about neutral, H⁺ extruded must be exchanged with other cations such as Na⁺ or K⁺. On the other hand, Na⁺ extrusion by the primary Na⁺ pump leads to the generation of energy without alkalization. This may be especially important at alkaline pH, where the cytoplasmic pH is even more acidic than the external pH. Furthermore, since the proton motive force is composed of $\Delta\Psi$ alone at alkaline pH, the Na⁺/H⁺ antiport system must be electrogenic and requires the influx of more than a stoichiometric amount of H⁺ for the extrusion of 1 mol Na⁺. In terms of energy efficiency in the generation of the sodium motive force, the primary Na⁺ pump seems to be superior to the coupled systems, especially under alkaline Na⁺-rich conditions.

ACKNOWLEDGMENT

I would like to thank Iyoko Sugihara for her secretarial support.

REFERENCES

1. **Mitchell, P.,** Coupling of phosphorylation to electron and hydrogen transfer by a chemiosmotic type of mechanism, *Nature (London),* 191, 144, 1961.
2. **Tokuda, H. and Kaback, H. R.,** Sodium-dependent methyl-1-thio-β-D-galactopyranoside transport in membrane vesicles isolated from *Salmonella typhimurium, Biochemistry,* 16, 2130, 1977.
3. **Tsuchiya, T., Raven, J., and Wilson, T. H.,** Co-transport of Na$^+$ and methyl-β-D-thiogalactopyranoside mediated by the melibiose transport system of *Escherichia coli, Biochem. Biophys. Res. Commun.,* 76, 26, 1977.
4. **Schuldiner, S. and Fishkes, H.,** Sodium-proton antiport in isolated membrane vesicles of *Escherichia coli, Biochemistry,* 17, 706, 1978.
5. **Unemoto, T., Tokuda, H., and Hayashi, M.,** Primary sodium pumps and their significance in bacterial energetics, in *The Bacteria,* Vol. 12, Bacterial Energetics, Krulwich, T. A., Ed., Academic Press, San Diego, 1990, 33.
6. **Tokuda, H., Nakamura, T., and Unemoto, T.,** Potassium ion is required for the generation of pH-dependent membrane potential and ΔpH by the marine bacterium *Vibrio alginolyticus, Biochemistry,* 20, 4198, 1981.
7. **Tokuda, H.,** Sodium translocation by NADH oxidase of *Vibrio alginolyticus:* isolation and characterization of the sodium pump-defective mutants, *Methods Enzymol.,* 125, 520, 1986.
8. **Nakamura, T., Tokuda, H., and Unemoto, T.,** K$^+$/H$^+$ antiporter functions as a regulator of cytoplasmic pH in a marine bacterium, *Vibrio alginolyticus, Biochim. Biophys. Acta,* 776, 330, 1984.
9. **Tokuda, H. and Unemoto, T.,** A respiration-dependent primary sodium extrusion system functioning at alkaline pH in the marine bacterium *Vibrio alginolyticus, Biochem. Biophys. Res. Commun.,* 102, 265, 1981.
10. **Tokuda, H. and Unemoto, T.,** Characterization of the respiration-dependent Na$^+$ pump in the marine bacterium *Vibrio alginolyticus, J. Biol. Chem.,* 257, 10007, 1982.
11. **Tokuda, H. and Unemoto, T.,** Growth of a marine *Vibrio alginolyticus* and moderately halophilic *V. costicola* becomes uncoupler-resistant when the respiration-dependent Na$^+$ pump functions, *J. Bacteriol.,* 156, 636, 1983.
12. **Tokuda, H.,** Isolation of *Vibrio alginolyticus* mutants defective in the respiration-coupled Na$^+$ pump, *Biochem. Biophys. Res. Commun.,* 114, 113, 1983.
13. **Tokuda, H., Asano, M., Shimamura, Y., Unemoto, T., Sugiyama, S., and Imae, Y.,** Roles of the respiratory Na$^+$ pump in bioenergetics of *Vibrio alginolyticus, J. Biochem.,* 103, 650, 1988.
14. **Unemoto, T. and Hayashi, M.,** NADH:quinone oxidoreductase as a site of Na$^+$-dependent activation in the respiratory chain of marine *Vibrio alginolyticus, J. Biochem.,* 85, 1461, 1977.
15. **Tokuda, H. and Unemoto, T.,** Na$^+$ is translocated at NADH:quinone oxidoreductase segment in the respiratory chain of *Vibrio alginolyticus, J. Biol. Chem.,* 259, 7785, 1984.
16. **Tokuda, H., Udagawa, T., and Unemoto, T.,** Generation of the electrochemical potential of Na$^+$ by the Na$^+$-motive NADH oxidase in inverted membrane vesicles of *Vibrio alginolyticus, FEBS Lett.,* 183, 95, 1985.
17. **Tokuda, H.,** Solubilization and reconstitution of the Na$^+$-motive NADH oxidase activity from the marine bacterium *Vibrio alginolyticus, FEBS Lett.,* 176, 125, 1984.
18. **Hayashi, M. and Unemoto, T.,** FAD and FMN flavoproteins participate in the sodium transport respiratory chain NADH:quinone reductase of a marine bacterium, *Vibrio alginolyticus, FEBS Lett.,* 202, 327, 1986.
19. **Hayashi, M. and Unemoto, T.,** Subunit component and their roles in the sodium-transport NADH:quinone reductase of a marine bacterium *Vibrio alginolyticus, Biochim. Biophys. Acta,* 890, 47, 1987.

20. **Tokuda, H.**, Respiratory Na⁺ pump and Na⁺-dependent energetics in *Vibrio alginolyticus, J. Bioenerg. Biomembr.*, 21, 693, 1989.

21. **Asano, M., Hayashi, M., Unemoto, T., and Tokuda, H.**, Ag⁺-sensitive NADH dehydrogenase in the Na⁺-motive respiratory chain of the marine bacterium *Vibrio alginolyticus, Agric. Biol. Chem.*, 49, 2813, 1985.

22. **Tokuda, H., Udagawa, T., Asano, M., Yamamoto, T., and Unemoto, T.**, Conjugation-dependent recovery of the Na⁺ pump in a mutant of *Vibrio alginolyticus* lacking three subunits of the Na⁺ pump, *FEBS Lett.*, 215, 335, 1987.

23. **Ragan, C. I.**, Structure of NADH-ubiquinone reductase (complex), *Curr. Top. Bioenerg.*, 15, 1, 1987.

24. **Wikström, M. and Krab, K.**, Respiration-linked H⁺ translocation in mitochondria: stoichiometry and mechanism, *Curr. Top. Bioenerg.*, 10, 51, 1980.

25. **Udagawa, T., Unemoto, T., and Tokuda, H.**, Generation of Na⁺ electrochemical potential by the Na⁺-motive NADH oxidase and Na⁺/H⁺ antiport system of a moderately halophilic *Vibrio costicola, J. Biol. Chem.*, 261, 2616, 1986.

26. **Ken-Dror, S., Preger, R., and Avi-Dor, Y.**, Functional characterization of the uncoupler insensitive Na⁺ pump of the halotolerant bacterium, Ba₁, *Arch. Biochem. Biophys.*, 244, 122, 1986.

27. **Ken-Dror, S., Lanyi, J. K., Schobert, B., Silver, B., and Avi-Dor, Y.**, An NADH:quinone oxidoreductase of the halotolerant bacterium Ba₁ is specifically dependent on sodium ions, *Arch. Biochem. Biophys.*, 244, 766, 1986.

28. **Tsuchiya, T. and Shinoda, S.**, Respiration-driven Na⁺ pump and Na⁺ circulation in *Vibrio parahaemolyticus, J. Bacteriol.*, 162, 794, 1985.

29. **Tokuda, H. and Kogure, K.**, Generalized distribution and common properties of Na⁺-dependent NADH:quinone oxidoreductases in Gram-negative marine bacteria, *J. Gen. Microbiol.*, 135, 703, 1989.

30. **Kogure, K. and Tokuda, H.**, Respiration-dependent primary Na⁺ pump in halophilic marine bacterium *Alcaligenes* strain 201, *FEBS Lett.*, 256, 147, 1989.

31. **Dimroth, P. and Thomer, A.**, A primary respiratory Na⁺ pump of an anaerobic bacterium: the Na⁺-dependent NADH:quinone oxidoreductase of *Klebsiella pneumoniae, Arch. Microbiol.*, 151, 439, 1989.

32. **Tokuda, H., Sugasawa, M., and Unemoto, T.**, Roles of Na⁺ and K⁺ in α-aminoisobutyric acid transport by the marine bacterium *Vibrio alginolyticus, J. Biol. Chem.*, 257, 788, 1982.

33. **Kakinuma, Y. and Unemoto, T.**, Sucrose uptake is driven by the Na⁺ electrochemical potential in the marine bacterium *Vibrio alginolyticus, J. Bacteriol.*, 163, 1293, 1985.

34. **Chernyak, B. V., Dibrov, P. A., Glagolev, A. N., Sherman, M. Yu., and Skulachev, V. P.**, A novel type of energetics in a marine alkalitolerant bacterium, *FEBS Lett.*, 164, 38, 1983.

35. **Yoshida, S., Sugiyama, S., Hojo, Y., Tokuda, H., and Imae, Y.**, Intracellular Na⁺ kinetically interferes with the rotation of the Na⁺-driven flagellar motors of *Vibrio alginolyticus, J. Biol. Chem.*, 265, 20346, 1990.

36. **Dibrov, P. A., Lazarova, R. L., Skulachev, V. P., and Verkhovskaya, M. L.**, The sodium cycle. II. Na⁺-coupled oxidative phosphorylation in *Vibrio alginolyticus* cells, *Biochim. Biophys. Acta*, 850, 458, 1986.

37. **Dibrov, P. A., Skulachev, V. P., Sokolov, M. V., and Verkhovskaya, M. L.**, The ATP-driven primary Na⁺ pump in subcellular vesicles of *Vibrio alginolyticus, FEBS Lett.*, 233, 355, 1988.

38. **Smirnova, I. A., Vaghina, M. L., and Kostyrko, V. A.**, ΔΨ and ΔpH generation by the H⁺ pumps of the respiratory chain and ATPase in subcellular vesicles from marine bacterium *Vibrio alginolyticus, Biochim. Biophys. Acta*, 1016, 385, 1990.

39. **Skulachev, V. P.**, Bacterial Na⁺ energetics, *FEBS Lett.*, 250, 106, 1989.

40. **Laubinger, W. and Dimroth, P.**, Characterization of the ATP synthase of *Propionigenium modestum* as a primary sodium pump, *Biochemistry*, 27, 7531, 1988.

41. **Laubinger, W. and Dimroth, P.**, The sodium ion translocating adenosine triphosphatase of *Propionigenium modestum* pumps protons at low sodium ion concentrations, *Biochemistry*, 28, 7194, 1989.

42. **Krumholz, L. R., Esser, U., and Simoni, R. D.**, Characterization of the H^+-pumping F_1F_0 ATPase of *Vibrio alginolyticus*, *J. Bacteriol.*, 172, 6809, 1990.

43. **Sakai, Y., Moritani, C., Tsuda, M., and Tsuchiya, T.**, A respiratory-driven and an artificially driven ATP synthesis in mutants of *Vibrio parahaemolyticus* lacking H^+-translocating ATPase, *Biochim. Biophys. Acta*, 973, 450, 1989.

44. **Kim, Y. J., Mizushima, S., and Tokuda, H.**, Fluorescence quenching studies on the characterization of energy generated at the NADH:quinone oxidoreductase and quinol oxidase segments of marine bacteria, *J. Biochem.*, 109, 616, 1991.

45. **Tokuda, H., Kim, Y. J., and Mizushima, S.**, *In vitro* protein translocation into inverted membrane vesicles prepared from *Vibrio alginolyticus* is stimulated by the electrochemical potential of Na^+ in the presence of *Escherichia coli* SecA, *FEBS Lett.*, 264, 10, 1990.

46. **Tokuda, H. and Unemoto, T.**, The Na^+-motive respiratory chain of marine bacteria, *Microbiol. Sci.*, 2, 65, 1985.

Chapter IG

THE Na$^+$-TRANSLOCATING ATP-SYNTHETASE FROM *PROPIONIGENIUM MODESTUM*

Peter Dimroth

TABLE OF CONTENTS

0-8493-6982-7/93/$0.00 + $.50

I. INTRODUCTION

ATP synthesis by oxidative phosphorylation in bacteria, mitochondria, or chloroplasts involves the formation of an electrochemical proton gradient ($\Delta\bar{\mu}_{H^+}$) by an electron transport chain and its conversion into the high-energy phosphoric anhydride bond of ATP by an F_1F_0-type ATPase.[1-3] More recently, the general picture of ATP synthesis was extended by the mechanism termed decarboxylation phosphorylation which, although related to oxidative phosphorylation, includes a number of distinct and unique properties.[4-7] The energy for membrane energization does not derive from oxidation but from a decarboxylation event and is not converted to ($\Delta\bar{\mu}_{H^+}$) but to $\Delta\bar{\mu}_{Na^+}$ (see Chapter ID). The electrochemical Na^+ ion gradient ($\Delta\bar{\mu}_{Na^+}$) is used directly for ATP formation by a Na^+-translocating F_1F_0-ATPase.

This Na^+-dependent ATP synthesis mechanism accounts for all the ATP synthesized in the strictly anaerobic bacterium *Propionigenium modestum*.[8-11] The existence of a similar Na^+-coupled F_1F_0-ATPase in other bacteria, albeit likely, has not been unequivocally demonstrated until now.[12]

This chapter will focus on the ATP synthesis mechanism in *P. modestum*. Also included will be a short review of ATP synthesis in bacteria that have been considered candidates for containing a Na^+-translocating ATPase, such as *Vibrio alginolyticus*, *Bacillus alcalophilus*, and *Mycoplasma gallisepticum*.

II. ENERGY METABOLISM OF *PROPIONIGENIUM MODESTUM*

P. modestum is a strictly anaerobic bacterium that gains its total energy for growth from the fermentation of succinate to propionate and CO_2 according to the equation: succinate $+ H_2O \rightarrow$ propionate $+ HCO_3^-$.[13] The small free energy change of the decarboxylation reaction ($\Delta G_0' = -20.6$ kJ/mol) does not allow ATP synthesis by substrate-linked phosphoryl group transfer, and no redox reactions occur which could drive ATP synthesis by electron-transport phosphorylation.

The energy metabolism of *P. modestum* is shown in Figure 1.[8] Succinate taken up into the cells is activated by CoA transfer from propionyl-CoA to yield succinyl-CoA. Rearrangement of the carbon skeleton with a vitamin B_{12}-containing enzyme yields (R)-methylmalonyl-CoA. This isomerizes to (S)-methylmalonyl-CoA and is then decarboxylated to propionyl-CoA. Under the catalysis of a CoA-transferase, propionate is formed and a new succinate molecule is coverted to succinyl-CoA. The only step of the fermentation pathway that is sufficiently exergonic to be used for energy transduction is the decarboxylation of (S)-methylmalonyl-CoA. The bacteria take advantage of this reaction and use the free energy of methylmalonyl-CoA decarboxylation to generate an electrochemical Na^+ ion gradient across the membrane. The responsible decarboxylase is a membrane-bound and biotin-containing enzyme

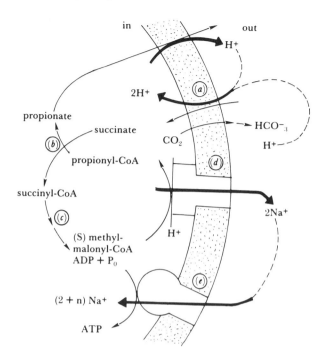

FIGURE 1. Energy metabolism of *Propionigenium modestum* with a Na^+ cycle coupling the exergonic decarboxylation of (*S*)-methylmalonyl-CoA to endergonic ATP synthesis.[5] A hypothetical proton circuit could couple succinate uptake with the extrusion of propionate and CO_2 *a*, succinate uptake system; *b*, succinate propionyl-CoA:CoA transferase; *c*, methylmalonyl-CoA mutase and methylmalonyl-CoA epimerase; *d*, methylmalonyl-CoA decarboxylase; *e*, ATPase. (From Dimroth, P., *Philos. Trans. R. Soc. London Ser. B*, 326, 465, 1990. With permission.)

that is closely related to other sodium ion-translocating decarboxylases found in anaerobic bacteria, i.e., oxaloacetate decarboxylase and glutaconyl-CoA decarboxylase (see Chapter ID).[14]

It is conceivable that uptake of succinate into the cells occurs in symport with protons, which could recycle by the proton-coupled secretion of the metabolic end products propionate and bicarbonate. The electrochemical Na^+ ion gradient derived from the decarboxylation of methylmalonyl-CoA is the only available energy source by which *P. modestum* can synthesize ATP. In principle, two different mechanisms could account for ATP synthesis in this anaerobic bacterium. Either the Na^+ gradient is first converted into a proton gradient which then energizes a conventional H^+-translocating F_1F_0-ATPase, or a novel type of ATPase could use the Na^+ gradient directly.

III. THE SODIUM CYCLE

The first indication for a Na^+-coupled ATP synthesis mechanism in *P. modestum* was obtained when it was observed that the ATPase solubilized from the bacterial membrane with the detergent N-octylglucoside was specifically activated 24-fold by Na^+ ions (5 mM).[8] Inverted bacterial vesicles catalzyed an active transport of Na^+ into their interior volume upon ATP hydrolysis. The transport was completely abolished by the Na^+-translocating ionophore monensin, but was not significantly affected by the uncoupler carbonylcyanide-p-trifluoromethoxyphenylhydrazone (CCFP), indicating that a proton gradient is not an intermediate for the accumulation of Na^+. The vesicles also catalyzed Na^+ uptake upon decarboxylation of malonyl-CoA which was abolished by monensin but unaffected by CCFP. It was also demonstrated that the decarboxylation of malonyl-CoA was energetically linked to ATP synthesis via a Na^+ circuit across the bacterial membrane.[8] The vesicles synthesized ATP from ADP and P_i upon decarboxylation of malonyl-CoA and synthesized malonyl-CoA from acetyl-CoA upon ATP hydrolysis. Both chemical syntheses were dependent on sealed membranes and on the $\Delta\bar{\mu}_{Na^+}$ developed in the exergonic reaction since dissipation of the Na^+ gradient with monensin abolished the synthesis of either ATP or malonyl-CoA.

IV. PURIFICATION AND SUBUNIT COMPOSITION OF THE ATPase

The ATPase activity was firmly bound to the bacterial membranes. Analogous to other F_1F_0-ATPases, the ATPase activity could be released from the membranes by treatment with EDTA at low ionic strength (1 mM Tris buffer) at pH 9.0[9] The solubilized enzyme was purified to homogeneity by fractionation with poly(ethylene glycol) and gel chromatography. Analysis of the enzyme by SDS gel electrophoresis showed the typical pattern of an F_1-type ATPase with subunits α, β, γ, δ, ϵ with apparent molecular masses of 58, 56, 37.6, 22.7, and 14 kDa, respectively. Interestingly, the specific activation by Na^+ of the detergent-extracted ATPase was not retained in the purified F_1-ATPase. The stimulation by Na^+ was recovered upon reconstitution of the ATPase enzyme complex (F_1F_0) from the purified F_1-ATPase and F_1-depleted membranes.[9] These results suggest that the ATPase complex has a binding site for Na^+ located on the membrane-bound F_0 moiety that is responsible for the activation by this alkali ion.

The ATPase complex (F_1F_0) was extracted from the bacterial membranes with Triton® X-100 and purified by fractional precipitation with poly(ethylene glycol).[10] Inspection of purified enzyme samples by SDS-gel electrophoresis indicated that, in addition to the five subunits of the F_1-ATPase, subunits with apparent molecular masses of 26, 23, and 7.5 kDa had been purified, which are termed a, b, and c according to the analogous F_0-ATPase subunits from other bacteria. A remarkable feature of subunit c is its strong aggregation, probably to a hexamer, which resists boiling in SDS and dissociates only in an autoclave (121°C) or with chloroform/methanol. The purified F_1F_0-ATPase was specifically activated by Na^+ ions, half-maximal activation being obtained with about 1 mM NaCl. Strong inhibition of the *P. modestum* ATPase was observed in the presence of the prokaryotic F_1F_0-ATPase inhibitors azide, tributyltin chloride, venturicidin, and dicyclohexylcarbodiimide (DCCD), whereas vanadate was not inhibitory. On incubation with [^{14}C]-DCCD, subunit *c* of the ATPase became specifically labeled,[10] presumably at the conserved glutamic acid residue within the C-terminal membrane-spanning α-helix (see below).[15]

V. RECONSTITUTION INTO PROTEOLIPOSOMES

A. ATP-DRIVEN Na$^+$ TRANSPORT

To investigate the transport function of the ATPase, the purified enzyme was reconstituted into proteoliposomes. These proteoliposomes catalyzed the rapid accumulation of Na$^+$ ions by ATP hydrolysis, as shown in Figure 2.[10] Also shown is the effect of certain ionophores on Na$^+$ uptake into the proteoliposomes. The transport of Na$^+$ ions was completely abolished by monensin and stimulated about four- to fivefold by either valinomycin (in the presence of K$^+$) or the uncoupler carbonylcyanide-*m*-chlorophenylhydrazone (CCCP). These results are indicative of an electrogenic Na$^+$ transport: a membrane potential which limits the transport rate is released by increasing the conductance for K$^+$ or H$^+$ across the membrane. The results also exclude the possibility that a proton gradient is an intermediate required to establish a Na$^+$ gradient by the *P. modestum* ATPase, but rather demonstrate the functioning of this enzyme as a primary Na$^+$ pump.

B. ATP-DRIVEN PROTON TRANSPORT

The ATPase activity of F_1F_0 of *P. modestum* was specifically activated by Na$^+$ ions with a K_m of about 0.8 mM.[10] Some ATPase activity was always found in the absence of Na$^+$ ions, the extent of activation being strongly influenced by pH. The activation of the ATPase by 5 mM NaCl was 1.5-fold at pH 5.0, 3.5-fold at pH 6.0, and 12-fold at pH 7.0. The pH optimum of the Na$^+$-stimulated enzyme was 7.0 and that of the Na$^+$-free enzyme was 6.0.[11] These results suggest that the enzyme is activated not only by Na$^+$ but also by H$^+$, albeit less efficiently. It is conceivable, therefore, that the ATPase catalyzes not only Na$^+$ but also H$^+$ translocation. The proton pump activity was demonstrated with reconstituted proteoliposomes by the ATP-dependent fluorescence quenching of 9-amino-6-chloro-2-methoxyacridine (ACMA),[11] which is indicative of the formation of a ΔpH (internally acidic) across the membrane. No ΔpH was formed in the presence of the uncoupler CCCP or by blocking the ATPase with DCCD. The effect of Na$^+$ ions on the quenching of ACMA fluorescence is shown in Figure 3.[11] Rate and extent of ΔpH formation were maximal in the absence of Na$^+$ and declined gradually to zero by increasing the Na$^+$ concentration from 0 to 1 mM. Half-maximal proton pumping activity was observed at 0.1 mM NaCl. The half-maximal

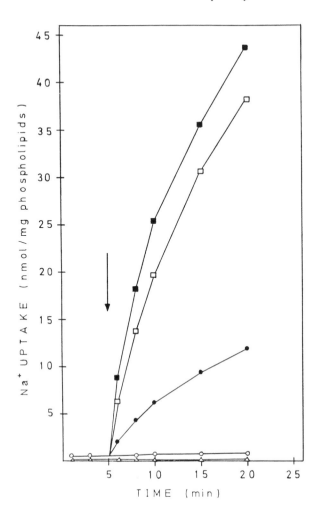

FIGURE 2. Kinetics of Na$^+$ transport into ATPase-containing proteoliposomes and effect of ionophores on Na$^+$ uptake.[10] The Na$^+$ transport was initiated by the addition of 2.5 mM K-ATP (arrow). Ionophores were contained as follows: (●) control; (■) 30 μM valinomycin; (□) 50 μM CCCP; (△) 50 μM monensin; (○) control in the absence of ATP. (Reprinted with permission form Laubinger, W. and Dimroth, P., *Biochemistry*, 27, 7531, 1988. Copyright 1988 American Chemical Society.)

rate of Na$^+$ transport was at 0.2 mM NaCl. These results suggest competition between Na$^+$ and H$^+$ for a common binding site on the enzyme and a switch from H$^+$ to Na$^+$ pumping as the Na$^+$ concentration increases. While Na$^+$ and H$^+$ are translocated at Na$^+$ concentrations of about 10^{-4} M at pH 7.0, Na$^+$ is the exclusive coupling ion at Na$^+$ concentrations of 10^{-3} M or above. Therefore, at physiological concentrations for *P. modestum* ($\sim$0.35 M NaCl), Na$^+$ will be the only coupling ion of the ATPase. It should be noted that the

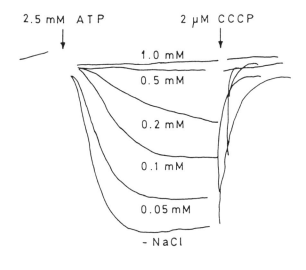

FIGURE 3. Proton pumping ATPase-containing proteoliposomes and effect of Na⁺ on this activity.[11] ACMA fluorescence quenching was initiated by adding 2.5 mM K-ATP (arrow) to reaction mixtures containing the concentrations of NaCl indicated. The quenching was released by adding 2 μM CCCP. (Reprinted with permission from Laubinger, W. and Dimroth, P., *Biochemistry*, 28, 7194, 1989. Copyright 1989 American Chemical Society.)

affinity of the ATPase for protons exceeds its affinity for Na⁺ by about three orders of magnitude. The V_{max} for Na⁺ transport is, however, about ten times the V_{max} for H⁺ transport. The efficiency of the *P. modestum* ATPase to function as a Na⁺ pump is therefore based primarily on a high V_{max} and not on a low K_m for this alkali cation.[11]

C. ATP-DRIVEN Na⁺/H⁺ TRANSPORT BY ATPase HYBRIDS

The relationship between the *P. modestum* ATPase and other F_1F_0-ATPases was further supported by the formation of functional hybrids with F_1 and F_0 components from different species. A reconstituted proteoliposomal system consisting of F_0 from *P. modestum* and F_1 from *Escherichia coli* or the thermophilic bacterium PS3 was capable of either H⁺ or Na⁺ pumping, and Na⁺ prevented H⁺ pumping at about the same concentration (>1 mM) as in the reconstituted system with the homologous *P. modestum* ATPase.[16] The ATPase of *E. coli*, on the other hand, was unable to catalyze Na⁺ translocation. A scheme summarizing these results is shown in Figure 4. We conclude from these data that the F_0 part defines the specificity for the coupling cation and that ion translocation through F_0 triggers ATP synthesis within F_1,

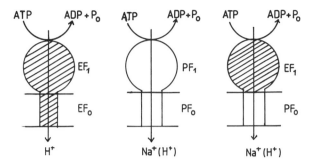

FIGURE 4.   Schematic drawing of the specificities in H^+ and Na^+ translocation by the homologous *E. coli* and *P. modestum* ATPases and an EF_1PF_0-ATPase hybrid. The abbreviations are E for *Escherichia coli* and P for *Propionigenium modestum*. F_1 and F_0 are the extrinsic and the intrinsic membrane-bound parts of the ATPase, respectively.

independent of whether the translocated ion is Na^+ or H^+. The change in ion specificity could be caused by relatively minor alterations at the cation binding site, not significantly affecting the recognition site for F_1 and, thus, the interaction of the F_1 and F_0 subcomplexes.

D. Na$^+$ (H$^+$) TRANSLOCATION THROUGH F$_0$

Proteoliposomes containing only F_0 of *P. modestum* catalyzed Na^+ or H^+ translocation (in the absence of Na^+) in response to a potassium diffusion potential induced by valinomycin.[49] The proton transport was impaired in the presence of Na^+ ions, and both Na^+ and H^+ translocation were abolished by DCCD treatment or after reconstitution of the holoenzyme with F_1. The maximal rate of Na^+ translocation by F_0 was similar to that of ATP-dependent Na^+ uptake catalyzed by F_1F_0. Unless much of our enzyme was denatured during the preparation of the F_0-containing proteoliposomes, these results argue against a highly active Na^+ (H^+) channel for *P. modestum* F_0. It is suggested that the F_0 moiety contains a cation binding site which in a distinct conformation is accessible from one side of the membrane only, and that the translocation involves a conformational change by which the cations can move to the other surface where they dissociate. In isolated F_0, a membrane potential may trigger the conformational transition since very low Na^+ conductivity is observed in its absence. In contrast, the Na^+ carrier monensin catalyzed a very rapid efflux of Na^+ ions from liposomes even in the absence of a membrane potential. Additional clear evidence was obtained recently that F_0 of *P. modestum* functions by a carrier mechanism and does not provide a channel or a pore. The protein catalyzed Na^+-counterflow independent of a membrane potential and unidirectional Na^+ flux in the direction of a negative membrane potential.[50]

VI. STRUCTURAL STUDIES

Like in *E. coli* and many other bacteria, the genes of the *P. modestum* ATPase seem to be organized in an operon-like structure. The amino acid sequences of the β-subunits of *P. modestum* and *E. coli* are 69% identical,[17] while the identities between the corresponding subunits *a, b,* and *c* are 18, 11, and 17%, respectively.[15,18] Site-directed mutagenesis studies with the *E. coli* ATPase have shown that the conserved amino acid residues of the *a* subunit Arg-210, Glu-219, and His-245 are essential for the function of H$^+$ translocation.[19,20] The conserved Arg-210 residue is found in all known ATPase sequences, including that of *P. modestum*. Position 219 contains either aspartate or histidine. ATPases from mitochondria, which have histidine at position 219, have glutamate at position 245, while ATPases with Glu-219 may have histidine, glycine, proline, or serine at position 245. Results from site-directed mutagenesis studies with the *E. coli* enzyme indicated partial retainment of H$^+$ translocation activity by replacing Glu-219 by aspartate or histidine, but no activity in the mutant with a leucine substitution. A mutation *a* His-245 → Glu produced a severe defect in the F$_0$-mediated proton translocation, but interestingly the double mutant (*a* Glu-219 → His, His-245 → Glu) yielded an ATPase complex with improved proton translocation as compared to the single mutants.[20] These results may indicate a close interaction of residues 219 and 245 of the *a* subunit and a requirement for an acidic residue in either of these positions in order to catalyze proton translocation.

The *P. modestum* ATPase has methionine at position 219 and aspartate at position 245 of the *a* subunit (*E. coli* numbering).[18] This enzyme is therefore related to the other ATPases by the retainment of an acidic amino acid in one of these positions. A methionine at position 219, however, has never been found before in an ATPase sequence. Whether this substitution is essential for the change in ion specificity is not known.

Subunit *c* contains a conserved aspartate or glutamate residue within the C-terminal membrane-spanning α-helix (Asp-62 in *E. coli*) that is specifically modified with low concentrations of DCCD.[21] The modification abolishes the ATPase activity and the translocation of protons (Na$^+$ ions in case of the *P. modestum* ATPase) through F$_0$. Asp-61 is therefore a good candidate to participate in the pathway of ion translocation. Subunit *c* of the *P. modestum* ATPase contains a glutamate residue at this position.[15] The highly conserved residues in the polar loop region between the two membrane-spanning helices, Ala-40, Arg-41, Glu-42, Pro-43 and Asp-44, are also found in the *P. modestum* ATPase.[15] This region may be important for the interaction with the F$_1$ subunits, which would explain its resistance to mutagenesis.

In summary, these results strongly support the phylogenetic relationship of the Na$^+$-translocating ATPase of *P. modestum* with the H$^+$-translocating ATPases from other organisms.

VII. GENERAL IMPLICATIONS FOR THE ATPase MECHANISM

The results presented have important implications for the mechanism of ATP synthesis and hydrolysis via F_1F_0-type enzymes. Thus, all models in which protons have a specific role that cannot be performed by another cation as well are unlikely to be correct. Our experiments clearly demonstrate that the cation binding site is situated on the F_0 sector of the molecule, not on F_1. We suppose that, in agreement with the model of Boyer,[22] the binding of the cations to these sites on F_0 from the side of high electrochemical potential triggers a conformational change which exposes the cations to the other surface of the membrane and brings about ATP synthesis on the F_1 moiety of the enzyme complex. After dissociation of the cations, the enzyme returns to its original conformation. It is also apparent that cation conduction through F_0 cannot proceed via a network of hydrogen-bonded chains ("proton wire"),[23] but more likely requires specific binding sties. As shown for crown ethers, such sites could bind either Na^+ or H_3O^+, indicating that H_3O^+ rather than H^+ may be the transported species.[24]

VIII. ATP SYNTHESIS BY *BACILLUS ALCALOPHILUS*

Alkaliphilic bacteria have the peculiar property of growing at high (>10) alkaline pH values.[25-27] The cytoplasm of these bacteria is one to two pH units less alkaline in order to protect the alkali-sensitive compounds in the cell. An acidification of the cytoplasm relative to the environment is the opposite of what is encountered in neutrophilic bacteria and has severe consequences on the bioenergetics of these cells. In neutrophilic bacteria the driving forces contributed by ΔpH and $\Delta\Psi$ are additive, but in alkaliphilic bacteria these forces are directed against each other so that the resulting $\Delta\bar{\mu}_{H^+}$ is low even at high $\Delta\Psi$ values. The system used for acidification of the cytoplasm is an electrogenic Na^+/H^+ antiport which at the same time generates a Na^+ gradient (ΔpNa^+) from the outside to the inside, adding driving force to the $\Delta\Psi$. The $\Delta\bar{\mu}_{Na^+}$ of these cells is therefore considerably higher than the $\Delta\bar{\mu}_{H^+}$ and could more appropriately drive endergonic membrane reactions. Numerous solute transport systems and the movement of the flagellar motor are, in fact, powered by the $\Delta\bar{\mu}_{Na^+}$, but surprisingly the synthesis of ATP is driven by $\Delta\bar{\mu}_{H^+}$[12,28-31] (see also Chapter IE).

A. PURIFICATION AND PROPERTIES OF THE ATPase

Inspection of the purified ATPase of *B. alcalophilus* by SDS-gel electrophoresis revealed the typical pattern of an F_1F_0-ATPase composed of eight different subunits. The α- and β-subunits of the *B. alcalophilus* ATPase reacted with antibodies raised against the corresponding subunits of the *E. coli* ATPase. Evidence for the presence of subunit c was obtained by specific labeling of this polypeptide with $[^{14}C]$-DCCD as in other F_1F_0-ATPases.[29]

A peculiar property of the *B. alcalophilus* ATPase is the very low ATPase activity with the physiological substrate Mg^{++}-ATP. The enzyme catalyzes substantial ATP hydrolysis either by substituting Ca^{++} for the Mg^{++} or after the addition of methanol. With 25% methanol the hydrolysis of Mg^{++}-ATP increases about 100-fold, indicating that the high potential of this enzyme to act as an ATP hydrolase is essentially blocked under physiological conditions.[29] The rationale for this property of the ATPase may be to protect the cells from ATP hydrolysis and concomitant proton extrusion at some transient drop of the $\Delta\bar{\mu}_{H^+}$ that would be detrimental at high environmental pH.

The kinetics of ATP hydrolysis with Ca^{++}-ATP revealed the cooperativity of three catalytic ATP binding sites.[29] These results are in accord with a model of the *E. coli* ATPase with three cooperative catalytic sites.[32]

B. RECONSTITUTION OF THE ATPase INTO PROTEOLIPOSOMES

The purified ATPase was reconstituted into proteoliposomes and the specificity for the coupling ion investigated with transport studies. The proteoliposomes were unable to catalyze Na^+ translocation to the inside under various conditions, but readily performed proton pumping, as shown by the ACMA fluorescence quenching assay.[30] The transprot was strictly dependent on Mg^{++}-ATP, not Ca^{++}-ATP, in accord with the supposition that Mg^{++} is the physiological metal ion of the ATPase, although the ATP hydrolysis activity in its presence is very low (see above).

The proteoliposomes catalyzed an ATP/[^{32}P] phosphate exchange that was insensitive to monensin but abolished by CCCP.[30] These results thus lead to the conclusion that the ATPase of *B. alcalophilus* uses H^+ and not Na^+ as the coupling ion. The same conclusion has been drawn for the F_1F_0-ATPase of *B. firmus* OF4.[31]

C. ELECTROCHEMICAL PROTON POTENTIAL OF *BACILLUS ALCALOPHILUS*

Because alkaliphilic bacilli have to synthesize ATP by a proton-coupled mechanism (see above), they encounter a severe bioenergetic problem if the $\Delta\bar{\mu}_{H^+}$ is as low as ~ -50 mV at high pH, as has been reported in the literature.[26,27,33,34] While the magnitude of the $\Delta\bar{\mu}_{H^+}$ is decisive for potential ATP synthesis mechanisms, there are a number of possible pitfalls in its accurate determination, especially in the determination of the $\Delta\Psi$ component of $\Delta\bar{\mu}_{H^+}$.[35] We therefore decided to reinvestigate the bioenergetic parameters of *B. alcalophilus* cells growing at high pH.

The bioenergetic parameters for two different strains of *B. alcalophilus* resulting from these studies are listed in Table 1.[36] While the ΔpH values are in the range of the previously reported values, our $\Delta\Psi$ and, therefore, $\Delta\bar{\mu}_{H^+}$ were considerably larger. Erroneously low $\Delta\Psi$ determined in previous investigations probably resulted from the high triphenylphosphonium bromide (TPP$^+$) concentrations (1 to 4 μM) used in these studies,[27,33,34] which may

TABLE 1
Bioenergetic Parameters of Two Different *B. alcalophilus* Strains (DSM 485 and ATCC 27647) at Alkaline pH[36]

Strain	pH$_{out}$	ΔpH	zΔpH (mV)	$\Delta\Psi$	$\Delta\bar{\mu}_{H^+}$	ATP/ADP	ΔGp kJ/mol (mV)
ATCC	10.3	1.9	+ 110	− 213	− 103	3.9	− 43.7 (− 453)
DSM	10.1	1.7	+ 97	− 206	− 109	4.5	− 44.1 (− 457)

Note: The values were obtained from harvested cells, which were suspended at high absorbance in fresh medium of pH 10.8 (strain ATCC 2764) or pH 10.5 (strain DSM 485). The phosphorylation potential (ΔGp) was converted from kilojoules per mole into millivolts by division through Faraday's constant (F = 96.5 kJ/V · mol).

From Hoffmann, A. and Dimroth, P., *Eur. J. Biochem.*, 201, 467, 1991. With permission.

lead to an underestimation of $\Delta\Psi$, especially at high values.[35] Our apparent $\Delta\Psi$ values were similar to those reported previously if determined with 1 μM TPP$^+$, but were considerably higher if the probe concentration was reduced to 10 to 50 nM.[36]

The proton motive force of − 103 mV resulting from these measurements is certainly on the lower edge of the bioenergetic scale, but could still account for ATP synthesis by conventional chemiosmosis at reasonable H$^+$/ATP stoichiometries. For *B. alcalophilus* strain ATCC 27647, the H$^+$/ATP ratio at pH 10.3 can be calculated from the phosphorylation potential of − 43 kJ/mol (− 453 mV) and $\Delta\bar{\mu}_{H^+}$ of − 103 mV to be 4.4.[36] Although an H$^+$/ATP stoichiometry of 3 is more commonly encountered, a ratio of 4 may exist in bacteria living under low-energy conditions, such as the alkaliphiles or certain anaerobes.[3] Examples are the methanogenic bacteria, which grow and synthesize ATP at $\Delta\bar{\mu}_{H^+}$ of − 120 mV,[37] and *P. modestum*, which at maximum can generate $\Delta\bar{\mu}_{Na^+}$ values of − 120 mV for theoretical reasons.[36] These low proton (sodium) motive forces allow ATP synthesis only at H$^+$(Na$^+$)/ATP coupling ratios >3.

Krulwich and colleagues[27] have proposed ATP synthesis in the alkaliphiles by a localized proton-translocation pathway within the membrane between respiratory chain components and the ATPase at high pH, but by conventional chemiosmosis at pH 8 to 9, assuming a pH-regulated gate that causes the switch between these two mechanisms.[27] If this mechanism was correct, the membrane potential should drop at a more alkaline pH, when the gate opens and the protons are short-circuited through an internal membrane pathway. In addition, the ATP synthesis should not be significantly affected by protonophorous ion channels traversing the membrane. Our experimental evidence is not in accord with this prediction: the membrane potential increases with increasing pH, and gramicidin abolishes growth and ATP synthesis at pH > 10.0.[36]

IX. MECHANISM OF ATP SYNTHESIS IN *VIBRIO ALGINOLYTICUS*

A peculiar property of *V. alginolyticus* is the expression at a slightly alkaline pH ($\sim$8.5) of a sodium ion-pumping NADH:ubiquinone oxidoreductase. The cells can grow in the presence of the protonophore CCCP if the Na$^+$ pump is functioning (pH 8.0), but become uncoupler sensitive if membrane energization depends on respiratory proton pumping (pH 6.0)[38,39] (see Chapter IF).

The electrochemical Na$^+$ gradient established by the respiratory Na$^+$ pump is taken advantage of for the movement of the flagellar motor[40] and as the driving force for various Na$^+$/solute cotransport systems.[41] In addition, marine *Vibrio* cells might use the Na$^+$ gradient established under the alkaline conditions for ATP synthesis. Supporting evidence for this hypothesis was obtained with whole cell studies with *V. alginolyticus*[42] and related *Vibrio* species.[43,44] It is puzzling, however, that this evidence finds no corroboration in experiments with membrane vesicles or proteoliposomes containing the purified ATPase of *V. alginolyticus*. Proteoliposomes prepared from the purified ATPase of *V. alginolyticus*[45] or from an *E. coli* clone expressing the gene for this enzyme[46] catalyzed an ATP-dependent proton translocation that was not influenced by NaCl concentrations as high as 500 mM. If there was competition between Na$^+$ and H$^+$ for a common binding site, as suggested for the *P. modestum* ATPase, H$^+$ translocation should be abolished when this amount of NaCl is added. In addition, direct measurement of ^{22}Na$^+$ uptake produced no positive signal, and a specific activation of ATP hydrolysis by Na$^+$, expected for a Na$^+$-translocating ATPase, was not observed.[46]

X. ATPase OF *MYCOPLASMA GALLISEPTICUM*

Sodium ions play a vital role in cell volume regulation by *M. gallisepticum*. These cells possess a Na$^+$-stimulated ATPase, suggested to function as a primary Na$^+$ pump. The ATPase was stimulated by Na$^+$ threefold at pH 8.5, but only very little at pH 5.5.[47,48] These data are reminiscent of the results obtained with the *P. modestum* ATPase and could indicate a switch between Na$^+$ and H$^+$ translocation in response to the concentrations of these two cations.

XI. CONCLUDING REMARKS

The ATPase of *P. modestum* has been the only definite example for a Na$^+$-translocating F$_1$F$_0$-ATPase until now. The only other primary pump in these bacteria is the Na^+-translocating methylmalonyl-CoA decarboxylase, and an electrochemical Na^+ ion gradient is, therefore, the only energy source by which ATP can be synthesized. In contrast, *B. alcalophilus* has no primary Na^+ pump, but pumps H^+ by the respiratory chain. A Na^+ gradient is formed by the secondary Na^+/H^+ antiporter and is used as the driving force for solute uptake and flagellar motion. ATP synthesis in this organism, however, clearly is independent of Na^+, but rather uses H^+ as the coupling ion. *V. alginolyticus* has primary pumps for H^+ and Na^+ ions and uses the Na^+ electrochemical potential to drive the flagellar motor and Na^+/solute symport processes. Additional evidence for Na^+-coupled ATP synthesis at alkaline pH was obtained from whole cell studies, but until now could not be verified on the enzymic level. Taken together, these results indicate that an overwhelming majority of bacteria use an H^+-translocating ATPase and that the Na^+-translocating ATPase may be restricted to a few species that mandatorily must synthesize ATP with a Na^+ gradient generated by a primary Na^+ ion pump.

REFERENCES

1. **Mitchell, P.**, *Chemiosmotic Coupling and Energy Transduction*, Glynn Research, Bodmin, Cornwall, England, 1968.
2. **Nicholls, D. G.**, *Bioenergetics. An Introduction to the Chemiosmotic Theory*, Academic Press, London, 1982.
3. **Harold, F. M.**, *The Vital Force. A Study of Bioenergetics*, W. H. Freeman, New York, 1986.
4. **Dimroth, P.**, Sodium ion transport decarboxylases and other aspects of sodium ion cycling in bacteria, *Microbiol. Rev.*, 51, 320, 1987.
5. **Dimroth, P.**, Mechanisms of sodium transport in bacteria, *Philos. Trans. R. Soc. London Ser. B*, 326, 465, 1990.
6. **Dimroth, P.**, Energy transductions by an electrochemical gradient of sodium ions, in *41st Colloquium Mosbach: The Molecular Basis of Bacterial Metabolism*, Hauska, G. and Thauer, R., Eds., Springer-Verlag, Berlin, 1990, 114.
7. **Dimroth, P.**, Na^+-coupled alternate to H^+-coupled primary transport systems in bacteria, *BioEssays*, 13, 463, 1991.
8. **Hilpert, W., Schink, B., and Dimroth, P.**, Life by a new decarboxylation-dependent energy conservation mechanism with Na^+ as coupling ion, *EMBO J.*, 3, 1665, 1984.
9. **Laubinger, W. and Dimroth, P.**, Characterization of the Na^+-stimulated ATPase of *Propionigenium modestum* as an enzyme of the F_1F_0 type, *Eur. J. Biochem.*, 168, 475, 1987.
10. **Laubinger, W. and Dimroth, P.**, Characterization of the ATP synthase of *Propionigenium modestum* as a primary sodium pump, *Biochemistry*, 27, 7531, 1988.
11. **Laubinger, W. and Dimroth, P.**, The sodium ion translocating adenosinetriphosphatase of *Propionigenium modestum* pumps protons at low sodium ion concentrations, *Biochemistry*, 28, 7194, 1989.
12. **Hoffmann, A., Laubinger, W., and Dimroth, P.**, Na^+-coupled ATP synthesis in *Propionigenium modestum*: is it a unique system?, *Biochim. Biophys. Acta*, 1018, 206, 1990.
13. **Schink, B. and Pfennig, N.**, *Propionigenium modestum* gen. nov. sp. nov. a new strictly anaerobic, nonsporing bacterium growing on succinate, *Arch. Microbiol.*, 133, 209, 1982.

14. **Dimroth, P.,** Na⁺-extrusion coupled to decarboxylation reactions, in *Alkali Cation Transport Systems in Prokaryotes,* Bakker, E. P., Ed., CRC Press, Boca Raton, FL, 1992, chap. ID.

15. **Ludwig, W., Kaim, G., Laubinger, W., Dimroth, P., Hoppe, J., and Schleifer, K. H.,** Sequence of subunit c of the sodium ion translocating adenosine triphosphate synthase of *Propionigenium modestum, Eur. J. Biochem.,* 193, 395, 1990.

16. **Laubinger, W., Deckers-Hebestreit, G., Altendorf, K., and Dimroth, P.,** A hybrid adenosinetriphosphatase composed of F_1 of *Escherichia coli* and F_0 of *Propionigenium modestum* is a functional sodium ion pump, *Biochemistry,* 29, 5458, 1990.

17. **Amann, R., Ludwig, W., Laubinger, W., Dimroth, P., and Schleifer, K. H.,** Cloning and sequencing of the gene encoding the beta subunit of the sodium ion translocating ATP synthase of *Propionigenium modestum, FEMS Microbiol. Lett.,* 56, 253, 1988.

18. **Kaim, G., Ludwig, W., Dimroth, P., and Schleifer, K. H.,** Sequence of subunits a and b of the sodium ion translocating adenosine triphosphate synthase of *Propionigenium modestum, Nucleic Acids Res.,* 18, 6697, 1990.

19. **Vik, S. B., Cain, B. D., Chun, K. T., and Simoni, R. D.,** Mutagenesis of the a subunit of the F_1F_0 ATPase from *Escherichia coli, J. Biol. Chem.,* 263, 6599, 1988.

20. **Cain, B. and Simoni, R. D.,** Interaction between Glu-219 and His-245 within the a subunit of F_1F_0 ATPase in *Escherichia coli, J. Biol. Chem.,* 263, 6606, 1988.

21. **Fillingame, R. H.,** Molecular mechanics of ATP synthesis by F_1F_0-type H^+-transporting ATP synthase, *Bacteria,* 12, 345, 1990.

22. **Boyer, P. D.,** A model for conformational coupling of membrane potential and proton translocation to ATP synthesis and to active transport, *FEBS Lett.,* 58, 1, 1975.

23. **Nagle, J. F. and Morowitz, H. J.,** Molecular mechanisms of proton transport in membranes, *Proc. Natl. Acad. Sci. U.S.A.,* 75, 298, 1978.

24. **Boyer, P. D.,** Bioenergetic coupling to protonmotive force, *Trends Biochem. Sci.,* 13, 5, 1988.

25. **Horikoshi, K. and Akiba, T.,** *Alcalophilic Microorganisms. A New Microbiol World.* Springer-Verlag, Berlin, 1982.

26. **Krulwich, T. A., Hicks, D. B., Seto-Young, D., and Guffanti, A. A.,** The bioenergetics of alkalophilic bacilli, *Crit. Rev. Microbiol.,* 16, 15, 1988.

27. **Ivey, D. M., Hicks, D. B., Guffanti, A. A., Sobel, G., and Krulwich, T. A.,** The problem of the electrochemical proton potential in alkaliphilic bacteria in *41st Colloquium Mosbach: The Molecular Basis of Bacterial Metabolism,* Hauska, G. and Thauer, R., Eds., Springer-Verlag, Berlin, 1990, 105.

28. **Guffanti, A. A. and Krulwich, T. A.,** ATP synthesis is driven by an imposed ΔpH or $\Delta\bar{\mu}_{H^+}$ but not by an imposed ΔpNa⁺ or $\Delta\bar{\mu}_{Na^+}$ in alkalophilic *Bacillus firmus* OF4 at high pH, *J. Biol. Chem.,* 263, 14748, 1988.

29. **Hoffmann, A. and Dimroth, P.,** The ATPase of *Bacillus alcalophilus.* Purification and properties of the enzyme, *Eur. J. Biochem.,* 194, 423, 1990.

30. **Hoffmann, A. and Dimroth, P.,** The ATPase of *Bacillus alcalophilus.* Reconstitution of energy transducing functions, *Eur. J. Biochem.,* 196, 493, 1991.

31. **Hicks, D. B. and Krulwich, T. A.,** Purification and reconstitution of the F_1F_0-ATP-synthase from alkaliphilic *Bacillus firmus* OF4, *J. Biol. Chem.,* 265, 20547, 1990.

32. **Senior, A. E.,** The proton-translocating ATPase of *Escherichia coli, Annu. Rev. Biochem.,* 59, 7, 1990.

33. **Guffanti, A. A., Susman, P., Blanco, R., and Krulwich, T. A.,** The protonmotive force and α-aminobutyric acid transport in an obligately alkaliphilic bacterium, *J. Biol. Chem.,* 253, 708, 1978.

34. **Guffanti, A. A., Finkelthal, O., Hicks, D. B., Falk, L., Sidhu, A., Garro, A., and Krulwich, T. A.,** Isolation and characterization of new facultatively alkalophilic strains of *Bacillus* species, *J. Bacteriol.,* 167, 766, 1986.

35. **Bakker, E. P.,** The role of alkali-cation transport in energy coupling of neutrophilic and acidophilic bacteria: an assessment of methods and concepts, *FEMS Microbiol. Rev.,* 75, 319, 1990.

36. **Hoffmann, A. and Dimroth, P.,** The electrochemical proton potential of *Bacillus alcalophilus, Eur. J. Biochem.,* 201, 467, 1991.

37. **Bott, M. and Thauer, R.,** Proton-motive-force-driven formation of CO from CO_2 and H_2 in methanogenic bacteria, *Eur. J. Biochem.,* 168, 407, 1987.

38. **Tokuda, H. and Unemoto, T.,** Characterization of the respiration-dependent Na^+ pump in the marine bacterium *Vibrio alginolyticus, J. Biol. Chem.,* 257, 10007, 1982.

39. **Tokuda, H. and Unemoto, T.,** Na^+ is translocated at NADH: quinone oxidoreductase segment in the respiratory chain of *Vibrio alginolyticus, J. Biol. Chem.,* 259, 7785, 1984.

40. **Dibrov, P. A., Kostyrko, V. A., Lazarova, R. L., Skulachev, V. P., and Smirnova, I. A.,** The sodium cycle. I. Na^+-dependent motility and modes of membrane energization in the marine alkalotolerant *Vibrio alginolyticus, Biochim. Biophys. Acta,* 850, 449, 1986.

41. **Unemoto, T., Tokuda, H., and Hayashi, M.,** Primary sodium pumps and their significance in bacterial energetics, *Bacteria,* 12, 33, 1990.

42. **Dibrov. P. A., Lazarova, R. L., Skulachev, V. P., and Verkhovskaya, M. L.,** The sodium cycle. II. Na^+-coupled oxidative phosphorylation in *Vibrio alginolyticus* cells, *Biochim. Biophys. Acta,* 850, 458, 1986.

43. **Sakai-Tomita, Y., Tsuda, M., and Tsuchiya, T.,** Na^+-coupled ATP synthesis in a mutant of *Vibrio parachaemolyticus* lacking H^+-translocating ATPase activity, *Biochem. Biophys. Res. Commun.,* 179, 224, 1991.

44. **Takada, Y., Fukunaga, N., and Sasaki, S.,** Na^+-driven ATP synthesis of a psychrophilic marine bacterium, *Vibrio* sp. strain ABE-1, *FEMS Microbiol. Lett.,* 82, 225, 1991.

45. **Dimitriev, O.Yu., Krasnoselskaya, I. A., Papa, S., and Skulachev, V. P.,** F_0F_1-ATPase from *Vibrio alginolyticus.* Subunit composition and proton pumping activity, *FEBS Lett.,* 284, 273, 1991.

46. **Krumholz, L. R., Esser, U., and Simoni, R. D.,** Characterization of the H^+-pumping F_1F_0 ATPase of *Vibrio alginolyticus, J. Bacteriol.,* 172, 6809, 1990.

47. **Shirvan, M. H., Schuldiner, S., and Rottem, S.,** Role of Na^+ cycle in cell volume regulation of *Mycoplasma gallisepticum, J. Bacteriol.,* 171, 4410, 1989.

48. **Shirvan, M. H., Schuldiner, S., and Rottem, S.,** Volume regulation in *Mycoplasma gallisepticum*: evidence that Na^+ is extruded via a primary Na^+ pump, *J. Bacteriol.,* 171, 4417, 1989.

49. **Kluge, C., Lambinger, W., and Dimroth, P.,** The Na^+-translocating ATPase of *Propionigenium modestum, Trans. Biochem. Soc.,* 20, 572, 1992.

50. **Kluge, C. and Dimroth, P.,** Studies on Na^+ and H^+ translocation through the F_0 part of the Na^+-translocating F_1F_0 ATPase from *Propionigenium modestum:* discovery of a membrane potential dependent step, *Biochemistry,* submitted.

Chapter IH

Na$^+$ TRANSLOCATION IN THE COURSE OF METHANOGENESIS FROM METHANOL OR FORMALDEHYDE

Volker Müller and Gerhard Gottschalk

TABLE OF CONTENTS

0-8493-6982-7/93/$0.00 + $.50

I. INTRODUCTION

Methanogenic bacteria are a phylogenetically diverse but nutritionally rather uniform group of strictly anaerobic bacteria. They are the last limb in the anaerobic food chain converting fermentation end products such as H_2 + CO_2, acetate, formate, methanol, and methylamines to methane and carbon dioxide; the latter compounds diffuse into aerobic zones and are then susceptible to aerobic oxidation.[1] Although the production of methane in anaerobic environments has been known for centuries, the organisms responsible for methanogenesis have been known only for a few decades. Their high sensitivity towards oxygen prevented for a long time the isolation of these organisms in pure culture in the laboratory. After the development of the Hungate technique[2] for the isolation and maintenance of strictly anaerobic organisms, methanogenic bacteria could be isolated and studied. The biochemical and genetic characterization of these organisms led to the discovery of a number of unusual features observed only in this group of bacteria; this finally resulted in the description of a new kingdom, the Archaea.[3] After the discovery of the pathway of methanogenesis[4] the question arose how this pathway is coupled to energy transduction. Therefore, the bioenergetics of these organisms was studied extensively in recent years in several laboratories, leading to the discovery of primary proton and sodium pumps coupled to methanogenesis.[5] It is the purpose of this chapter to summarize the results which led to the description of a sodium motive system coupled to the reversible conversion of formaldehyde to the redox level of methanol. (For a discussion of the sodium motive system coupled to the first step of methanogenesis from H_2 + CO_2, the reader is referred to Chapter IJ). Before this topic will be discussed, it is important to describe the pathway of methanogenesis.

II. PATHWAY OF METHANOGENESIS FROM DIFFERENT SUBSTRATES*

For an understanding of the bioenergetics it is important to stress that all substrates of methanogens are channeled into one central pathway. This path-

---

* $\Delta G_0'$ values used throughout the text are taken from Keltjens and van der Drift unless stated otherwise.[6]

way involves a number of unique coenzymes; some of them are found only in methanogenic bacteria. Methanofuran (MF),[7] tetrahydromethanopterin (H_4MPT),[8] and coenzyme M (HS-CoM)[9] (Figure 1) are used as C_1 carriers, the deazaflavin F_{420}[10] serves as a low-potential hydrogen carrier ($E_0' = -340$ to -350 mV), and the recently identified 7-mercaptoheptanoylthreonine phosphate (HS-HTP)[11] plays an essential role in the mechanism of energy coupling. The factor F_{430},[12] a nickel porphinoid, is the prosthetic group of the methyl-S-CoM reductase.

The sequence of reactions leading to methane from various substrates can be described on the basis of a central pathway. This pathway involves the binding of a C_1 moiety to the carriers MF, H_4MPT, and HS-CoM. The coenzyme-bound C_1 moiety is reduced from the formyl (methenyl) level over the methylene level to the methyl level and is oxidized vice versa (Figure 2). From the most oxidized intermediate, formyl-methanofuran, CO_2 is liberated, this reaction is reversible and also serves as the first step in methanogenesis from CO_2. From the most reduced intermediate, methyl-S-CoM, methane and a heterodisulfide are produced. The heterodisulfide can be reduced by either $F_{420}H_2$ or H_2 (Reactions 8 and 9 in Figure 2; see below, Section III). Depending on their redox level, different methanogenic substrates are fed into the central chain at different sites. CO_2 enters the pathway at the most oxidized level, formaldehyde at the methylene level, and methanol at the methyl level. By using different substrates and substrate combinations one can then investigate individual reactions in the overall pathway of methanogenesis. Before the bioenergetics of methanogenesis is discussed, a short description of the pathway of methanogenesis from formaldehyde and methanol will be given.

A. METHANOGENESIS FROM FORMALDEHYDE

$$HCHO + H_2O \rightarrow CO_2 + 4[H] \qquad (1)$$

$$\frac{HCHO + 4[H] \rightarrow CH_4 + H_2O}{2HCHO \rightarrow CH_4 + CO_2} \qquad \begin{matrix} (2) \\ \Delta G_0' = -184.2 \text{ kJ/mol} \quad (3) \end{matrix}$$

$$HCHO + 2H_2 \rightarrow CH_4 + H_2O \qquad \Delta G_0' = -157.3 \text{ kJ/mol} \qquad (4)$$

Formaldehyde enters the central pathway of methanogenesis at the level of methylene-H_4MPT, whereby formaldehyde reacts spontaneously in a non-enzymatic reaction with H_4MPT to yield the methylene derivative.[13] Under N_2 (Equation 3), formaldehyde has to be disproportionated: 50% is oxidized in the central pathway to CO_2 (Equation 1) to yield reducing equivalents in the form of $F_{420}H_2$ necessary to reduce the other 50% to methane (Equation 2). Under a hydrogen atmosphere (Equation 4), formaldehyde is completely reduced to methane by reducing equivalents generated via a hydrogenase; this reaction includes only exergonic reactions catalyzed by the methylene-H_4MPT

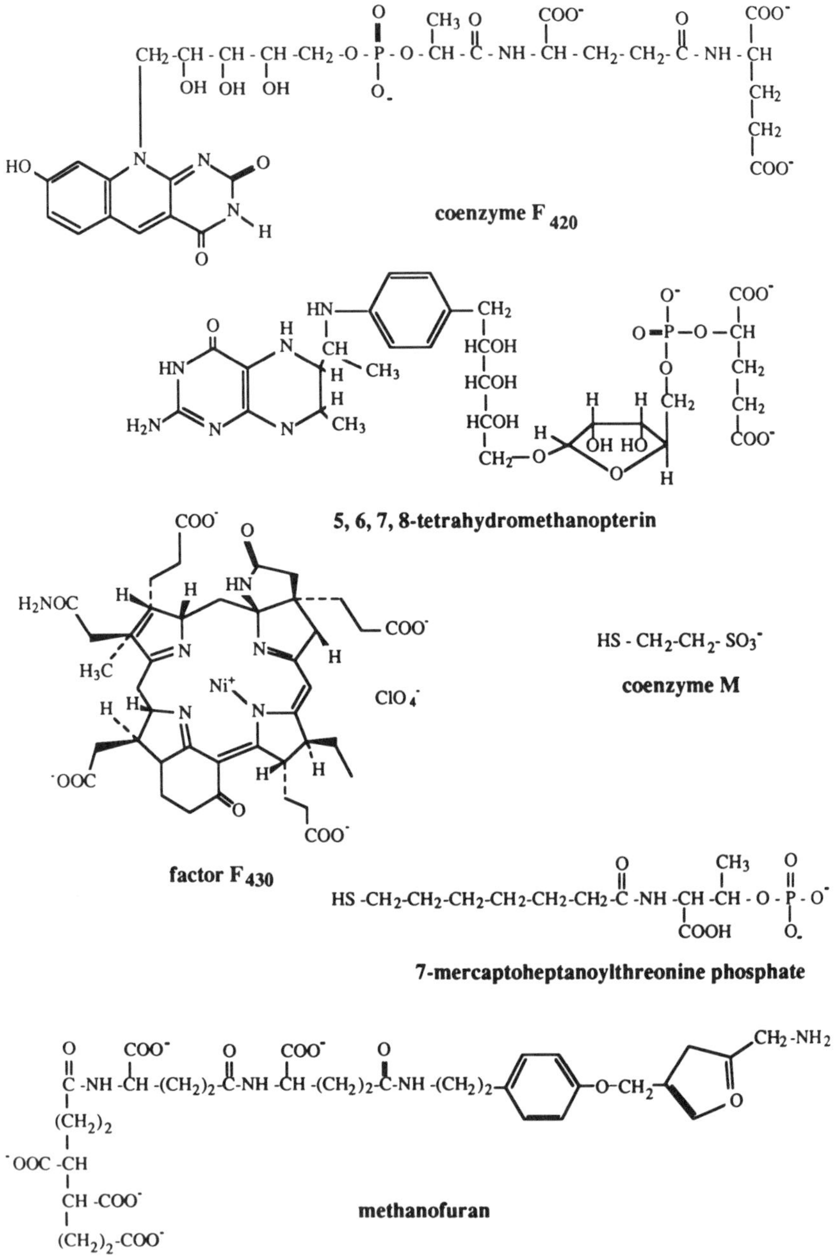

coenzyme F $_{420}$

5, 6, 7, 8-tetrahydromethanopterin

HS - CH_2-CH_2- SO_3^-

coenzyme M

factor F $_{430}$

ClO_4^-

7-mercaptoheptanoylthreonine phosphate

methanofuran

FIGURE 1.   Coenzymes involved in methanogenesis.

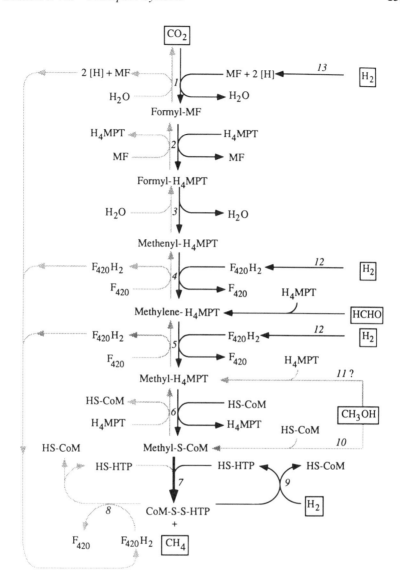

FIGURE 2. Pathway of methanogenesis. The reduction reactions are shown by black lines, the oxidation reactions by gray lines, and common reactions by thick lines. Substrates and products are shown in boxes. Enzymes involved: *1*, formylmethanofuran dehydrogenase; *2*, formylmethanofuran:H_4MPT formyltransferase; *3*, methenyl-H_4MPT cyclohydrolase; 4, methylene-H_4MPT dehydrogenase; *5*, methylene-H_4MPT reductase; *6*, methyl-H_4MPT:HS-CoM methyltransferase; *7*, methyl-S-CoM reductase; *8*, $F_{420}H_2$:heterodisulfide oxidoreductase; *9*, H_2:heterodisulfide oxidoreductase; *10*, methanol:HS-CoM methyltransferase; *11*, hypothetical methanol:H_4MPT methyltransferase; *12*, F_{420}-dependent hydrogenase; *13*, reaction not established.

reductase plus the methyl-H_4MPT:HS-CoM methyltransferase (Equation 5) and the methyl-S-CoM reductase (Equation 6):

$$Methylene\text{-}H_4MPT + H_2 + HS\text{-}CoM \rightarrow CH_3\text{-}S\text{-}CoM + H_4MPT$$

$$\Delta G_0' = -49.5 \text{ kJ/mol} \quad (5)$$

$$CH_3\text{-}S\text{-}CoM + H_2 \rightarrow CH_4 + HS\text{-}CoM \qquad \Delta G_0' = -85 \text{ kJ/mol} \qquad (6)$$

As we shall see later, both reactions are coupled to ion extrusion across the membrane.

B. METHANOGENESIS FROM METHANOL AND METHYLAMINES

$$CH_3OH + H_2O \rightarrow CO_2 + 6[H] \qquad \Delta G_0' = +17.9 \text{ kJ/mol} \qquad (7)$$

$$\underline{3CH_3OH + 6[H] \rightarrow 3CH_4 + 3H_2O \qquad \Delta G_0' = -112.5 \text{ kJ/mol}} \qquad (8)$$
$$4CH_3OH \rightarrow 3CH_4 + CO_2 + 2H_2O \qquad \Delta G_0' = -106.5 \text{ kJ/mol} \qquad (9)$$

$$CH_3OH + H_2 \rightarrow CH_4 + H_2O \qquad \Delta G_0' = -112.5 \text{ kJ/mol} \quad (10)$$

Depending on the gas atmosphere applied, methanol is disproportionated to methane and carbon dioxide (Equation 9) or exclusively reduced to methane (Equation 10). The disproportionation reaction takes place in the absence of molecular hydrogen. Of the methanol added, 25% is oxidized via the central pathway yielding the reduced form of F_{420} ($F_{420}H_2$), (Equation 7), which reduces the other 75% of the methanol to methane (Equation 8). Although it is not known whether the methyl group enters the central pathway at the level of HS-CoM or H_4MPT in the course of the oxidation sequence, recent studies indicate that the methyl group is not oxidized via free intermediates, but is channeled into the central pathway and then oxidized via the coenzyme-bound intermediates.[16,17] The conversion of the methyl group to methylene-H_4MPT (the reverse of Equation 5) is endergonic and has to be energetically connected to an exergonic reaction such as the methyl-S-CoM reductase reaction.

In the presence of H_2, methanol is exclusively reduced to methane with reducing equivalents obtained from the oxidation of hydrogen. This reaction is carried out by resting[18] or growing cells of *Methanosarcina barkeri*[19] and is the only substrate combination which supports growth of *Methanosphaera stadtmanae*,[20] underlining its importance for the bioenergetics of methanogens. Furthermore methanogenesis from methanol + H_2 is the simplest type of methane formation. Besides a methyltransferase reaction[21] which channels the methyl group in to the central pathway at the level at HS-CoM, it includes only one exergonic redox reaction catalyzed by the methyl-S-CoM reductase (Equation 6) and has been, therefore, an important substrate for the elucidation of the mechanism of energy coupling. The methyl-S-CoM reductase reaction is common to all methanogenic substrates and is accompanied by a free energy change of -85 kJ/mol, which is sufficiently large for the synthesis of ATP by electron transport phosphorylation.

III. THE REDUCTION OF METHYL-S-CoM AND THE MECHANISM OF ENERGY COUPLING

The elucidation of the mechanism of energy coupling in methanogenic bacteria came from a study with resting cells of *Methanosarcina barkeri* using methanol $+ H_2$ as a substrate.[18] Inhibitor studies have led to the conclusion that this reaction is coupled to ATP synthesis by electron transport phosphorylation with H^+ as the coupling ion. After the identification of the so-called factor B of the methylreductase system as HS-HTP, Reaction 6 was further resolved into two reactions. In the first step, methyl-S-CoM and HS-HTP are condensed, giving rise to methane and a heterodisulfide of HS-CoM and HS-HTP[22,23] which is reduced in a second step to the corresponding thiols:[24]

$$CH_3-S-CoM + HS-HTP \rightarrow CoM-S-S-HTP + CH_4 \tag{11}$$

$$CoM-S-S-HTP + 2[H] \rightarrow HS-CoM + HS-HTP \tag{12}$$

$$\Delta G_0' = -40 \text{ kJ/mol with } H_2 \text{ as reductant}^{25}$$

$$\Delta G_0' = -27 \text{ kJ/mol with } F_{420}H_2 \text{ as reductant}^{25}$$

After the development of an energetically intact inverted vesicle system of the methanogenic strain Göl (this strain is nutritionally and bioenergetically identical to *Methanosarcina barkeri*, but has the advantage of a proteinaceous cell wall which can be disrupted easily; this is a prerequisite to obtain vesicles which cannot be obtained from *M. barkeri*), the partial reactions could be analyzed for their ability to couple product formation with the synthesis of ATP.

Everted membrane vesicles are able to couple the reduction of methyl-S-CoM[26] (Equation 6) as well as the reduction of the heterodisulfide[27] (Equation 12) with the generation of a membrane potential and the synthesis of ATP. These experiments identify the heterodisulfide as the electron acceptor of the proton-motive electron transport chain; correspondingly, the enzyme was shown to be membrane bound in strain Göl.[28] As electron donors, an F_{420}-dehydrogenase[29] and a hydrogenase[30] have been identified; the

corresponding enzymes in Gö1 are membrane bound and have been purified.[31,32] The rationale for the presence of two electron donor systems is the
use of different electron carrier systems depending on the substrate utilized:
in the course of methanogenesis from methanol, reduced F_{420} is generated in
the oxidative branch and $F_{420}H_2$ serves as a reductant for the heterodisulfide
reductase; during methanogenesis form $H_2 + CO_2$, electrons for the heterodisulfide reduction are channeled directly into the electron transport chain
by a hydrogenase without the intermediary formation of $F_{420}H_2$. A scheme
of the mechanism of energy coupling in Gö1 is depicted in Figure 3.

IV. THE ROLE OF SODIUM IONS IN THE PATHWAY OF METHANOGENESIS FROM METHANOL

From the study of Perski et al.[33] it has been known since 1981 that growth
of methanogens as well as methanogenesis by resting cells is strictly dependent
on the presence of sodium ions in the medium. The analogy with other systems
suggested a participation of Na^+ in the mechanism of energy coupling, solute
transport, or intracellular enzyme catalysis. On the basis of recent bioenergetic
studies the question about the role of sodium in methanogens can be answered
at least partially.

A. SODIUM IONS ARE NOT INVOLVED IN ATP SYNTHESIS AS COUPLED TO THE METHYL-S-CoM REDUCTASE REACTION

As outlined above, methanogenesis from methanol + H_2 is coupled to
ATP synthesis by electron transport phosphorylation and is, therefore, a suitable substrate with which to address the question of the role of Na^+ in ATP
synthesis. Methanogenesis from methanol + H_2 as well as the corresponding
increase in the intracellular ATP content was independent of Na^+.[34] The same
holds true for methanogenesis from trimethylamine + H_2.[35,36] These are the
first substrates known to yield methane in the absence of Na^+. Later on it
was shown more directly that protons are extruded in the course of methanogenesis from methanol + H_2.[37] These experiments ruled out an involvement of Na^+ in ATP synthesis as coupled to the methyl-S-CoM reductase
reaction.

The disproportionation of methanol is sodium dependent; since sodium
ions are not involved in the reductive branch of methanogenesis from methanol, their site of action must be in the oxidative branch. The sodium-dependent step was identified by adding substrates and substrate combinations
with different redox levels to resting cells of *Methanosarcina barkeri*. Thus,
the sodium dependence could be overcome by addition of formaldehyde.[34]
Radiolabeling studies revealed that, in the absence of Na^+ and in the presence
of both methanol and formaldehyde, methanol is exclusively reduced and
formaldehyde exclusively oxidized according to the equation

$$2*CH_3OH + H°CHO \rightarrow 2*CH_4 + °CO_2 \qquad (13)$$

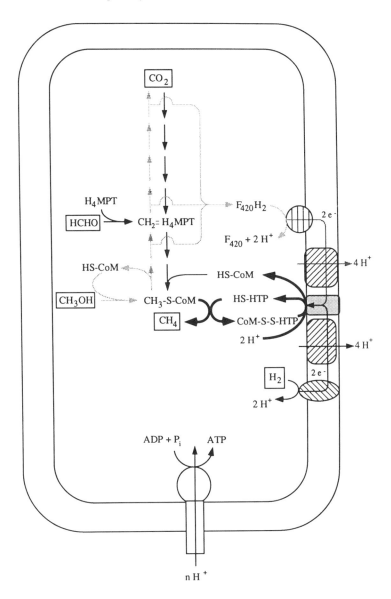

FIGURE 3. Tentative scheme of energy transduction coupled to the reduction of the heterodisulfide. The reduction reactions are shown by black lines, the oxidation reactions by gray lines, and common reactions by thick lines. During methanogenesis from methanol, $F_{420}H_2$ is generated in the oxidative branch and reoxidized in an electron transport chain with the heterodisulfide as electron acceptor (see Reaction 8 in Figure 2). During methanogenesis from methanol $+ H_2$, formaldehyde $+ H_2$, or $CO_2 + H_2$, electrons for the reduction of the heterodisulfide are provided by the hydrogenase and fed directly into the electron transport chain (see Reaction 9 in Figure 2). The proton motive force established is used to generate ATP via an ATP synthase. The nature of the membrane-integral electron carrier is not known.

These experiments indicate that the conversion of methanol to methylene-H_4MPT is the sodium-dependent step.

The conversion of the methyl group of methanol to methylene-H_4MPT (the reverse of Equation 5) is an endergonic process and must be coupled to exergonic reactions. That the coupling involves an energized membrane rather than ATP can be concluded from the fact that this reaction is not observed in a cell-free extract even in the presence of ATP and that it is impaired in whole cells in the absence of a membrane potential.[34] From the dependence of the reaction on sodium and an energized membrane the idea arose that this reaction might be driven by an electrochemical sodium ion gradient. However, at that time a sodium ion gradient across the membrane had not been demonstrated.

B. THE GENERATION OF A SECONDARY SODIUM ION GRADIENT DURING METHANOGENESIS FROM METHANOL

As a prerequisite for the validity of the hypothesis that $\Delta\bar{\mu}_{Na^+}$ serves as the driving force for the endergonic conversion of methanol to the level of formaldehyde, the presence and mode of generation of a transmembrane sodium ion gradient had to be shown. This was done by measuring the substrate-dependent extrusion of $^{22}Na^+$ from preequilibrated cell suspensions.[38] Upon addition of methanol or methanol + H_2 to resting cells,[22] Na^+ was actively extruded from the cytoplasm, thereby generating a transmembrane sodium ion gradient. Methanogenesis from methanol + H_2 was stimulated by the protonophore tetrachlorosalicylanilide, but sodium extrusion was inhibited; this is in accordance with a secondary mechanism of sodium ion translocation.

A secondary sodium ion translocation is most commonly achieved by the action of a Na^+/H^+ antiporter. Such a cation exchanger with interesting inhibitor features was previously described in *Methanobacterium thermoautotrophicum*; this exchanger was inhibited by amiloride and harmaline,[39] compounds known to inhibit the electroneutral antiporter in eukaryotes[40] (see also Chapter IJ). By measuring the ΔpNa-induced acidification of a weakly buffered medium, such an antiporter with the same inhibitor features was also shown to be present in *Methanosarcina barkeri*.[38] Using the specific inhibitor amiloride it was demonstrated that the sodium extrusion during methanogenesis from methanol results from the activity of a Na^+/H^+ antiporter energized by the primary electrochemical proton potential generated in the last step of methanogenesis. Sodium extrusion was largely inhibited by amiloride; simultaneously, the intracellular pH increased.

C. THE ELECTROCHEMICAL SODIUM ION GRADIENT AS THE DRIVING FORCE FOR THE CONVERSION OF METHANOL TO THE FORMAL REDOX LEVEL OF FORMALDEHYDE

After the elucidation of the formation of a secondary sodium ion gradient during methanogenesis from methanol, the hypothetical $\Delta\bar{\mu}_{Na^+}$ driven conversion of methanol to the level of formaldehyde was verified by different approaches.[41,42] First, an active Na^+/H^+ antiporter was shown to be required for methanogenesis from methanol. The Na^+/H^+ antiporter is responsible for the extrusion of Na^+; inhibition of the exchanger resulted in an inhibition of methanol oxidation, but not of methanol reduction. Second, a sodium ion gradient is required for methanogenesis from methanol; artificially imposed inversed Na^+ gradients ($Na^+_i > Na^+_e$) impaired methanogenesis. Furthermore the presence of the synthetic Na^+/H^+ antiporter monensin led to a dissipation of ΔpNa but had no effect on $\Delta \Psi$, thus lowering $\Delta \bar{\mu}_{Na^+}$ but not affecting $\Delta \bar{\mu}_{H^+}$ correspondingly, methanogenesis from methanol — but not from methanol + H_2 — was inhibited. In view of the presence of a natural antiporter in the membrane of methanogenic bacteria, the inhibitory effect of the synthetic antiporter monensin might be explained by different regulation patterns and by the fact that the natural antiporter catalyzes an electrogenic exchange, whereas monensin acts electroneutral (see also Chapter IJ). Third, a Na^+ influx associated with the oxidation of methanol has been demonstrated. The inhibition of the Na^+/H^+ antiporter, which is responsible for Na^+ extrusion during methanol disproportionation, led to a rapid influx of Na^+ during methanol disproportionation; simultaneously, the $\Delta \psi$ decreased, indicating an inwardly directed electrogenic translocation of Na^+. These effects were not observed with the substrate combination methanol + H_2. Fourth, methanol oxidation in the absence of a $\Delta \bar{\mu}_{Na^+}$ is impaired, but can be restored by generating an artificial Na^+ gradient across the membrane.

In summary, these experiments show that the $\Delta \bar{\mu}_{Na^+}$ is the driving force for the conversion of methanol to methylene-H_4MPT (see Figure 4). However, they do not answer the question — which of the reactions involved is the $\Delta \bar{\mu}_{Na^+}$-driven reaction. The answer came recently from a study of the reverse reaction sequence.

V. THE CONVERSION OF FORMALDEHYDE TO THE FORMAL REDOX LEVEL OF METHANOL IS ACCOMPANIED BY A PRIMARY ELECTROGENIC Na⁺ TRANSLOCATION

Most of the reactions in the central pathway of methanogenesis are reversible. If the conversion of methanol to methylene-H_4MPT is driven by $\Delta \bar{\mu}_{Na^+}$, the reverse reaction, the conversion of methylene-H_4MPT to the redox level of methanol, should be accompanied by a primary sodium extrusion. Methylene-H_4MPT is produced nonenzymatically from formaldehyde + H_4MPT present in the cells. Under an atmosphere of molecular hydrogen, formaldehyde is exclusively reduced to methane via methyl-S-CoM according to Equation 4. Thus, by comparing the substrates methanol + H_2 and formaldehyde + H_2 the reactions leading from methylene-H_4MPT to the formal redox level of methanol (methyl-H_4MPT or methyl-S-CoM) can be analyzed.

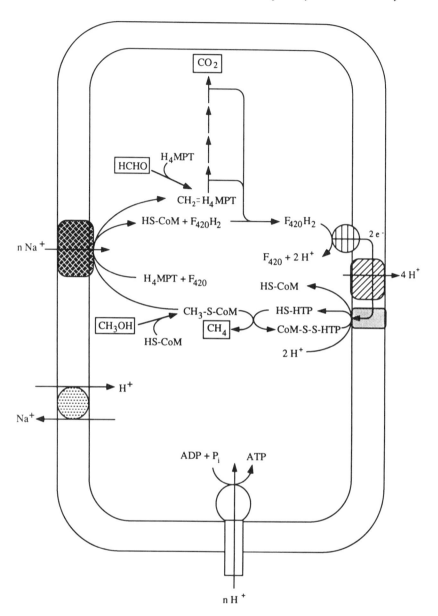

FIGURE 4. Tentative scheme of Na^+ translocation during methanogenesis from methanol. The electrochemical proton potential generated during the heterodisulfide reduction is used to generate a secondary sodium ion gradient via the Na^+/H^+ antiporter. The $\Delta\bar{\mu}_{Na^+}$ then drives the endergonic conversion of methanol to the formal redox level of formaldehyde (e.g., methylene-H_4MPT). The stoichiometry of the Na^+/H^+ antiporter is not known.

In the first step, the sodium dependence of this reaction was analyzed.[43] Methanogenesis from formaldehyde $+ H_2$, but not from methanol $+ H_2$, was strictly dependent on the presence of sodium ions, with an apparent K_m for Na^+ of 0.8 ± 0.2 mM. To elucidate the type of sodium dependence, experiments with $^{22}Na^+$ were performed.[43] Upon addition of formaldehyde $+ H_2$ to cell suspensions of *Methanosarcina barkeri*, sodium ions were actively extruded from the cytoplasm, resulting in the generation of a transmembrane Na^+ gradient of at least 1:20 (this value represents the minimal value for ΔpNa because it is based on the lowest level of Na^+_i which can be measured accurately under the conditions used). Using inhibitors, the type of sodium translocation was determined. Extrusion of Na^+ was not inhibited by protonophores or inhibitors of the Na^+/H^+ antiporter, indicating a primary mechanism. Na^+ translocation resulted in the generation of a protonophore-resistant membrane potential of -60 mV; correspondingly, a protonophore-mediated reversed ΔpH (inside acidic) of the same magnitude as the $\Delta \Psi$ was formed. Sodium extrusion was not inhibited by inhibitors of the F_1F_0-ATP synthase present in this organism. These results clearly demonstrate a primary electrogenic sodium ion translocation coupled to methanogenesis from formaldehyde $+ H_2$. Experiments with *Methanobacterium thermoautotrophicum* gave the same result.[44] A scheme of the sodium ion translocation during the conversion of formaldehyde is depicted in Figure 5.

The demonstration of this primary sodium pump has two implications for the bioenergetics of methanogenesis. First, the presence of a $\Delta \tilde{\mu}_{Na^+}$-driven methanol oxidation and a $\Delta \tilde{\mu}_{Na^+}$-generating formaldehyde conversion to the level of methanol is in accordance with a reversible Na^+ motive system connected to the central pathway. Second, the conversion of methylene-H_4MPT to the methanol level is part of the central pathway of methanogenesis and, therefore, it becomes clear that methane formation from $H_2 + CO_2$ also involves the primary sodium pump. During methanogenesis from $H_2 + CO_2$ the primary sodium pump coupled to the conversion of methylene-H_4MPT represents a second site of energy transduction. The energy stored in the electrochemical sodium ion gradient was shown to drive ATP synthesis via a secondary proton gradient built up by the Na^+/H^+ antiporter[45] (see also Chapter IJ).

VI. BIOCHEMISTRY OF THE REVERSIBLE $\Delta \tilde{\mu}_{Na^+}$-GENERATING/UTILIZING SYSTEM

The experiments described so far showed that the reversible conversion of formaldehyde to the formal redox level of methanol led to sodium transport across the membrane; however, this reaction sequence is complex, involves several enzymes (Figure 6) and has not been resolved in detail. With regard to the methanol disproportionation, the reactions leading from methanol to methylene-H_4MPT are not known. There are two enzymes known which

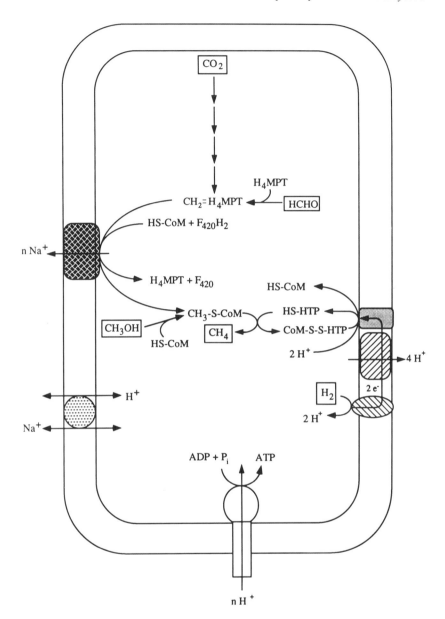

FIGURE 5. Tentative scheme of Na^+ translocation during methanogenesis from formaldehyde. The conversion of methylene-H_4MPT to the formal redox level of methanol is coupled to a primary electrogenic Na^+ translocation. Also shown is the proton-motive electron transport coupled to the heterodisulfide reduction.

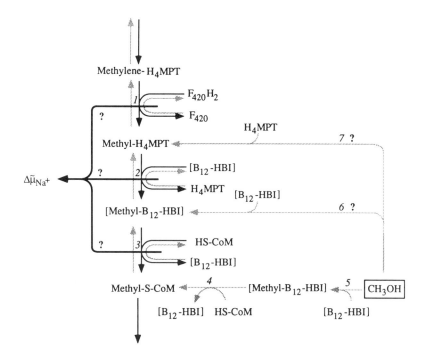

FIGURE 6. Reactions involved in the reversible conversion of methylene-H$_4$MPT to the formal redox level of methanol, possible sites for the entry of the methyl group from methanol into the central path, and possible sites of Na$^+$ translocation. (The methylene-H$_4$MPT reductase as the site of Na$^+$ translocation was ruled out recently; see text.) The reduction reactions are shown by black lines and the oxidation reactions by gray lines. [B$_{12}$-HBI], enzyme-bound 5-hydroxy-benzimidazolylcobamide; [methyl-B$_{12}$-HBI], enzyme-bound methyl-5-hydroxybenzimidazolyl-cobamide; *1*, methylene-H$_4$MPT reductase; *2*, methyl-H$_4$MPT:B$_{12}$-HBI methyltransferase (MT); *3*, MT-bound B$_{12}$-HBI:HS-CoM methyltransferase; *4*, MT$_1$-bound methyl-B$_{12}$-HBI:HS-CoM methyltransferase (MT$_2$); *5*, methanol:B$_{12}$-HBI methyltransferase (MT$_1$); *6*, hypothetical methanol:B$_{12}$-HBI methyltransferase; *7*, hypothetical methanol:H$_4$MPT methyltransferase.

transfer the methyl group from methanol to HS-CoM:[21] a methanol:5-hydroxybenzimidazolylcobamide (B$_{12}$-HBI) methyltransferase, also known as MT$_1$, which becomes methylated and a methyl-B$_{12}$-HBI:HS-CoM methyltransferase (MT$_2$), which transfers the methyl group from MT$_1$ to HS-CoM. There is no doubt that these enzymes are involved in the reductive branch; since the reductive branch is sodium independent, a function of MT$_1$ and MT$_2$ in sodium ion translocation can be excluded. With regard to the oxidative branch, there is an ongoing controversy about the site of entry of the methyl group into the central pathway. Although there is good evidence that the oxidation of the methyl group involves the C$_1$ carriers H$_4$MPT and MF,[16,17] it is not clear whether it enters the pathway at the level of HS-CoM by the action of MT$_1$ and MT$_2$ or at the level of B$_{12}$-HBI or H$_4$MPT (Figure 6).

More is known about the reverse reaction, the conversion of formaldehyde to the level of methanol, which includes two reactions: a methylene-H_4MPT reductase reaction and a methyl-H_4MPT:HS-CoM methyltransferase reaction (Figure 6):

$$\text{Methylene–}H_4MPT + F_{420}H_2 \rightarrow \text{Methyl–}H_4MPT + F_{420}$$

$$\Delta G_0' = -5.2 \text{ kJ/mol}^{46} \tag{14}$$

$$\text{Methyl–}H_4MPT + \text{HS–CoM} \rightarrow \text{Methyl–S–CoM} + H_4MPT$$

$$\Delta G_0' = -29.7 \text{ kJ/mol} \tag{15}$$

The reductase has been purified from *Methanobacterium thermoautotrophicum* strain Marburg[46] and ΔH[47] as well as from *Methanosarcina barkeri*.[48] The isolated enzyme consists of either four identical subunits with a molecular mass of 36 kDa (strain Marburg and *Methanosarcina barkeri*) or one subunit of 35 kDa (strain ΔH). The enzyme is not stimulated by Na^+ and has been isolated from the soluble fraction. In contrast to methylene-tetrahydrofolate (H_4F) reductases isolated from different organisms, it does not contain flavins or iron-sulfur centers. The enzyme is strictly specific for the physiological electron donor F_{420}; viologen dyes, NADH, or FADH do not serve as electron donors. The standard free energy change was calculated to be −5.2 kJ/mol (E_0' methylene-H_4MPT/methyl-H_4MPT = −323 mV; E_0' $F_{420}/F_{420}H_2$ = −350 mV).

Comparatively little is known about the methyl-$H_{w6.54}MPT$:HS-CoM methyltransferase. This reaction has a standard free energy change of −29.7 kJ/mol, which is sufficient to translocate approximately 2 mol of Na^+ per mole of methyl-H_4MPT converted. However, the enzyme has not yet been purified. Attempts have been made to investigate the reaction by analyzing the overall reaction from formaldehyde to methyl-S-CoM in the presence of bromoethanesulfonate, a compound which inhibits the demethylation of methyl-S-CoM. Using this approach Kengen et al.[49] have shown that this sequence of reactions proceeds in the cytoplasm, whereas Sauer[50] found that activity in the membrane fraction of *Methanobacterium thermoautotrophicum*. The methyltransferase reaction involves a corrinoid protein:[51] during the conversion of formaldehyde to methyl-S-CoM, methyl-B_{12}-HBI is formed as an intermediate. Based on these results, the presence of a methyl-H_4MPT:B_{12}-HBI methyltransferase has been proposed which accepts the methyl group from methyl-H_4MPT. In the next step, methyl-S-CoM is formed by a methyl-B_{12}-HBI:HS-CoM methyltransferase (Figure 6). By the concerted action of the two enzymes, the methyl group of H_4MPT is transferred to HS-CoM; this reaction sequence is analogous to the methyltransferases involved in the conversion of methanol. The presence of a methylcobalamin:HS-CoM methyltransferase had been reported earlier to be present not only in methylotrophic methanogens, but also in hydrogenotrophic methanogens;[52] however, the function of the enzyme in the central pathway was unclear. This enzyme might be involved in the conversion of both methanol and formaldehyde and be responsible for the transfer of the methyl group from a methylated intermediate to HS-CoM. If this is correct, the branching point for the reduction and oxidation of the methyl group might be methyl-B_{12}-HBI.

In view of a corrinoid-containing methyl-H_4MPT:HS-CoM methyltransferase, the presence of a redox-active membrane-bound corrinoid isolated from *Methanobacterium thermoautotrophicum*[53] is interesting. Although a function in the central pathway could not be assigned to this protein, antibodies against it cross-react with the soluble methyltransferase;[54] these experiments can be taken as evidence for a membrane-bound corrinoid protein which functions as a methyltransferase (see also below). This is corroborated by the work of Sauer,[50] who found the methyl-H_4MPT:methyl-S-CoM methyltransferase in the membrane fraction of *Methanobacterium thermoautotrophicum*. Preliminary experiments obtained in our laboratory also indicate that, after a gentle disruption of the cells, the methylene-H_4MPT reductase as well as the methyltransferase from strain Göl are in the particulate fraction.[55]

VII. THE METHYL-H_4MPT:HS-CoM METHYLTRANSFERASE IS A PRIMARY SODIUM PUMP IN STRAIN GÖL

Even a close look at the known biochemical data and features of the enzymes involved does not allow conclusions to be drawn as to the identity of the sodium-translocating reaction. The solution to this question came very recently from a study with inverted membrane vesicles, using the intermediates methylene- and methyl-H_4MPT as substrates for a Na⁺ translocation.[56] Upon addition of methylene-H_4MPT to washed inverted vesicles of strain Göl incubated in the presence of HS-CoM, methyl-S-CoM was formed; simultaneously, ^{22}Na⁺ was transported into the lumen of the vesicles. Inhibitor studies revealed that the sodium transport was primary and electrogenic. Sodium transport was strictly dependent on H-CoM. Interestingly, the conversion of methyl-H_4MPT to methyl-S-CoM also led to sodium transport; methyl-H_4MPT could be substituted for by methyl-H_4F. That the methyl transfer reaction involves a corrinoid protein is demonstrated by the fact that methyl-S-CoM formation as well as sodium transport is inhibited by propyliodide; upon illumination, methyl-S-CoM formation and, simultaneously, Na⁺ transport are restored.[55] The methyl-H_4MPT:HS-CoM methyltransferase was shown to be localized in the membrane fraction and its activity is strictly dependent on Na⁺.[55]

Although it is currently not known whether the methyl transfer from methyl-H_4MPT to B_{12}-HBI or from methyl-B_{12}-HBI to HS-CoM is the sodium-dependent step (Reactions 2 and 3 in Figure 6), the experiments described above demonstrate that the methyl-H_4MPT:methyl-S-CoM methyltransferase is a membrane-bound corrinoid-containing enzyme and that if functions as a primary sodium pump. This is, at least to our knowledge, the first demonstration of a methyltransferase functioning as an ion pump. The methyltransferase as a sodium-dependent site also readily explains the observed sodium dependence of methanogenesis from acetate.[57] Methanogenesis from acetate is accompanied by the generation of a transmembrane Na^+ gradient, but the actual sodium-dependent reaction is not known.[58] Recent experiments demonstrated that the methyl group of acetate enters the central pathway at the level of H_4MPT.[16] The subsequent methyltransferase reaction could then be the sodium-dependent site. However, more data are needed for a final conclusion.

VIII. RELATED OBSERVATIONS IN ACETOGENIC BACTERIA

The finding of a primary sodium pump connected to the conversion of methylene-H_4MPT to the formal redox level of methanol initiated a study on the role of Na^+ in acetogenic bacteria. These organisms are also strictly anaerobic and grow at the expense of $H_2 + CO_2$, which is converted via the acetyl-CoA pathway to acetate.[60] This pathway does not allow net ATP formation by substrate-level phosphorylation and involves the conversion of CO_2 via formyl, methenyl, methylene, and methyl intermediates. The C_2 carrier used is H_4F, which is structurally and functionally analogous to H_4MPT. In view of the sodium-motive methylene-H_4MPT conversion in methanogens, the idea arose that the corresponding reaction in acetogens might also be associated with sodium extrusion. Furthermore, membrane-bound corrinoids are found not only in a number of different methanogenic bacteria, but also in acetogenic bacteria;[59] therefore, it is tempting to speculate that the Na^+-translocating methyltransferase might be widespread in anaerobic bacteria using the acetyl-CoA or analogous pathways.

Growth as well as acetate formation by *Peptostreptococcus productus*,[61] *Acetogenium kivui*,[62] and *Acetobacterium woodii*[63] is strictly dependent on sodium ions. In experiments with resting cells of *Acetobacterium woodii* it was demonstrated that, as in methanogens, the conversion of formaldehyde to the formal redox level of methanol is the sodium-dependent step. Furthermore, acetogenesis is accompanied by the generation of a transmembrane sodium gradient of -90 mV.[63] Recent experiments obtained in our laboratory indicate that, during the conversion of formaldehyde $+$ CO to acetate by inverted vesicles of *Acetobacterium woodii*, Na^+ is extruded by a primary mechanism.[64]

IX. CONCLUDING REMARKS

The coupling of exergonic and endergonic reactions during methanogenesis from methanol involves an electrochemical sodium ion gradient which is used as the driving force for the endergonic conversion of the methyl group of methanol or trimethylamine to methyl-H$_4$MPT. This is the first example of a $\Delta\bar{\mu}_{Na^+}$ as the driving force of an endergonic reaction other than ATP synthesis or transport.

The sodium-dependent reaction sequence is reversible and functions as a primary sodium pump in the course of the conversion of methylene-H$_4$MPT to the redox level of methanol; since this reaction is part of the central pathway of methanogenesis, it represents a second site of energy transduction during methanogenesis from H$_2$ + CO$_2$. The primary sodium pump is also present in acetogens and, therefore, seems to be common in anaerobic bacteria dependent on the conversion of C$_1$-compounds via the acetyl-CoA or analogous pathways. The recent identification of the methyl-H$_4$MPT:HS-CoM methyltransferase as the primary sodium pump is the first demonstration of a methyl transfer reaction involved in ion translocation; this reaction deserves further investigation as to the nature of the components involved and the biochemical mechanism of sodium ion translocation.

ACKNOWLEDGMENT

The work from the authors' laboratory was supported by the Deutsche Forschungsgemeinschaft.

REFERENCES

1. **Wolfe, R. S.**, Microbial formation of methane, *Adv. Microbiol. Physiol.*, 6, 107, 1971.
2. **Hungate, R. E.**, in *Methods in Microbiology*, Vol. 3B, Norris, J. R. and Ribbons, D. W., Eds., Academic Press, New York, 1969, 117.
3. **Woese, C. R., Kandler, O., and Wheelis, M. L.**, Towards a natural system of organisms: proposal for the domains Archaea, Bacteria and Eucarya, *Proc. Natl. Acad. Sci. U.S.A.*, 87, 4576, 1990.
4. **Rouvière, P. and Wolfe, R. S.**, Novel biochemistry of methanogenesis, *J. Biol. Chem.*, 263, 7913, 1988.
5. **Blaut, M., Müller, V., and Gottschalk, G.**, Energetics of methanogens, in *The Bacteria, a Treatise on Structure and Function*, Vol. 12, Krulwich, T. A., Ed., Academic Press, New York, 1989, 505.
6. **Keltjens, J. T. and van der Drift, C.**, Electron transfer reactions in methanogens, *FEMS Microbiol. Rev.*, 39, 259, 1986.

7. **Leigh, J. A., Rinehart, K. L., and Wolfe, R. S.**, Methanofuran (carbon dioxide reducing factor), a formyl carrier in methane production from carbon dioxide in *Methanobacterium*, *Biochemistry*, 24, 995, 1985.

8. **Van Beelen, P., Stassen, P. M., Bosch, J. W. G., Vogels, G. D., Guijt, W., and Haasnoot, A. G.**, Elucidation of the structure of methanopterin, a coenzyme from *Methanobacterium thermoautotrophicum*, using two-dimensional nuclear-magentic-resonance techniques, *Eur. J. Biochem.*, 138, 563, 1984.

9. **Taylor, C. D. and Wolfe, R. S.**, Structure and methylation of coenzyme M ($HSCH_2CH_2SO_3$), *J. Biol. Chem.*, 249, 4879, 1974.

10. **Eirich, L. D., Vogels, G. D., and Wolfe, R. S.**, Proposed structure for coenzyme F_{420} from *Methanobacterium*, *Biochemistry*, 17, 4583, 1978.

11. **Noll, K. M., Rinehart, K. L., Jr., Tanner, R. S., and Wolfe, R. S.**, Structure of component B (7-mercaptoheptanoylthreonine phosphate) of the methylcoenzyme M methylreductase system of *Methanobacterium thermoautotrophicum*, *Proc. Natl. Acad. Sci. U.S.A.*, 83, 4238, 1986.

12. **Pfaltz, A., Jaun, B., Fässler, A., Eschenmoser, A., Jaenchen, R., Gilles, H. H., Diekert, G., and Thauer, R. K.**, F_{430} aus methanogenen Bakterien: Struktur des porphinoiden Ligandensystems, *Helv. Chim. Acta*, 65, 828, 1984.

13. **Escalante-Semerena, J. C. and Wolfe, R. S.**, Formaldehyde activation factor, tetrahydromethanopterin, a coenzyme of methanogenesis, *Proc. Natl. Acad. Sci. U.S.A.*, 81, 1976, 1984.

14. **Hippe, H., Caspari, D., Fiebig, K., and Gottschalk, G.**, Utilization of trimethylamine and other methyl compounds for growth and methane formation by *Methanosarcina barkeri*, *Proc. Natl. Acad. Sci. U.S.A.*, 76, 494, 1979.

15. **Naumann, E., Fahlbusch, K., and Gottschalk, G.**, Presence of a trimethylamine:HS-coenzyme M methyltransferase in *Methanosarcina barkeri*, *Arch. Microbiol.*, 138, 79, 1984.

16. **Fischer R. and Thauer, R. K.**, Methyltetrahydromethanopterin as an intermediate in methanogenesis from acetate in *Methanosarcina barkeri*, *Arch. Microbiol.*, 151, 459, 1989.

17. **Mahlmann, A., Deppenmeier, U., and Gottschalk, G.**, Methanofuran-b is required for CO_2 formation from formaldehyde by *Methanosarcina barkeri*, *FEMS Microbiol. Lett.*, 61, 115, 1989.

18. **Blaut, M. and Gottschalk, G.**, Coupling of ATP synthesis and methane formation from methanol and molecular hydrogen in *Methanosarcina barkeri*, *Eur. J. Biochem.*, 141, 217, 1984.

19. **Müller, V., Blaut, M. and Gottschalk, G.**, Utilization of methanol plus hydrogen by *Methanosarcina barkeri* for methanogenesis and growth, *Appl. Environ. Microbiol.*, 52, 269, 1986.

20. **Miller, T. L. and Wolin, M. J.**, *Methanosphaera stadtmaniae*, gen. nov, sp. nov.: a species that forms methane by reducing methanol with hydrogen, *Arch. Microbiol.*, 141, 116, 1985.

21. **Van der Meijden, P., Heythuysen, H. J., Pouwels, F. P., Houwen, F. P., van der Drift, C., and Vogels, G. D.**, Methyltransferase involved in methanol conversion by *Methanosarcina barkeri*, *Arch. Microbiol.*, 134, 238, 1983.

22. **Ellermann, J., Hedderich, R., Böcher, R., and Thauer, R. K.**, The final step in methane formation — investigations with highly purified methyl-CoM reductase (component C) from *Methanobacterium thermoautotrophicum* (strain Marburg), *Eur. J. Biochem.*, 171, 669, 1988.

23. **Bobik, T. A., Olson, K. D., Noll, K. M., and Wolfe, R. S.**, Evidence that the heterodisulfide of coenzyme M and 7-mercaptoheptanoylthreonine phosphate is a product of the methylreductase reaction in *Methanobacterium*, *Biochem. Biophys. Res. Commun.*, 149, 455, 1987.

24. **Hedderich, R. and Thauer, R. K.**, *Methanobacterium thermoautotrophicum* contains a soluble enzyme system that specifically catalyzes the reduction of the heterodisulfide of coenzyme M and 7-mercaptoheptanoylthreonine phosphate with H_2, *FEBS Lett.*, 234, 23, 1988.

25. **Thauer, R. K.**, Energy metabolism of methanogenic bacteria, *Biochim. Biophys. Acta*, 1018, 256, 1990.

26. **Peinemann, S., Blaut, M., and Gottschalk, G.**, ATP synthesis coupled to methane formation from methyl-CoM and H_2 catalyzed by vesicles of the methanogenic bacterial strain Göl, *Eur. J. Biochem.*, 186, 175, 1989.

27. **Peinemann, S., Hedderich, R., Blaut, M., Thauer, R. K., and Gottschalk, G.**, ATP synthesis coupled to electron transfer from H_2 to the heterodisulfide of 2-mercaptoethanesulfonate and 7-mercaptoheptanoylthreonine phosphate in vesicle preparations of the methanogenic bacterium strain Göl, *FEBS Lett.*, 263, 57, 1990.

28. **Deppenmeier, U., Blaut, M., Mahlmann, A., and Gottschalk, G.**, Membrane-bound $F_{420}H_2$-dependent heterodisulfide reductase in methanogenic bacterium strain Göl and *Methanolobus tindarius*, *FEBS Lett.*, 1, 199, 1990.

29. **Deppenmeier, U., Blaut, M., Mahlmann, A., and Gottschalk, G.**, Reduced coenzyme F_{420}:heterodisulfide oxidoreductase, a proton-translocating redox system in methanogenic bacteria, *Proc. Natl. Acad. Sci. U.S.A.*, 87, 9449, 1990.

30. **Deppenmeier, U., Blaut, M., and Gottschalk, G.**, H_2:heterodisulfide oxidoreductase, a second energy-conserving system in the methanogenic strain Göl, *Arch. Microbiol.*, 155, 272, 1991.

31. **Haase, P., Deppenmeier, U., Blaut, M., and Gottschalk, G.**, Purification and characterization of a $F_{420}H_2$-dehydrogenase from *Methanolobus tindarius*, *Eur. J. Biochem.*, 203, 527, 1992.

32. **Deppenmeier, U., Blaut, M., and Gottschalk, G.**, Purification and characterization of a F_{420} non-reactive membrane-bound hydrogenase from the methanogenic strain Göl, *Arch. Microbiol.*, 157, 505, 1992.

33. **Perski, H. J., Moll, J., and Thauer, R. K.**, Sodium dependence of growth and methane formation in *Methanobacterium thermoautotrophicum*, *Arch. Microbiol.*, 130, 319, 1981.

34. **Blaut, M., Müller, V., Fiebig, K., and Gottschalk, G.**, Sodium ions and an energized membrane required by *Methanosarcina barkeri* for the oxidation of methanol to the level of formaldehyde, *J. Bacteriol.*, 164, 95, 1985.

35. **Müller, V.**, Der Na⁺-Kreislauf in *Methanosarcina barkeri*: Aufbau und Funktion eines transmembranen elektrochemischen Na⁺-Gradienten, Ph.D. thesis, University of Göttingen, Göttingen, Germany, 1987.

36. **Müller, V., Kozianowski, G., Blaut, M., and Gottschalk, G.**, Methanogenesis from trimethylamine + H_2 by *Methanosarcina barkeri* is coupled to ATP formation by a chemiosmotic mechanism, *Biochem. Biophys. Acta*, 892, 207, 1987.

37. **Blaut, M., Müller, V., and Gottschalk, G.**, Proton translocation coupled to methanogenesis from methanol + hydrogen in *Methanosarcina barkeri*, *FEBS Lett.*, 215, 53, 1987.

38. **Müller, V., Blaut, M., and Gottschalk, G.**, Generation of a transmembrane gradient of Na⁺ in *Methanosarcina barkeri*, *Eur. J. Biochem.*, 162, 461, 1987.

39. **Schönheit, P. and Beimborn, D. B.**, Presence of a Na⁺/H⁺ antiporter in *Methanobacterium thermoautotrophicum* and its role in Na⁺ dependent methanogenesis, *Arch. Microbiol.*, 142, 354, 1985.

40. **Krulwich, T. A.**, Na⁺/H⁺ antiporters, *Biochim. Biophys. Acta*, 726, 245, 1983.

41. **Müller, V., Blaut, M., and Gottschalk, G.**, The transmembrane electrochemical gradient of Na⁺ as driving force for methanol oxidation in *Methanosarcina barkeri*, *Eur. J. Biochem.*, 172, 601, 1988.

42. **Winner, C. and Gottschalk, G.**, H_2 and CO_2 production from methanol or formaldehyde by the methanogenic bacterium strain Göl treated with 2-bromoethansulfonic acid, *FEMS Microbiol. Lett.*, 65, 259, 1989.

43. **Müller, V., Winner, C., and Gottschalk, G.,** Electron transport-driven sodium extrusion during methanogenesis from formaldehyde + H$_2$ by *Methanosarcina barkeri, Eur. J. Biochem.,* 178, 519, 1989.

44. **Kaesler, B. and Schönheit, P.,** The sodium cycle in methanogenesis. CO$_2$ reduction to the formaldehyde level in methanogenic bacteria is driven a primary electrochemical potential of Na$^+$ generated by formaldehyde reduction to CH$_4$, *Eur. J. Biochem.,* 186, 309, 1989.

45. **Kaesler, B. and Schönheit, P.,** The role of sodium ions in methanogenesis. Formaldehyde oxidation to CO$_2$ and 2 H$_2$ in methanogenic bacteria is coupled with primary electrogenic Na$^+$ translocation at a stoichiometry of 2–3 Na$^+$/CO$_2$, *Eur. J. Biochem.,* 184, 223, 1989.

46. **Ma, K. and Thauer, R. K.,** Purification and properties of N^5,N^{10}-methylenetetrahydromethanopterin reductase from *Methanobacterium thermoautotrophicum* (strain Marburg), *Eur. J. Biochem.,* 191, 187, 1990.

47. **teBrömmelstroet, B. W., Hensgens, C. M. H., Keltjens, J. T., Van der Drift, C., and Vogels, G. D.,** Purification and properties of 5,10-methylenetetrahydromethanopterin reductase, a coenzyme F$_{420}$-dependent enzyme, from *Methanobacterium thermoautotrophicum* strain ΔH, *J. Biol. Chem.,* 265, 185, 1990.

48. **Ma, K. and Thauer, R. K.,** N^5,N^{10}-methylenetetrahydromethanopterin reductase from *Methanosarcina barkeri, FEMS Microbiol. Lett.,* 70, 119, 1990.

49. **Kengen, S. W., Mosterd, J. J., Nelissen, R. L. H., Keltjens, J. T., van der Drift, C., and Vogels, G. D.,** Reductive activation of the methyl-tetrahydromethanopterin:coenzyme M methyltransferase from *Methanobacterium thermoautotrophicum* strain ΔH, *Arch. Microbiol.,* 150, 405, 1988.

50. **Sauer, F. D.,** Tetrahydromethanopterin methyltransferase, a component of the methane synthesizing complex of *Methanobacterium thermoautotrophicum, Biochem. Biophys. Res. Commun.,* 133, 177, 1986.

51. **Poirot, C. M., Kengen, S. W., Valk, E., Keltjens, J. T., van der Drift, C., and Vogels, G. D.,** Formation of methylcoenzyme M from formaldehyde by cell free extracts of *Methanobacterium thermoautotrophicum*. Evidence for the involvement of a corrinoid-containing methyltransferase, *FEMS Microbiol. Lett.,* 40, 7, 1987.

52. **Taylor, C. D. and Wolfe, R. S.,** A simplified assay for coenzyme M (HSCH$_2$CH$_2$SO$_3$). Resolution of methylcobalamin — coenzyme M methyltransferase and use of sodium borohydride, *J. Biol. Chem.,* 249, 4886, 1974.

53. **Schulz, H., Albracht, S. P. J., Coremans, J. M. C., and Fuchs, G.,** Purification and some properties of the corrinoid-containing membrane protein from *Methanobacterium thermoautotrophicum, Eur. J. Biochem.,* 171, 589, 1988.

54. **Stupperich, E., Juza, A., Eckerskorn, C., and Edelmann, L.,** An immunological study of corrinoid proteins from bacteria revealed homologous antigenic determinants of a soluble corrinoid-dependent methyltransferase and corrinoid-containing membrane proteins from *Methanobacterium* species, *Arch. Microbiol.,* 155, 28, 1990.

55. **Becher, B., Müller, V., and Gottschalk, G.,** unpublished data, 1991.

56. **Becher, B., Müller, V., and Gottschalk, G.,** The methyl-tetrahydromethanopterin:coenzyme M methyltransferase of *Methanosarcina* strain Göl is a primary sodium pump, *FEMS Microbiol. Lett.,* 91, 239, 1992.

57. **Perski, H. J., Schönheit, P., and Thauer, R. K.,** Sodium dependence of methane formation in methanogenic bacteria, *FEBS Lett.,* 143, 323, 1982.

58. **Peinemann, S., Müller, V., Blaut, M., and Gottschalk, G.,** Bioenergetics of methanogenesis from acetate by *Methanosarcina barkeri, J. Bacteriol.,* 170, 1369, 1988.

59. **Dangel, W., Schulz, H., Diekert, G., König, H., and Fuchs, G.,** Occurrence of corrinoid-containing membrane proteins in anaerobic bacteria, *Arch. Microbiol.,* 148, 52, 1987.

60. **Wood, H. G., Ragsdale, S. W., and Pezacka, E.,** The acetyl-CoA pathway of autotrophic growth, *FEMS Microbiol. Rev.,* 39, 345, 1986.

61. **Geerligs, G., Schönheit, P., and Diekert, G.,** Sodium dependent acetate formation from CO₂ in *Peptostreptococcus productus* (strain Marburg), *FEMS Microbiol. Lett.,* 57, 253, 1989.
62. **Yang, H. and Drake, H. L.,** Differential effects of sodium on hydrogen- and glucose-dependent growth of the acetogenic bacterium *Acetogenium kivui, Appl. Environ. Microbiol.,* 56, 81, 1990.
63. **Heise, R., Müller, V., and Gottschalk, G.,** Sodium dependence of acetate formation by the acetogenic bacterium *Acetobacterium woodii, J. Bacteriol.,* 171, 5473, 1989.
64. **Heise, R., Müller, V., and Gottschalk, G.,** unpublished data, 1992.

Chapter IJ

THE ROLE OF Na$^+$ IN THE FIRST STEP OF CO$_2$ REDUCTION TO METHANE IN METHANOGENIC BACTERIA

Peter Schönheit

TABLE OF CONTENTS

0-8493-6982-7/93/$0.00 + $.50

I. INTRODUCTION

Methanogenic bacteria are strictly anaerobic archaebacteria that are defined by their ability to form methane as the end product of their energy metabolism. This group of organisms had been studied extensively because of their phylogenetic status as archaebacteria (Archaea)[1-3] and because of their important ecological role in the anaerobic carbon cycle.[4] The study of the biochemistry of methanogenesis is of particular interest because of the participation of several unique types of cofactors and a number of unusual types of enzymes.[5-8] All methanogens belong to the euryarchaeotal branch of the Archaea forming phylogenetically rather distantly related major groups, the Methanococcales (e.g., *Methanococcus voltae*), the Methanobacteriales (e.g., *Methanobacterium thermoautotrophicum*), and the Methanomicrobiales (e.g., *Methanosarcina barkeri*). The phylogenetic diversity is also reflected by a great diversity in morphology, cell wall structure, and metabolic and physiological properties (for recent reviews see References 9 to 11). Most methanogens can be cultivated on mineral salt media containing various energy substrates (see below), and ammonia and sulfide generally serve as nitrogen and sulfur sources, respectively. Various trace elements proved to be essential for growth of methanogens, most notably nickel, cobalt, and molybdenum,[12-14] which are components of various enzymes and coenzymes involved in methanogenesis, and Na^+, which is involved in energy coupling (see below).

This chapter deals with some energetic aspects of methanogenesis. Methanogens gain energy by coupling the exergonic formation of methane from various substrates with the synthesis of ATP (for recent reviews on the energetics of methanogenesis see References 7 and 15 to 17). Methanogens can use only a limited number of energy substrates for growth and methane formation. Methane can be formed either from C_1-compounds such as CO_2/H_2, formate, methanol, and methylamines or from acetate.[1,9,11] Very recently pyruvate was shown to be a methanogenic substrate.[25]

Most methanogenic species form CH_4 by the reduction of CO_2 with H_2 as electron donor; about 50% of those can also use formate. Recently a few species capable of utilizing primary and secondary alcohols as electron donors for CO_2 reduction have been described.[19,20] Perski et al.[21] have observed that growth of *Methanobacterium thermoautotrophicum* on CO_2/H_2 as the sole carbon and energy source requires Na^+ ions. Subsequent studies[22] have shown that, in addition, CH_4 formation from CO_2/H_2 in resting cells of various methanogens depends on Na^+ ions. Furthermore, CH_4 formation from other energy substrates such as methanol or acetate were found to be stimulated by Na^+.[22] Thus, a specific role of the cation in the energy metabolism was likely. The function of Na^+ in the formation of CH_4 from CO_2 has been identified by the analysis of the partial reactions of the CO_2 reduction pathway with respect to their energetics and their requirement for Na^+ (see below and

Chapter IH). As a result of these studies the pathway of CO_2 reduction by H_2 to CH_4 has been divided into exergonic and endergonic reactions. The first step of the pathway, the reduction of CO_2 to a formyl moiety, which is bound as a formamide to the methanogenic coenzyme methanofuran, is an endergonic reaction ("CO_2 activation") that has to be coupled *in vivo* with the exergonic reactions of the pathway. This chapter deals with the mechanism of that coupling process and the involvement of Na^+ transport. Available evidence is summarized indicating that in methanogens the endergonic activation of CO_2 is driven by the electrochemical potential of Na^+ ions ($\Delta\bar{\mu}_{Na^+}$), which is generated in the exergonic parts of the CO_2 reduction pathway. In particular, the following items are discussed: (1) the enzymology and thermodynamics of the partial reaction involved in CO_2 reduction to methane, (2) the mechanism of energy coupling of endergonic CO_2 reduction to formyl-methanofuran and the role of Na^+ ions, and (3) the role of the Na^+/H^+ antiporter in CO_2 reduction to CH_4.

II. CO₂ REDUCTION TO CH₄

Methanogenic substrates can be divided into two groups: CH_4 is formed either by the reduction of CO_2 or by the reduction of a preformed methyl group with different electron donors. CO_2 can be reduced to CH_4 by H_2, formate, and primary or secondary alcohols as electron donors. The $\Delta G_0'$ values given were calculated from ΔGf_0 (Gibbs free energy of formation) values given by Thauer et al.;[23] CO_2 is considered to be in the gaseous state.

Almost all methanogens form methane by the reduction of CO_2 with molecular hydrogen (Equation 1):

$$4H_2 + CO_2 \rightarrow CH_4 + 2H_2O; \Delta G_0' = -131 \text{ kJ/mol } CH_4 \qquad (1)$$

Formate conversion to CH_4 involves the complete oxidation of formate to CO_2 by formate dehydrogenase, generating reducing equivalents which are subsequently used to reduce CO_2 to CH_4 (Equations 2 to 4):

$$4HCOO^- + 4H^+ \rightarrow 4CO_2 + 8[H] \qquad (2)$$

$$8[H] + 1CO_2 \rightarrow CH_4 + 2H_2O \qquad (3)$$

$$\overline{4HCOO^- + 4H^+ \rightarrow CH_4 + 3CO_2 + 2H_2O; \Delta G_0' = -145 \text{ kJ/mol } CH_4} \qquad (4)$$

Primary and secondary alcohols, such as ethanol and propanol or 2-propanol, 2-butanol, and cyclopentanol, have been shown to serve as electron donors for CO_2 reduction to CH_4. The primary alcohols (e.g., ethanol) are oxidized to their corresponding acids (Equation 5) and secondary alcohols (e.g., 2-propanol) are oxidized to their corresponding ketones (Equation 6):

$$2CH_3CH_2OH + CO_2 \rightarrow 2CH_3COO^- + 2H^+ + CH_4; \quad \Delta G_0' = -112 \text{ kJ/mol } CH_4 \quad (5)$$

$$4(CH_3)_2CHOH + CO_2 \rightarrow 4CH_3COCH_3 + CH_4 + 2H_2O; \Delta G_0' = -32 \text{ kJ/mol } CH_4 \quad (6)$$

Almost all of the alcohol-utilizing methanogens can in addition use H_2 or formate as the electron donor for CO_2 reduction to CH_4.

The second group of methanogenic substrates, e.g., methanol, methyl-amines, acetate, and pyruvate, all contain a preformed methyl group which is reduced to methane by various electron donors. The energetics of meth-anogenesis from methanol, as well as the role of Na^+ in this pathway, is discussed in Chapter IH. For the energetics of methane formation from acetate see References 24 and 25.

Formaldehyde has been shown to be a methanogenic substrate for cell suspensions of methanogens rather than a growth substrate. This is probably due to the high toxicity of this compound. In cell suspensions formaldehyde can either be reduced to methane with molecular H_2 (Equation 7) or, under a N_2 atmosphere, be disproportionated to CH_4 and CO_2 (Equation 8). Upon inhibition of methanogenesis by specific inhibitors, under a N_2 atmosphere, formaldehyde is oxidized to CO_2 and $2H_2$ (Equation 9). Studies on the en-ergetics of these various modes of formaldehyde conversion gave new insights into the mechanism of energy transduction and the role of Na^+ during CH_4 formation from CO_2 or methanol (see below and Chapter IH).

$$HCHO + 2H_2 \rightarrow CH_4 + H_2O; \quad \Delta G_0' = -158 \text{ kJ/mol } CH_4 \quad (7)$$

$$2HCHO \quad \rightarrow CH_4 + CO_2; \quad \Delta G_0' = -185 \text{ kJ/mol } CH_4 \quad (8)$$

$$HCHO + H_2O \rightarrow CO_2 + 2H_2; \quad \Delta G_0' = -27 \text{ kJ/mol } CO_2 \quad (9)$$

A. ENZYMOLOGY

The reduction of CO_2 to CH_4 proceeds in four two-electron steps via coenzyme-bound intermediates of the formal redox states of formate, for-maldehyde, and methanol. The C_1-carrying coenzymes have been identified as methanofuran (MFR), tetrahydromethanopterin (H_4MPT), and coenzyme M (H-S-CoM).[5-8] The structures of the coenzymes involved in CO_2 reduction to CH_4 are given in Figure 1 of Chapter IH. In this chapter, Figure 1 shows partial reactions, intermediates, and enzymes involved in CH_4 formation from CO_2 reduction as well as the partial reactions that require Na^+ ions. The figure also includes the site of entry of formaldehyde and methanol. There is evidence that the disproportionation of both formaldehyde and methanol to CO_2 and CH_4 is catalyzed by the same set of enzymes as are involved in

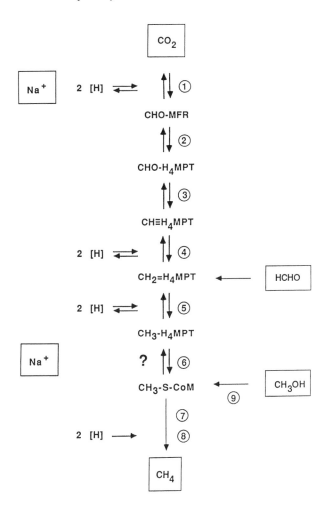

FIGURE 1. Proposed pathway of methanogenesis from CO_2, methanol, and formaldehyde: intermediates, enzymes, and sites of Na^+ dependence. There is evidence that the oxidation of methanol and formaldehyde proceeds via the reversal of reactions involved in CO_2 reduction (see arrows indicating reversibility). It is assumed that methanol binds first to coenzyme M prior to oxidation. Formaldehyde (HCHO) binds nonenzymatically to tetrahydromethanopterin. MFR, methanofuran; CHO-MFR, formyl-MFR; H_4MPT, tetrahydromethanopterin; CHO-H_4MPT, formyl-H_4MPT; CH≡H_4MPT, methenyl-H_4MPT; CH_2=H_4MPT, methylene-H_4MPT; CH_3-H_4MPT, methyl-H_4MPT; H-S-CoM, coenzyme M; CH_3-S-CoM, methyl-coenzyme M. Numbers in circles refer to enzymes involved: ① formyl-MFR dehydrogenase; ② formyl-MFR:H_4MPT formyl-transferase; ③ methenyl-H_4MPT cyclohydrolase; ④ methylene-H_4MPT dehydrogenase; ⑤ methylene-H_4MPT reductase; ⑥ methyl-H_4MPT:H-S-CoM methyltransferase; question mark indicates that reversibility has not yet been demonstrated; ⑦ methyl-coenzyme M reductase; ⑧ heterodisulfide reductase; ⑨ methanol:H-S-CoM methyltransferases. For the structures of coenzymes involved in CO_2 reduction to CH_4 see Figure 1 of Chapter IH.

CO_2 reduction to CH_4. In Table 1 the $\Delta G_0{}'$ values of the partial reactions of CO_2 reduction to CH_4 (Reactions 1 to 8) are given.

The reduction of CO_2 to formyl-MFR (CHO-MFR; formate level) is catalyzed by formyl-MFR dehydrogenase. It is assumed that first an N-substituted carbamate, carboxy-MFR, is formed from CO_2 and MFR (Equation 10), which is then reduced to formyl-MFR by formyl-MFR dehydrogenase (Equation 11; see Fig. 2). The enzyme has been purified from *Methanosarcina barkeri*[26,27] and from *Methanoautotrophicum thermoautotrophicum*.[28,29] Both enzymes are iron-sulfur proteins containing different molybdopterin derivatives.[27,29]

The reduction of formyl-MFR to N^5,N^{10}-methylene-H_4MPT (formaldehyde level) involves N^5-formyl-H_4MPT and N^5,N^{10}-methenyl-H_4MPT as intermediates. The enzymes catalyzing Reaction 2, formyl-MFR:H_4MPT formyltransferase, and Reaction 3, N^5,N^{10}-methenyl-H_4MPT cyclohydrolase, have been purified (see References 5 and 30). Reduction of N^5,N^{10}-methenyl-H_4MPT to N^5,N^{10}-methylene-H_4MPT (Reaction 4) is catalyzed in *Methanosarcina barkeri* by a coenzyme F_{420}-dependent methylene-H_4MPT dehydrogenase activity and a coenzyme F_{420}-reducing hydrogenase. In *Methanobacterium thermoautotrophicum* and other *Methanobacteriales* the reduction is catalyzed by an F_{420}-dependent enzyme and an F_{420}-independent enzyme (see Reference 3). The latter protein constitutes a novel type of hydrogenase that reduces protons to H_2 with methylene-H_4MPT as electron donor.[32]

N^5,N^{10}-methylene-H_4MPT is further converted to methyl-coenzyme M (CH_3-S-CoM; methanol level) via N^5-methyl-H_4MPT. Reduction of methylene-H_4MPT to N^5-methyl-H_4MPT (Reaction 5) is catalyzed by coenzyme F_{420}-dependent methylene-H_4MPT reductase.[8,33] The subsequent methyl-group transfer from H_4MPT to coenzyme M (H-S-CoM) (Reaction 6) is catalyzed by a methyl-H_4MPT:H-S-CoM methyltransferase.[8] This enzyme has not yet been purified.

The final step in methanogenesis is the reductive demethylation of CH_3-S-CoM to CH_4. This reduction involves two reactions. CH_3-S-CoM is reduced with N-7-mercaptoheptanoyl-threonine-phosphate (H-S-HTP); (see Figure 1) as electron donor to yield CH_4 and a heterodisulfide of H-S-CoM and H-S-HTP (CoM-S-S-HTP; Reaction 7). This reaction is catalyzed by CH_3-S-CoM reductase, which contains a nickel porphinoid, factor F_{430}, as prosthetic group (for a recent review see Reference 34). The subsequent reduction of the heterodisulfide with H_2 to yield H-S-HTP and H-S-CoM (Reaction 8) is catalyzed by CoM-S-S-HTP-dependent heterodisulfide reductase.[35]

Reducing equivalents required for CO_2 reduction are generated from the oxidation of either H_2 or formate, or primary and secondary alcohols. The activation of molecular H_2 (in Reactions 1, 2, 5, 8; Table 1) is catalyzed in methanogens by three hydrogenases,[36-39] a coenzyme F_{420}-reducing hydrogenase, a viologen dye-reducing hydrogenase, and a hydrogenase that specifically reduces protons at the expense of methylene-H_4MPT oxidation.

TABLE 1
Free Energy Changes ($\Delta G_0'$) of Reactions and Redox Potentials (E_0') of C-Intermediates Involved in CH_4 Formation from H_2 and CO_2[a]

	Reaction		$\Delta G_0'$ (kJ/reaction)	E_0' (mV)
1.	$CO_2 + MFR + H_2$	$\rightarrow$ Formyl-MFR + H_2O	+16	−497
2.	Formyl-MFR + H_4MPT	$\rightarrow$ N^5-formyl-H_4MPT + MFR	−5	
3.	N^5-methyl-H_4MPT + H^+	$\rightarrow$ N^5,N^{10}-methenyl-H_4MPT^+ + H_2O	−2	
4.	N^5,N^{10}-methenyl-H_4MPT^+ + H_2	$\rightarrow$ N^5,N^{10}-methylene-H_4MPT + H^+	−5	−386
5.	N^5,N^{10}-methylene-H_4MPT + H_2	$\rightarrow$ N^5-methyl-H_4MPT	−20	−323
6.	N^5-methyl-H_4MPT + H-S-CoM	$\rightarrow$ Methyl-S-CoM + H_4MPT	−30	
7.	Methyl-S-CoM + H-S-HTP	$\rightarrow$ CH_4 + CoM-S-S-HTP	−45	
8.	CoM-S-S-HTP + H_2	$\rightarrow$ H-S-CoM + H-S-HTP	−40	−210
1–8.	$CO_2 + 4H_2$	$\rightarrow$ $CH_4 + 2H_2O$	−131	−245

Note: MFR, methanofuran; H_4MPT, tetrahydromethanopterin; H-S-CoM, coenzyme M; H-S-HTP, *N*-7-mercaptoheptanoylthreonine phosphate.

[a] $\Delta G_0'$ and E_0' values were taken from Reference 7; E_0' (H^+/H_2) = $^-$414 mV.

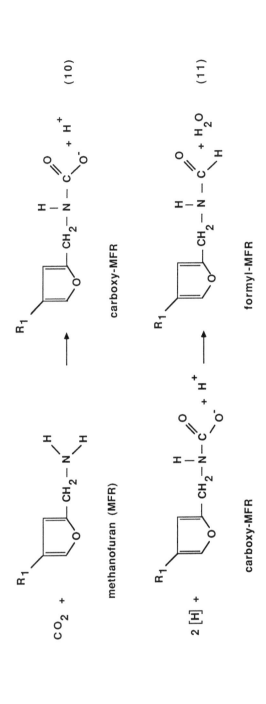

FIGURE 2. Proposed reactions involved in formyl-methanofuran synthesis from CO_2, 2[H], and methanofuran in methanogenic bacteria (see Reference 7). For the structure of R see Figure 1 of Chapter IH.

Oxidation of formate is catalyzed by formate dehydrogenase, which uses coenzyme F_{420} as a physiological electron acceptor.[40] Primary or secondary alcohols are oxidized by alcohol dehydrogenases which are specific for either NADP$^+$ or coenzyme F_{420}.

B. SITES OF ENERGY COUPLING

The reduction of CO_2 by H_2 to CH_4 (Equation 1) is an exergonic process that is coupled with the synthesis of ATP. Under standard conditions the free energy change of the reaction is -131 kJ/mol CH_4. However, in natural habitats methanogens grow at H_2 partial pressures between 1 and 10 Pa (10^{-4} to 10^{-5} atm). Under these conditions the free energy change of the reaction is about -30 to -40 kJ/mol CH_4. The energy requirement for ATP synthesis from ADP and P_i under cellular conditions is about 60 to 80 kJ/mol assuming an intracellular phosphate potential of -50 kJ/mol and a thermodynamic efficiency of ATP synthesis of 60 to 70%.[23,41] Thus, *in vivo* less than 1 mol ATP per mole CO_2 reduced to CH_4 can be formed.

On the basis of thermodynamic data (Table 1) the pathway of CO_2 reduction to CH_4 can be divided into three parts (Figure 3). First, the reduction of CO_2 to methylene-H_4MPT (CH_2=H_4MPT, formaldehyde level) is endergonic. This part is followed by two exergonic reactions, the conversion of methylene-H_4MPT to methyl-coenzyme M (CH_3-S-CoM, methanol level) and the reduction of CH_3-S-CoM to CH_4. The mode of energy coupling of these reactions has been studied in whole cells by comparing the reduction of CO_2 (Equation 1), formaldehyde (Equation 7), and methanol (Equation 12) by H_2 to CH_4 in different methanogens.[42-45]

$$H_2 + CH_3OH \rightarrow CH_4 + H_2O; \Delta G_0' = -112 \text{ kJ/mol } CH_4 \qquad (12)$$

Formaldehyde binds nonenzymatically to H_4MPT to form N^5,N^{10}-methylene-H_4MPT, CH_2=H_4MPT;[46] methanol is transferred to H-S-CoM by two methyltransferases (see Reference 9) to yield CH_3-S-CoM (Figures 1 and 3). Thus, formaldehyde and methanol could be used as substitutes for the coenzyme-bound intermediates, and the modes of energy coupling and the role of Na$^+$ in the two exergonic parts of the CO_2 reduction pathway could be studied. Furthermore, if methanogenesis is inhibited, cell suspensions catalyze the conversion of formaldehyde to CO_2 and $2H_2$ (Equation 9). This reverse reaction has been used to study the mechanism of energy coupling of the endergonic part of the pathway (Figure 3).

1. Methyl-Coenzyme M Reduction to CH_4 — Site of Primary $\Delta\bar{\mu}_{H^+}$ Generation and ATP Synthesis

The most exergonic part of the CO_2 reduction pathway is the terminal reaction, the reduction of CH_3-S-CoM to CH_4 (Table 1, Reactions 7 and 8; CH_3-S-CoM + $H_2 \rightarrow CH_4$ + H-S-CoM, $\Delta G_0 = -85$ kJ/mol). Gottschalk

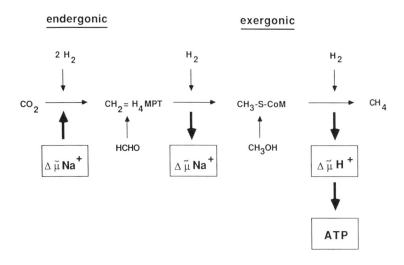

FIGURE 3. Modes of energy coupling of endergonic and exergonic partial reactions involved in methane formation from H_2 and CO_2. CH_2=H_4MPT, methylene-H_4MPT; CH_3-S-CoM, methylcoenzyme M; $\Delta\tilde{\mu}_{H^+}$, transmembrane electrochemical potential of H^+; $\Delta\tilde{\mu}_{Na^+}$, transmembrane electrochemical potential of Na^+.

and co-workers showed that this reaction was coupled with primary proton translocation, generating an electrochemical proton potential ($\Delta\tilde{\mu}_{H^+}$) which then drives the synthesis of ATP according to a chemiosmotic mechanism.[44,46] The reaction is independent of Na^+ ions, thus excluding the involvement of Na^+ translocation in both CH_3-S-CoM reduction to CH_4 and the process of ATP synthesis[48] (for details see Chapter IH).

2. Methylene-H_4MPT Conversion to Methyl-Coenzyme M — A First Site of Primary Na^+ Translocation

The mode of energy coupling of the second exergonic site, the conversion of methylene-H_4MPT to CH_3-S-CoM (Table 1, Reactions 5 and 6), has been studied by comparing the energetics of CH_4 formation from H_2/formaldehyde and H_2/methanol in cell suspensions of *Methanosarcina barkeri*. The conversion of methylene-H_4MPT to CH_3-S-CoM has been shown to be Na^+ dependent and to be coupled to primary electrogenic Na^+ extrusion by the cells. This topic is discussed in Chapter IH.

III. CO_2 REDUCTION TO THE FORMALDEHYDE LEVEL — A SECOND SITE OF PRIMARY Na^+ TRANSLOCATION

The first reaction of the CO_2 reduction pathway, the formation of formyl-MFR from CO_2, H_2(2[H]), and MFR (Reaction 1), has a $\Delta G_0'$ value of $+16$ kJ/mol. Under conditions of a low H_2 partial pressure (10 Pa) this reaction is even more endergonic $\Delta G' = +41$ kJ/mol). Since the subsequent reactions, catalyzed by formyltransferase, cyclohydrolase, and methylene-H$_4$MPT dehydrogenase, are only slightly exergonic (Figure 1; Table 1), the overall reduction of CO$_2$ to methylene-H$_4$MPT is endergonic. CH$_4$ formation from CO$_2$/H$_2$, rather than from formaldehyde/H$_2$, is sensitive to uncouplers.[42,43] This result suggests that CO$_2$ reduction to the formaldehyde level (methylene-H$_4$MPT; "CO$_2$ activation") is also endergonic *in vivo* and can proceed only when coupled to the exergonic formaldehyde reduction to CH$_4$.[42,49] Furthermore, ion movement across the membrane could play a role in this coupling process.

In principle, two mechanisms of coupling can be envisaged: (1) activation of CO$_2$ occurs at the level of the substrate at the expense of ATP hydrolysis ("substrate activation", as in acetogenic bacteria; see below); or (2) the redox potentials (E') of the electrons required for CO$_2$ reduction are pushed toward more negative values at the expense of electrochemical gradients of either H$^+$ or Na$^+$ by the mechanism of reverse electron transport ("redox activation"). Since ATP-consuming synthetases are not involved in CO$_2$ reduction to methylene-H$_4$MPT (Table 1, Reactions 1 to 4), the latter mechanism is more likely.

In the following sections, experiments are described which indicate that CO$_2$ reduction to methylene-H$_4$MPT (formaldehyde level) is driven by the primary electrochemical Na$^+$ potential generated during the reduction of formaldehyde to CH$_4$.

A. FORMALDEHYDE OXIDATION TO CO$_2$ — COUPLING TO PRIMARY $\Delta\bar{\mu}_{Na^+}$ GENERATION

Cell suspensions of various methanogens catalyze the exergonic oxidation of formaldehyde to CO$_2$ and 2H$_2$ (Equation 9) if methanogenesis is inhibited by the coenzyme M analogue 2-bromoethanesulfonate.[50,51] The following findings indicate that in *Methanosarcina barkeri* formaldehyde conversion to CO$_2$ and H$_2$ is coupled to primary electrogenic Na$^+$ extrusion, resulting in the formation of an electrochemical Na$^+$ potential $(\Delta\bar{\mu}_{Na^+})$:[50]

1. Formaldehyde oxidation was coupled to the extrusion of Na$^+$ ions; Na$^+$ extrusion was a primary process since it was inhibited by Na$^+$ ionophores and was not affected by protonophores or Na$^+$/H$^+$ antiporter inhibitors.

2. Formaldehyde oxidation was associated with the generation of a membrane potential of the order of 100 mV (inside negative) concomitant with electrogenic Na$^+$ extrusion. The membrane potential could be dissipated by sodium ionophores rather than by protonophores.

3. Formaldehyde oxidation was coupled to the synthesis of ATP in an indirect process since it was inhibited by Na$^+$/H$^+$ antiporter inhibitors, by protonophores, and by the H$^+$-ATPase inhibitor dicyclohexylcarbodiimide.

These findings indicate the following sequences of reactions: formaldehyde oxidation is coupled to the generation of primary transmembrane Na^+ potential, $\Delta\bar{\mu}_{Na^+}$, which is converted into a transmembrane proton potential, $\Delta\bar{\mu}_{Na^+}$, via Na^+/H^+ antiporter. Subsequently, $\Delta\bar{\mu}_{Na^+}$ drives the synthesis of ATP via H^+-ATP synthase (Figure 4). ATP synthesis, as measured by the increase of the intracellular ATP concentration, showed a lag which may be due to a rate-limiting step in the transformation of $\Delta\bar{\mu}_{Na^+}$ into a $\Delta\bar{\mu}_{H^+}$ (see also Reference 52). Such a lag was not observed if ATP synthesis was coupled to CH_3OH reduction to CH_4, a reaction which generates $\Delta\bar{\mu}_{H^+}$ directly.

The possibility that primary Na^+ extrusion was driven by a Na^+-translocating ATPase was excluded; the protonophore tetrachlorosalicylanilide was found to uncouple formaldehyde oxidation from ATP synthesis without affecting Na^+ extrusion. Thus, ATP cannot be the driving force for Na^+ extrusion in *Methanosarcina barkeri*. This situation may be different in *Methanococcus voltae*, which has been reported to contain a Na^+-translocating ATPase.[53,54] The function of this enzyme is, however, far from clear.

The site of Na^+ translocation during formaldehyde conversion to CO_2 and $2H_2$ is not known. After formaldehyde has reacted with H_4MPT, it is probably oxidized to CO_2 via the reverse of Reactions 1 to 4 (Table 1; Figure 1) since the oxidation of formaldehyde to CO_2 in cell extracts of *Methanosarcina barkeri* requires both methanofuran and tetrahydromethanopterin.[55,56] There is also evidence that methanol oxidation via methylene-H_4MPT to CO_2 is mechanistically and energetically the reverse of CO_2 reduction to the redox level of methanol; e.g., *Methanosarcina barkeri* contains the same set of enzymes after growth on either H_2/CO_2 or methanol (see Reference 31). Since the formyl-MFR dehydrogenase reaction is the most exergonic step of the sequence (formyl-MFR + $H_2O \rightarrow CO_2$ + H_2 + MFR, $\Delta G_0' = -16$ kJ/mol) it is assumed that electron transport from formyl-methanofuran ($E_0' = -497$ mV) as electron donor to protons ($E_0' = -414$ mV) as electron acceptors yielding CO_2 and H_2, is coupled to electrogenic Na^+ translocation. Formyl-MFR dehydrogenase is apparently membrane associated.[28] The components of this electron transport chain are not known (Figures 5 and 7).

Thus, in methanogens formaldehyde oxidation to CO_2 represents a second site of a primary $\Delta\bar{\mu}_{Na^+}$ generation. The energy stored in the gradient can be used, via $\Delta\bar{\mu}_{Na^+}$, for the synthesis of ATP (see above), or $\Delta\bar{\mu}_{Na^+}$ can drive endergonic reactions directly (see Chapter IH).

B. CO_2 REDUCTION TO METHYLENE-H_4MPT — COUPLING TO PRIMARY $\Delta\bar{\mu}_{Na^+}$ CONSUMPTION

If Reactions 1 to 4 (Table 1; Figure 1) are reversible, it seems probable that CO_2 reduction to methylene-H_4MPT is driven by an electrochemical Na^+

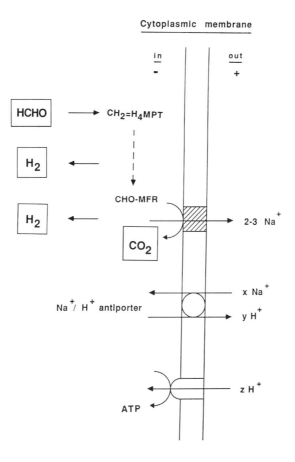

FIGURE 4. Proposed mechanism of ATP synthesis coupled to formaldehyde oxidation to CO_2 and $2H_2$. The hatched box indicate an electron transport chain catalyzing primary Na^+ extrusion (see Figure 5); $CH_2=H_4MPT$, methylene-tetrahydromethanopterin; CHO-MFR, formyl-methanofuran. The Na^+/H^+ antiporter converts $\Delta\bar{\mu}_{Na^+}$ into $\Delta\bar{\mu}_{H^+}$. The stoichiometry of Na^+ translocation was taken from Reference 50; x, y, z = unknown stoichiometric factors.

potential. Evidence for this notion has been obtained from experiments with cell suspensions of *Methanosarcina barkeri*. The effects of ionophores on CH_4 formation from H_2/CO_2 and from H_2/formaldehyde were compared and the energetic parameters of the cells were measured.[57]

1. CO_2 reduction to CH_4 was insensitive to protonophores when the activity of Na^+/H^+ antiporter was inhibited; under these conditions the electrochemical Na^+ potential ($\Delta\bar{\mu}_{Na^+}$) was -120 mV and was composed of a membrane potential of -80 mV and a chemical transmembrane Na^+ gradient of -40 mV. The electrochemical proton potential ($\Delta\bar{\mu}_{H^+}$) was almost absent (-10 mV) since an inverse pH gradient

Cytoplasmic membrane

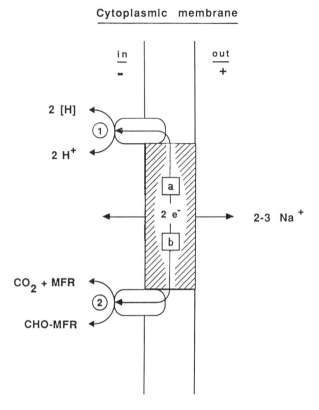

FIGURE 5. Proposed mechanism of energy coupling of CO_2 reduction to formyl-methanofuran (CHO-MFR): a reversible primary Na^+-translocating site involved in the first step of CO_2 reduction to CH_4. The hatched box indicate an electron transport chain catalyzing Na^+ translocation. ① Membrane-associated dehydrogenase; 2[H] can be H_2 or reduced coenzyme F_{420}; ② formyl-methanofuran dehydrogenase; a, b: unknown electron carriers. The stoichiometry of Na^+ translocation was taken from Reference 50.

(inside acidic) of $+70$ mV was present. Thus, almost no driving force for H^+ ions existed under these conditions and, accordingly, the cellular ATP content was low (<1 nmol/mg). This result clearly indicates that $\Delta\bar{\mu}_{Na^+}$, rather than $\Delta\bar{\mu}_{H^+}$ or ATP hydrolysis, forms the driving force for CO_2 reduction to methylene-H_4MPT; in the absence of the Na^+/H^+ antiporter inhibitor, protonophores were found to inhibit CH_4 formation from H_2/CO_2.[42,57] This inhibition is probably due to the activity of the Na^+/H^+ antiporter, which dissipates the primary $\Delta\bar{\mu}_{H^+}$ in the presence of protonophores.

2. CO_2 reduction to CH_4, rather than formaldehyde reduction to CH_4, was sensitive toward Na^+ ionophores, which dissipated $\Delta\bar{\mu}_{Na^+}$; this result also suggested CO_2 activation driven by $\Delta\bar{\mu}_{Na^+}$.

C. STOICHIOMETRIES OF PRIMARY Na$^+$ TRANSLOCATION

Further support for the conclusion that CO_2 activation is driven by $\Delta\tilde{\mu}_{Na^+}$ has come from determination of stoichiometries of primary Na$^+$ translocation coupled to CH_4 formation from H_2/CO_2 and from H_2/formaldehyde. The transport experiments were performed with whole cells of *Methanosarcina barkeri* equilibrated with ^{22}Na$^+$; the stoichiometry of Na$^+$ export was determined from the initial rate of Na$^+$ efflux and the corresponding electron transport rate, i.e., methanogenesis.[57] In the presence of protonophores and Na$^+$/H$^+$ antiporter inhibitors, CO_2 reduction to CH_4 was coupled to the extrusion of 1 to 2 mol Na$^+$ per mol CH_4; formaldehyde reduction to CH_4 was coupled to the extrusion of 3 to 4 mol Na$^+$ per mol CH_4. Thus, during CO_2 reduction to the formaldehyde level, 2 to 3 mol Na$^+$ ions must have been taken up. In accordance, the reverse reaction, formaldehyde oxidation to CO_2 and 2 H_2, was coupled to the extrusion of 2 to 3 mol Na$^+$ per mol CO_2.[50] Furthermore, the disproportionation of formaldehyde, which includes both formaldehyde reduction to CH_4 and formaldehyde oxidation to CO_2, was coupled to the primary translocation of 5 to 7 mol Na$^+$ per 2 mol formaldehyde, which equals the sum of the amounts of Na$^+$ ions extruded by the two partial reactions.[57] These stoichiometries were determined at low, but equal, rates of CH_4 or CO_2 formation. During preincubation of the cells, required for Na$^+$ equilibration across the membrane, the rates of the catabolic reactions decreased by more than 90%.[57] It is assumed that the stoichiometries of Na$^+$ translocation at high rates of CH_4 or CO_2 formation are the same as low reaction rates.

In summary, the data of Section III.A to C strongly suggest that the driving force for the endergonic reduction of CO_2 to methylene-H_4MPT is the transmembrane electrochemical sodium gradient which is generated during formaldehyde reduction to CH_4. The mechanism of coupling between the exergonic and endergonic reactions is that of a reverse electron transport: e.g., electrons coming from H_2 (E' at pH$_2$ of 10 Pa = -300 mV) have to be pushed to more negative values for the reduction of CO_2 to formyl-MFR (E_0' = -497 mV). The energy stems from the electrochemical sodium potential (Figure 7). The subsequent reduction of formyl-MFR to methylene-H_4MPT (formaldehyde level) is slightly exergonic. Thus, CO_2 is activated in methanogens by a novel mechanism which is different from that of acetogens (see below).

These conclusions have been drawn from whole cell studies. Evidently, conclusive evidence for primary Na$^+$ translocation coupled to CO_2 reduction to formyl-MFR will require the purification and reconstitution of the components involved and the demonstration of a direct role of Na$^+$ in the reactions catalyzed by the reconstituted system.

IV. ROLE OF THE Na^+/H^+ ANTIPORTER IN CO_2 REDUCTION TO CH_4 — THE Na^+ CYCLE

Perski et al.[21,22] have shown that growth and CH_4 formation from $H_2/$CO_2$ in methanogens is dependent on Na^+ ions. Moreover, CH_4 formation from other substrates, methanol and acetate, was found to require Na^+ ions.[22] Thus, a specific role of the cation in the coupling mechanism of ATP synthesis was envisaged. Later it was found that ATP synthesis driven by a potassium diffusion potential in *Methanobacterium thermoautotrophicum* was stimulated by Na^+.[58] However, a direct role of the cation in the mechanism of ATP synthesis could be excluded since CH_4 formation from H_2 and CH_3OH, as well as ATP synthesis coupled to this reaction, was not stimulated by Na^+ ions.[80] Two Na^+-dependent sites involved in CO_2 reduction to CH_4 have been identified: the conversion of methylene-H_4MPT to CH_3-S-CoM (Chapter IH) and the reduction of CO_2 to formyl-MFR (see above).

In order to explain the Na^+ stimulation of ATP synthesis driven by a diffusion potential, the presence of a Na^+/H^+ antiporter was proposed.[58] In this artificial system the acidification of the cytoplasm, which occurs in response to electrogenic potassium efflux, could be prevented by the activity of the antiporter. Subsequently, Na^+/H^+ antiporter activity has been demonstrated in both *Methanobacterium thermoautotrophicum*[59] and *Methanosarcina barkeri*.[60] An important result of these studies was that the Na^+/H^+ antiporter was inhibited by amiloride and harmaline, which have been described as inhibitors of eukaryotic Na^+/H^+ antiporters.[61] Using these inhibitors it has been shown that an active antiporter is essential for methanogenesis from H_2/CO_2.[59,62] The antiporter also accepts Li^+ instead of Na^+ since Li^+ stimulates CH_4 formation from H_2/CO_2 in the absence of Na^+.[59] In subsequent studies the use of amiloride and its more potent derivative, ethyl-isopropyl-amiloride,[63] permitted the differentiation between primary and secondary Na^+ movements during partial reactions of the CO_2 reduction pathway (see above).

The Na^+/H^+ exchange mediated by the antiporter is probably electrogenic. A Na^+/H^+ stoichiometry has been calculated in whole cells of *Methanosarcina barkeri* by measuring the membrane potential ($\Delta\Psi$) and the transmembrane chemical gradients of Na^+ and H^+ in the steady state during CH_4 formation from H_2/CO_2:[64] the $\Delta\Psi$ was about -120 mV, a ΔpH could not be detected, and the measured Na^+ gradient ($[Na^+]_{outside}/[Na^+]_{inside}$) was at least ten (higher gradients could not be quantitated correctly). The stoichiometry of an electrogenic Na^+/H^+ antiport (y Na^+/x H^+) can be estimated from Equation 13 (see Reference 65):

$$-Z \log([Na^+]_{in}/[Na^+]_{out}) = -[(x - y)\Delta\psi - xZ\Delta pH]; \quad Z = 60 \text{ mV} \quad (13)$$

On the basis of the measured values a stoichiometry of about 1 Na^+ to 1.5 H^+ was calculated.

The antiporters of both *Methanobacterium thermoautotrophicum* and *Methanosarcina barkeri* are currently being purified in our lab using a ^{3}H-labeled inhibitor that binds covalently to the Na^+/H^+ antiporter.

The role of Na^+/H^+ antiporter — In principle, an electrogenic Na^+/H^+ antiporter converts transmembrane electrochemical potentials of protons ($\Delta\bar\mu_{H^+}$) into those of Na^+ ions ($\Delta\bar\mu_{Na^+}$) and vice versa. The direction of the exchange is determined by the magnitude of the prevailing ion gradients. During methane formation from H_2/CO_2 the antiporter is involved in the conversion of a $\Delta\bar\mu_{Na^+}$ into $\Delta\bar\mu_{H^+}$ as indicated by the following findings obtained with *Methanosarcina barkeri*:[57,64] (1) in the steady state of CH_4 formation from H_2/CO_2 the value of $\Delta\bar\mu_{Na^+}$ was higher (-180 mV) than that of $\Delta\bar\mu_{H^+}$ (-120 mV); (2) Na^+/H^+ antiporter inhibitors inhibited all reactions that are coupled with primary Na^+ extrusion, the degree of the inhibition being roughly proportional to the amount of Na^+ ions translocated — for example, formaldehyde oxidation to CO_2 and $2H_2$ (2 to 3 mol Na^+ per mole CO_2), formaldehyde reduction to CH_4 (3 to 4 mol Na^+ per mole CH_4), or formaldehyde disproportionation to CO_2 and CH_4 (5 to 7 mol Na^+ per 2 mol formaldehyde) were inhibited by either 30, 50, or 80%, respectively. This inhibition was reversed by Na^+ ionophores. These data indicate that the Na^+/H^+ antiporter is involved in catalyzing Na^+ backflow into the cell, in exchange for protons, thereby closing the Na^+ cycle. Accordingly, CO_2 reduction to CH_4, which is coupled to the net primary extrusion of only 1 to 2 mol Na^+ per mol CH_4, is partially inhibited (ca. 30%) by Na^+/H^+ antiporter inhibitors.[57]

The Na^+ cycle — The data described above fit into a scheme showing how the transmembrane transport cycles of H^+ and Na^+ couple exergonic and endergonic reactions to each other during CH_4 formation from CO_2 (Figure 6): the exergonic reactions are the conversion of methylene-H_4MPT (CH_2=H_4MPT, formaldehyde level) to methyl-coenzyme M (CH_3-S-CoM, methanol level) and the reduction of CH_3-S-CoM to CH_4. The former reaction is coupled to primary extrusion of Na^+ ions; the latter is coupled to primary extrusion of protons. During methylene-H_4MPT conversion to CH_3-S-CoM, more Na^+ ions are extruded than are taken up during CO_2 reduction to methylene-H_4MPT. These "extra" Na^+ ions exchange against H^+ via the Na^+/H^+ antiporter. The resulting $\Delta\bar\mu_{H^+}$ can be used for ATP synthesis. Thus, the necessity of a Na^+/H^+ antiporter in CH_4 formation from H_2/CO_2 follows directly from the imbalance of the different Na^+ stoichiometries of the partial reactions involved. In summary, coupling of exergonic and endergonic reactions in CH_4 formation from H_2 and CO_2 involves both primary Na^+ and H^+ cycles, both of which are linked by the electrogenic Na^+/H^+ antiporter.

According to this scheme the antiporter has a stoichiometric function in CH_4 formation from CO_2. Since it generates $\Delta\bar\mu_{H^+}$ from a primary $\Delta\bar\mu_{Na^+}$ it is involved in ATP synthesis. Thus, the antiporter, in combination with the H^+-translocating ATP synthase, allows the complete conversion of the

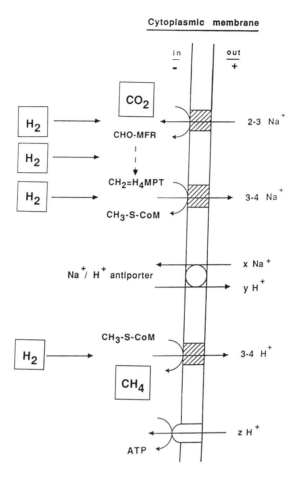

FIGURE 6. Proposed function of electrochemical H^+ and Na^+ potentials in energy conservation coupled to CH_4 formation from H_2/CO_2. The Na^+/H^+ antiporter is involved in the generation of $\Delta\bar{\mu}_{H^+}$ from $\Delta\bar{\mu}_{Na^+}$. CHO-MFR, formyl-methanofuran; $CH_2=H_4MPT$, methylene H_4MPT; CH_3-H_4MPT, methyl-tetrahydromethanopterin; CH_3-S-CoM, methyl-coenzyme M. The hatched boxes indicate membrane-bound electron transport chains catalyzing Na^+ or H^+ translocation. The stoichiometries of Na^+ and H^+ translocation were taken from References 47, 50, and 57. x, y, z = unknown stoichiometric factors.

electrochemical Na^+ potential for the formation of ATP via electron transport phosphorylation. This is important since the ATP yield of methanogenesis from CO_2 is low (see above).

One can only speculate about the question why gradients of both H^+ and Na^+ are involved in driving endergonic reactions during CO_2 reduction to CH_4. Such a situation would allow an independent regulation of $\Delta\bar{\mu}_{Na^+}$-driven electron transport (methanogenesis) and of $\Delta\bar{\mu}_{H^+}$-driven ATP synthesis. This

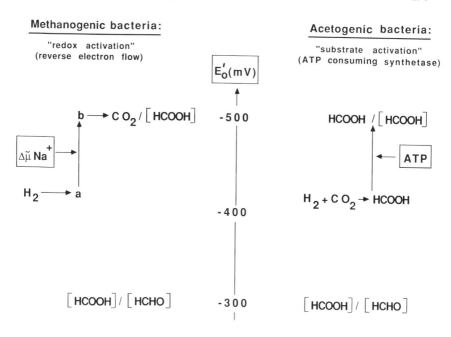

FIGURE 7. Different mechanisms of endergonic CO_2 activation in methanogens vs. acetogens. [HCOOH], coenzyme-bound formate (in methanogens, formyl-methanofuran or formyl-tetrahydromethanopterin; in acetogens, formyl-tetrahydrofolate); [HCHO], coenzyme-bound formaldehyde (in methanogens, methylene-tetrahydromethanopterin; in acetogens, methylene-tetrahydrofolate). It is assumed that both the redox potentials of the CO_2/[HCOOH] couple and of the [HCOOH]/[HCHO] couple in methanogens and acetogens are equal. $\Delta\tilde{\mu}_{Na^+}$, transmembrane electrochemical potential of Na^+; a, b = unknown electron carriers involved in $\Delta\tilde{\mu}_{Na^+}$-driven reverse electron flow.

might be important in certain metabolic situations in nature, e.g., when growth of the bacteria is initiated after periods of energy starvation.

Besides a stoichiometric function, a role of the methanogenic Na^+/H^+ antiporter in pH regulation (for reviews see References 61 and 66 and Chapter IA) has also been proposed[59] to explain Na^+-dependent ΔpH formation (inside alkaline at pH 5, inside acidic at pH 9) in cell suspensions of *Methanobacterium thermoautotrophicum*. However, the data are difficult to interpret because of interference of Na^+-dependent methanogenesis from H_2/CO_2. (For a discussion of pH regulation in methanogens see Reference 16.)

Role of the Na^+/H^+ antiporter in methanol disproportionation — During methane formation from methanol (see Chapter IH) 1 mol of methanol is oxidized to CO_2 to generate 6[H], which reduces 3 mol of methanol to CH_4. There is evidence that the oxidation of methanol to CO_2 is mechanistically and energetically the reverse of CO_2 reduction to the methanol level (see Figure 1). Thus, during methanol oxidation to CO_2 the Na^+-translocating reactions and the Na^+/H^+ antiporter operate in the direction opposite of that described for the CO_2 reduction pathway. The stoichiometries of Na^+ translocation can explain, for example, why methanol oxidation requires secondary transport of Na^+ ions (see Chapter IH). During methanol oxidation to the formaldehyde level (methylene-H_4MPT) three to four Na^+ are taken up; formaldehyde oxidation to CO_2 and $2H_2$ is coupled to the extrusion of only two to three Na^+. Thus, the "missing" Na^+ ions have to be provided by the Na^+/H^+ antiporter via transformation of a primary $\Delta\bar{\mu}_{H^+}$ into a secondary $\Delta\bar{\mu}_{Na^+}$. The endergonic methanol oxidation is therefore energetically linked to CH_3OH reduction, which generates a primary $\Delta\bar{\mu}_{H^+}$. Thus, the role of the Na^+/H^+ antiporter in methanol disproportionation is to generate $\Delta\bar{\mu}_{Na^+}$ from $\Delta\bar{\mu}_{H^+}$ by mediating Na^+ extrusion.

V. MECHANISM OF CO_2 ACTIVATION IN METHANOGENS VS. ACETOGENS

Several reactions of CO_2 reduction to CH_4 by methanogens resemble those of CO_2 reduction to acetate by acetogenic bacteria. Most anaerobic acetogenic eubacteria form acetate from CO_2 and H_2 according to Equation 14:

$$4H_2 + 2CO_2 \rightarrow CH_3COO^- + H^+ + 2H_2O;\ \Delta G_0' = -95\ \text{kJ/mol acetate} \quad (14)$$

Acetate is synthesized via the carbon monoxide/acetyl-CoA pathway (for recent reviews see References 24 and 67 to 69): one molecule of CO_2 is reduced via formyl-tetrahydrofolate (formyl-H_4F) and methylene-H_4F to methyl-H_4F, and a second CO_2 is reduced to an enzyme-bound carbonyl. The methyl group is condensed with the carbonyl group and coenzyme A (CoA) to form acetyl-CoA. Acetyl-CoA is then converted to acetate via acetyl phosphate, and ATP is formed in the acetate kinase reaction.

Both the reduction of CO_2 to acetate and that of CO_2 to CH_4 start with the endergonic CO_2 reduction to coenzyme-bound formate. In acetogens, CO_2 is converted to formyl-H_4F (formate level), which is energetically similar to CO_2 conversion to formyl-H_4MPT or formyl-MFR in methanogens. However, the mode of energy coupling is different. In acetogens, this endergonic reaction is driven at the expense of ATP hydrolysis: CO_2 is reduced to free formate via formate dehydrogenase; formate is then activated in the ATP-consuming formyl-H_4F synthetase reaction to yield formyl-H_4F. Thus, in acetogens, CO_2 activation is accomplished by the mechanism of "substrate activation" rather than by reverse electron flow ("redox activation") as shown for methanogens (Figure 7). Why these two completely different pathways and mechanisms of CO_2 activation have evolved is presently not understood.

ACKNOWLEDGMENTS

The work performed in the author's laboratory was supported by grants from the Deutsche Forschungsgemeinschaft and from the Fonds der Chemischen Industrie.

REFERENCES

1. **Balch, W. E., Fox, G. E., Magrum, L. J., Woese, C. R., and Wolfe, R. S.,** Methanogens: reevaluation of a unique biological group, *Microbiol. Rev.,* 43, 260, 1979.
2. **Woese, C. R.,** Bacterial evolution, *Microbiol. Rev.,* 51, 221, 1987.
3. **Woese, C. R., Kandler, O., and Wheelis, M. L.,** Towards a natural system of organisms: proposal for the domains archaea, bacteria, and eucarya, *Proc. Natl. Acad. Sci. U.S.A.,* 87, 4576, 1990.
4. **Oremland, R. S.,** Biogeochemistry of methanogenic bacteria, in *Biology of Anaerobic Microorganisms,* Zehnder, A. J. B., Ed., John Wiley & Sons, New York, 1988, 641.
5. **Rouvière, P. E. and Wolfe, R. S.,** Novel biochemistry of methanogenesis, *J. Biol. Chem.,* 263, 7913, 1988.
6. **DiMarco, A. A., Bobik, T. A., and Wolfe, R. S.,** Unusual coenzymes of methanogenesis, *Annu. Rev. Biochem.,* 59, 355, 1990.
7. **Thauer, R. K.,** Energy metabolism of methanogenic bacteria, *Biochim. Biophys. Acta,* 1018, 256, 1990.
8. **Keltjens, J. T., te Brömmelstroet, B. W., Kengen, S. W. M., van der Drift, C., and Vogels, G. D.,** 5,6,7,8-Tetrahydromethanopterin-dependent enzymes involved in methanogenesis, *FEMS Microbiol. Rev.,* 87, 327, 1990.
9. **Vogels, G. D., Keltjens, J. T., and van der Drift, C.,** Biochemistry of methane production, in *Biology of Anaerobic Microorganisms,* Zehnder, A. J. B., Ed., John Wiley & Sons, New York, 1988, 707.
10. **Fuchs, G. and Stupperich, E.,** Carbon assimilation pathways in archaebacteria, *Sys. Appl. Microbiol.,* 7, 364, 1977.
11. **Jones, W. J., Nagle, D. P., and Whitman, W. B.,** Methanogens and the diversity of archaebacteria, *Microbiol. Rev.,* 51, 135, 1987.
12. **Schönheit, P., Moll, J., and Thauer, R. K.,** Nickel, cobalt, and molybdenum requirement for growth of Methanobacterium thermoautotrophicum, *Arch. Microbiol.,* 123, 105, 1979.
13. **Scherer, P. and Sahm, H.,** Effect of trace elements and vitamins on the growth of *Methanosarcina barkeri, Acta Biotechnol.,* 1, 57, 1981.
14. **Whitman, W. B., Ankwanda, E., and Wolfe, R. S.,** Nutrition and carbon metabolism of *Methanococcus voltae, J. Bacteriol.,* 149, 852, 1982.
15. **Keltjens, J. T. and van der Drift, C.,** Electron transfer reactions in methanogens, *FEMS Microbiol. Rev.,* 39, 259, 1986.
16. **Blaut, M., Müller, V., and Gottschalk, G.,** Energetics of methanogens, in *The Bacteria,* Vol. 12, Krulwich, T. A., Ed., Academic Press, New York, 1990, 505.

17. **Gottschalk, G. and Blaut, M.,** Generation of proton and sodium motive forces in methanogenic bacteria, *Biochim. Biophys. Acta,* 1018, 263, 1990.

18. **Bock, A.-K. and Schönheit, P.,** unpublished data.

19. **Widdel, F.,** Growth of methanogenic bacteria in pure culture with 2-propanol and other alcohols as hydrogen donors, *Appl. Environ. Microbiol.,* 51, 1056, 1986.

20. **Zellner, G. and Winter, J.,** Secondary alcohols as hydrogen donors for CO_2-reduction by methanogens, *FEMS Microbiol. Lett.,* 44, 323, 1987.

21. **Perski, H. J., Moll, J., and Thauer, R. K.,** Sodium dependence of growth and methane formation in *Methanobacterium thermoautotrophicum, Arch. Microbiol.,* 130, 319, 1981.

22. **Perski, H. J., Schönheit, P., and Thauer, R. K.,** Sodium dependence of methane formation in methanogenic bacteria, *FEBS Lett.,* 143, 323, 1982.

23. **Thauer, R. K., Jungermann, K., and Decker, K.,** Energy conservation in chemotrophic anaerobic bacteria, *Bacteriol. Rev.,* 41, 100, 1977.

24. **Thauer, R. K., Möller-Zinkhan, D., and Spormann, A. M.,** Biochemistry of acetate catabolism in anaerobic chemotrophic bacteria, *Annu. Rev. Microbiol.,* 43, 43, 1989.

25. **Peinemann, S., Müller, V., Blaut, M., and Gottschalk, G.,** Bioenergetics of acetate by *Methanosarcina barkeri, J. Bacteriol.,* 170, 1369, 1988.

26. **Karrasch, M., Börner, G., Enssle, M., and Thauer, R. K.,** The molybdoenzyme formylmethanofuran dehydrogenase from *Methanosarcina barkeri* contains a pterin cofactor, *Eur. J. Biochem.,* 194, 367, 1990.

27. **Karrasch, M., Börner, G., and Thauer, R. K.,** The molybdenum cofactor of formylmethanofuran dehydrogenase from *Methanosarcina barkeri* is a molybdopterin guanine dinucleotide, *FEBS Lett.,* 274, 48, 1990.

28. **Börner, G., Karrasch, M., and Thauer, R. K.,** Formylmethanofuran dehydrogenase activity in cell extracts of *Methanobacterium thermoautotrophicum* and of *Methanosarcina barkeri, FEBS Lett.,* 244, 21, 1989.

29. **Börner, G., Karrasch, M., and Thauer, R. K.,** Molybdopterin adenine dinucleotide and molybdopterin hypoxanthine dinucleotide in formylmethanofuran dehydrogenase from *Methanobacterium thermoautotrophicum, FEBS Lett.,* 290, 31, 1991.

30. **Breitung, J. and Thauer, R. K.,** Formylmethanofuran: tetrahydromethanopterin formyltransferase from *Methanosarcina barkeri.* Identification of N^5-formyl-tetrahydromethanopterin as the product, *FEBS Lett.,* 275, 226, 1990.

31. **Schwörer, B. and Thauer, R. K.,** Activities of formylmethanofuran dehydrogenase, methylenetetrahydromethanopterin dehydrogenase, methylenetetrahydromethanopterin reductase, and heterodisulfide reductase in methanogenic bacteria, *Arch. Microbiol.,* 155, 459, 1991.

32. **Zirngibl, C., Hedderich, R., and Thauer, R. K.,** N^5,N^{10}-Methylenetetrahydromethanopterin dehydrogenase from *Methanobacterium thermoautotrophicum* has hydrogenase activity, *FEBS Lett.,* 261, 112, 1990.

33. **Ma, K. and Thauer, R. K.,** Purification and properties of N^5,N^{10}-methylenetetrahydromethanopterin reductase from *Methanobacterium thermoautotrophicum (strain Marburg), Eur. J. Biochem.,* 191, 187, 1990.

34. **Friedmann, H. C., Klein, A., and Thauer, R. K.,** Structure and function of the nickel porphinoid, coenzyme F_{430}, and of its enzyme, methyl coenzyme M reductase, *FEMS Microbiol. Rev.,* 87, 339, 1990.

35. **Hedderich, R., Berkessel, A., and Thauer, R. K.,** Purification and properties of heterodisulfide reductase from *Methanobacterium thermoautotrophicum* (strain Marburg), *Eur. J. Biochem.,* 193, 255, 1990.

36. **Graf, E. G. and Thauer, R. K.,** Hydrogenase from *Methanobacterium thermoautotrophicum,* a nickel-containing enzyme, *FEBS Lett.,* 136, 165, 1981.

37. **Fox, J. A., Livingston, D. J., Orme-Johnson, W. H., and Walsh, C. T.,** 8-Hydroxy-5-deazaflavin-reducing hydrogenase from *Methanobacterium thermoautotrophicum, Biochemistry,* 26, 4219, 1987.

38. **Fiebig, K. and Friedrich, B.**, Purification of the F_{420}-reducing hydrogenase from *Methanosarcina barkeri* (strain Furaso), *Eur. J. Biochem.*, 184, 79, 1989.
39. **Reeve, J. N. and Beckler, G. S.**, Conservation of primary structure in prokaryotic hydrogenases, *FEMS Microbiol. Rev.*, 87, 419, 1990.
40. **Ferry, J. G.**, Formate dehydrogenase, *FEMS Microbiol. Rev.*, 87, 377, 1990.
41. **Thauer, R. K. and Morris, J. G.**, Metabolism of chemotrophic anaerobes: old view and new aspects, in *The Microbe*, Part II, Prokaryotes and Eukaryotes, Kelly, D. P. and Carr, N. G., Eds., University Press, Cambridge, 1984.
42. **Blaut, M. and Gottschalk, G.**, Protonmotive force-driven synthesis of ATP during methane formation from molecular hydrogen and formaldehyde or carbon dioxide in *Methanosarcina barkeri*, *FEMS Microbiol. Lett.*, 24, 103, 1984.
43. **Kaesler, B. and Schönheti, P.**, Methanogenesis and ATP synthesis in methanogenic bacteria at low electrochemical proton potentials. An explanation for the apparent uncoupler insensitivity of ATP synthesis, *Eur. J. Biochem.*, 174, 189, 1988.
44. **Blaut, M. and Gottschalk, G.**, Coupling of ATP synthesis and methane formation from methanol and molecular hydrogen in *Methanosarcina barkeri*, *Eur. J. Biochem.*, 141, 217, 1984.
45. **Miller, T. L. and Wolin, M. J.**, *Methanosphaera stadtmaniae* gen. nov., sp. nov.: a species that forms methane by reducing methanol with hydrogen, *Arch. Microbiol.*, 141, 116, 1985.
46. **Escalante-Semarena, J. C. and Wolfe, R. S.**, Formaldehyde oxidation and methanogenesis, *J. Bacteriol.*, 158, 721, 1984.
47. **Blaut, M., Müller, V., and Gottschalk, G.**, Proton translocation coupled to methanogenesis from methanol + hydrogen in *Methanosarcina barkeri*, *FEBS Lett.*, 215, 53, 1987.
48. **Blaut, M., Müller, V., Fiebig, K., and Gottschalk, G.**, Sodium ions and an energized membrane required by *Methanosarcina barkeri* for the oxidation of methanol to the level of formaldehyde, *J. Bacteriol.*, 164, 95, 1985.
49. **Kell, D. B., Doddema, H. J., Morris, J. G., and Vogels, G. D.**, Energy coupling in methanogens, in *Proc. 3rd Int. Symp. Microbiol Growth on C-1 Compounds*, Dalton, H., Ed., Hyden, London, 1981, 159.
50. **Kaesler, B. and Schönheit, P.**, The role of sodium ions in methanogenesis. Formaldehyde oxidation to CO_2 and 2 H_2 in methanogenic bacteria is coupled with primary electrogenic Na^+ translocation at a stoichiometry of 2–3 Na^+/CO_2, *Eur. J. Biochem.*, 184, 223, 1989.
51. **Winner, C. and Gottschalk, G.**, H_2 and CO_2 production from methanol or formaldehyde by the methanogenic bacterium strain Göl treated with 2-bromoethanesulfonic acid, *FEMS Microbiol. Lett.*, 65, 259, 1989.
52. **Müller, V., Winner, C., and Gottschalk, G.**, Electron-transport-driven sodium extrusion during methanogenesis from formaldehyde and molecular hydrogen by *Methanosarcina barkeri*, *Eur. J. Biochem.*, 178, 519, 1988.
53. **Dharmavaram, R. M. and Konisky, J.**, Characterization of a P-type ATPase of archaebacterium *Methanococcus voltae*, *J. Biol. Chem.*, 264, 14085, 1989.
54. **Carper, S. W. and Lancaster, J. R., Jr.**, An electrogenic sodium-translocating ATPase in *Methanococcus voltae*, *FEBS Lett.*, 200, 177, 1986.
55. **Fischer, R. and Thauer, R. K.**, Methyltetrahydromethanopterin as an intermediate in methanogenesis from acetate in *Methanosarcina barkeri*, *Arch. Microbiol.*, 151, 459, 1989.
56. **Mahlmann, A., Deppenmeier, U., and Gottschalk, G.**, Methanofuran-b is required for CO_2 formation from formaldehyde by *Methanosarcina barkeri*, *FEBS Microbiol. Lett.*, 61, 115, 1989.
57. **Kaesler, B. and Schönheit, P.**, The sodium cycle in methanogenesis. CO_2 reduction to the formaldehyde level in methanogenic bacteria is driven by a primary electrochemical potential of Na^+ generated by formaldehyde reduction to CH_4, *Eur. J. Biochem.*, 186, 309, 1989.

58. **Schönheit, P. and Perski, H. J.**, ATP synthesis driven by a potassium diffusion potential in *Methanobacterium thermoautotrophicum* is stimulated by sodium, *FEMS Microbiol. Lett.*, 20, 263, 1983.

59. **Schönheit, P. and Beimborn, D. B.**, Presence of a Na^+/H^+ antiporter in *Methanobacterium thermoautotrophicum* and its role in Na^+ dependent methanogenesis, *Arch. Microbiol.*, 142, 354, 1985.

60. **Müller, V., Blaut, M., and Gottschalk, G.**, Generation of a transmembrane gradient of Na^+ in *Methanosarcina barkeri, Eur. J. Biochem.*, 162, 461, 1987.

61. **Krulwich, T. A.**, Na^+/H^+ antiporters, *Biochim. Biophys. Acta*, 726, 245, 1983.

62. **Schönheit, P. and Beimborn, D. B.**, Monensin and gramicidin stimulate CH_4 formation from H_2 and CO_2 in *Methanobacterium thermoautotrophicum* at low external Na^+ concentration, *Arch. Microbiol.*, 146, 181, 1986.

63. **Kleyman, T. R. and Cragoe, E. J., Jr.**, Amiloride and its analogs as tool in the study of ion transport, *J. Membr. Biol.*, 105, 1, 1989.

64. **Kaesler, B. and Schönheit, P.**, unpublished data.

65. **Harold, F. M.**, *The Vital Force. A Study of Bioenergetics*, W. H. Freeman, San Francisco, 1986.

66. **Booth, I. R.**, Regulation of cytoplasmic pH in Bacteria, *Microbiol. Rev.*, 49, 359, 1985.

67. **Wood, H. G., Ragsdale, S. W., and Pezacka, E.**, The acetyl-CoA pathway of autotrophic growth, *FEMS Microbiol. Rev.*, 39, 345, 1986.

68. **Fuchs, G.**, CO_2 fixation in acetogenic bacteria: variations on the theme, *FEMS Microbiol. Rev.*, 39, 181, 1986.

69. **Diekert, G.**, CO_2 reduction to acetate in anaerobic bacteria, *FEMS Microbiol. Rev.*, 87, 391, 1990.

Section II: K^+ Transport Systems

Chapter IIA

CELL K$^+$ AND K$^+$ TRANSPORT SYSTEMS IN PROKARYOTES

Evert P. Bakker

TABLE OF CONTENTS

I. INTRODUCTION AND PERSPECTIVE

Prokaryotes contain relatively high concentrations of K^+ in their cytoplasm. This concentration is determined by the sum of the activities of the K^+ transport systems located in the cell membrane of these organisms. *Escherichia coli*[1-3] and probably also all other bacteria possess several K^+ uptake and K^+ efflux systems. The activity of these systems is coordinated in such a way that the cells take up enough K^+ to develop a negative turgor pressure. Once this situation has been reached, the cells shut off the activity of their K^+ uptake systems and thereby circumvent unnecessary energy losses due to rapid and futile cycling of K^+ across the cell membrane.[1,4] The second important function of cell K^+ in bacterial physiology is its role in the homeostasis of cytoplasmic pH.[5] Transmembrane K^+ transport processes play an important role in this process too.

This overview introduces the different aspects of K^+ transport described in the subsequent chapters of this section. It also contains a description of the role of cell K^+ in osmoregulation of prokaryotes. Many of the data have been obtained with *E. coli*. However, with a few exceptions, there is no reason to believe that other prokaryotes contain K^+ transport systems essentially different from those of *E. coli*. This chapter avoids extensive overlap with previous reviews. More details, in particular about older data, can be found in a number of excellent reviews covering K^+ transport in bacteria[1,4,6-10] and in the chapters of this section that follow the overview.

II. CELL K^+

All cells extrude Na^+ and accumulate K^+.[1,11-13] The reason for this preference of K^+ in the cytoplasm is not clear. Wiggins[12] has discussed this phenomenon recently. She proposes that due to its somewhat larger ionic radius the interaction of K^+ with the water molecules of the cytoplasm and with the anionic groups of the cellular macromolecules is different from that of Na^+ and that the effect of K^+ is more beneficial than that of Na^+ for allowing enzymes to possess native conformations and to be active.[12] Alternatively, it can be argued that life occurs in a world which is relatively rich in Na^+ and low in K^+ (see Reference 13) and that exclusion of Na^+ from the cytoplasm allows the cells to generate an electrochemical Na^+ gradient across their cell membrane ($\Delta\bar{\mu}_{Na^+}$) which can be used to drive other processes across this membrane (see Section I of this book). According to this hypothesis, K^+, the second most abundant alkali cation in nature,[13] is accumulated by the cells in order to enable them to develop a negative turgor pressure. The cytoplasmic enzymes must then have adapted their properties to the relatively high K^+ concentration in the cytoplasm.

A. MEASUREMENT OF CELL K⁺

K⁺ can be detected either by flame photometry or by atomic absorption spectroscopy. Both methods are sufficiently sensitive and accurate to give reliable results. Alternatively, the amount of ⁴²K⁺ in cells loaded with this isotope can be determined. However, this method is not very convenient since ⁴²K⁺ is a strong γ-ray emitter with a half-life of only 12.4 h. In addition, the isotope ⁴²K⁺ is not commercially available anymore in Europe. Finally, cell K⁺ can be determined by ³⁹K-NMR.[14]

The problem with the determination of the K⁺ content of bacterial cells lies in the way the cells are treated before the actual measurement is done. In particular, the method by which extracellular K⁺ is removed is crucial. Centrifugation and washing of the cells can give unreliable results since the cells may have released part of their K⁺ during these steps. For instance, K⁺ leaks out from the pellet of *Bacillus acidocaldarius* cells before the supernatant is removed and the cells in the pellet fraction are denatured.[15]

In Chapter IIIA Kleiner recommends centrifugation of cells through silicone oil as the only method that prevents loss of NH_4^+ from the cell pellet after the centrifugation step. Here I advocate the same method as being highly reliable for the preparation of cell samples used for the K⁺ determination. The first advantage is that the cell pellet contains very little extracytoplasmic water. Hence, in most situations the cells do not have to be washed free of external K⁺ before they are centrifuged through the oil. Second, K⁺ released from the cells in the pellet is retained in this fraction since K⁺ does not diffuse back through the oil to the medium. The oil that should be used depends on the density of the medium and on the type of cells. In most cases, silicone oil AR200 (D = 1.04) or mixtures of this oil with AR20 (D = 1.01) give satisfactory results. Cells of *Halobacterium halobium* have been centrifuged through CR500 oil[16] (all of these oils are from Wacker Chemie, Munich, Germany). After centrifugation the supernatant and most of the oil are removed by suction and the pellet fraction is treated with f.i. trichloroacetic acid in order to denature cell proteins. K⁺ is then determined in this extract after denatured protein has been removed by centrifugation.[15,17]

B. CONCENTRATION AND ACTIVITY OF K⁺ IN THE CYTOPLASM

Most bacteria that grow at low values of medium osmolarity contain K⁺ concentrations of the order of 0.2 to 0.5 mol/l of cytoplasmic H_2O, and cell K⁺ increases with the osmolarity of the medium[1,11,14,18,19] (Figure 1 shows this phenomenon for *E. coli*). It is, however, a matter of debate to what extent cytoplasmic K⁺ is osmotically active. The macromolecules in the cytoplasm contain an excess of negative groups, which amounts to 0.2 and 0.6 mol/l of cell water in eukaryotes[12] and prokaryotes (my own calculation), respectively.

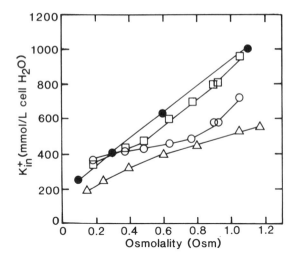

FIGURE 1. K$^+$ content of *Escherichia coli* as a function of medium osmolality. Data from different groups are compared. Open symbols, growing cells; closed circles, stationary phase cells;[19] diamonds, data from Epstein and Schultz.[19] The latter values might be too low since the cytoplasmic volume of the cells has been overestimated.[99] The open squares and circles represent the maximal and the steady-state values, respectively, after growing *E. coli* cells have been submitted to hyperosmotic shock[35] (see Figure 3 for details).

K$^+$ is likely to interact with these fixed negative groups by forming ion pairs with them. This strong binding would render these K$^+$ ions osmotically inactive.[12]

However, besides K$^+$, prokaryotes also contain large amounts of Mg^{2+} and polyamines (in *E. coli* both are about 50 to 100 mmol/l of total cell water).[20,21] Polyamines[21] and at least part of the cell magnesium[20,22] are not released after permeabilization of the cell membrane, suggesting that they bind irreversibly to cellular components. By contrast, both *Halobacterium cutirubrum*[22] and *E. coli*[20] can exchange their complete K$^+$ content against Na$^+$ rapidly, suggesting that even if K$^+$ binding to anionic groups was to occur in the cytoplasm it is reversible.

One would expect that ^{39}K-NMR measurements with *E. coli* cells might give some information about the extent or strength of ion pair formation between K$^+$ and the macromolecules in the cytoplasm. However, Richey et al.[14] who have done such measurements, conclude that their data do not allow any conclusion about K$^+$ binding to cellular components. Thus, it is impossible to give an exact answer to the question about the osmotic activity of K$^+$ in the bacterial cytoplasm. Some arguments are in favor of the notion that K$^+$ ions in these cells are involved in ion pairing,[12] but there exist no experimental data that support this hypothesis. To the contrary, the evidence that K$^+$ is involved in the regulation of the turgor pressure in prokaryotes cell is overwhelming (see Section II.C). Hence, K$^+$ ions must exert some osmotic activity in the cytoplasm, which contradicts the notion that in cells grown at low osmolarity almost all of the K^+ ions are osmotically inactive.[12]

C. CELL K^+ AND TURGOR PRESSURE

Some cytoplasmic enzymes are specifically activated by K^+ ions. However, the K^+ concentration of 1 to 10 mM required for this effect is much smaller than the at least several hundred millimoles present in the cytoplasm of most prokaryotes.[1] At such concentrations K^+ is the major solute of the cytoplasm, and cell K^+ is therefore considered to play an important role in the process of turgor pressure regulation in prokaryotes.[1,9,11] Turgor pressure is defined as the difference between the osmotic pressure exerted by the solutes in the medium surrounding the cells and that of the solutes in the cytoplasm. In the few eubacteria in which it has been measured, turgor pressure amounts to -0.1 to -0.4 mPa.[11,23-25] Turgor brings the cells into a swollen state, which is believed to be essential for growth.[9,11]

The notion that in prokaryotes cell K^+ is crucial to maintenance of turgor pressure is based on the observation that the K^+ concentration of the cells varies with the osmotic pressure of the medium.[9,11,14,18,19] When cells are put under chronic osmotic stress by growth at elevated medium osmolarities they increase their K^+ contents. Organic anions like glutamate are synthesized concomitantly and serve as counterions to the extra K^+ taken up from the medium.[26-29] In general, internal K^+ concentrations increase monotonously with medium osmolarity (as shown in Figure 1 for *E. coli*), but in halophiles and halotolerants the pattern changes at a critical value of the latter parameter.[11,30] Above this value the cells stop accumulating additional K^+ and start to take up or synthesize neutral organic solutes like the zwitterions proline, glycine betaine, or ectoine[27,30-34] and/or carbohydrates like trehalose.[28,29,33] The reason for this switch is that at high concentrations K^+ requires so many water molecules for hydration that in the cytoplasm it will pull away water from the surface of the proteins, leading to denaturation of these proteins. The zwitterionic solutes that accumulate in the cytoplasm at high medium osmolarities hardly interfere with the H_2O structure of the cytoplasm[12] and are therefore called "compatible solutes".[11]

With *E. coli* cells a critical value of medium osmolarity above which the cells do not take up additional K^+ and start to synthesize compatible solutes[11,30] is not observed. Rather, in growing cells of this organism the concentration of K^+ (Figure 1, open circles) increases monotonously with medium osmolarity. Under these conditions the cells also synthesize trehalose.[35,36] Above 0.8 Osm the cytoplasmic K^+ concentration rises more steeply with medium osmolality (Figure 1, open circles), probably because the cells are not able to synthesize additional trehalose.[36] Stationary phase cells contain more K^+ than do growing cells (Figure 1, closed and open symbols, respectively), possibly because the trehalose concentration in the former type of cells is much lower than that of the latter.[36] In *E. coli*, which is hardly a halotolerant organism, the maximum concentration of trehalose in the cytoplasm amounts to about 0.4 M.[35,36] This concentration is very similar to that observed with the halophilic phototrophic eubacterium *Ectothiorhodospira halochloris*.[33] For the latter organism it has been proposed that an internal concentration of trehalose higher than 0.4 M is harmful to the cells.[33]

A high internal K^+ concentration does not seem to be required for growth of *E. halochloris* at 2.7 M.[31] However, under these conditions the sum of betaine, ectoine, and trehalose in the cytoplasm equals only about 3 M.[33] Assuming that these compounds exert activities of one, it can be concluded that they cannot be exclusively responsible for the maintenance of cell turgor. The same is true for the ectoine plus betaine concentration in the cytoplasm of the moderately halophilic eubacterium Ba.[37] One wonders whether the K^+ content of the two organisms has been underestimated.

The extremely halophilic archaebacteria have developed a strategy completely different from that of halophilic eubacteria. Instead of compatible solutes, *Halobacteriaceae* accumulate about 4 to 5 M K^+, 1 M Na^+, and 5 to 6 M Cl^- in their cytoplasm.[38] These organisms have developed the following strategy for circumventing dehydration of their soluble proteins: (1) the proteins possess many negatively charged residues that cluster at their surface, (2) these residues are involved in the binding of solvated K^+ and Na^+ ions to the proteins, and (3) halophilic proteins have a very compact structure.[38-40]

The Na^+ concentration in prokaryotes also may rise with external osmolarity. However, one expects (1) that the internal Na^+ concentration remains lower than that of the medium and (2) that the $[K^+]/[Na^+]$ ratio in the cytoplasm is larger than one. Data that show the opposite should be taken with caution. For instance, the eubacterium *Halobacteroides acetoethylicus* has been reported to contain 0.24 M K^+ and 0.92 M Na^+ when grown at 1.7 M NaCl.[41] However, these cells have been washed with hypotonic buffer.[41] In Section IV of this overview and in Chapter IIG it is shown that such an "osmotic downshock" may lead to extremely rapid and extensive loss of K^+ from the cells.

In prokaryotes the turgor pressure decreases with increasing hyperosmotic stress.[11,24,25] In Halobacteriaceae it approaches zero.[42,43] Only under these conditions can some of these organisms develop a square[44] or triangular[45] cell form.[11]

Cell-wall-free bacteria also develop small turgor pressures. Nevertheless, *Mycoplasma mycoides* var. *Capri* contains 0.3 M K^+ when grown in a medium of relatively low osmolarity,[46] suggesting that these cells possess at least some turgor. The archaebacterium *Thermoplasma acidophilum* has been reported to contain only 18 mM K^+ when grown at a medium osmolality of 50 mOsm.[47] This internal K^+ concentration appears to be too low. I suggest that in the time span that had elapsed between the formation of the cell pellet in the centrifuge and the handling of the sample some K^+ may have leaked out from the cell pellet into the medium (see Section II.A regarding this problem).

III. K+ TRANSPORT SYSTEMS

Figure 2 gives a survey of the K+ transport systems described for different bacteria. All bacteria in which K+ transport has been studied show the same pattern of K+ transport systems as *E. coli*. K+ uptake and K+ efflux take place through separate systems. Uptake is electrogenic, which means that it is greatly facilitated by the internally negative membrane potential ($\Delta\Psi$) present in most bacteria. K+ efflux occurs via antiporters and is probably electroneutral. This arrangement of separate systems enables the cells to lose K+ under conditions at which $\Delta\Psi$ remains high. Several of the K+ uptake and K+ efflux systems can transport K+ with very high rates. Uncontrolled transport would therefore lead to rapid futile cycling of K+ across the cell membrane, which would cost energy. The cells avoid this situation by regulating the activity of the different systems separately. Under most conditions the systems are hardly active. K+ uptake is stimulated by low turgor pressure.[9,48,49] Efflux is stimulated by high cytoplasmic pH or by high turgor pressure.[2,9] Under the latter conditions the cells also might release K+ via stretch-activated channels (see Chapter IIH).

A. MEASUREMENT OF K+ TRANSPORT

Before the different transport systems are discussed, the methods by which transports are measured will be described. The experiments are carried out as outlined in Section II.A, except that cell samples are taken as a function of time. Net K+ transport also can be determined with a K+-specific electrode.[50-52] With respiring cell suspensions care should be taken to avoid anaerobiosis during the measurements.

For the measurement of net K+ uptake, the cells must first be depleted of this cation. With *E. coli* this is done by incubating the cells with 2,4-dinitrophenol in a Na+-containing buffer[53] or by treating them with Tris-EDTA followed by washing the cells in Na+-containing buffer.[17] Some stationary-phase cells of *Enterococcus hirae* already contain little K+. Alternatively, this organism also can be depleted of K+ by treatment with 2,4-dinitrophenol.[54] The K+ content of *B. acidocaldarius* has been reduced by washing the cells with a Na+-containing buffer.[15] In all of these cells K+ has been replaced by Na+ and H+; cell K+ also can be replaced by weak bases like NH4+ (Chapter IIIC), triethanolamine (Chapter IF and Reference 51), or even Tris.[50] During its net uptake K+ replaces these other cations in the cytoplasm.

The K+-depleted cells are preincubated for 5 to 10 min with an energy source (almost always glucose). K+ is then added to the suspension, and samples of 0.1 to 0.5 mg cell protein are withdrawn from the suspension as a function of time. The cells are separated from the medium by filtration[53,54] or by centrifugation through silicone oil.[15,17] The first method cannot be used under special conditions, like after phage infection, since then the cells release

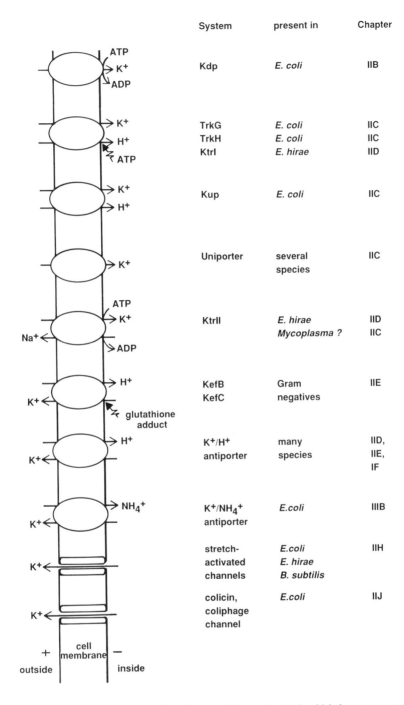

System	present in	Chapter
Kdp	*E. coli*	IIB
TrkG	*E. coli*	IIC
TrkH	*E. coli*	IIC
KtrI	*E. hirae*	IID
Kup	*E. coli*	IIC
Uniporter	several species	IIC
KtrII	*E. hirae*	IID
	Mycoplasma ?	IIC
KefB	Gram	IIE
KefC	negatives	
K^+/H^+ antiporter	many species	IID, IIE, IF
K^+/NH_4^+ antiporter	*E.coli*	IIIB
stretch-activated channels	*E.coli* *E. hirae* *B. subtilis*	IIH
colicin, coliphage channel	*E.coli*	IIJ

FIGURE 2. K^+ transport systems in prokaryotes. The organism(s) in which the system occur(s) and the chapter in which it (they) is (are) discussed are indicated at the right-hand part of the figure.

K$^+$ and Na$^+$ during the few seconds it takes to wash the cells on the filter.[20] The disadvantages of the second method are that it is loud and that it is not suitable for measuring rapid K$^+$ movement since it has a dead time of at least 15 s.[17]

K$^+$ transport in the exchange mode can only be measured with ^{42}K$^+$.[48,49] For this purpose cells are first "loaded" with either ^{42}K$^+$ or nonradioactive K$^+$. After a steady state has been reached, ^{42}K$^+$-loaded cells are diluted with a medium of identical composition, but without radioactivity; K$^+$-loaded cells receive a trace amount of ^{42}K$^+$. The radioactivity in the cells is measured as a function of time.

K$^+$ efflux experiments also start with K$^+$-loaded cells. Efflux is initiated, for instance, by increasing the cytoplasmic pH or the turgor pressure of the cells[2,50] (see also Chapter IIE).

B. K$^+$ UPTAKE SYSTEMS

This topic is described in detail in Chapters IIA, IIC, IID, and IIG. Therefore, only some general features of these systems will be described here. *E. coli* and probably all other bacteria contain several types of K$^+$ uptake systems. These systems have different affinities for K$^+$, which allows the cells to take up the cation at a wide range of external K$^+$ concentrations.[1] The activity of most of the systems cannot be traced with ^{86}Rb$^+$ since this alkali cation is accepted poorly[55] (see also Chapters IIB and IIC).

1. The K$^+$-Translocating Kdp-ATPase

Kdp is the best-studied bacterial K$^+$ uptake system (for reviews, see References 1, 10, and 56 and Chapter IIB). It transports K$^+$ with a high affinity into the cell (the K_m for K$^+$ uptake is about 2 μM).[53] The transport complex consists of three inner-membrane proteins, KdpA, KdpB, and KdpC, which are encoded by the genes of the *kdpABC* operon. This operon is expressed under conditions at which the cells possess a low turgor pressure, which is sensed by a two-component system formed by the KdpD and KdpE proteins.[57] KdpD is a membrane protein and is believed to be the sensor that transduces the signal to the DNA-binding protein KdpE, which controls transcription of the *kdpABC* operon.[57] In addition to its stimulatory effect on transcription, low turgor pressure also increases the activity of the Kdp system.[49]

Kdp belongs to the P-type of ion-translocating ATPases.[58] The KdpB protein forms the catalytic subunit. Its primary sequence shows homology with the large subunit of eukaryotic ion-translocating ATPases,[59] indicating that P-type ATPases have been preserved well during evolution. *kdp* genes or Kdp proteins have been identified in a wide variety of Gram-negative[60] and in one Gram-positive organism[61] (see Chapter IIB).

2. Other ATP-Driven K$^+$ Uptake Systems

Two ATP-driven systems have reported to be present in *Enterococcus hirae* (formerly *Streptococcus faecalis* ATCC 9790). However, the claim that the KtrI system of this organism is a K^+-translocating ATPase of the P-type[62] has been retracted[63] (see also Reference 10 and Chapters IIB and IID). In the latter chapter, Kakinuma states that the KtrII system of *E. hirae* represents a novel type of ion-translocating ATPase that exchanges K^+ against Na^+ across the cell membrane of that organism. A system similar to that of KtrII has been proposed to exist in *M. mycoides* var. *Capri*.[64] However, Chapter IIC will argue that the evidence is incomplete for the notion that in this organism K^+ uptake is coupled to ATP hydrolysis.

3. Low-Affinity Uptake Systems

During growth at K^+ concentrations above 1 mM prokaryotes take up K^+ via low-affinity systems (Chapter IIC). *Escherichia coli* contains at least three of these systems: TrkG, TrkH, and Kup[3,65] (the latter is also called TrkD[3]). The TrkG and TrkH systems are the more important ones since they transport K^+ ten times faster than does Kup.[53] All three systems are constitutive.

TrkG and TrkH possess almost identical properties. The reason that *E. coli* contains two systems might be that the *trkG* gene has entered the cells with a prophage.[3,66,67] The TrkG/H systems have the following properties:

1. The transport complex contains several kinds of subunits.
2. The systems require both ATP and $\Delta\bar{\mu}_{H^+}$ for activity;[68] ATP may regulate transport, and $\Delta\bar{\mu}_{H^+}$ appears to drive this process.
3. Low turgor pressure activates the systems;[48,49] thus, together with Kdp the TrkG/H systems are responsible for additional K^+ uptake after hyperosmotic shock (see Sections II.C and IV).
4. The systems transport K^+ with very high turnover numbers (about $10^4 \cdot s^{-1} \cdot$ transport complex^{-1}).[69]

The Kup system consists of only one type of subunit. It probably functions as a secondary porter. In contrast to Kdp, TrkG, and TrkH, Kup is not involved in osmoadaptation after a hyperosmotic shock. It is difficult to understand why cells of *E. coli* have preserved the relative sluggish Kup system, but the same question arises about the TrkG system.

It is not yet clear how these three *E. coli* systems translocate K^+ across the membrane. The best guess is that they are all K^+/H^+ symporters. Systems with a similar mode of action also exist in other bacteria. The activity of some of these systems appears to be regulated by ATP and/or turgor pressure as well. Another group of low-affinity K^+ uptake systems transports K^+ by uniport.[70,71] More details about all of these systems can be found in Reference 1 and Chapter IIC.

C. K⁺ EFFLUX SYSTEMS

1. K⁺/H⁺ Antiporters

E. coli possesses at least three systems[2] (Chapter IIE and Figure 2). Two of those systems, KefB and KefC, underly an unusual regulation in which an adduct of glutathione stimulates activity.[72,73] In contrast to the third *E. coli* system, KefB and KefC are probably not involved in the homeostatic processes of pH and turgor-pressure regulation. Rather, they serve as emergency systems for the situation in which the glutathione in the cells has reacted with sulfhydryl reagents (see Chapter IIE). Except that it is present, nothing is known about the third K⁺/H⁺ antiport system of *E. coli*.[2] K⁺/H⁺ antiporters have been described for many bacteria. The KefB/KefC type of system is confined to Gram-negative bacteria[74] (see Chapter IIE).

2. K⁺/NH₄⁺ Antiporter

E. coli takes up NH_4^+ exchange for K⁺ ions in a process that may be mediated by a K⁺/NH_4^+ antiporter.[75] The system is described in Chapter IIIB. It is not known whether other bacteria contain a similar system.

3. Stretch-Activated Channels

After osmotic downshock, *E. coli* releases almost all of its K⁺ within seconds (see Section IV and Chapter IIH). This rate of K⁺ loss is too fast to be mediated by a secondary transport system. The possibility that this K⁺ efflux occurs through stretch-activated channels is discussed in Chapter IIH. Recently, this kind of channel has been described for both Gram-negative[76,77] and Gram-positive bacteria.[78]

IV. K⁺ TRANSPORT DURING OSMOTIC STRESS IN *E. COLI*

In this section recent data from my group illustrate the role that K⁺ transport plays in the adaptation of *E. coli* to osmotic shock. The most striking effects occur during acute osmotic stress, but net transmembrane K⁺ movement is important even for adaptation to chronic stress. Growing cells of a strain that do not contain functional Kdp, KefB, and KefC systems were submitted to a hyperosmotic shock by the addition of 0.4 *M* NaCl. After the cells had adapted to this condition, they were subjected to an osmotic downshock by dilution of the suspension 1:1 with H_2O (Figure 3).

A. HYPEROSMOTIC SHOCK

Within minutes after hyperosmotic shock the cells more than doubled their K⁺ concentration by taking up extra K⁺ via the Trk (or Kdp) system

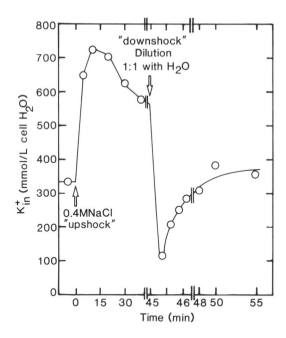

FIGURE 3. Transmembrane K^+ movements after hyperosmotic and hypoosmotic shock. The K^+ concentration in the cells is given as a function of time. Cells of *E. coli* strain LB322 ($\Delta kdpABC5\ kefB$::Tn10 *kefC121-1*)[35] were grown at 37°C in a minimal medium of 0.2 Osm with glucose as the carbon source ("K5 medium"; see Reference 100). At zero time the osmolarity of the medium was increased by the addition of 0.4 *M* NaCl. At t = 45 min the medium was diluted 1:1 with water of 37°C.[35]

(Figure 3). Concomitantly the cells synthesized glutamate, which accumulated to about 200 m*M* in the cytoplasm.[29,35] The high intracellular K^+-glutamate concentration stimulates the transcription of the *proU* operon, which encodes a high-affinity uptake system for glycine betaine[79-81] (see Chapter IIF). In the presence of external glycine betaine (or proline or taurine, which give similar effects), cells growing at high osmolarity will take up large amounts of this compatible solute, which replaces part of the previously accumulated "extra" K^+ from the cytoplasm.[2,29,79,79a] This K^+ release probably occurs via $K^+/$ H^+ antiporters,[2] although it cannot be excluded that stretch-activated channels are also involved.

In the absence of glycine betaine or proline, the cells entered a second phase of adaptation about 15 min after they had been submitted to hyperosmotic shock (Figure 3). Part of the extra K^+ was extruded (Figure 3). Concomitantly the cells degraded part of their glutamate, and these two ions were replaced by trehalose.[28,29,35,36] In the steady state, growing cells contained about 550 mmol of K^+, about 80 mmol of glutamate, and about 360 mmol of trehalose per liter of cytoplasmic water.[35] With the assumption that these three compounds all have activity coefficients of 1.0 and that they are the only osmotically active species of the cytoplasm, one calculates for these *E. coli* cells a turgor pressure of -0.5 MPa. This value is only slightly higher than the -0.1 to -0.4 MPa measured in gas-vacuolated prokaryotes.[23-25]

B. HYPOOSMOTIC SHOCK

After a hypoosmotic shock, cell turgor may become so great that the cells break. After a combined hypoosmotic and cold shock, *E. coli* cells lose large amounts of solutes.[82] For a long time this effect has been considered to be due to a short, reversible rupture of the cell wall and membranes. The observation that bacteria contain stretch-activate channels[76-78] has shed a new light on this phenomenon (see Chapter IIH). It now appears that solute efflux after osmotic downshock is a specific process and cannot be due to a short rupture of the cell membrane.[35,83] The right-hand part of Figure 3 shows the effect on cell K⁺ after growing cells that had adapted to high osmolarity were diluted 1:1 with water of 37°C. Within seconds the cells released 90% of their K⁺ (Figure 3) and glutamate and 60% of their trehalose.[35] This rapid release of solutes allowed the cells to reduce their turgor pressure to a low value and thereby avoid rupture. The solute efflux process is specific since after osmotic downshock sucrose did not enter the cytoplasm and the cells released neither ATP nor alanine.[35] Independently, it has been shown that after hypoosmotic shock osmoadapted cells loaded with glycine betaine release this solute rapidly and completely.[84] The specific efflux of compounds as diverse as trehalose, K⁺, glutamate, and glycine betaine must take place through more than one efflux system. However, these systems have not yet been characterized, and it is not clear whether they are stretch-activated channels.

The cells recovered quickly after their turgor pressure had collapsed, primarily by accumulating K⁺ (Figure 3) and glutamate (not shown) from the medium, leading to restoration of the cell turgor. This reaccumulation of K⁺ occurs via the Trk or Kdp systems, but not via Kup (Figure 2 and Section III.B).[35]

Ectothiorhodospira halochloris loses 35% of its glycine betaine pool after the NaCl concentration in the medium is reduced. Subsequently, the cells reabsorb 38% of this previously released solute.[85] This suggests that *E. halochloris* employs a mechanism similar to that of *Escherichia coli* for reducing its turgor pressure after hypoosmotic shock. Nothing is known about how other prokaryotes behave under these conditions.

V. K⁺ TRANSPORT AND pH HOMEOSTASIS

Bacteria maintain their cytoplasmic pH at a value close to neutrality.[5,86-88] In Section I of this book the role of Na⁺ transport in this process has been discussed. However, transmembrane movement of K⁺ contributes too.[3,5,88] First, via its uptake, K⁺ is involved in the alkalinization of the cytoplasm. K^+ uptake is electrogenic and decreases the internally negative membrane potential generated by H^+ extrusion via the proton pumps of the cells. The depolarization allows the cells to extrude more protons. Hence, uptake of K^+ ions leads to an interconversion of the $\Delta\Psi$ into a sizable transmembrane pH difference (ΔpH) and allows the cells to maintain a slightly alkaline pH under conditions at which the medium is more acidic[5,17,89,90] (see also Chapter IIG for *Rhodobacter sphaeroides*). An extreme situation occurs in acidophilic bacteria. At pH values around 2 the electrogenic uptake of K^+ is so rapid that the cells develop reversed membrane potentials (internally positive). The cells have to do this because at the existing $\Delta\bar{\mu}_{H^+}$ values of 220 mV they would otherwise not be able to maintain their internal pH value close to six.[5,88]

At the other end of the pH scale a second type of K^+ transport process is involved in the prevention of overalkalinization of the cytoplasm: H^+ uptake takes place via both Na^+/H^+ and K^+/H^+ antiporters.[5] In *E. coli*[3,50] and in *Vibrio alginolyticus* (Chapter IF) the activity of K^+/H^+ antiporter(s) increases at values of $pH_{in} > 7.6$, resulting in release of K^+ and acidification of the cytoplasm.

VI. ARTIFICIALLY INDUCED K^+ RELEASE

A number of compounds reduce the K^+ content of bacteria, particularly under conditions at which the cells maintain large ratios of $[K^+]_{in}/[K^+]_{out}$ across their cell membranes. Membrane-active bacteriocins and many bacteriophages form channels in the plasma membrane of sensitive cells (see Chapter IIJ). K^+ release from these cells is electrically compensated for by the uptake of monovalent cations (in most cases Na^+) from the medium[91] and is accompanied by a large depolarization of the cell membrane.[91,92]

Bactericidal aminoglycosides (e.g., streptomycin or gentamicin) also cause net K^+ release, but this effect is much smaller than that of the channel-forming bacteriocins. Treatment of the cells with antibiotic leads to an increase in the passive permeability of the *E. coli* cell membrane toward several ions, including streptomycin. Under these conditions the membrane potential remains relatively high.[93,94] This effect is not due to the direct interaction of the antibiotics with the membrane. Rather, the bactericidal aminoglycosides cause errors in the process of translation, followed by the incorporation of some of the mistranslated proteins into the cell membrane, leading to the observed permeability changes.[93-95] Although cell death is not the direct consequence of membrane permeabilization, the latter is nevertheless essential for the bactericidal action of these antibiotics.[93-95]

Finally, K^+ uptake mutants that contain an efflux system for tetracycline show accelerated K^+ uptake.[96] The effect probably has little to do with the mechanism of tetracycline efflux since experiments with truncated proteins have shown that K^+ uptake activity is mediated by the N-terminal portion of the protein,[97] that for tetracyclin efflux the C-terminus is also required, and that the protein even induces aminoglycoside uptake.[97,98] I propose that the efflux protein increases the passive permeability of the membrane, leading to additional, $\Delta\Psi$-driven uptake of K^+ and streptomycin.

ACKNOWLEDGMENTS

I thank W. Epstein, I. R. Booth, A. Siebers, and K. Altendorf for discussion and Manfred Schleyer for carrying out the experiments on osmoadaptation of *E. coli*. The work of my group has been supported by grants from the Deutsche Forschungsgemeinschaft (SFB171, project C1), the Fonds der Chemischen Industrie, and the European Community (contract SC-1-0334-C(A)).

REFERENCES

1. **Walderhaug, M. O., Dosch, D. C., and Epstein, W.,** Potassium transport in bacteria, in *Ion Transport in Prokaryotes,* Rosen, B. P. and Silver, S., Eds., Academic Press, New York, 1987, 84.
2. **Bakker, E. P., Booth, I. R., Dinnbier, U., Epstein, W., and Gajewska, A.,** Evidence for multiple K^+ export systems in *Escherichia coli, J. Bacteriol.,* 169, 3743, 1987.
3. **Dosch, D. C., Helmer, G. L., Sutton, S. H., Salvacion, F. F., and Epstein, W.,** Genetic analysis of potassium transport loci in *Escherichia coli:* evidence for three constitutive systems mediating uptake of potassium, *J. Bacteriol.,* 173, 687, 1991.
4. **Bakker, E. P.,** Control of futile transmembrane potassium cycling in *Escherichia coli,* in *The Roots of Biochemistry,* Kleinkauf, H., von Doehren, H., and Jaenicke, L., Eds., Walter de Gruyter, Berlin, 1988, 699.
5. **Booth, I. R.,** Regulation of cytoplasmic pH in bacteria, *Microbiol. Rev.,* 49, 359, 1985.
6. **Harold, F. M. and Altendorf, K.,** Cation transport in bacteria: K^+, Na^+, and H^+, *Curr. Top. Membr. Transp.,* 5, 1, 1974.
7. **Silver, S.,** Transport of cations and anions, in *Bacterial Transport,* Rosen, B. P., Ed., Marcel Dekker, New York, 1978, 222.
8. **Epstein, W. and Laimins, L.,** Potassium transport in *Escherichia coli:* diverse systems with common control by osmotic forces, *Trends Biochem. Sci.,* 5, 21, 1980.
9. **Epstein, W.,** Osmoregulation by potassium transport in *Escherichia coli, FEMS Microbiol. Rev.,* 39, 73, 1986.
10. **Epstein, W.,** Bacterial transport ATPases, in *Bacterial Energetics,* Vol. 12, The Bacteria: A Treatise on Structure and Function, Academic Press, Orlando, 1990, 87.
11. **Brown, A. D.,** *Microbial Water Stress Physiology. Principles and Perspectives,* John Wiley & Sons, Chichester, England, 1990.
12. **Wiggins, P. M.,** Role of water in some biological processes, *Microbiol. Rev.,* 54, 432, 1990.
13. **Kernan, R. P.,** *Cell Potassium,* John Wiley & Sons, New York, 1980.

14. **Richey, B., Cayley, D. S., Mossing, M. C., Kolka, C., Anderson, C. F., Farrer, T. C., and Record, M. T., Jr.,** Variability of the intracellular ionic environment of *Escherichia coli.* Differences between *in vitro* and *in vivo* effects of ion concentrations on protein-DNA interactions and gene expression, *J. Biol. Chem.,* 262, 7157, 1987.

15. **Michels, M. and Bakker, E. P.,** Low affinity potassium uptake system in *Bacillus acidocaldarius, J. Bacteriol.,* 169, 4341, 1987.

16. **Michel, H. and Oesterhelt, D.,** Light-induced changes of the pH gradient and the membrane potential in *H. halobium, FEBS Lett.,* 65, 175, 1976.

17. **Bakker, E. P. and Mangerich, W. E.,** Interconversion of components of the bacterial proton motive force by electrogenic potassium transport, *J. Bacteriol.,* 147, 820, 1981.

18. **Christian, J. H. B. and Waltho, J. A.,** The sodium and potassium content of non-halophilic bacteria in relation to salt tolerance, *J. Gen. Microbiol.,* 25, 97, 1961.

19. **Epstein, W. and Schultz, S. G.,** Cation transport in *Escherichia coli.* V. Regulation of cation content, *J. Gen. Physiol.,* 49, 221, 1965.

20. **Keweloh, H. and Bakker, E. P.,** Permeability changes in the cytoplasmic membrane of *Escherichia coli* K-12 early after infection with bacteriophage T1, *J. Bacteriol.,* 160, 347, 1984.

21. **Kashigawa, K. H., Kobayashi, H., and Igarashi, K.,** Apparently unidirectional polyamine transport by proton motive force in polyamine-deficient *Escherichia coli, J. Bacteriol.,* 165, 972, 1986.

22. **Lanyi, J. K. and Silverman, M. P.,** The state of binding of intracellular K$^+$ in *Halobacterium cutirubrum, Can. J. Microbiol.,* 18, 993, 972.

23. **Walsby, A. E.,** The water relationships of gas vacuolate prokaryotes, *Proc. R. Soc. London Ser. B,* 208, 73, 1980.

24. **Reed, R. H. and Walsby, A. E.,** Changes in cell turgor pressure in response to increases in external NaCl concentration in the gas-vacuolate cyanobacterium *Microcystis* sp., *Arch. Microbiol.,* 143, 290, 1985.

25. **Koch, A. L. and Pinette, M. F. S.,** Nephelometric determination of turgor pressure in growing Gram-negative bacteria, *J. Bacteriol.,* 169, 3654, 1987.

26. **Christian, J. H. B. and Waltho, J. A.,** Solute concentrations within cells of halophilic and non-halophilic bacteria, *Biochim. Biophys. Acta,* 65, 506, 1962.

27. **Measures, J. C.,** Role of amino acids in osmoregulation of nonhalophilic bacteria, *Nature (London),* 257, 1975, 398.

28. **Larsen, P. I., Sydnes, L. K., Landfald, B., and Strøm, A. R.,** Osmoregulation in *Escherichia coli* by accumulation of organic osmolytes: betaines, glutamic acid, and trehalose, *Arch. Microbiol.,* 147, 1, 1987.

29. **Dinnbier, U., Limpinsel, E., Schmid, R., and Bakker, E. P.,** Transient accumulation of potassium glutamate and its replacement by trehalose during adaptation of growing cells of *Escherichia coli* K-12 to elevated sodium chloride concentrations, *Arch. Microbiol.,* 150, 348, 1988.

30. **Kushner, D. J. and Kamekura, M.,** Physiology of halophilic bacteria, in *Halophilic Bacteria,* Vol. 1, Rodriguez-Valera, F., Ed., CRC Press, Boca Raton, FL, 1988, 109.

31. **Imhoff, J. F.,** Halophilic phototrophic bacteria, in *Halophilic Bacteria,* Vol. 1, Rodriguez-Valera, F., Ed., CRC Press, Boca Raton, FL, 1988, 85.

32. **Galinski, E. A., Pfeiffer, H.-P., and Trüper, H. G.,** 1,4,5,6-Tetrahydro-2-methyl-4-pyrimidinecarboxylic acid. A novel cyclic amino acid from halophilic phototrophic bacteria of the genus *Ectothiorhodospira, Eur. J. Biochem.,* 149, 135, 1985.

33. **Galinski, E. A. and Herzog, R. M.,** The role of trehalose as a substrate for nitrogen-containing compatible solutes *(Ectothiorhodospira halochloris), Arch. Microbiol.,* 153, 607, 1990.

34. **Whatmore, A. M., Chudek, J. A., and Reed, R. H.,** The effects of osmotic upshock on the intracellular solute pools of *Bacillus subtilis, J. Gen. Microbiol.,* 136, 2527, 1990.

35. **Schleyer, M.,** Osmoadaptation von *Escherichia coli* nach hypoosmotischem Schock, M.Sc. thesis, University of Osnabrück, Osnabrück, Germany, 1991.

36. **Welsch, D. T., Reed, R. H., and Herbert, R. A.,** The role of trehalose in osmoadaptation of *Escherichia coli* NCIB 9484: the interaction of trehalose, K⁺ and glutamate during osmoadaptation in continuous culture, *J. Gen. Microbiol.,* 137, 745, 1991.

37. **Regev, R., Gilboa, H., and Avi-Dor, Y.,** ¹³C-NMR study of the interrelation between synthesis and uptake of compatible solutes in two moderately halophilic eubacteria, *Arch. Biochem. Biophys.,* 278, 106, 1990.

38. **Grant, W. D.,** General view of Halophiles, in *Superbugs. Microorganisms in Extreme Environments,* Horikoshi, K. and Grant, W. D., Eds., Springer-Verlag, Berlin, 1990, 37.

39. **Zaccai, G. and Eisenberg, H.,** A model for the stabilization of halophilic proteins, in *Life Under Extreme Conditions. Biochemical Adaptation,* di Prisco, G., Ed., Springer-Verlag, Berlin, 1991, 125.

40. **Baldacci, G., Guinet, F., Tillet, J., Zaccai, G., and de Recondo, A.-M.,** Functional implications related to the gene structure of the elongation factor EF$_{Tu}$ from *Halobacterium marismortui, Nucleic Acids Res.,* 18, 507, 1990.

41. **Rengipat, S., Lowe, S. E., and Zeikus, J. G.,** Effect of extreme salt concentrations on physiology and biochemistry of *Halobacteroides acetoethylicus, J. Bacteriol.,* 170, 3065, 1988.

42. **Walsby, A. E.,** The pressure relationships of gas vacuoles, *Proc. R. Soc. London Ser. B,* 178, 301, 1971.

43. **Walsby, A. E.,** The pressure relations of halophilic and nonhalophilic prokaryotic cells determined using gas vacuoles as pressure probes, *FEMS Microbiol. Rev.,* 39, 45, 1986.

44. **Walsby, A. E.,** A square bacterium, *Nature (London),* 283, 69, 1980.

45. **Otozai, K., Takashina, T., and Grant, W. D.,** A novel triangular archaebacterium, *Haloarcula japonica,* in *Superbugs. Microorganisms in Extreme Environments,* Horikoshi, K. and Grant, W. D., Eds., Springer-Verlag, Berlin, 1990, 63.

46. **Leblanc, G. and Le Grimellec, C.,** Active K⁺ transport in *Mycoplasma mycoides* var. *Capri.* Net and unidirectional K⁺ movements, *Biochim. Biophys. Acta,* 554, 156, 1979.

47. **Searcy, D. G.,** *Thermoplasma acidophilum:* intracellular pH and K⁺ concentration, *Biochim. Biophys. Acta,* 451, 278, 1976.

48. **Meury, J. and Kepes, A.,** The regulation of potassium fluxes for the adjustment and maintenance of potassium levels in *Escherichia coli, Eur. J. Biochem.,* 119, 165, 1981.

49. **Rhoads, D. B. and Epstein, W.,** Cation transport in *Escherichia coli.* IX. Regulation of K transport, *J. Gen. Physiol.,* 72, 283, 1978.

50. **Yamasaki, K., Moriyami, Y., Futai, M., and Tsuchiya, T.,** Uptake and extrusion of K⁺ regulated by extracellular pH in *Escherichia coli, FEBS Lett.,* 120, 125, 1980.

51. **Bakker, E. P., Kroll, R. G., and Booth, I. R.,** Potassium transport in *Escherichia coli:* sodium is not a substrate of the potassium uptake system TrkA, *FEMS Microbiol. Lett.,* 23, 293, 1984.

52. **Boulanger, P. and Letellier, L.,** Characterization of ion channels involved in the penetration of phage T4 DNA into *Escherichia coli* cells, *J. Biol. Chem.,* 263, 9767, 1988.

53. **Rhoads, D. B., Waters, F. B., and Epstein, W.,** Cation transport in *Escherichia coli.* VIII. Potassium transport mutants, *J. Gen. Physiol.,* 67, 325, 1976.

54. **Bakker, E. P. and Harold, F. M.,** Energy coupling to potassium transport in *Streptococcus faecalis.* Interplay of ATP and the protonmotive force, *J. Biol. Chem.,* 255, 433, 1980.

55. **Rhoads, D. B., Woo, A., and Epstein, W.,** Discrimination between Rb⁺ and K⁺ by *Escherichia coli, Biochim. Biophys. Acta,* 469, 45, 1977.

56. **Epstein, W.,** The Kdp system: a bacterial K⁺ transport ATPase, *Curr. Top. Membr. Transp.,* 23, 153, 1985.

57. **Walderhaug, M. O., Polarek, J. W., Voelkner, P., Daniel, J. M., Hesse, J. E., Altendorf, K., and Epstein, W.,** KdpD and KdpE, proteins that control expression of the *kdpABC* operon, are members of the two component sensor-effector class of regulators, *J. Bacteriol.,* 174, 2152, 1992.

58. **Pedersen, P. L. and Carafoli, E.**, Ion motive ATPases. I. Ubiquity, properties and significance to cell function, *Trends Biochem. Sci.*, 12, 146, 1987.

59. **Hesse, J. E., Wieczorek, L., Altendorf, K., Reicin, A. S., Dorus, E., and Epstein, W.**, Sequence homology between two membrane transport ATPases, the Kdp-ATPase of *Escherichia coli* and the Ca^{2+}-ATPase of sarcoplasmic reticulum, *Proc. Natl. Acad. Sci. U.S.A.*, 81, 4746, 1984.

60. **Walderhaug, M. O., Litwack, E. D., and Epstein, W.**, Wide distribution of homologs of *Escherichia coli* Kdp K^+-ATPase among Gram-negative bacteria, *J. Bacteriol.*, 171, 1192, 1989.

61. **Hafer, J., Siebers, A., and Bakker, E. P.**, The high-affinity K^+-translocating ATPase complex from *Bacillus acidocaldarius* consists of three subunits, *Mol. Microbiol.*, 3, 487, 1989.

62. **Solioz, M., Mathews, S., and Fuerst, P.**, Cloning of the K^+-ATPase of *Streptococcus faecalis*. Structural and evolutionary implications of its homology to the KdpB-protein of *Escherichia coli*, *J. Biol. Chem.*, 262, 7358, 1987.

63. **Apell, H.-J. and Solioz, M.**, Electrogenic transport by the *Enterococcus hirae* ATPase, *Biochim. Biophys. Acta*, 1017, 221, 1990.

64. **Benyoucef, M., Rigaud, J.-L., and LeBlanc, G.**, Cation transport in *Mycoplasma mycoides* var Capri cells. The nature of the link between K^+ and Na^+ transport, *Biochem. J.*, 208, 539, 1982.

65. **Bossemeyer, D., Schlösser, A., and Bakker, E. P.**, Specific cesium transport via the *Escherichia coli* Kup (TrkD) K^+ uptake system, *J. Bacteriol.*, 171, 2219, 1989.

66. **Schlösser, A., Kluttig, S., Hamann, A., and Bakker, E. P.**, Subcloning, nucleotide sequence and expression of trkG, a gene that encodes an integral membrane protein involved in potassium uptake via the Trk system of *Escherichia coli*, *J. Bacteriol.*, 173, 3170, 1991.

67. **Schlösser, A., Hamann, A., Schleyer, M., and Bakker, E. P.**, The K^+-uptake systems TrkG and TrkH from *Escherichia coli*: a pair of unusual transport systems involved in osmoregulation, in *Molecular Mechanisms of Transport*, Quagliariello, E. and Palmieri, F., Eds., Elsevier, Amsterdam, 1992, 51.

68. **Rhoads, D. B. and Epstein, W.**, Energy coupling to net K^+ transport in *Escherichia coli* K-12, *J. Biol. Chem.*, 252, 1394, 1977.

69. **Bossemeyer, D., Borchard, A., Dosch, D. C., Helmer, G. C., Epstein, W., Booth, I. R., and Bakker, E. P.**, K^+ transport protein TrkA of *Escherichia coli* is a peripheral membrane protein that requires other *trk* gene products for attachment to the cytoplasmic membrane, *J. Biol. Chem.*, 264, 16403, 1989.

70. **Wagner, G., Hartmann, R., and Oesterhelt, D.**, Potassium uniport and ATP synthesis in *Halobacterium halobium*, *Eur. J. Biochem.*, 89, 169, 1978.

71. **Golby, P., Cotton, N. P. J., Carver, M., and Jackson, J. B.**, Evidence that the low-affinity K^+ transport system of *Rhodobacter capsulatus* is a uniporter. The effects of ammonium on the transporter, *Arch. Microbiol.*, 157, 125, 1992.

72. **Meury, J. and Kepes, A.**, Glutathione and the gated potassium channels of *E. coli*, *EMBO J.*, 1, 339, 1982.

73. **Elmore, M. J., Lamb, A. J., Ritchie, G. Y., Douglas, R. M., Munro, A., Gajewska, A., and Booth, I. R.**, Activation of potassium efflux from *Escherichia coli* by glutathione metabolites, *Mol. Microbiol.*, 4, 405, 1990.

74. **Douglas, R. M., Roberts, J. A., Munro, A. W., Ritchie, G. Y., Lamb, A. J., and Booth, I. R.**, The distribution of homologues of the *Escherichia coli* KefC K^+-efflux system in other bacterial species, *J. Gen. Microbiol.*, 137, 1999, 1991.

75. **Jayakumar, A., Epstein, W., and Barnes, E. M.**, Characterization of ammonium (methylammonium) potassium antiport in *Escherichia coli*, *J. Biol. Chem.*, 260, 7528, 1985.

76. **Martinac, B., Buechner, M., Delcour, A. H., Adler, J., and Kung, C.**, Pressure-sensitive ion channel in *Escherichia coli*, *Proc. Natl. Acad. Sci. U.S.A.*, 84, 2297, 1987.

77. **Berrier, C., Coulombe, A., Houssin, C., and Ghazi, A.,** A patch clamp study of ion channels of inner and outer membranes and of contact zones of *E. coli,* fused into giant liposomes. Pressure-activated channels are localized in the inner membrane, *FEBS Lett.,* 259, 27, 1989.

78. **Zoratti, M. and Petronilli, V.,** Ion-conducting channels in a Gram-positive bacterium, *FEBS Lett.,* 240, 105, 1988.

79. **Sutherland, L., Cairney, L., Elmore, M., Booth, I. R., and Higgins, C. F.,** Osmotic regulation of transcription: induction of the *proU* betaine transport gene is dependent on accumulation of intracellular potassium, *J. Bacteriol.,* 168, 805, 1986.

79a. **McLaggan, D. and Epstein, W.,** *Escherichia coli* accumulates the eukaryotic osmolyte taurine at high osmolarity, *FEMS Microbiol. Lett.,* 81, 209, 1991.

80. **Higgins, C. F., Dorman, C. J., Stirling, D. A., Waddell, L., Booth, I. R., May, G., and Bremer, E.,** A physiological role for DNA supercoiling in the osmotic regulation of gene expression in *S. typhimurium* and *E. coli, Cell,* 52, 569, 1988.

81. **Prince, W. S. and Villarejo, M. R.,** Osmotic control of *proU* transcription is mediated through direct action of potassium glutamate on the transcription complex, *J. Biol. Chem.,* 265, 17673, 1990.

82. **Tsapis, A. and Kepes, A.,** Transient breakdown of the permeability barrier of the membrane of *Escherichia coli* upon hypoosmotic shock, *Biochem. Biophys. Acta,* 469, 1, 1977.

83. **Ghazi, A.,** personal communication, 1991.

84. **Koo, S.-P., Higgins, C. F., and Booth, I. R.,** Regulation of compatible solute accumulation in *Salmonella typhimurium:* evidence for a glycine betaine efflux system, *J. Gen. Microbiol.,* 137, 2617, 1991.

85. **Tschichholz, I. and Trüper, H. G.,** Fate of compatible solutes during dilution stress in *Ectothiorhodospira halochloris, FEMS Microbiol. Ecol.,* 73, 181, 1991.

86. **Padan, E., Zilberstein, D., and Schuldiner, S.,** pH homeostasis in bacteria, *Biochim. Biophys. Acta,* 650, 131, 1982.

87. **Macnab, R. M. and Castle, A. M.,** A variable stoichiometry model for pH homeostasis in bacteria, *Biophys. J.,* 52, 637, 1987.

88. **Bakker, E. P.,** The role of alkali-cation transport in energy coupling of neutrophilic and acidophilic bacteria: an assessment of methods and concepts, *FEMS Microbiol. Rev.,* 75, 319, 1990.

89. **Kashket, E. R. and Barker, S. L.,** Effects of potassium ions on the electrical and pH gradients across the membrane of *Streptococcus lactis* cells, *J. Bacteriol.,* 130, 1017, 1977.

90. **Kroll, R. G. and Booth, I. R.,** The role of potassium transport in the generation of a pH gradient in *Escherichia coli, Biochem. J.,* 198, 691, 1981.

91. **Keweloh, H. and Bakker, E. P.,** Increased permeability and subsequent resealing of the host cell membrane early after infection of *Escherichia coli* with bacteriophage T1, *J. Bacteriol.,* 160, 1984, 354.

92. **Labedan, B. and Letellier, L.,** Membrane potential changes during first steps of coliphage infection, *Proc. Natl. Acad. Sci. U.S.A.,* 78, 215, 1981.

93. **Busse, H.-J., Wöstmann, C., and Bakker, E. P.,** The bactericidal action of streptomycin:membrane permeabilization caused by the insertion of mistranslated proteins into the cytoplasmic membrane of *Escherichia coli* and subsequent caging of the antibiotic inside the cell due to degradation of these proteins, *J. Gen. Microbiol.,* 138, 551, 1992.

94. **Davis, B. D.,** Mechanism of bactericidal action of aminoglycosides, *Microbiol. Rev.,* 51, 341, 1987.

95. **Bakker, E. P.,** Aminoglycoside and aminocyclitol antibiotics: hygromycin B is an atypical compound that exerts effects on cells of *Escherichia coli* characteristic for bacteriostatic aminocyclitols, *J. Gen. Microbiol.,* 138, 563, 1992.

96. **Dosch, D. C., Salvacion, F. F., and Epstein, W.,** Tetracycline resistence element of pBR322 mediates potassium transport, *J. Bacteriol.,* 160, 1188, 1984.

97. **Griffith, J. K., Kogoma, T., Corvo, D. L., Anderson, W. L., and Kazim, A. L.,**
An N-terminal domain of the tetracycline resistance protein increases susceptibility to
aminoglycosides and complements potassium uptake defects in *Escherichia coli, J. Bacteriol.,* 170, 598, 1988.
98. **Merlin, T. L., Davis, G. E., Anderson, W. L., Moyzis, R. K., and Griffith, J. K.,**
Aminoglycoside uptake increased by *tet* gene expression, *Antimicrob. Agents Chemother.,*
33, 1549, 1989.
99. **Epstein, W.,** personal communication, 1991.
100. **Epstein, W. and Kim, B. S.,** Potassium transport loci in *Escherichia coli* K-12, *J. Bacteriol.,* 108, 639, 1971.

Chapter IIB

K+-TRANSLOCATING Kdp-ATPases AND OTHER BACTERIAL P-TYPE ATPases

Annette Siebers and Karlheinz Altendorf

TABLE OF CONTENTS

0-8493-6982-7/93/$0.00 + $.50

I. ION-MOTIVE ATPases

Membrane ion-motive ATPases catalyze the hydrolysis of ATP to ADP and inorganic phosphate and utilize the free energy change of this reaction for the vectorial transmembrane movement of ions against their electrochemical gradient. These ion-translocating ATPases are classified into three groups, denoted F_1F_0- or F-type, vacuolar or V-type, and E_1E_2- or P-type, respectively.[1] The F-type ATPases are involved in oxidative phosphorylation in bacteria and mitochondria, in photophosphorylation in chloroplasts, and in membrane energization in fermentative bacteria. These enzymes consist of a peripheral catalytic F_1 moiety responsible for ATP synthesis or ATP hydrolysis and an integral F_0 moiety mediating the translocation of H^+ or Na^+ ions across the membrane (see also Chapter IG). The vacuolar or V-type ATPases are H^+ pumps promoting the acidification of intracellular compartments such as the vacuoles of fungi and plants, secretory vesicles, or the organelles of the exo- or endocytotic pathway. The subunit composition of V-type ATPases has not yet been completely established, but it appears to resemble that of the F-type enzymes. Sequence similarity (25% identical residues in the catalytic subunits) suggests that these two ATPases share a common ancestor.[2] Also, recent results show that the proton pumps of the archaebacteria *Sulfolobus acidocaldarius* and *Methanococcus thermolithothrophicus* are only distantly related to the F-type ATPases, whereas similarity to the eukaryotic V-type ATPases is pronounced (50% sequence identity in the catalytic subunits).[3] Enzymes of the third group, the P-type ATPases, transport Na^+, K^+, H^+, Ca^{2+}, Mg^{2+}, or Cd^{2+} across membranes of pro- and eukaryotes. During the catalytic process these ATPases cycle through two main conformational states, E_1 and E_2, and form a phosphorylated intermediate (β-aspartylphosphate). P-type ATPases are inhibited by micromolar concentrations of the phosphate analogue orthovanadate. They may be composed of up to three subunits, one of which bears the catalytic center. Despite large variations in their relative molecular masses (60 to 140 kDa), all of these catalytic subunits exhibit a similar topology in the membrane and contain regions with significant sequence similarity.[4]

There are several bacterial ion-motive ATPases that do not fit into any of the three classes of enzymes presented above — for instance, the ArsABC arsenate-resistance ATPase[5,6] and a putative Na^+-ATPase of *Methanococcus voltae* (see Section IV).[7] As more information on ion-motive ATPases becomes available, the current classification system may have to be extended.

This chapter focuses on the Kdp system from *Escherichia coli,* which was the first P-type ATPase found in bacteria. Models for the structure and function of the Kdp subunits are presented, and current knowledge about the distribution of Kdp K^+-ATPases and other P-type ATPases in prokaryotes is summarized. Bacterial transport ATPases,[8] K^+ transport in bacteria,[9] and the Kdp-ATPase itself[8,10-12] have been reviewed extensively by Epstein and his

TABLE 1
Inhibitors of Ion-Motive ATPases

Inhibitor	ATPase classes			Kdp-ATPase (K_i values, μM)
	F-type	**V-type**	**P-type**	
Vanadate	−	−	+ +	1.5
FITC	/	/	+ +	3.5
DCCD	+ +	+ +	+	60
NEM	−	+	+	100
Bafilomycin A_1	−	+ +	+	2.0

Note: To abstract from enzyme-, tissue-, or species-dependent variations within one ATPase class, the full spectrum of inhibitory values reported for one specific inhibitor in the ATPase literature has been subdivided into three rough inhibition categories. The symbols mean: + +, strong inhibition (effective in the nanomolar or micromolar range); +, moderate inhibition (effective in the micromolar and millimolar range); −, no inhibition; /, no information available. Exact K_i- values are only listed for the Kdp-ATPase. FITC = fluorescein isothiocyanate; DCCD = *N,N'*-dicyclohexylcarbodiimide; NEM = *N*-ethylmaleimide.

co-workers. For information not included in this chapter the reader is referred to this comprehensive literature and to the other chapters of this book (Section II).

II. THE Kdp-ATPase OF *ESCHERICHIA COLI*

Potassium transport across the cytoplasmic membrane of *E. coli* is a complex process: constitutive and inducible systems, influx and efflux systems, a primary pump, as well as secondary porters and stretch-activated channels are involved (for an overview, see Chapter IIA of this book). Apart from the constitutive K⁺ uptake systems Kup, TrkH, and TrkG,[13,14] the cells possess the inducible Kdp system. Since its description in 1970,[15] Kdp has been the subject of genetic,[12] physiological,[11,16] and biochemical[17,18] studies. Its intricate regulation at the levels of both transcription and enzyme activity, its extraordinarily high affinity (K_m for K⁺ uptake = 2 μM),[16] and its specificity for K⁺ ions[19] all suggest that Kdp functions as an emergency system. Upon deprivation of intracellular K⁺ or at low cell turgor pressure, Kdp scavenges traces of K⁺ from the medium down to an external K⁺ concentration of 50 nM. The transport complex is composed of three inner membrane proteins: KdpA (59, 189 Da), KdpB (72, 112 Da) and KdpC (20, 267 Da).[20,21] An ATPase activity reflecting the properties of the transport system was found to be associated with membranes of Kdp-derepressed cells, confirming that ATP is the driving force for K⁺ uptake via the Kdp system.[22] This section gives a state-of-the-art report on the Kdp-ATPase and ends with the presentation of models for structure and function of the single Kdp subunits and their integration in the complex.

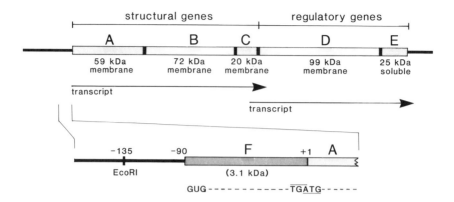

FIGURE 1. Relationship between genes, polypeptides, and mRNA of *E. coli* Kdp-ATPase. The gene order and the molecular weights are taken from References 21 and 30. Information about the transcripts is given in References 27 and 28. The *kdpF* gene has been described in Reference 28.

A. REGULATION

Intracellular K^+ concentration is regulated in such a manner that the turgor pressure remains relatively constant.[11] Within this framework, K^+ uptake through the Kdp system is regulated at two levels. At low cellular turgor pressure the Kdp-ATPase is switched on.[11] The resulting increase in internal K^+ concentration is accompanied by the synthesis of glutamate, which serves as the major counterion. Thus, K^+ and glutamate restore the turgor pressure of the cell.[11,23,24] The mechanism by which cells sense turgor pressure and translate it into variations in K^+ transport is not known. However, it is quite conceivable that the membrane exerts a direct mechanical effect on the Kdp-ATPase (see Section II.E.4). As shown by a transcriptional *kdpA::lacZ* fusion, the *kdpABC* genes are only expressed at low turgor pressure; expression is not determined by either external or internal K^+ per se.[25]

Five *kdp* genes, clustered at minute 16 on the *E. coli* chromosome,[26] are required for Kdp activity (Figure 1). Three of these genes, *kdpA, kdpB,* and *kdpC,* form the *kdpABC* operon[21] that codes for the three membrane-bound subunits of the Kdp-ATPase.[20] An adjacent operon containing the *kdpD* and *kdpE* genes codes for two proteins that regulate the expression of the *kdpABC* operon.[27] Sequence analysis of the *kdpABC* region has revealed an additional open reading frame *(kdpF)* located upstream of, and overlapping by one base pair with, *kdpA* (Figure 1).[28] The nucleotide sequence indicates that the *kdpF* product should be quite hydrophobic. The *kdpF* gene has been cloned; upon expression in minicells a protein was synthesized with a molecular weight similar to that predicted from the nucleotide sequence (3,090). However, evidence is still lacking that *kdpF* is also expressed *in vivo*.[29]

The *kdpDE* operon codes for a large membrane-bound protein (KdpD) and a smaller soluble protein (KdpE)[30] which together constitute a typical

High (Normal) Turgor

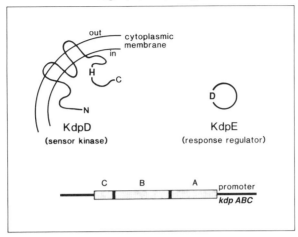

Low Turgor

FIGURE 2. Turgor-regulated *kdp* expression. A suggested model illustrating how the KdpD and KdpE proteins may act to regulate expression of the *kdpABC* operon. Low turgor is assumed to cause a conformational change in KdpD so that its phosphotransferase and kinase are activated, resulting in the phosphorylation of KdpE, which in turn binds to the *kdpABC* promoter to activate transcription. H (histidine) in KdpD and D (aspartate) in KdpE represent possible phosphorylation sites. (Modified from Walderhaug, M. O. et al., *J. Bacteriol.*, 1992.)

sensor kinase/response regulator system (Figure 2).[31] A decrease in turgor presumably leads to autophosphorylation of KdpD (sensor), which in turn acts as a phosphotransferase for KdpE (regulator). In its phosphorylated form KdpE should interact with the promoter region of the *kdpABC* operon to stimulate transcription. A hydrophobicity plot suggests that the KdpD protein (mol wt 98,656) is largely cytoplasmic, with four membrane-spanning segments near its middle. The C-terminal domain and, by inference, the N-terminal domain are probably cytoplasmic since the former contains a region homologous to other kinases. The intimate association of KdpD with the cytoplasmic membrane suggests that this protein senses turgor through changes in membrane tension produced within the bilayer. The analysis of mutants in which regulation by KdpD is abolished and/or expression is constitutive should lead to a better understanding of the sensing mechanism. Although KdpD is drawn as a monomer in Figure 2, the occurrence of intracistronic complementation[32] indicates that KdpD is present as an oligomer in the membrane.

Sequence comparison shows that KdpD and KdpE are homologous to other pairs of sensor/regulator proteins such as CreC/CreB,[33] PhoR/PhoB,[34] and EnvZ/OmpR.[35] Similarity with KdpD is not extensive, but there are several regions of closely related or identical residues, including His-673, which corresponds to the phosphorylated His-213 in PhoR[34] and His-139 in NrII (NtrB).[36] Most (six out of seven) of the mutations leading to constitutive expression of Kdp reside within the first 158 residues of KdpD.[27] Thus, it is tempting to speculate that the N-terminal part of the protein, together with the four membrane-spanning helices, plays an important role in sensing turgor. The resulting signal is then transferred to the C-terminal half and converted into kinase activity.

As predicted from its sequence, KdpE is a soluble protein with a molecular weight of 25,240. Similarity of KdpE with other regulators is much larger than that of KdpD with the other sensors.[30] The protein with the best similarity (36% identical residues) to KdpE is the PhoB protein, closely followed by that to several other regulators.[30] Of particular interest is the conservation of Asp-9 and Asp-52 in KdpE because the corresponding residues in CheY, a regulator of chemotaxis, have been shown to be important for its phosphorylation.[37,38]

In agreement with the notion that KdpD and KdpE are members of the sensor/effector group of regulators, we have shown both phosphorylation of KdpD in everted membrane vesicles from *E. coli* and transfer of the phosphate residue from KdpD to KdpE in a cytoplasmic fraction.[39]

Presumably, the phosphorylated form of KdpE enhances transcription by interacting with the promoter region of the *kdpABC* operon. The nucleotide sequence upstream of the *kdpABC* operon comprising the promoter region suggests that the RNA polymerase recognition sites (the -35 and -10 regions) strongly deviate from the consensus sequence.[40] This agrees with data from other genes which are also under positive control.[41,42] For OmpR, a transcriptional activator of the genes for the outer membrane porins OmpC and OmpF, binding sites have been found in the region from -40 to -100.[43,44] Further analyses are necessary to show whether this is also the case for KdpE.

B. PURIFICATION

Our original protocol for the isolation of the KdpABC complex was based on its solubilization by the detergent Aminoxid WS 35 (C_{7-17}-1-alkoylamino-3-dimethylaminopropane-3-N-oxide) from membranes of *kdp*-induced cells followed by several chromatographic steps.[17,45] The resulting preparation was more than 90% pure, but obviously because of the inherent instability and extreme sensitivity of the KdpB subunit to proteolytic attack[17] it had a relatively low specific ATPase activity of about 700 μmol $\times$ g^{-1} $\times$ min^{-1} and a modest activity yield (6.5%). The purified native Kdp complex and the single SDS-denatured subunits KdpA, B, and C have been used to raise monospecific polyclonal antibodies.[17]

The following intrinsic features of the Kdp transport system are preserved in the isolated Kdp-ATPase: (1) its high affinity for K$^+$ (K_m app. = 10 μM) and Mg^{2+}-ATP (K_m = 80 μM), (2) its narrow specificity for monovalent cations (neither Na$^+$, Cs$^+$, Li$^+$, Rb$^+$, nor NH$_4^+$ acts as a stimulator) and nucleotides, (3) its requirement for divalent cations (Co^{2+} and Mn^{2+} can substitute for Mg^{2+}, whereas Ca^{2+} acts as an inhibitor), and (4) its inhibitor sensitivity.

The pattern by which inhibitors affect the enzyme activity of the isolated Kdp complex is similar to that of other P-type ATPases (Table 1): Vanadate inhibits Kdp with a K_i of 1.5 μM. Fluorescein isothiocyanate (FITC) affects all eukaryotic P-type ATPases by binding to a specific lysine residue that is supposed to form part of the ATP binding site.[46] The Kdp-ATPase is protected by ATP against inactivation by FITC.[17] *N,N'*-dicyclohexylcarbodiimide (DCCD), which affects ion pumps of all three groups of ATPases,[1] also impairs Kdp function. The effect of *N*-ethylmaleimide (NEM) on Kdp and other ATPases points to the involvement of essential cysteine residues in the catalytic process. The macrolide antibiotic bafilomycin A$_1$[47] has turned out to be a valuable tool for distinguishing among the three different types of ATPases.[48] This compound represents the first highly potent and relatively specific inhibitor of V-type ATPases. P-type ATPases exhibit K_i values over a wide concentration range, whereas none of the F-type ATPases investigated so far is affected by bafilomycins. Surprisingly, the archaebacterial *S. acidocaldarius* ATPase, although very closely related to the V-type ATPases,[2,3] is not inhibited by bafilomycin A$_1$.[49]

We have recently established a more efficient purification scheme for Kdp based on affinity chromatography with the triazine dye matrix Fractogel TSK AF-Red.[50] This method yields a Kdp preparation of similar purity but with a severalfold higher recovery of ATPase activity than did the conventional method. The reduction in the preparation time minimizes KdpB degradation and results in a high specific ATPase activity (up to 2000 μmol $\times$ g^{-1} $\times$ min^{-1}).

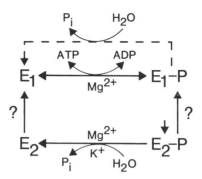

FIGURE 3. Minimum reaction scheme of the Kdp-ATPase (for details see text). (From Siebers, A. and Altendorf, K., *J. Biol. Chem.*, 264, 5831, 1989. With permission.)

C. PHOSPHORYLATED INTERMEDIATE

E_1E_2-ATPases undergo cyclic conformational changes between "high-energy" form E_1 and "low energy" form E_2 and their phosphorylated equivalents. These conversions are linked with the binding, occlusion, or release of the transported ion(s).[8] The Kdp-ATPase forms an unstable phosphointermediate during the ATPase reaction cycle.[18,51] Indirect evidence for an acylphosphate linkage has been provided by the alkali lability and the sensitivity of the enzyme-bound phosphate toward hydroxylamine. Based on studies with effectors and inhibitors, we have proposed a scheme (Figure 3) that includes (1) Mg^{2+}-dependent phosphorylation leading to the formation of the phosphorylated intermediate, (2) Mg^{2+}-dependent and K^+-stimulated dephosphorylation promoting the breakdown of the phosphointermediate, and (3) K^+-uncoupled basal dephosphorylation. For the Kdp-ATPase, experimental proof of E_1E_2 interconversions is lacking, but they are likely to occur in view of many mechanistic analogies to the eukaryotic K^+-translocating E_1E_2-ATPases.

Experiments designed to identify the phosphorylated subunit and the modified amino acid residue are under way. Upon labeling purified KdpABC complex with [γ-^{32}P]ATP, radioactivity is associated with KdpB, corroborating that this protein is the catalytic subunit of the complex.[18] By protein microsequencing of KdpB-derived cyanogen bromide fragments the phosphorylation site has been localized within a region of 100 amino acids, between Leu-283 and Met-383.[52] This agrees with sequence comparisons which predict Asp-307 to be the phosphorylated residue in Kdp.[4,21] Other data suggestive of an essential catalytic function of Asp-307 come from site-directed mutagenesis:[53] exchange of Asp-307 for a series of other amino acids completely abolishes both K^+ transport activity and phosphoenzyme formation. By contrast, mutation of the nearby Asp-300 does not cause these effects.

D. TOPOLOGY

The topology of the Kdp proteins in the membrane has been investigated with membrane vesicles of defined orientation that were exposed to membrane-impermeable reagents such as proteases[54,55] and the protein-modifying reagent [^{125}I]diazoiodosulfanilic acid.[55] From these studies, KdpB appears to be chemically modified and proteolytically digested from both membrane sides, although the latter process is more efficient from the cytoplasmic side. KdpA is almost resistant to proteases and does not react with the surface label from either side. The results with KdpC are contradictory. Whereas this subunit is not iodinated from either side of the membrane, degradation by proteases is observed from the cytoplasmic side, but only under relatively harsh conditions. We explain these data by inferring that KdpC is shielded by the extended intracellular parts of KdpB (see the model for the Kdp complex in Figure 7). This protection of KdpC is effective against surface labeling, but breaks down after digestion of the protruding part of KdpB by proteases.[55]

E. STRUCTURAL AND FUNCTIONAL MODELS

In this section models on the structure and function of the single Kdp subunits and their integration in the KdpABC complex are presented. They are based on (1) experimental data on phosphorylation and topology (see Sections II.C and II.D); (2) computer predictions about protein hydrophobicity and secondary structure, calculated from the nucleotide sequence of the *kdpABC* genes;[21] and (3) literature data on sequence comparisons and homology within the group of P-type ATPases.[4,21,56,57]

1. KdpB

KdpB is the catalytic subunit. The model of KdpB membrane organization (Figure 4) depicts an integral membrane protein with six membrane-spanning α-helices, a small and a large cytoplasmic protein domain, and only small periplasmic protein loops. This structure matches the key structural features of all P-type ATPases[58] and is also compatible with the proposed headpiece/stalk/membrane part morphology of this group of enzymes.[4] According to three-dimensional image reconstruction of the Ca^{2+}-ATPase, the cytoplasmic domains constitute a globule that is connected by a narrow stalk to the hydrophobic membrane-anchoring segment of the enzyme. In KdpB, the most probable candidates for the formation of the stalk are the extramembraneous α-helices adjacent to the membrane-anchored helices 1, 2, 3, and 5. The putative stalk helices encompass amino acid residues 13 to 22, 83 to 94, 199 to 210, and 588 to 608, respectively.

Serrano[4] has described ten regions of homology that are common to all P-type ATPases (sequences a to j in Figure 4). The only conserved region with hydrophobic character (region e) is located in transmembrane segment

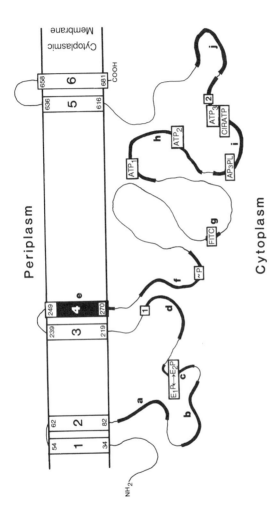

FIGURE 4. Membrane model of the KdpB protein. The model is drawn to scale, corresponding to the length of the primary sequence but neglecting secondary structure folding. Transmembraneous α-helical segments are indicated as boxes; extramembraneous regions are drawn as lines. Ten regions (a to j) known to be conserved in all P-type ATPases[4] are highlighted by blackening of membrane boxes or by thickening of nonmembrane lines. Essential residues or sequences to which a function has been assigned in other P-type ATPases have been transferred to corresponding Kdp regions; the homologous or analogous position within the Kdp sequence is conserved in Kdp and by numbers if it is not conserved. For explanation of the tags see the text.

4, adjacent to the catalytic site. A role of this stretch in energy transduction has been proposed.[59]

Insights into the molecular mechanism of cation transport by P-type ATPases come from work on the yeast plasma membrane H^+-ATPase[4,60] and the Ca^{2+}-ATPase of sarcoplasmic reticulum. With the latter enzyme, the sites of phosphorylation[57,61] ATP binding,[57,62] cation binding,[63,64] and conformational E_1P-E_2P transitions[65] have been investigated by site-directed mutagenesis. Apart from a rough identification of the site of phosphorylation (Section II.C), direct experimental evidence about functional residues of the Kdp-ATPase is lacking. However, the close relatedness between P-type ATPases allows an informed guess as to the Kdp counterparts of functional residues characterized in other members of the family.

The catalytic site — The phosphorylation site ("~P" in Figure 4) has the consensus sequence DKTGT[L/I]T in all known P-type ATPases.[57] The corresponding residue leading this sequence in Kdp is Asp-307 (see Section II.C).

The FITC binding site — The binding site for fluorescein isothiocyanate ("FITC" in Figure 4) has been proposed to be part of, or at least close to, the ATP binding region because all FITC-sensitive ATPases can be protected by addition of ATP.[17,46] A lysine has been identified as the modified residue. It is conserved in all eukaryotic P-type ATPases and the bacterial Kdp-ATPase (Lys-395), but not in the other prokaryotic cation pumps. Site-directed mutagenesis of this lysine does not support a direct involvement of this residue in ATP binding.[57]

The ATP binding site — Taylor and Green[66] have suggested a detailed model of the spatial structure of the nucleotide binding domain in cation pumps, in which conserved amino acids were grouped around a postulated ATP binding cleft. Three conserved aspartic acid residues (corresponding to Asp-447, Asp-473, and Asp-522 in KdpB; "ATP₁", "ATP₂", and "ATP₃" in Figure 4) are located within the predicted ATP binding loops. Mutagenesis of the respective residues in the yeast H^+-ATPase alters the nucleotide specificity of the enzymes, pointing to their participation in nucleotide binding.[60] Two further putative ATP binding sites have been identified by biochemical work with ATP analogues. Adenosine-triphosphopyridoxal ("AP₃PL" in Figure 4) binds to a lysine residue in the Ca^{2+}-ATPase of sarcoplasmic reticulum[67] (corresponding to Lys-499 in KdpB), and the alkylating ATP analogue γ-[4-(N-2-chloroethyl-N-methylamino)]benzylamide ("CIRATP" in Figure 4) reacts with an aspartic acid residue in the Na^+/K^+-ATPase (corresponding to Asp-518 in KdpB).[68]

Conformational changes — The sequence motif TGE is located within the small cytoplasmic domain of P-type ATPases. Site-specific mutations in the Ca^{2+}-ATPase have revealed that these three residues participate in the conformational change associated with the E_1P-E_2P transition ("$E_1P{\leftrightarrow}E_2P$" in Figure 4).[65] The corresponding motif in Kdp is Thr-159–Gly–Glu.

Certain essential residues are conserved in some cation pumps, but not in the Kdp-ATPase. These include: (1) The glycine residue that determines the sensitivity of the *Schizosaccharomyces pombe* H^+-ATPase toward vanadate[69] and seems to be involved in E_1-E_2 conformational changes of the phosphate site in the Ca^{2+}-ATPase[70] (the corresponding Kdp position is labeled "1" in Figure 4) and (2) The lysine residue (position "2" in Figure 4) that represents one of the binding sites for the ATP analogue $5'p$-fluorosulfonylbenzoyladenosine (FSBA) in the Na^+/K^+-ATPase;[71] this lysine residue is not conserved in any of the prokaryotic ATPases.[57]

2. KdpA

The KdpA subunit is a hydrophobic protein that is predicted to span the membrane with 12 α-helical segments (Figure 5). Genetic studies suggest that KdpA is involved in K^+ binding and translocation.[8] A total of 12 different *kdpA* mutants have been isolated with markedly reduced affinity for K^+, but virtually unaltered rates of transport (K_m mutants). These mutations cluster in the extramembraneous loops between transmembrane helices 2 and 3, 4 and 5, 10 and 11.[8,72] This may indicate that these loops constitute the periplasmic K^+ binding pocket. A fourth cluster of mutation sites is located within helices 7, 8, and 9. This hints to the existence of membrane-embedded K^+ binding sites which may be part of an ion channel formed by the central helices 7 to 9. Intriguingly, three negatively charged amino acids are present in helices 5, 6, and 8 of the KdpA membrane part. These residues might form a K^+ transport pathway through the membrane (Glu-302 $\rightarrow$ Glu-370 $\rightarrow$ Glu-272; see arrows in Figure 5).

KdpA does not show significant sequence similarity with any other protein. However, its overall architecture is similar to that established for several secondary membrane porters.[55,58] In particular, the assembly from 12 membrane-spanning α-helices seems to be the structural paradigm of carriers such as those for sugars, citrate, or tetracycline that function in symport or antiport with H^+ or Na^+.[73] Also, there are striking analogies of KdpA to the lactose carrier of *E. coli*[74] beyond the unifying 12-helix structural motif: the occurrence of 7 to 9 Pro residues, 3 Glu residues, an excess of positive charge within the membrane part, and, in the exposed regions, a positive net charge on the inside and a negative net charge on the outside.[75]

3. KdpC

KdpC (Figure 6) appears to contain only one membrane-spanning α-helix close to the N-terminus. The residual part of the protein is presumably oriented toward the cell interior, but is not easily accessible in the KdpABC complex (see Section II.D). Further characteristics of the secondary structure are two antiparallel β-sheets and an amphipathic α-helix at the C-terminus. The function of KdpC is still subject to debate. Mutations in the *kdpC* gene render the complex defective in both K^+ transport and ATPase activity.

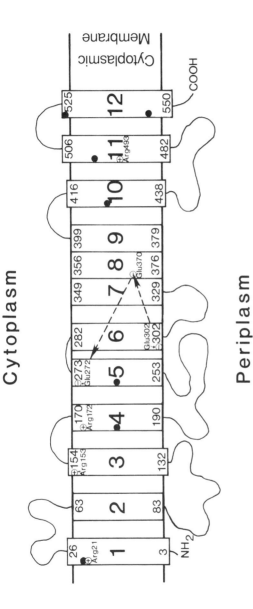

FIGURE 5. Membrane model of the KdpA protein. In the membrane-spanning α-helical segments (boxes 1 to 12) the localization of negatively charged (Glu-272, Glu-302, Glu-370) and positively charged (Arg-21, Arg-153, Arg-172, Arg-493) amino acids is indicated by the respective charge symbols; intramembraneous proline residues (position 22, 180, 263, 425, 499, 525, 545) are symbolized by filled circles. The broken-lined arrows mark a hypothetical transport pathway for K⁺ ions across the membrane. The model is drawn to scale.

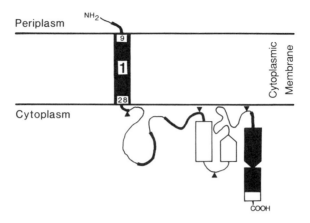

FIGURE 6. Membrane model of the KdpC protein. In this model that is drawn to scale, characteristics of the secondary structure are indicated as follows: intra- or extramembraneous α-helices are represented as boxes, β-sheets as pointed boxes, and β-turns as filled triangles. Regions of homology shared with the β_1-subunit of the Na^+/K^+-ATPase[76] are highlighted by blackening of the boxes (in regions with a distinct secondary structure) or by thickening of the lines (in regions lacking an obvious secondary structure).

The structure of KdpC is similar to that of the β_1- and β_2-isoforms of the Na^+/K^+-ATPase[76] and the β-subunit of the H^+/K^+-ATPase[77] in that all four proteins possess only one transmembrane helix leading into a large extramembraneous domain. However, this resemblance may be accidental since in the β-subunits, in contrast to the KdpC protein, the extramembraneous protein parts with the attached oligosaccharide chains are clearly oriented toward the extracellular space. In addition, for optimal alignment of the four sequences many gaps must be introduced, and the similarities within the detected blocks of homology are weak at most. Strictly conserved residues are concentrated within the membrane anchor, a glycine-rich structure at the end of the first one third of the molecule and the C-terminal region of KdpC. Neither the N-linked glycosylation sites nor the six cysteines forming three disulfide linkages in the β-subunits are conserved in KdpC.

We have recently obtained evidence for the existence of a (presumably regulatory) binding site for ATP on KdpC. Specifically, upon photoaffinity labeling of the Kdp complex with [^{32}P]8-azido-ATP not only KdpB was labeled; an even stronger incorporation of radioactivity was observed in the N-terminal cyanogen bromide fragment (Met-1 to Met-75) of KdpC.[78] The notion that ATP may play a regulatory role in addition to being a substrate has arisen independently from enzymatic studies on the complex.[18]

It has been established that the β-subunits are important for the functional assembly of eukaryotic P-type ATPases.[79] Possibly, KdpC plays a similar role in the assembly of the KdpABC complex.

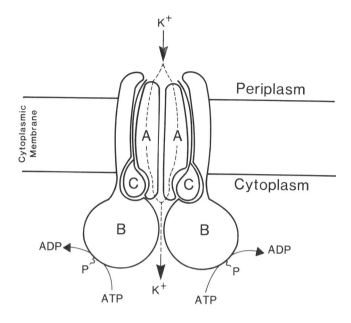

FIGURE 7. Membrane model of the KdpABC complex. Putative arrangement of subunits A, B, and C of the *E. coli* Kdp complex in the membrane (for details see text).

4. KdpABC

From genetic studies it appears that the KdpABC complex is oligomeric with respect to the KdpA protein.[10] Preliminary data on the stoichiometries of the subunits within the complex suggest that all three are present in equimolar amounts $(KdpA_2B_2C_2)$.[54] The picture emerging for the Kdp complex is that of a membrane-embedded core of two KdpA molecules surrounded by the intramembraneous segments of KdpC and KdpB. The large extramembraneous parts of KdpB assemble into a stalk-attached globule protruding into the cytoplasm and covering most of the nonmembrane regions of KdpC (Figure 7).

In comparison to other P-type K⁺-translocating ATPases the Kdp system is unique because of both its high affinity and its extremely high specificity for K⁺. During evolution, a single type of subunit may not have sufficed to reach this degree of specialization. Rather, the precursor of KdpB may have specialized in ATP hydrolysis and delegated transport to the precursor of KdpA, which originated from the family of secondary ion porters (see Section II.E.2).[55,58] This division of labor may have necessitated a third component, KdpC, to mediate between the two. Specifically, it may transmit conformational changes between the energy-producing KdpB and the energy-consuming KdpA subunit. Experimental evidence for an integrative function of KdpC comes from studies with *kdpC* or *kdpB* amber mutants.[20] Solubilization and purification experiments revealed that KdpB and KdpA fall apart in the absence of KdpC. By contrast, a stable KdpAC complex is formed in the absence of KdpB.[80]

In the Ca^{2+}-ATPase of sarcoplasmic reticulum the high-affinity binding sites for Ca^{2+} ions are localized within the transmembrane domain.[63,64] These binding sites involve four negatively charged and two polar amino acid residues (three glutamic acids, one aspartic acid, one asparagine, and one threonine). Alignment of P-type ATPase sequences shows a high degree of conservation of these six residues in all eukaryotic members of this family as well as in the prokaryotic MgtB-ATPase.[81] According to our model for KdpB (Figure 4), four of the six analogous residues are localized outside the membrane. The only acceptable membrane-integrated counterpart in KdpB would be Asp-673 (analogous to a glutamic acid residue in the Ca^{2+}-ATPase). By contrast, the KdpA subunit possesses three glutamic acid residues in the membrane part that might constitute an ion pathway through the membrane (see Section II.E.2 and Figure 5). These observations confirm the notion (see above) that KdpB has abandoned its transport function in favor of the better-adapted KdpA protein.

In analogy with the model proposed for the Ca^{2+}-ATPase,[63] K^+ transport through the KdpABC complex might proceed as follows. Upon binding of K^+ ions to extramembraneous loops of KdpA, K^+ ions are channeled to the high-affinity binding site in the membrane. Simultaneously, ATP is bound to the nucleotide-binding domain of KdpB in the E_1 conformation. Occupancy of K^+ binding sites leads to conformational changes which are transmitted to the nucleotide and phosphorylation domains to bring the bound ATP and the Asp-307 residue into a reactive configuration. After phosphate transfer, the energy stored in the E_1P form of the enzyme is converted into a conformational motion (transition E_1P into E_2P), possibly resulting in rotation or tilting of one or more of the KdpA transmembrane segments. A channel leading to the cytoplasmic side of the membrane is thereby opened, allowing diffusion of the K^+ ions along negatively charged and polar residues. KdpC might act as an additional, ATP-regulated gating device controlling the release of K^+ from the channel. The idea that KdpA functions as a gated channel is challenging; this way, phosphorylation/dephosphorylation and the superimposed regulatory ATP control could determine the open/closed state of the channel, whereas the turgor pressure could modulate the transport efficiency by influencing the width of the gate.

III. Kdp-TYPE ATPases OF OTHER BACTERIA

A priori, it would seem that high-affinity K^+ uptake systems are particularly important for free-living bacteria since these have to put up with the most drastic changes in their ionic environment.

The first report of a Kdp-like system was for *Rhizobium* sp. UMKL 20.[82] This organism possesses a membrane-bound K^+-stimulated and vanadate-sensitive ATPase activity that is repressed by growth in media containing high concentrations of K$^+$. In *R. meliloti,* a P-type ATPase has been discovered by sequencing of the *fixGHI* operon.[83] This FixI-ATPase was predicted to be a cation pump specifically required for symbiontic nitrogen fixation (Table 2). It is not known whether the two *Rhizobium* ATPases are identical.

A variety of bacteria have been screened by DNA-DNA hybridization for the presence of *kdp* genes.[84] All enterobacteria tested gave positive results with all *kdp* probes. Diverse genera of cyanobacteria *(Anabaena, Anacystis, Synechococcus)* and *Pseudomonas aeruginosa* revealed much weaker similarity. No hybridization under moderate stringency was detected to the DNA of any Gram-positive organism, to *Acholeplasma laidlawii* (see Table 2), or to two archaebacterial species. In general, the *kdpB* gene, encoding the catalytic subunit of the complex, hybridized most strongly, whereas the *kdpA* and *kdpD* sequences gave weak or negative results.[84] Independent studies with a *kdpABC* probe have revealed positive hybridization signals also in phototrophic bacteria[85] and in *Spiroplasma citri*[86] (see Table 2). The detection of a Kdp-type system in cyanobacteria[84] did not come as a surprise since transport studies with cells of *Anabaena variabilis*[87] had already documented the existence of a high-affinity K$^+$ transport system ($K_m = 40$ μM) that discriminates against Rb$^+$. At present, it is not clear whether there exists any relation between this K$^+$ uptake system and the partially sequenced cyanobacterial P-type ATPase in *Synechococcus* 6301[88] (see Table 2).

Particularly for organisms with large differences in G + C content, DNA-DNA hybridization is of limited use. For the identification of Kdp systems, antibodies have proved to be more powerful tools. For example, DNA probes failed to detect a Kdp-type sequence in either of the Gram-positive bacteria examined.[84] However, with antisera a 70-kDa homologue of KdpB was readily detected in *Bacillus acidocaldarius.*[89] Subsequently, immunological cross-reactivity of the polyclonal complex- or subunit-specific antibodies raised against *E. coli* Kdp[17] was observed with a 70-kDa protein in *Acinetobacter calcoaceticus,*[90] a 75-kDa protein in *Paracoccus denitrificans,*[89] and with proteins in the 70-kDa range in several phototrophic bacteria: *Rhodobacter sphaeroides, Rhodobacter capsulatus,* and *Rhodospirillum rubrum* (Table 2).[91] In general, the immunoreaction was only observed with the anti-KdpABC or anti-KdpB sera, indicating that the catalytic subunit had been preserved best during evolution. After the *E. coli* enzyme, the best-studied Kdp-type ATPases are those from *B. acidocaldarius* and *Rhodobacter sphaeroides.* These two systems are described in more detail below.

In *B. acidocaldarius, in vivo* transport studies identified a high-affinity K$^+$ uptake system. This system strongly resembles the *E. coli* Kdp system in that it discriminates against Rb$^+$ and is expressed only in cells grown with low external K$^+$ concentrations.[89] A K$^+$-stimulated and vanadate-sensitive ATPase activity was found to be associated with membranes of this organism. The corresponding ATPase was solubilized and partially purified by ion exchange chromatography.[92] The properties of this enzyme with respect to substrate affinity, specificity, dependence on divalent cations, and inhibitor

TABLE 2
Bacterial P-Type ATPases

ATPase	Transported ion	Organism	Molecular mass (kDa)[a]	Basis of classification					Function
				Physiological data[b]	Formation of phospho-intermediate	Immunological cross-reaction[c]	DNA hybridization[d]	DNA sequence	
KdpABC-ATPase	K^+	*Escherichia coli*	72	+	+	/	/	+	Regulation of cell turgor
		Bacillus acidocaldarius	70	+	+	+	/	/	
		Paracoccus denitrificans	75	+	/	+	/	/	
		Phototrophic bacteria	70–73	+	/	+	+	/	
H^+-ATPase	H^+	*Enterococcus hirae*	78	+	+	/	$-^d$	/	Regulation of cell pH
?-ATPase	?	*Enterococcus hirae*	63	/	/	/		+	?
FixI-ATPase	?	*Rhizobium meliloti*	80	/	/	/	$-^d$	+	Symbiotic nitrogen fixation
CadA-ATPase	Cd^{2+}	*Staphylococcus aureus*	78	/	/	/	/	+	Export of toxic Cd^{2+} ions
MgtB-ATPase	Mg^{2+}	*Salmonella typhimurium*	101	/	/	/	/	+	Uptake of Mg^{2+} ions
		Mollicutes							
?-ATPase	?	*Acholeplasma laidlawii*	80, 60/68	/	+	+	$-^d$	/	?
?-ATPase	?	*Mycoplasma gallisepticum*	68	+	/	+	/	/	?

Enzyme	Transported ion	Organism	M_r	a	b	c	d	Physiological role
Ca²⁺-ATPase	Ca²⁺	*Spiroplasma citri*	?	/	+	/	/	Maintenance of a low cytoplasmic Ca²⁺ concentration
		Streptococci	37	/	+	/	−[d]	
?-ATPase	?	Cyanobacteria *Synechococcus* 6301	?	/	/	+[d]	(+)	?

Note: The symbols used in this table are: +, positive result; −, negative result; /, no data available; ?, transported ion unknown.

a   The molecular mass of the catalytic subunit is given: in boldface, if calculated from the DNA sequence; in regular print, if derived from the migration behavior in SDS polyacrylamide gels.

b   Physiological data can result from transport studies and/or ATPase activity measurements in cells, membranes, or proteoliposomes and from the characterization of (partially) purified enzyme. An important criterion for classification as an P-type ATPase is sensitivity to vanadate.

c   Immunological methods comprise immunoblotting or immunoprecipitation experiments with antibodies specific for another member of the P-type ATPase family.

d   Data marked by this superscript come from Reference 84. In this study the genomes of a wide variety of distantly related bacteria were screened for heterologous hybridization using the *kdp* genes of *E. coli* as a probe. References for independent hybridization studies are given in the text.

sensitivity are virtually identical to those of the *E. coli* Kdp-ATPase. Furthermore, the *Bacillus* enzyme also forms a phosphorylated intermediate.[93] The enzyme appears to consist of three subunits with molecular masses of 70, 44, and 23 kDa.[92] The 70-kDa and 44-kDa proteins cross-react with antisera generated against the *E. coli* KdpB and KdpA subunits, respectively.[93] Taken together, these results argue strongly for a close relatedness of the ATPases from the two organisms.

With the photoheterotrophic purple nonsulfur bacterium *Rhodobacter sphaeroides,* transport studies have revealed the existence of a high-affinity K^+ uptake system in cells grown at low external K^+ concentrations.[94] The ATPase activity of membranes is stimulated by K^+ and is sensitive to vanadate.[95] The solubilized enzyme was purified by dye-ligand affinity chromatography[95] essentially according to the one-column procedure described for *E. coli*[50] (see Section II.B). The enzyme from *R. sphaeroides* consists of three polypeptides with molecular masses of 70, 43.5, and 23.5 kDa. There is one important difference with the *E. coli* enzyme. In the latter KdpB tends to decompose,[17] whereas the catalytic subunit in the *R. sphaeroides* enzyme is stable.[95] This inherent stability may make the *R. sphaeroides* enzyme the candidate of choice for crystallization attempts.

So far we have restricted the discussion in this section to eubacteria. An interesting question is whether P-type ATPases exist in archaebacteria as well. For *Methanobacterium thermoautotrophicum* it has been shown that K^+ can be accumulated 50,000-fold through a nonuniport mechanism.[96] This result prompted us to search for a Kdp-type system in this organism. To this end, cells were grown anaerobically in mineral media containing high or low K^+ concentrations. Only one protein, a 32-kDa polypeptide, is additionally synthesized by K^+-limited cells. This protein, which can be extracted from the membranes by Aminoxid, did not exhibit a K^+- or Na^+-stimulated ATPase activity, and immunoblotting revealed no specific cross-reactivity of any *M. thermoautotrophicum* protein with the anti-*E. coli* Kdp sera.[97] These data suggest that *M. thermoautotrophicum* may possess an inducible K^+ uptake system that is not of the Kdp type.[97]

IV. OTHER BACTERIAL P-TYPE ATPases

Until recently, common opinion has held that P-type ATPases are a typical feature of eukaryotes. Prokaryotes were considered to regulate the ionic composition of their cytoplasm mainly via secondary symporters or antiporters. However, since the Kdp system from *E. coli* was identified as a P-type ATPase,[21] the family of bacterial P-type ATPases has increased steadily (Table 2).

Solioz and co-workers[98,99] have isolated an ion-translocating P-type ATPase from *Enterococcus hirae* (formerly *Streptococcus faecalis*). The original suggestion that the enzyme pumps K^+ has been revised in favor of a role in H^+ export.[100] The same group has determined the nucleotide sequence of a P-type ATPase gene.[101] This gene does not encode the protein that has been isolated,[8] indicating that *E. hirae* must contain at least two different P-type ATPases (see Table 2). Note that in this organism the extrusion of Na$^+$ and of Ca^{2+} is also mediated by primary pumps (see also Chapter IID). For *Streptococcus lactis* and *Streptococcus sanguis* there is a good evidence that P-type ATPases are involved in Ca^{2+} export (see Table 2).[102]

Cd^{2+} resistance in *Staphylococcus aureus* is encoded by a plasmid and based on active extrusion of this toxic ion. The Cd^{2+}-ATPase involved in this activity exhibits homology to P-type cation pumps, especially those of *E. hirae* and *Escherichia coli*.[5,103] So far, *in vitro* measurements of ATPase activity or protein phosphorylation have not been successful due to difficulties in overexpression of the *cadA* gene.[103]

Salmonella typhimurium contains three distinct transport systems for Mg^{2+}: CorA, MgtA, and MgtB.[104] MgtB, an uptake system, is repressed at high extracellular Mg^{2+} concentrations. The amino acid sequence deduced from the nucleotide sequence of the *mgtb* locus indicates that the MgtB protein is a P-type ATPase.[81] Strikingly, this enzyme is more similar to the Ca^{2+}-ATPases of yeast and mammalian sarcoplasmic reticulum than to the P-type ATPases of other prokaryotes.[8,81] Another gene, coding for a membrane-integrated 22.5-kDa protein designated MgtC, is organized with *mgtb* in the same operon.[81] MgtC may constitute an accessory subunit of the Mg^{2+}-ATPase; it shows no similarity to any protein recorded in the data bases.[81]

Reports on energy-driven ion movements in Mollicutes point to the existence of multiple membrane ATPases of the F-type as well as of the P-type. *Acholeplasma laidlawii* contains a Mg^{2+}-stimulated Na$^+$-ATPase[105] that has been purified from the plasma membrane[106] and probably classifies as an F-type enzyme. Two studies provide evidence for the occurrence of P-type ATPases in *A. laidlawii*. First, phosphorylation of crude plasma membranes with [γ-^{32}P]ATP yields acylphosphate polypeptides of 80 and 60 kDa.[107] It is unclear whether these represent two separate phosphoenzymes of whether the smaller is the result of proteolytic degradation of the larger. Second, *A. laidlawii* membranes were examined for immunological cross-reactivity with an antiserum against the yeast plasma membrane H$^+$-ATPase.[108] The immunoblotting assay identified a strongly reacting 68-kDa protein. Also, *Mycoplasma gallisepticum* appears to possess an electrogenic, ATP-driven Na$^+$ pump.[108] Two distinct ATPases are associated with the membranes as monitored by differences in pH profile, Na$^+$ stimulation, and inhibitor sensitivity. The coexistence of F- and P-type ATPases in *M. gallisepticum* has been substantiated by immunoblotting.[108] Membranes of a third representative of the Mollicutes, *Spiroplasma citri*, were solubilized and the extract separated by chromatofocusing.[86] Here also, two distinct ATPase activities differing in pH optima and sensitivity to vanadate could be distinguished. One of these corresponds to an F-type ATPase and the other presumably to a P-type ATPase.[86] A 37-kDa polypeptide was the major compound in the vanadate-sensitive fraction. The ultimate proof that the 37-kDa protein is the authentic

ATPase and not a contaminant or a breakdown product is still lacking. The minimal size of a functional P-type ATPase lies around 60 kDa (Table 2), and it seems unlikely that a single monomeric 37-kDa protein can fulfill all of the transport-linked functions.

In the archaebacterium *Methanococcus voltae,* an ATPase of unknown function has been purified and characterized.[109] This ATPase was originally classified as a P-type ATPase because it exhibited characteristic features of this ATPase category such as sensitivity to vanadate and formation of an acylphosphate intermediate.[109] Surprisingly, its nucleotide sequence revealed that the protein represents a novel type of peripheral membrane ATPase. Data bank searches have failed to identify a relationship to any other ATPase.[7]

V. CONCLUDING REMARKS

The current picture of the origin of ion-motive ATPases[4,110] suggests that, in primitive cells, simple versions of H^+-pumping ATPases, the ancestors of F- and V-type ATPases, evolved as a device to balance the osmotic net influx of extracellular ions and to extrude protons generated by fermentative metabolism. P-type ATPases are regarded as a later invention; they appear to be assembled from preexisting elements.

Whereas the F- and V-type ATPases remained dedicated o the transport of H^+ and occasionally Na^+ (see Chapter IG), the P-type ATPases developed transport abilities for a wide range of mono- and divalent cations: H^+, Na^+, K^+, Ca^{2+}, Mg^{2+}, and Cd^{2+}. The proteins originating from a common P-type ATPase ancestor underwent modifications, evolving the capability to transport different cations inwardly and outwardly through different membranes according to the needs for the survival of the cell. As noted by Snavely et al.,[81] it is striking that the simplest P-type ATPases, i.e., those consisting of one single subunit, are all involved in export of ions from the cytoplasm either to the extracellular space or to intracellular compartments (for instance, the H^+- or Ca^{2+}-ATPases of protozoa, fungi, and plant/animal cells, the H^+-ATPase of *Enterococcus hirae,* or the Cd^{2+}-ATPase from *Staphylococcus aureus*). By contrast, pumps mediating the inward movement of cations (Kdp, MgtB) or combining influx and export functions (Na^+/K^+- or H^+/K^+-ATPase) appear to require accessory subunits. Possibly, the asymmetry of functional sites, i.e., extracellular sites for cation binding and intracellular sites for ATP binding and hydrolysis, demand more sophisticated protein structures. The accessory glycoproteins (the β-subunits) of the Na^+/K^+- and H^+/K^+-ATPase might be an adaptation to the protein export machinery of the animal cell, and the glycosylation meets the surface protein requirement of the mammalian cells. The function of MgtC, the putative accessory subunit of the Mg^{2+}-ATPase from *S. typhimurium* is completely unclear.[81] The Kdp-ATPase, with its two accessory subunits, is the most specialized P-type ATPase. This complex possesses an outstanding functionality in terms of ion selectivity and transport capacity. The role of K^+ as a preferred and essential osmoregulatory ion in bacteria explains the widespread distribution of the Kdp-type ATPase among such distantly related eubacterial groups as Gram-positive and Gram-negative organisms, cyanobacteria, phototrophic, and enteric bacteria.

K+ translocation by the KdpABC complex is the coordinated work of three subunits. This makes the enzyme an excellent model system for the study of subunit assembly and of communication between components of a hetero-oligomeric enzyme complex. Another challenging feature of the Kdp system is its regulation by turgor pressure. An environmental stimulus evoking mechanical forces in the membrane is transmitted into a signal acting at the level of gene expression and, at the same time, exerts control on the activity of already synthesized and membrane-integrated Kdp transport complexes. The investigation of the underlying intricate regulatory network is the subject of our current research.

ACKNOWLEDGMENTS

We thank Drs. Tilly Bakker-Grunwald and Evert Bakker for invaluable help in preparing the manuscript. We thank Johanna Petzold for skillfully typing several successive versions of the manuscript. Research in our laboratory on Kdp-ATPase is funded by the Deutsche Forschungsgemeinschaft (SFB 171, B5) and the Fonds der Chemischen Industrie.

REFERENCES

1. **Pedersen, P. L. and Carafoli, E.,** Ion motive ATPases. I. Ubiquity, properties, and significance to cell function, *Trends Biochem. Sci.,* 12, 146, 1987.
2. **Nelson, N. and Taiz, L.,** The evolution of H+-ATPases, *Trends Biochem. Sci.,* 14, 113, 1989.
3. **Gogarten, J. P., Kibak, H., Dittrich, P., Taiz, L., Bowman, E. J., Bowman, B. J., Manolson, M. F., Poole, R. J., Date, T., Oshima, T., Konishi, J., Denda, K., and Yoshida, M.,** Evolution of the vacuolar H+-ATPase: implications for the origin of eukaryotes, *Proc. Natl. Acad. Sci. U.S.A.,* 86, 6661, 1989.
4. **Serrano, R.,** Structure and function of proton translocating ATPase in plasma membranes of plants and fungi, *Biochim. Biophys. Acta,* 947, 1, 1988.
5. **Silver, S., Nucifora, G., Chu, L., and Misra, T. K.,** Bacterial resistance ATPases: primary pumps for exporting toxic cations and anions, *Trends Biochem. Sci.,* 14, 76, 1989.
6. **Rosen, B. P., Hsu, C.-M., Karkaria, C. E., Owolabi, J. B., and Tisa, L. S.,** Molecular analysis of an ATP-dependent ion pump, *Philos. Trans. R. Soc. London Ser. B,* 326, 455, 1990.

7. **Dharmavaram, R., Gillevet, P., and Konisky, J.**, Nucleotide sequence of the gene encoding the vanadate-sensitive membrane-associated ATPase of *Methanococcus voltae, J. Bacteriol.*, 173, 2131, 1991.

8. **Epstein, W.**, Bacterial transport ATPases, in *Bacterial Energetics*, Vol. 12, The Bacteria: A Treatise on Structure and Function, Krulwich, T. A., Ed., Academic Press, Orlando, 1990, 87.

9. **Walderhaug, M. O., Dosch, D. C., and Epstein, W.**, Potassium transport in bacteria, in *Ion Transport in Prokaryotes*, Rosen, B. P. and Silver, S., Eds., Academic Press, New York, 1987, 84.

10. **Epstein, W.**, The Kdp system: a bacterial K^+ transport ATPase, *Curr. Top Membr. Transp.*, 23, 153, 1985.

11. **Epstein, W.**, Osmoregulation by potassium transport in *Escherichia coli, FEMS Microbiol. Rev.*, 39, 73, 1986.

12. **Polarek, J. W., Walderhaug, M. O., and Epstein, W.**, Genetics of Kdp, the K^+-transport ATPase of *Escherichia coli, Methods Enzymol.*, 157, 655, 1988.

13. **Bossemeyer, D., Schlösser, A., and Bakker, E. P.**, Specific cesium transport via the *Escherichia coli* Kup (TrkD) K^+ uptake system, *J. Bacteriol.*, 171, 2219, 1989.

14. **Dosch, D. C., Helmer, G. L., Sutton, S. H., Salvacion, F. F., and Epstein, W.**, Genetic analysis of potassium transport loci in *Escherichia coli:* evidence for three constitutive systems mediating uptake of potassium, *J. Bacteriol.*, 173, 687, 1991.

15. **Epstein, W. and Davies, M.**, Potassium-dependent mutants of *Escherichia coli* K-12, *J. Bacteriol.*, 101, 836, 1970.

16. **Rhoads, D. B., Waters, F. B., and Epstein, W.**, Cation transport in *Escherichia coli.* VIII. Potassium transport mutants, *J. Gen. Physiol.*, 67, 325, 1976.

17. **Siebers, A. and Altendorf, K.**, The K^+-translocating Kdp-ATPase from *Escherichia coli:* purification, enzymatic properties and production of complex- and subunit-specific antisera, *Eur. J. Biochem.*, 178, 131, 1988.

18. **Siebers, A. and Altendorf, K.**, Characterization of the phosphorylated intermediate of the K^+-translocating Kdp-ATPase from *Escherichia coli, J. Biol. Chem.*, 264, 5831, 1989.

19. **Rhoads, D. B., Woo, A., and Epstein, W.**, Discrimination between Rb^+ and K^+ by *Escherichia coli, Biochim. Biophys. Acta*, 469, 45, 1977.

20. **Laimins, L. A., Rhoads, D. B., Altendorf, K., and Epstein, W.**, Identification of the structural proteins of an ATP-driven potassium transport system in *Escherichia coli, Proc. Natl. Acad. Sci. U.S.A.*, 75, 3216, 1978.

21. **Hesse, J. E., Wieczorek, L., Altendorf, K., Reicin, A. S., Dorus, E., and Epstein, W.**, Sequence homology between two membrane transport ATPases, the Kdp-ATPase of *Escherichia coli* and the Ca^{2+}-ATPase of sarcoplasmic reticulum, *Proc. Natl. Acad. Sci. U.S.A.*, 81, 4746, 1984.

22. **Epstein, W., Whitelaw, V., and Hesse, J.**, A K^+ transport ATPase in *Escherichia coli, J. Biol. Chem.*, 253, 6666, 1978.

23. **Larsen, P. I., Sydnes, L. K., Landfald, B., and Strøm, A. R.**, Osmoregulation in *Escherichia coli* by accumulation of organic osmolytes: betaines, glutamic acid, and trehalose, *Arch. Microbiol.*, 147, 1, 1987.

24. **Dinnbier, U., Limpinsel, E., Schmid, R., and Bakker, E. P.**, Transient accumulation of potassium glutamate and its replacement by trehalose during adaptation of growing cells of *Escherichia coli* K12 to elevated sodium chloride concentrations, *Arch. Microbiol.*, 150, 348, 1988.

25. **Laimins, L. A., Rhoads, D. B., and Epstein, W.**, Osmotic control of *kdp* operon expression in *Escherichia coli, Proc. Natl. Acad. Sci. U.S.A.*, 78, 464, 1981.

26. **Bachmann, B. J.**, Linkage map of *Escherichia coli* K-12, edition 8, *Microbiol. Rev.*, 54, 130, 1990.

27. **Polarek, J. W., Williams, G., and Epstein, W.**, The products of the *kdpDE* operon are required for the expression of the Kdp ATPase of *Escherichia coli, J. Bacteriol.*, 174, 2145, 1992.

28. **Polarek, J. W.,** The Identification and Analysis of the Regulatory Genes of the Kdp System of *Escherichia coli,* Ph.D. thesis, University of Chicago, Chicago, IL, 1986.

29. **Möllenkamp, T. and Altendorf, K.,** unpublished data, 1991.

30. **Walderhaug, M. O., Polarek, J. W., Voelkner, P., Daniel, J. M., Hesse, J. E., Altendorf, K., and Epstein, W.,** KdpD and KdpE, proteins that control expression of the *kdpABC* operon, are members of the two-component sensor-effector class of regulators, *J. Bacteriol.,* 174, 2152, 1992.

31. **Stock, J. B., Ninfa, A. J., and Stock, A. M.,** Protein phosphorylation and regulation of adaptive responses in bacteria, *Microbiol. Rev.,* 53, 450, 1989.

32. **Rhoads, D. B., Laimins, L., and Epstein, W.,** Functional organization of the *kdp* genes of *Escherichia coli* K12, *J. Bacteriol.,* 135, 445, 1978.

33. **Wanner, B. L.,** Phosphorus assimilation and its control of gene expression in *Escherichia coli,* in *The Molecular Basis of Bacterial Metabolism,* 41st Colloquium Mosbach, Hauska, G. and Thauer, R., Eds., Springer-Verlag, Heidelberg, Germany, 1990, 152.

34. **Makino, K., Shinagawa, H., Amemura, M., Kawamoto, T., Yamada, M., and Nakata, A.,** Signal transduction in the phosphate regulon of *Escherichia coli* involves phosphotransfer between PhoR and PhoB proteins, *J. Mol. Biol.,* 210, 551, 1989.

35. **Mizuno, T. and Mizushima, S.,** Signal transduction and gene regulation through the phosphorylation of two regulatory components: the molecular basis for the osmotic regulation of the porin genes, *Mol. Microbiol.,* 4, 1077, 1990.

36. **Ninfa, A. J. and Bennett, R. L.,** Identification of the site of autophosphorylation of the bacterial protein kinase/phosphatase Nr_{II}, *J. Biol. Chem.,* 266, 6888, 1991.

37. **Bourret, R. B., Hess, J. F., and Simon, M. I.,** Conserved aspartate residues and phosphorylation in signal transduction by the chemotaxis protein CheY, *Proc. Natl. Acad. Sci. U.S.A.,* 87, 41, 1990.

38. **Sanders, D. A., Gillece-Castro, B. L., Stock, A. M., Burlingame, A. L., and Koshland, D. E., Jr.,** Identification of the site of phosphorylation of the chemotaxis response regulator protein, CheY, *J. Biol. Chem.,* 246, 21770, 1989.

39. **Voelkner, P. and Altendorf, K.,** unpublished data, 1991.

40. **Möllenkamp, T. and Altendorf, K.,** unpublished data, 1992.

41. **Mizuno, T., Chou, M.-Y., and Inouye, M.,** DNA sequence of the promoter region of the *ompC* gene and the amino acid sequence of the signal peptide of pro-OmpC protein of *Escherichia coli, FEBS Lett.,* 151, 159, 1983.

42. **Raibaud, O. and Schwartz, M.,** Positive control of transcription initiation in bacteria, *Annu. Rev. Genet.,* 18, 173, 1984.

43. **Norioka, S., Ramakrishnan, G., Ikenaha, K., and Inouye, M.,** Interaction of a transcriptional activator, OmpR, with reciprocally osmoregulated genes, *ompF* and *ompC,* of *Escherichia coli, J. Biol. Chem.,* 261, 17,113, 1986.

44. **Jo, Y.-L., Nara, F., Ichihara, S., Mizuno, T., and Mizushima, S.,** Purification and characterization of the OmpR protein, a positive regulator involved in osmoregulatory expression of the *ompF* and *ompC* genes in *Escherichia coli, J. Biol. Chem.,* 261, 15,252, 1986.

45. **Siebers, A., Wieczorek, L., and Altendorf, K.,** K⁺-ATPase from *Escherichia coli:* isolation and characterization, *Methods Enzymol.,* 157, 668, 1988.

46. **Farley, R. A. and Faller, L. D.,** The amino acid sequence of an active site peptide from the H,K-ATPase of gastric mucosa, *J. Biol. Chem.,* 260, 3899, 1985.

47. **Werner, G., Hagenmaier, H., Drautz, H., Baumgartner, A., and Zähner, H.,** Bafilomycins, a new group of macrolide antibiotics. Production, isolation, chemical structure and biological activity, *J. Antibiot.,* 37, 110, 1984.

48. **Bowman, E. J., Siebers, A., and Altendorf, K.,** Bafilomycins: a class of inhibitors of membrane ATPases from microorganisms, animal cells, and plant cells, *Proc. Natl. Acad. Sci. U.S.A.,* 85, 7972, 1988.

49. **Schäfer, G.,** personal communication, 1989.

50. **Siebers, A., Kollmann, R., Dirkes, G., and Altendorf, K.**, Rapid, high-yield purification and characterization of the K^+-translocating Kdp-ATPase from *Escherichia coli*, *J. Biol. Chem.*, 267, 12,717, 1992.

51. **Epstein, W., Laimins, L., and Hesse, J.**, A phosphorylated intermediate of the Kdp system, an ATP-driven K^+ transport system of *E. coli*, presented at the 11th Int. Congr. Biochemistry, Toronto, July 8 to 13, 1979, 449.

52. **Siebers, A. and Broermann, R.**, unpublished data, 1990.

53. **Puppe, W., Siebers, A., and Altendorf, K.**, The phosphorylation site of the Kdp-ATPase of *Escherichia coli*. Site-directed mutagenesis of aspartic acid residues 300 and 307 of the KdpB subunit, *Mol. Microbiol.*, in press.

54. **Laimins, L. A.**, Organization of the KDP Potassium Transport System of *Escherichia coli*, Ph.D. thesis, University of Chicago, Chicago, IL, 1981.

55. **Siebers, A.**, Kalium-Transport bei *Escherichia coli:* Funktionelle, immunologische und topographische Untersuchungen der Kdp-ATPase, Ph.D. thesis, Universität Osnabrück, Osnabrück, Germany, 1988.

56. **Jørgensen, P. L. and Andersen, J. P.**, Structural basis for E_1-E_2 conformational transitions in Na,K-pumps and Ca-pump proteins, *J. Membr. Biol.*, 103, 95, 1988.

57. **Maruyama, K., Clarke, D. M., Fujii, J., Inesi, G., Loo, T. W., and MacLennan, D. H.**, Functional consequences of alterations to amino acids located in the catalytic center (isoleucine 348 to threonine 357) and nucleotide-binding domain of the Ca^{2+}-ATPase of sarcoplasmic reticulum, *J. Biol. Chem.*, 264, 13038, 1989.

58. **Epstein, W., Walderhaug, M. O., Polarek, J. W., Hesse, J. E., Dorus, E., and Daniel, J. M.**, The bacterial Kdp K^+-ATPase and its relation to other transport ATPases, such as the Na^+/K^+- and Ca^{2+}-ATPases in higher organisms, *Philos. Trans. R. Soc. London Ser. B*, 326, 479, 1990.

59. **Shull, G. E., Schwartz, A., and Lingrel, J. B.**, Amino-acid sequence of the catalytic subunit of the $(Na^+ + K^+)$ATPase deduced from a complementary DNA, *Nature (London)*, 316, 691, 1985.

60. **Portillo, F. and Serrano, R.**, Dissection of functional domains of the yeast proton-pumping ATPase by directed mutagenesis, *EMBO J.*, 7, 1793, 1988.

61. **Maruyama, K. and MacLennan, D. H.**, Mutation of aspartic acid-351, lysine-352, and lysine-515 alters the Ca^{2+} transport activity of the Ca^{2+}-ATPase expressed in COS-1 cells, *Proc. Natl. Acad. Sci. U.S.A.*, 85, 3314, 1988.

62. **Clarke, D. M., Loo, T. W., and MacLennan, D. H.**, Functional consequences of alterations to amino acids located in the nucleotide binding domain of the Ca^{2+}-ATPase of sarcoplasmic reticulum, *J. Biol. Chem.*, 265, 22223, 1990.

63. **Clarke, D. M., Loo, T. W., Inesi, G., and MacLennan, D. H.**, Location of high affinity Ca^{2+}-binding sites within the predicted transmembrane domain of the sarcoplasmic reticulum Ca^{2+}-ATPase, *Nature (London)*, 339, 476, 1989.

64. **Clarke, D. M., Loo, T. W., and MacLennan, D. H.**, Functional consequences of alterations to polar amino acids located in the transmembrane domain of the Ca^{2+}-ATPase of sarcoplasmic reticulum, *J. Biol. Chem.*, 265, 6262, 1990.

65. **Clarke, D. M., Loo, T. W., and MacLennan, D. H.**, Functional consequences of mutations of conserved amino acids in the β-strand domain of the Ca^{2+}-ATPase of sarcoplasmic reticulum, *J. Biol. Chem.*, 265, 14088, 1990.

66. **Taylor, W. R. and Green, N. M.**, The predicted secondary structure of the nucleotide-binding sites of six cation-transporting ATPases lead to a probable tertiary fold, *Eur. J. Biochem.*, 179, 241, 1989.

67. **Yamamoto, H., Tagaya, M., Fukui, T., and Kawakita, M.**, Affinity labeling of the ATP-binding site of Ca^{2+}-transporting ATPase of sarcoplasmic reticulum by adenosine triphosphopyridoxal: identification of the reactive lysyl residue, *J. Biochem. (Tokyo)*, 103, 452, 1988.

68. **Ovchinnikov, Y. A., Dzhandzugazyan, K. N., Lutsenko, S. V., Mustayev, A. A., and Modyanov, N. N.,** Affinity modification of E_1-form of Na⁺,K⁺-ATPase revealed Asp-710 in the catalytic site, *FEBS Lett.*, 217, 111, 1987.

69. **Ghislain, M., Schlesser, A., and Goffeau, A.,** Mutation of a conserved glycine residue modifies the vanadate sensitivity of the plasma membrane H⁺-ATPase from *Schizosaccharomyces pombe, J. Biol. Chem.*, 262, 17549, 1987.

70. **Andersen, J. P., Vilsen, B., Leberer, E., and MacLennan, D. H.,** Functional consequences of mutations in the β-strand sector of the Ca²⁺-ATPase of sarcoplasmic reticulum, *J. Biol. Chem.*, 264, 21018, 1989.

71. **Ohta, T., Nagano, K., and Yoshida, M.,** The active site structure of Na⁺/K⁺-transporting ATPase: location of the 5′-(p-fluorosulfonyl)benzoyladenosine binding site and soluble peptides released by trypsin, *Proc. Natl. Acad. Sci. U.S.A.*, 83, 2071, 1986.

72. **Epstein, W.,** personal communication, 1988.

73. **Hendersen, P. J. F. and Maiden, M. C. J.,** Homologous sugar transport proteins in *Escherichia coli* and their relatives in both prokaryotes and eukaryotes, *Philos. Trans. R. Soc. London Ser. B,* 326, 391, 1990.

74. **Kaback, H. R., Bibi, E., and Roepe, P. D.,** β-Galactoside transport in *E. coli:* a functional dissection of *lac* permease, *Trends Biochem. Sci.*, 15, 309, 1990.

75. **von Heijne, G.,** The distribution of positively charged residues in bacterial inner membrane proteins correlates with the trans-membrane topology, *EMBO J.*, 5, 3021, 1986.

76. **Shull, G. E., Lane, L. K., and Lingrel, J. B.,** Amino-acid sequence of the β-subunit of the (Na⁺ + K⁺)ATPase deduced from a cDNA, *Nature (London)*, 321, 429, 1986.

77. **Shull, G. E.,** cDNA cloning of the β-subunit of the rat gastric H,K-ATPase, *J. Biol. Chem.*, 265, 12123, 1990.

78. **Dröse, S. and Siebers, A.,** unpublished results, 1991.

79. **McDonough, A. A., Geering, K., and Farley, R. A.,** The sodium pump needs its β subunit, *FASEB J.*, 4, 1598, 1990.

80. **Siebers, A., Epstein, W., and Altendorf, K.,** unpublished data, 1991.

81. **Snavely, M. D., Miller, C. G., and Maguire, M. E.,** The *mgtB* Mg²⁺ transport locus of *Salmonella typhimurium* encodes a P-type ATPase, *J. Biol. Chem.*, 266, 815, 1991.

82. **Lim, S. T.,** K⁺-ATPase from *Rhizobium* sp. UMKL 20, *Arch. Microbiol.*, 142, 393, 1985.

83. **Kahn, D., David, M., Domergue, O., Daveran, M.-L., Ghai, J., Hirsch, P. R., and Batut, J.,** *Rhizobium meliloti fixGHI* sequence predicts involvement of a specific cation pump in symbiotic nitrogen fixation, *J. Bacteriol.*, 171, 929, 1989.

84. **Walderhaug, M. O., Litwack, E. D., and Epstein, W.,** Wide distribution of homologs of *Escherichia coli* Kdp K⁺-ATPase among Gram-negative bacteria, *J. Bacteriol.*, 171, 1192, 1989.

85. **Abee, T. and Bakker, E. P.,** unpublished data, 1989.

86. **Simoneau, P. and Labarère, J.,** Evidence for the presence of two distinct membrane ATPases in *Spiroplasma citri, J. Gen. Microbiol.*, 137, 179, 1991.

87. **Reed, R. H., Rowell, P., and Stewart, W. D. P.,** Uptake of potassium and rubidium ions by the cyanobacterium *Anabaena variabilis, FEMS Microbiol. Lett.*, 11, 233, 1981.

88. **Cozens, A. L. and Walker, J. E.,** The organization and sequence of the genes for ATP synthase subunits in the cyanobacterium *Synechococcus* 6301, *J. Mol. Biol.*, 194, 359, 1987.

89. **Bakker, E. P., Borchard, A., Michels, M., Altendorf, K., and Siebers, A.,** High affinity potassium uptake system in *Bacillus acidocaldarius* showing immunological cross-reactivity with the Kdp system from *Escherichia coli, J. Bacteriol.*, 169, 4342, 1987.

90. **Hellingwerf, K. J. and Abee, T.,** unpublished data, 1991.

91. **Abee, T., Knol, J., Hellingwert, K. J., Bakker, E. P., Siebers, A., and Konings, W. N.**, A Kdp-like, high-affinity, K$^+$-ATPase is expressed during growth of *Rhodobacter sphaeroides* in low potassium media. Distribution of this K$^+$-ATPase among purple non-sulphur phototrophic bacteria, *Arch. Microbiol.*, in press.

92. **Hafer, J., Siebers, A., and Bakker, E. P.**, The high-affinity K$^+$-translocating ATPase complex from *Bacillus acidocaldarius* consists of three subunits, *Mol. Microbiol.*, 3, 487, 1989.

93. **Brockhage, B. and Bakker, E. P.**, unpublished data, 1990.

94. **Abee, T. and Hellingwerf, K. J.**, unpublished data, 1990.

95. **Abee, T., Siebers, A., Altendorf, K., and Konings, W. N.**, submitted for publication.

96. **Schönheit, P., Beimborn, D. B., and Perski, H.-J.**, Potassium accumulation in growing *Methanobacterium thermoautotrophicum* and its relation to the electrochemical proton gradient, *Arch. Microbiol.*, 140, 247, 1984.

97. **Siebers, A. and Schönheit, P.**, unpublished data, 1991.

98. **Hugentobler, G., Heid, I., and Solioz, M.**, Purification of a putative K$^+$-ATPase from *Streptococcus faecalis*, *J. Biol. Chem.*, 258, 7611, 1983.

99. **Fürst, P. and Solioz, M.**, The vanadate-sensitive ATPase of *Streptococcus faecalis* pumps potassium in a reconstituted system, *J. Biol. Chem.*, 261, 4302, 1986.

100. **Apell, H.-J. and Solioz, M.**, Electrogenic transport by the *Enterococcus hirae* ATPase, *Biochim. Biophys. Acta*, 1017, 221, 1990.

101. **Solioz, M., Mathews, S., and Fürst, P.**, Cloning of the K$^+$-ATPase of *Streptococcus faecalis*, *J. Biol. Chem.*, 262, 7358, 1987.

102. **Ambudkar, S. V., Lynn, A. R., Maloney, P. C., and Rosen, B. P.**, Reconstitution of ATP-dependent calcium transport from Streptococci, *J. Biol. Chem.*, 261, 15596, 1986.

103. **Nucifora, G., Chu, L., Misra, T. K., and Silver, S.**, Cadmium resistance from *Staphylococcus aureus* plasmid pI258 *cadA* gene results from a cadmium-efflux ATPase, *Proc. Natl. Acad. Sci. U.S.A.*, 86, 3544, 1989.

104. **Hmiel, S. P., Snavely, M. D., Florer, J. B., Maguire, M. E., and Miller, C. G.**, Magnesium transport in *Salmonella typhimurium:* genetic characterization and cloning of three magnesium transport loci, *J. Bacteriol.*, 171, 4742, 1989.

105. **Jinks, D. C., Silvius, J. R., and McElhaney, R. N.**, Physiological role and membrane lipid modulation of the membrane-bound (Mg^{2+}, Na$^+$)-adenosine triphosphatase activity in *Acholeplasma laidlawii*, *J. Bacteriol.*, 136, 1027, 1978.

106. **Lewis, R. N. A. H. and McElhaney, R. N.**, Purification and characterization of the membrane (Na$^+$ + Mg^{2+})-ATPase from *Acholeplasma laidlawii* B, *Biochim. Biophys. Acta*, 735, 113, 1983.

107. **Walderhaug, M. O., Post, R. L., Saccomani, G., Leonard, R. T., and Briskin, D. P.**, Structural relatedness of three ion-transport adenosine triphosphatases around their active sites of phosphorylation, *J. Biol. Chem.*, 260, 3852, 1985.

108. **Shirvan, M. H., Schuldiner, S., and Rottem, S.**, Volume regulation in *Mycoplasma gallisepticum:* evidence that Na$^+$ is extruded via a primary Na$^+$ pump, *J. Bacteriol.*, 171, 4417, 1989.

109. **Dharmavaram, R. M. and Konisky, J.**, Characterization of a P-type ATPase of the archaebacterium *Methanococcus voltae*, *J. Biol. Chem.*, 264, 14085, 1989.

110. **Maloney, P. C. and Wilson, T. H.**, The evolution of ion pumps, *Bioscience*, 35, 43, 1985.

Chapter IIC

LOW-AFFINITY K$^+$ UPTAKE SYSTEMS

Evert P. Bakker

TABLE OF CONTENTS

0-8493-6982-7/93/$0.00 + $.50

I. INTRODUCTION

The media in which bacteria are cultivated almost always contain K^+ concentrations above 1 mM. Under these conditions the expression of the *kdpABC* operon is inhibited (Chapter IIB), and the bacteria take up K^+ via low-affinity systems (K_m values for K^+ uptake are 0.2 to 2 mM). Most of these systems transport K^+ rapidly (V_{max} = 50 to 500 nmol K^+ taken up per minute per milligram cell dry weight or cell protein), and at least in *Escherichia coli* they are constitutive. The low-affinity systems also transport Rb^+ and some accept Cs^+ or Tl^+. However, in most bacteria these (alkali) cations are taken up with considerably lower rates and affinities than K^+. A detailed analysis of K^+ transport genes and K^+ transport systems has been made for only *E. coli*.[1,2] This bacterium contains several types of low-affinity K^+ uptake systems that act in parallel.[2,3] The *E. coli* data will be discussed below first (Section II). Subsequently, what is known about the low-affinity systems in species other than *E. coli* will be described, as far as this is not covered in Chapters IID and IIF or reviews by Epstein and co-workers.[1,4]

II. *ESCHERICHIA COLI*

A. HISTORICAL OVERVIEW

Until recently, the understanding of K^+ uptake via the low-affinity systems was poor. However, now it is clear that much of the existing confusion is taken away by assuming that *E. coli* contains three low-affinity systems.[2,3] Table 1 and this section summarize how this confusion has arisen.

In 1971 Epstein and Kim[5] identified three *trk* loci, *trkA, trkD,* and *trkE* (trk stands for <u>tr</u>ansport of <u>K</u>$^+$; mutations in *trk* genes affect K^+ uptake via the low-affinity systems). These three loci, as well as the later identified *trkG* and *trkH* genes, lie widely scattered on the chromosome (Figure 1).[2,5-8] An early analysis of K^+ uptake by strains mutated in *trkA, trkD,* or *trkE* led to the proposal that *E. coli* contains three constitutive K^+ uptake system, TrkA, TrkD, and TrkF.[9] The former two were thought to be encoded by *trkA* and *trkD,* respectively, and TrkF was supposed to mediate the very slow K^+ uptake by cells with an impaired Kdp, TrkA, and TrkD system.[9] Since at that time the probably erroneous observation was made that a mutation in *trkE* affects the activity of both the TrkA and TrkD systems, it was concluded that a functional *trkE* gene is required for full activity of K^+ uptake by both systems.[9] After the identification of the *trkG* gene,[6,7] the transport data were reinterpreted in 1981; all of the mutations in *trk* genes were thought to affect a single system, which was called TrkA[6] and later Trk.[1,7,10] This picture did not change with the identification of the *trkH* gene.[2,8] The dismantling of this "super system" started after it had been observed (1) that mutations in *trkA, trkE, trkG,* or *trkH* do not affect the activity of the *trkD* gene product[2,11,12] and (2) that only cells which contain a functional *trkD* gene show low-affinity

TABLE 1
Low-Affinity K⁺ Uptake Systems in *E. coli*

Year	Gene Identified	System(s)	Ref.
1971	*trkA* *trkD* *trkE*	—	5
1976		TrkA TrkD TrkF	9
1981	*trkG*	Single system (TrkA) TrkF	6,7
1982		Single system (Trk) TrkF	1,7,10
1984	*trkH*		2,8
1989		Trk Kup (TrkD) TrkF	2,11
1991		TrkG TrkH Kup (TrkD) TrkF	2,3,13

Note: A survey is given how K⁺ uptake data have been interpreted during the years. See the text for details.

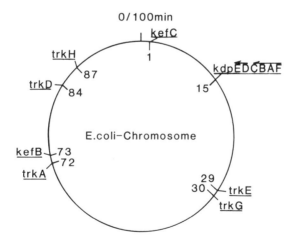

FIGURE 1. K⁺ transport genes in *Escherichia coli*. The data are from Epstein and his co-workers.[1,2,5,7,8,82,83] The numbers (in minutes) indicate the position of the genes on the *E. coli* chromosome.[55] (From Schlösser, A. et al., in *Molecular Mechanisms of Transport,* Quagliariello, E. and Palmieri, F., Eds., Elsevier, Amsterdam, 1992, 51. With permission.)

Cs[11] uptake.[11] These data show that TrkD represents a separate system. In order to distinguish it from the Trk system, TrkD has been renamed Kup, standing for K[+] uptake.[11] Subsequently, careful analysis of the effect of mutations in the *trk* genes on transport led to the proposal that Trk consists of two almost identical systems, TrkG and TrkH.[2] Recent molecular biological data support this notion.[3,13] The TrkG system is composed of the integral membrane protein TrkG[14] and the peripheral membrane protein TrkA.[12] The *trkE* gene product increases the affinity of the system for K[+].[2] The composition of the TrkH system is similar to that of TrkG, except that the TrkH protein replaces the TrkG protein[2,3] and a functional *trkE* gene product is absolutely required.[2]

B. THE TrkG AND TrkH SYSTEMS

Older work has shown that the ''Trk'' system possesses a number of remarkable features that distinguish it from all other known bacterial transport systems.[1] However, molecular biological and biochemical studies on the system started late, and the molecular basis for the unusual properties of the system(s) is not yet known. Molecular biology on the Trk systems is not easy, since (1) the *trk* genes had to be cloned separately (Figure 1), (2) the expression of *trk* genes is very low,[12,15] and (3) the Trk systems are composed of more than one type of subunit.[2,12] Epstein and his co-workers[1,2,7,8,10] have cloned the six *trk* genes, which has made possible the determination of the nucleotide sequence of these genes and has enabled us to overproduce some of the Trk proteins. The early results obtained with intact cells will be summarized first (Section II.B.1). Subsequently, the molecular properties of the two Trk systems will be described (Section II.B.2), and a model about their mode of action will be presented (Section II.B.3).

1. Intact Cells

All of the conclusions drawn from work with intact cells were reached before it was recognized that Trk(A) represents two systems, TrkG and TrkH.[2] In view of the large similarities between the systems (Section II.B.2), these conclusions are probably also valid for each of them. However, this assumption has not been verified. For historical reasons the combination of the TrkG and TrkH systems also will be referred to as ''Trk'', ''TrkA'', or ''Trk(A)''.

a. Energy Coupling

Rhoads and Epstein[16] have investigated how TrkA couples energy to K[+] uptake.[16] Surprisingly, TrkA does not belong to either of the two common groups of active transport systems in prokaryotes, i.e., primary pumps, which couple the free energy from ATP hydrolysis to the transport of ions or substrates across the cell membrane, and secondary porters, which couple the transmembrane movement of H[+] or Na[+] to solute transport. Rather, TrkA resembles both systems, since it requires for K^+ uptake activity both a high ATP level in the cytoplasm of the cells and a high transmembrane $\Delta\tilde{\mu}_{H^+}$.[16]

b. The Role of ATP

In the presence of ATP both the Kdp and the TrkA systems mediate the exchange of $^{42}K^+$ against nonradioactive K^+ across the cell membrane in a process that does not depend on $\Delta\tilde{\mu}_{H^+}$.[17] At the time of this finding it became apparent that Kdp functions as a K^+-translocating ATPase.[18] Hence, the $^{42}K^+$-K^+ exchange data were interpreted to mean that ATP hydrolysis drives K^+ uptake via TrkA too.[17] However, these data also do not contradict a different interpretation, namely that ATP activates the transport process and that the value of $\Delta\tilde{\mu}_{H^+}$ determines the direction of net transport.[19] Subsequent work has shown that during uptake of K^+ the TrkA system consumes at least 15 times less ATP per K^+ ion transported than does Kdp.[20] These data suggest that ATP regulates rather than drives K^+ transport via the TrkA system.[20]

c. The Role of $\Delta\tilde{\mu}_{H^+}$

Several groups have proposed that TrkA[21-23] as well as similar systems from other bacteria[19,24] function as K^+/H^+ symporters. This notion is based on a thermodynamic argument, showing that the ratio A, describing K^+ distribution across the membrane (Equation 1)

$$A = \log([K^+]_{in}/[K^+]_{out}) \tag{1}$$

exceeds the membrane potential ($\Delta\Psi$), as shown in Equation 2,

$$RTA/F \geq \Delta\psi \tag{2}$$

in which R is the gas constant, T, the absolute temperature, and F, the Faraday constant. However, except for *Bacillus acidocaldarius*[24] (see Section III.D), the value of A remains far below the theoretical limit for a H^+/K^+ symporter (Equation 3):

$$RTA/F = 2\Delta\psi - RT/F\Delta pH \tag{3}$$

in which ΔpH gives the transmembrane pH gradient.[19] There may be several reasons why K^+ accumulation via the *E. coli* Trk(A) system does not obey Equation 3. Most importantly, this accumulation ratio is determined under conditions at which $[K^+]_{out}$ is extremely low (1 to 5 μM) and three orders of magnitude smaller than the K_m value for K^+ uptake by the system (Table 2).[2,9] Thus, under these conditions K^+ accumulation is more likely to reflect a kinetic rather than thermodynamic control. In addition, *E. coli* contains K^+ efflux systems (Chapters IIA, IIE, and IIH). The activity of these systems will also reduce the value of A.

Except for the thermodynamic argument, no evidence has been presented in favor of a K^+/H^+ symport mechanism for any bacterial TrkA-like system. Most notably, it has not been possible to demonstrate that K^+ uptake via these systems is accompanied by alkalinization of the medium. It is not clear why these experiments have failed. Work with H^+-solute symporters like the lactose carrier from *E. coli* has shown that with these systems solute uptake by the cells is accompanied by alkalinization of the medium under conditions at which lipophilic ions prevent the formation of $\Delta\Psi$.[25-27] It may be that for K^+ uptake via TrkA the necessary cell conditions (the cells (1) must contain a high ATP level, (2) should not metabolize too rapidly for alkalinization to remain detectable, and (3) should have a low $\Delta\Psi$) are too much to be fulfilled simultaneously and that therefore the experiments have failed. However, it is clear that the present data are far from sufficient for concluding that the Trk systems function as K^+-H^+ symporters.

d. Electrogenicity of K^+ Uptake

K^+-loaded prokaryotes develop smaller $\Delta\Psi$ and larger ΔpH values than do K^+-depleted cells.[28-35] Especially with glycolyzing bacteria like *Enterococcus hirae* (formerly *Streptococcus faecalis*)[28,30] or *Mycoplasma mycoides* var. *Capri*,[32] this effect is so large that it is difficult to detect any membrane potential in K^+-loaded cells at all. As described in Section IV of Chapter IIA, this K^+-dependent interconversion of $\Delta\Psi$ and ΔpH is assumed to be due to electrogenic K^+ transport via the K^+ uptake systems and contributes to the process of pH homeostasis.[36] Cells of *E. coli* that contain either Kdp or Trk show the effect,[30,31] suggesting that K^+ uptake via both systems is electrogenic.[30] In Chapter IIG the K^+-dependent interconversion of $\Delta\Psi$ and ΔpH in *Rhodobacter sphaeroides* is discussed in detail.

e. Regulation by Cell Turgor Pressure; Circumvention of Futile Cycling

During net K^+ uptake via the Kdp or Trk systems, *E. coli* cells diminish their $^{42}K^+$-K^+ exchange activity when the internal K^+ concentration reaches its steady-state value.[17,37-39] By contrast, after hyperosmotic shock both this exchange activity[17,37,38] and net uptake[17,40] (Figure 3 in Chapter IIA) increase. These effects are due to a direct influence of cell turgor on the activity of these systems.[1,6] At low turgor pressure (i.e., plasmolysis) Trk and Kdp are active, and at high turgor pressure (i.e., in growing cells) the activity of these systems is low. For the Trk systems the difference between the two situations is large: at high turgor pressure, Trk has a V_{max} value for K^+ uptake that equals only 1 to 2% of that of K^+-depleted cells.[11,38,39]

From a physiological point of view the down-regulation of K^+ uptake activity makes sense since it prevents the rapid futile cycling of K^+ across the cell membrane and therefore an unnecessary waste of energy. In Chapter IIIC conditions are discussed where Kdp and Trk were believed to be involved in futile transmembrane K^+ cycling.[41] However, recent work has shown that futile NH_3-NH_4^+ cycling causes the increased energy demand of the cells and that the Trk systems are not involved in this process.[42]

2. Molecular Biology and Biochemistry

Attempts to combine the separately cloned *trk* genes into artificial operons have not been successful. The properties of these genes and of their products have therefore been studied separately.

a. trkA Gene and Its Product

Under most conditions the *trkA405* mutation or the deletion of the *trkA* gene abolishes the activity of both the TrkG and TrkH systems completely.[2,9,43] Hence, it has been thought that TrkA forms the core of the system(s) and that is an integral membrane protein.[7,10] However, immunological studies have shown that TrkA forms a peripheral membrane protein bound to the inner side of the cell membrane.[11] In disrupted cells only about 5% of the protein occurs in the membrane fraction. Repeated washing of the membranes does not lead to a substantial release of TrkA.[44] One way to explain this result is to assume that the membranes contain a small number of high-affinity binding sites for TrkA (only about two per cell equivalent).[44] Alternatively, these data could mean that soluble TrkA protein does not bind to the membrane since it is not in a binding-competent form (see below).

The membrane-bound form of TrkA is important for the K^+ transport process since, in a strain that carries point mutations in both *trkG* and *trkH* and does not take up K^+ via the TrkG and TrkH systems, TrkA occurs exclusively in soluble form.[12] Moreover, both membrane binding and K^+ uptake are restored by the introduction into the cells of a multiple-copy plasmid carrying a *trkG* or *trkH* gene.[12] These data suggest that TrkG and TrkH each have a function in the anchoring of the TrkA protein to the membrane.[12,44]

The amino acid sequence of TrkA (EMBL accession number X52114) supports the view that TrkA is not an integral membrane protein; 44% of its amino acid residues are polar, and it does not contain a stretch of hydrophobic residues that is long enough to traverse the membrane in the form of an α-helix.[43] The TrkA protein does not show significant similarity with any other known protein. However, the Diagon plot of its amino acid sequence indicates that TrkA forms an internal dimer.[43] In addition, intragenic complementation suggests that more than one copy of the TrkA protein is present per Trk complex.[43]

The N-terminus of each half TrkA contains a Rossmann fold,[45] GAGQVGGT at residues 7 to 13 and GGGNIGAGL at residues 239 to 247, respectively.[43] This kind of fold is important for the binding of nucleotides to enzymes.[46] Hence, we speculated that ATP exerts its activating effect on the Trk systems by binding to the TrkA protein. However, attempts to label membrane-bound TrkA with radioactively labeled ATP or 8-azido-ATP have failed.[43] This negative result may be due to the fact that the membranes contain extremely little TrkA protein. Thus, even if labeling had occurred, it might have been below the limit of detection.[43]

b. trkG, trkH, and Their Products

The *trkG* gene is located in the prophage *rac* region of the *E. coli* chromosome.[1,2,14] The role of the TrkG protein in K^+ transport was therefore at first unclear. However, recent data have shown that *trkG* has supplied *E. coli* K-12 with a second copy of a *trkH*-like gene.[2,3,13] The two genes have almost the same size,[3,14] and despite their 14% difference in $G + C$ content (37% and 51% for *trkG* and *trkH*, respectively) the aligned genes contain 49% identical nucleotides. The proteins TrkG (485 amino acid residues) and TrkH (483, not 421 amino acid residues[3,47]) contain 40% identical and 18% conservatively exchanged residues for a total of 58% similar residues. TrkG is an integral membrane protein,[13] and the closely similar structure of TrkG and TrkH suggests that the same is true for TrkH.[3] Except with each other, TrkG and TrkH do not share any significant similarity with other membrane proteins, not even with KdpA (see Chapter IIB).

TrkG and TrkH can span the membrane up to 12 times.[3,14] Figure 2 compares the transmembrane folding of the two proteins, assuming that they contain ten transmembrane helices. This model was chosen because it puts the positively charged residues (closed circles) of TrkG and TrkH at one side of the membrane, most likely the cytoplasm.[48] In Figure 2, hydrophilic regions with a similarity of more than 70% are indicated by thick lines. These regions are found at both sides of the membrane. They may be involved in K^+ binding and, at the cytoplasmic side, in the anchoring of TrkA to the membrane (see Section II.B.2.a). In particular, the loops between helices IV and V and between helices VIII and IX are strongly conserved and may be important in the latter process.

The putative transmembrane helices IV and X are 90 and 71% similar, respectively, a larger than average value. Helix IV consists exclusively of apolar residues. It might interact with the membrane phospholipids or with hydrophobic sides of other transmembrane helices. Helices VI and VIII are amphiphilic. The approximate locations of the polar residues and of tyrosine inside the apolar part of the membrane are indicated in Figure 2 with the one-letter amino acid code. The 15 positions in the two proteins at which these residues are identical or conservatively exchanged are indicated with circles or diamonds, respectively. These residues cluster around the middle part of the membrane (Figure 2). Unlike the family of Lac-type carriers, TrkG and TrkH do not contain a conserved histidine residue in the membrane[49] (see also Chapter IC). However, helix VIII contains a conserved cysteine residue (Figure 2). This residue might be responsible for the sensitivity of the Trk systems to *N*-ethylmaleimide (see Chapter IIE).

c. trkE and Its Product

Almost nothing is known about *trkE* and its product. Epstein and his co-workers are determining the nucleotide sequence of the gene. Their results

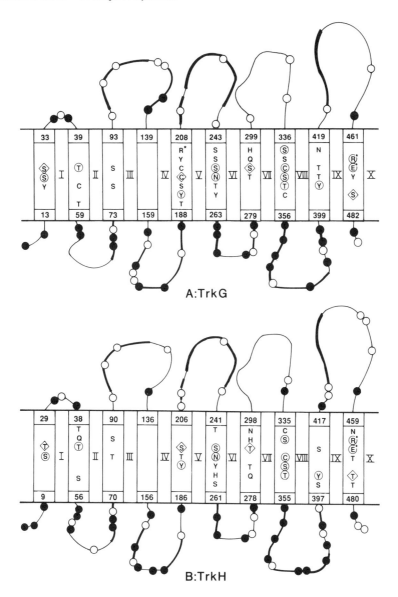

FIGURE 2. The *E. coli* TrkG and TrkH proteins have a similar transmembrane folding. It is assumed that both proteins span the membrane ten times. The closed and open circles indicate positively (K,R) and negatively charged residues (D,E), respectively. Thick lines show polar stretches in which the two proteins possess a larger than average similarity. Polar residues inside the membrane are indicated with one-letter symbols. Positions at which these residues are identical and conservatively exchanged are indicated by circles and diamonds, respectively. See the text for further details. (From Schlösser, A. et al., in *Molecular Mechanisms of Transport,* Quagliariello, E. and Palmieri, F., Eds., Elsevier, Amsterdam, 1992, 51. With permission.)

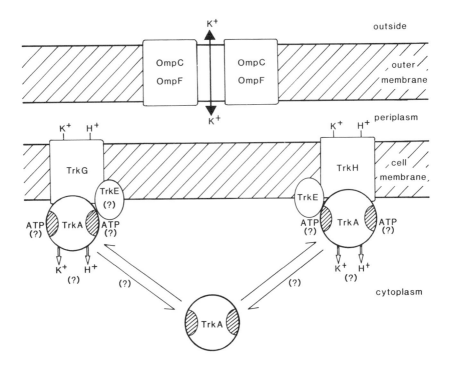

FIGURE 3. Model explaining how the TrkG and TrkH systems translocate K^+. See the text for details. (From Schlösser, A. et al., in *Molecular Mechanisms of Transport*, Quagliariello, E. and Palmieri, F., Eds., Elsevier, Amsterdam, 1992, 51. With permission.)

might give some clues about the puzzling observation that TrkE is indispensible for the TrkH system, but that TrkG is partially active without it. Studies with a strain that contains a *trkE-lacZ* fusion suggest that TrkE is a membrane protein.[44] In contrast to TrkG and TrkH, TrkE is not involved in the binding of TrkA to the membrane.[12,44]

3. Model for the Function of the TrkG and TrkH Systems

Figure 3 shows my working hypothesis about the mechanism of K^+ translocation via the TrkG and TrkH systems. It is based on the data discussed above. The apolar TrkG and TrkH proteins form the core of each system. The function of TrkE is not clear. It might support TrkG and TrkH in the K^+ uptake process, which is drawn to occur in symport with protons (see Section II.B.1.c), or it might play a role in turgor sensing. TrkG and TrkH bind the TrkA protein to the membrane. It is assumed that TrkA confers the activation by ATP to the complexes, despite the fact that experimental evidence in support of this notion is still lacking (see Section II.B.2.a). Activation might take place by binding of TrkA to TrkG and TrkH and nucleotide binding to

TrkA (Figure 3). Alternatively, TrkA might be a protein kinase and phosphorylate other components of the complexes. The first model explains why membrane binding of TrkA is essential for the K^+ transport process.[12,44] The latter notion is more compatible with the observation that a large portion of TrkA is found among the soluble proteins.[12]

It is not known how the TrkG and TrkH systems sense changes in the turgor pressure of the cells. My working hypothesis is that different values of cell turgor exert different lateral pressures on the transport complex, leading to altered interactions between the transmembrane helices or between the subunits of the complex. However, the structure of the TrkG and TrkH systems is too complex to allow at present an analysis of the effect of cell turgor on it. This question can be addressed more easily with a simpler turgor-sensing system like KdpD[50] (Chapter IIB).

Recently it has been reported that in a temperature-sensitive *dnaK* strain at high NaCl concentrations the Trk systems become inactive at the permissive temperature. This suggests that under these conditions at least one of the components of the TrkG and TrkH system cannot maintain a native conformation.[51] In cooperation with Meury it was tested whether, under conditions at which the Trk systems become inactive, TrkA dissociates from the membrane. However, at high salt concentrations no differences were observed between the amount of membrane-bound TrkA in the mutant and the wild type. Inactivation in the *dnaK* strain is specific for the Trk systems since under conditions at which they are inhibited both TrkF and Kdp remain active.[51,52] Future work will have to show how the chaperone protein DnaK[53] supports the activity of the Trk systems under conditions of salt stress.

A final comment concerns the turnover number of the Trk systems. The main Na^+/H^+ antiporter from *E. coli* (NhaA) transports 10^3 equivalents of $Na^+ \cdot s^{-1} \cdot complex^{-1}$ (Chapter IB). The turnover number of the Trk systems may even be higher. The copy number of the TrkA protein is of the order of 20 to 50,[12] and that of TrkG less than 200.[15] From the latter number and the V_{max} of the TrkG system for K^+ (Reference 2) one calculates that this system has a turnover number of about 10^4. Apparently, secondary transport systems for alkali cations can translocate these ions with a rate similar to that of small mobile carriers like valinomycin.[54]

C. THE Kup (TrkD) SYSTEM

The *trkD* gene encodes a third low-affinity K^+ uptake system, called TrkD[2] or Kup.[11] The gene is located at 84.5 min on the *E. coli* chromosome (Figure 1), just upstream of the *rbsD-R* gene cluster.[55] The region encoding K^+ uptake activity is confined to a 2-kilobase (kb) DNA fragment and comprises the *Kpn*I and *Pvu*II sites at 4010.25 and 4010.75 kb, respectively, on the *E. coli* restriction map.[56,57] The partial nucleotide sequence of *trkD* shows that it encodes a hydrophobic protein. On SDS-polyacrylamide gels the Kup protein has an apparent molecular weight of 45,000.[56] Its real molecular weight may be considerably higher since many membrane proteins migrate too rapidly on these gels (see Reference 14 and Chapters IB, IC, and IIB for other examples).

K^+ uptake activity of Kup varies from preparation to preparation,[9] but in wild-type cells its V_{max} never exceeds one tenth of that of the Trk systems.[2,9,12] Cells that carry a *trkD*-containing multiple-copy plasmid show a five- to eightfold gene-dose effect (V_{max} for K^+ uptake = 100 to 180 nmol $\cdot$ min^{-1} $\cdot$ mg^{-1} dry weight of cells).[11,12,56] K^+ uptake is almost independent of pH$_{out}$. At pH 6 the *trkD*-plasmid-containing cells accumulate K^+ via Kup about 20,000-fold under conditions at which $\Delta\Psi$ equals 170 mV.[56,58] Insertion of these numbers into Equation 2 shows that K^+ accumulation is too steep for Kup to function as a K^+ uniporter. Subsequent experiments have shown that Kup only requires a high $\Delta\bar{\mu}_{H^+}$ (or $\Delta\bar{\mu}_{Na^+}$) for activity.[56] It has not yet been checked whether Kup is active in the absence of Na$^+$. My working hypothesis is that Kup functions as a simple K^+/H^+ or K^+/Na^+ symporter.

In contrast to Trk and Kdp, Kup is not involved in osmoadaptation of *E. coli*.[40,56] At 0.3 M NaCl its activity is inhibited completely.[56] In this respect, Kup superficially resembles several bacterial carbohydrate transport systems that become inactive after osmotic upshock.[59] However, the activity of the latter is partially restored after the cells have regained their turgor;[59] this is not the case with Kup.[56] At low medium osmolarity (50 mOsm) the affinity of Kup for K^+ increases ($K_m \approx 80$ μM).[56] Under these conditions, Kup may become important to the cells since its presence will postpone expression of the *kdpABC* operon.

D. THE TrkF SYSTEM

Cells that lack functional Kdp, Trk, and Kup systems still grow at K^+ concentrations above 15 mM.[5,9] The very slow K^+ uptake observed under these conditions is thought to be mediated by the TrkF system.[1,9] Since a gene for TrkF has not been identified, it is not clear whether the system consists of a protein (complex) that mediates K^+ uptake or whether this uptake reflects the membrane-potential-driven, nonspecific permeation of K^+ through the cell membrane.[9,16] The observation that the rate of K^+ is proportional to the K^+ concentration in the medium supported the latter notion.[9] However, in Osnabrück it has been observed that K^+ uptake via TrkF becomes saturated at high K^+ concentrations (K_m and V_{max} values are 20 to 30 mM and 10 to 15 nmol $\cdot$ min^{-1} $\cdot$ mg^{-1} dry weight of cells, respectively).[56] TrkF does not discriminate between K^+, Rb$^+$, and Cs$^+$.[60] It might be renamed Krc, standing for $\underline{K}^+$, $\underline{R}b^+$, and $\underline{C}s^+$.[60]

III. OTHER BACTERIA

In the past few years, knowledge about low-affinity K^+ uptake systems from bacteria other than *E. coli* has hardly increased. The description that follows will mainly review new data and will indicate whether the K^+ uptake system described plays a role in osmoadaptation of the cells. Older results are discussed only if my interpretation of them is different from that given in recent reviews.[1,4] Table 2 gives a survey of the species for which low-affinity K^+ uptake systems have been described.

A. ENTEROBACTERIACEAE

K^+ uptake by low-affinity systems has not been studied in any member of the Enterobacteriaceae other than *E. coli*. However, with molecular biological and immunological methods it has been estimated how widespread *trk* genes and TrkA-like proteins are among other species of the family.[15,43] *trkA*- and *trkD*-like genes, as well as the TrkA protein, also occur in many other Enterobacteriaceae (Figure 4), but not in bacteria outside this group.[15,43] This result is different from that of Kdp, for which the same approach has shown that it occurs in many Gram-negatives[61] and in one Gram-positive[62] (see Chapter IIB). Hence, during evolution Kdp has been preserved better than Trk or Kup. The negative results obtained with Trk or Kup do not prove that bacteria outside the Enterobacteriaceae do not possess K^+ uptake systems that are functionally similar to the low-affinity systems of *E. coli*. Rather, the genes and proteins of these bacteria are so different from those of *E. coli* that cross-reactivity is not observed.

B. PHOTOTROPHS

K^+ uptake has been studied in several phototrophic bacteria (Table 2; Chapters IIB and IIG). *Rhodobacter capsulatus* (formerly *Rhodopseudomonas capsulata*),[35,63,64] Rhodobacter sphaeroides,[65] (formerly *Rhodopseudomonas sphaeroides*), and *Chromatium vinosum*[66] contain low-affinity uptake systems (Table 2). The reported V_{max} values vary from species to species[63-66] or with the growth conditions.[63,64] In addition, the about 20-fold differences observed in uptake rates (Table 2) may be due to the fact that in some situations K^+ transport has been measured in the net uptake mode and in others in the exchange mode. These rates can differ by a factor of 50 (Section II.B.1.e). Finally, with some cells K^+ uptake has been assayed by the use of $^{86}Rb^+$ [63-65] or $^{204}Tl^+$ [66] as a probe for K^+. In only one study has it been verified that the uptake parameters of Rb^+ resemble those of K^+.[63]

It is difficult to draw conclusions about the mechanism of energy coupling to K^+ uptake by phototrophs since (1) all of the work has been done with intact cells; (2) K^+ uptake mutants are not available, and it is therefore not known whether parallel systems are present; and (3) inhibitors have been used which may cause artifacts. The *C. vinosum* system appears to function as an ATPase since it requires only ATP for K^+ uptake.[66] By contrast, the *R. sphaeroides* system requires only a high $\Delta\bar{\mu}_{H^+}$ and may function as a K^+ uniporter.[65] Recent observations with *R. capsulatus* also suggest that this bacterium accumulates Rb^+ by uniport.[35,64] Originally it had been concluded

TABLE 2
Survey of Low-Affinity K$^+$ Uptake Systems in Bacteria

Species	System	K$_m$^{a}	V$_{max}$^{a}	Other ions transported	Involved in osmoadaptation	Ref.
Escherichia coli	TrkG	0.3–1	240	Rb$^+$, Tl$^{+ b}$	Yes[b]	2,3,67,85
	TrkH	2.2–3	310–450	Rb$^+$, Tl$^{+ b}$	Yes[b]	2,3,67,85
	Kup	0.3	30	Rb$^+$, Cs$^+$	no	2,11
	TrkF	20–30	15	Rb$^+$, Cs$^+$	?	9,56,60
Paracoccus denitrificans		1–2	270	Rb$^+$ (?)	?	86
Rhodobacter capsulatus		0.2	8	Rb$^+$, Cs$^+$	?	63
		1–2[c]	Fast[c]	K$^+$	?	64
Rhodobacter sphaeroides		?	6[d]	?	Yes (?)	65,68
Chromatium vinosum		1.3[e]	100[e]	?	Yes	66
Anabaena variabilis		4.5	?[f]	Rb$^+$	Yes	87,88
Synechocystis		?	?	?	Yes	88–90
Mycoplasma mycoides	I	0.3	80	Rb$^+$	?	71,73
	II	?	?	No Rb$^+$	?	73
Bacillus subtilis		0.2	26	?	Yes	74–76
Bacillus acidocaldarius		1	50	Rb$^+$	Hardly	24
Enterococcus hirae	KtrI	0.55	55	Rb$^+$	No	19, Chapter IID
	KtrII	0.5	20	Rb$^+$	?	Chapter IID
Methanospirillum hungatei		0.4[d]	29[d]	K$^+$	?	91
Halobacterium halobium		?	?	Rb$^+$	Yes	92–94

Note: Details can be found in Section III or in Reference 1.

[a] K$_m$ in mM; V$_{max}$ in nmol · min^{-1} · mg^{-1} dry weight or cell protein.

[b] These data have not yet been determined separately for the TrkG and TrkH systems.

[c] Data determined with ^{86}Rb$^+$; the V$_{max}$ of the system is stated to be two orders of magnitude larger than those reported in References 63 and 65.

[d] Determined with ^{86}Rb$^+$.

[e] Determined with ^{204}Tl$^+$; data recalculated.

[f] V$_{max}$ is expressed in nmol · m^{-2} · s^{-1}.

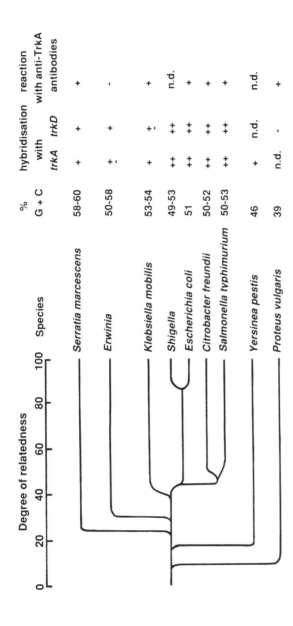

FIGURE 4. Distribution of *trkA*- and *trkD*-like genes and TrkA-like proteins among Enterobacteriaceae.[15,43] The three columns on the right give the results from Southern and Western blot experiments. Symbols: + +, strong cross-reactivity; +, modest cross-reactivity; ±, weak cross-reactivity; −, no cross-reactivity; n.d., not determined. (Tree redrawn after Reference 84.)

that the uptake system accepts K^+, Rb^+, and NH_4^+.[35] However, for NH_4^+ this has been retracted.[64] Due to the electrogenicity of K^+ or Rb^+ uptake, the cells are depolarized after these ions have been added to the medium (see Section II.A.1.d and Chapter IIA). From the rate of depolarization, as measured by changes in the carotenoid absorption spectrum of the chromatophores,[35] the current associated with alkali cation uptake has been calculated.[35,64] This current has been compared with the rate of $^{86}Rb^+$ uptake by the cells. The result shows that one unit of positive charge enters the cells per molecule of $^{86}Rb^+$ transported, suggesting that $^{86}Rb^+$ uptake takes place by uniport.[64] For the following reasons I hesitate to accept this hypothesis. First, K^+ uptake has been monitored with $^{86}Rb^+$ [64]; the experiments should be repeated with $^{42}K^+$ since at least in *E. coli* Rb^+ is poorly accepted by the major low-affinity K^+ uptake systems,[67] and Rb^+ uptake occurs with more "slip" than that of K^+ (see Figure 2B of Reference 58). Second, membrane current has been measured across the chromatophore membrane, and K^+ uptake takes place across the cell membrane. Although in intact cells these two membranes are interconnected, it is not clear whether they always possess an equal $\Delta\Psi$. If changes in $\Delta\Psi$ of the chromatophores lag behind those of the plasma membrane, the ion:charge ratio during transport will be underestimated. Clearly, more work is needed to establish whether the low-affinity K^+ (Rb^+) uptake system of *R. capsulatus* occurs via uniport.[64]

K^+ uptake via low-affinity systems also plays a role in osmoadaptation of phototrophs and cyanobacteria. The situation in *Ectothiorhodospira halochloris* has been discussed in Chapter IIA. In *C. vinosum* $[K^+]_{in}$ increases from 0.15 to 0.9 M after the osmolality of the medium has been increased from 0.06 to 0.42 Osm.[66] Results obtained with *R. sphaeroides* suggest that the same is true for this organism.[68] The form of the gas vacuoles of the cyanobacterium *Microcystis* sp. has been used to estimate changes in the turgor pressure.[69] Osmotic upshock with NaCl leads to a decrease in cell turgor, and K^+ uptake is required for partial recovery.[69] Thus, in contrast to *E. coli*, changes in cell turgor of *Microcystis* sp. can be compared directly with changes in K^+ uptake activity.[69]

C. *MYCOPLASMA MYCOIDES* VAR. CAPRI

Leblanc, Le Grimellec, and their co-workers[32,70-73] have characterized K^+ uptake by the Gram-negative, cell-wall-free bacterium *Mycoplasma mycoides* var. Capri (strain PG3). This organism ferments glucose to weak acids, with lactic acid being the main product (30%).[70] Energy coupling of strain PG3 resembles that of *Enterococcus hirae* (see Chapter IID). ATP is generated in the cytoplasm by substrate-level phosphorylation, and the cells develop $\Delta\Psi$ by coupling ATP hydrolysis to the extrusion of H^+ (or perhaps Na^+; see below) via an F_1F_0 type of ATPase.

K^+-depleted cells take up this cation via a low-affinity system (Table 2).[70] The factor by which the bacteria accumulate K^+ is too high to be supported by $\Delta\Psi$ alone,[32,71,72] suggesting that K^+ uptake does not occur by uniport (Equations 1 and 2). Dicyclohexylcarbodiimide, an inhibitor of F_1F_0-ATPases, prevents both the generation of $\Delta\Psi$ and K^+ uptake.[70,71] Other treatments that abolish $\Delta\Psi$ also affect K^+ uptake.[71] In addition, Na^+ ions stimulate K^+ uptake and K^+ ions stimulate Na^+ extrusion;[72,73] membrane-bound ATPase activity is stimulated twofold by Na^+ ions (half-maximal effect at 30 mM).[73] Finally, rapid K^+ uptake is accompanied by a temporary decrease in the ATP concentration of the cells.[72] These data have been interpreted to mean that at least part of the K^+ is taken up by an ATP-consuming Na^+/K^+ exchange system.[73] For the interpretation of the data, the observation that the presence of a 5 μM concentration of the protonophore carbonylcyanide p-trifluoromethoxyphenyl hydrazone (FCCP) does not inhibit K^+ uptake is crucial.[1,71] However, this concentration of FCCP also does not affect $\Delta\Psi$.[32,71,72] By contrast, other treatments that reduce $\Delta\Psi$ also affect K^+ uptake.[71] One explanation of these data is that the putative Na^+/K^+-ATPase is of new type since it is inhibited by low concentrations of N,N'-dicyclohexylcarbodiimide (DCCD) and requires $\Delta\Psi$. However, none of these data contradict the notion that K^+ uptake takes place via a system similar to KtrI of *Enterococcus hirae* (Chapter IID) or TrkG and TrkH of *Escherichia coli* (Section II) and that Na^+ extrusion occurs via an independent system. In strain PG3 as well as in *Mycoplasma gallisepticum* the latter process might be mediated by an F_1F_0-ATPase of the Na-translocating type[4] (see Chapter IG regarding this type of F_1F_0-ATPase).

D. *BACILLUS*

Bacillus subtilis contains a low-affinity K^+ uptake system.[74] In this organism, K^+ uptake plays a role in turgor regulation:[75,76] after a hyperosmotic shock with 0.4 M NaCl the cells double their K^+ concentration. Subsequently, they synthesize proline, which reaches a concentration of about 0.5 M in the cytoplasm.[75] This situation is different from that in *E. coli,* which only synthesizes proline after osmotic upshock in strains in which feedback inhibition of proline synthesis is impaired.[77]

The thermoacidophile *B. acidocaldarius* contains a low-affinity K^+ uptake system that also accepts Rb^+.[24] The system requires both ATP and the $\Delta\bar{\mu}_{H^+}$ for activity.[24] In this respect it resembles the Trk system of *Escherichia coli* (Section II) and KtrI from *Enterococcus hirae* (Chapter IID). *B. acidocaldarius* accumulates K^+ about 2000-fold at pH values between 2 and 4.[24] Under these conditions the cells maintain a cytoplasmic pH close to neutrality and develop membrane potentials that vary from a small internally positive value at pH 2 to a small negative value at pH 4.[78] Insertion of these data into Equation 3 shows that between pH 2 and 4 K^+ accumulation reaches 90% of the expected value if the system is a K^+/H^+ symporter. This indicates that symport with H^+ is a feasible mechanism for K^+ uptake by this organism.[24]

B. acidocaldarius grows poorly at high medium osmolarities.[24] It contains relatively little K^+ (120 mM) and, after addition of 0.3 M NaCl, its K^+

concentration increases to only 160 mM.[24] It is not known whether under the latter conditions *B. acidocaldarius* synthesizes proline, as does *B. subtilis*.[75]

E. *ENTEROCOCCUS HIRAE*

In Chapter IID Kakinuma writes that *Enterococcus hirae* can synthesize two low-affinity K$^+$ uptake systems, KtrI and KtrII. The main system is KtrI. It resembles the Trk systems of *Escherichia coli* closely.[19] The two major differences are (1) KtrI does not discriminate between K$^+$ and Rb$^+$; and (2) after a hypoosmotic shock cells do not take up additional K$^+$, suggesting that KtrI is not involved in osmoadaptation (Chapter IID). For some time KtrI has been believed to function as a K$^+$-translocating ATPase of the P-type. However, this view has now been revised[4,79] (see Chapters IIB and IID).

The KtrII system is synthesized under conditions at which the cells cannot remove sufficient Na$^+$ from their cytoplasm. Kakinuma explains that KtrII extrudes Na$^+$ in exchange for K$^+$ at the expense of ATP hydrolysis (Chapter IID). In contrast to the major K$^+$ uptake system of *Mycoplasma mycoides* var. Capri,[71] KtrII is active at low $\Delta\Psi$. Recently it has been shown that the major subunit of KtrII resembles the catalytic subunit of the vacuolar type of ATPases from eukaryotes.[80]

F. ARCHAEBACTERIA

Very little is known about K$^+$ uptake by archaebacteria. Walderhaug et al.[1] have reviewed the data obtained for *Halobacterium halobium* and methanogens up to 1988. Since then no further studies have been published about this topic in these organisms. K$^+$ stimulates H$^+$ extrusion by respiring cells of the thermoacidophilic archaebacterium *Sulfolobus acidocaldarius,* indicating that the organism takes up K$^+$ in response to H$^+$ extrusion via the respiratory chain.[81]

IV. CONCLUDING REMARKS

Since the last comprehensive review on K$^+$ transport in bacteria was written,[1] the understanding of the process of K$^+$ uptake via the low-affinity systems in *Escherichia coli* has improved considerably. Many of the observations that then seemed bizarre are explained by the concept that this organism contains three low-affinity systems instead of one and that the major two systems (TrkG and TrkH) share a common subunit (the TrkA protein; Figure 3). The mechanism by which these systems translocate K$^+$ is still far from clear. More work must be done to determine whether all three systems indeed function as K$^+$/H$^+$ symporters and to clarify the role of ATP in the transport of K$^+$ via the TrkG and TrkH systems.

The lack of progress in the field of low-affinity K$^+$ uptake systems in almost all bacteria other than *E. coli* is only partially due to fact that K$^+$ uptake mutants are not available for these organisms. The basic problem is that K$^+$ transport experiments have been done traditionally by investigators primarily interested in bacterial physiology and bioenergetics. Most of these people have not attacked the problem with molecular biological methods and have left the field after having made some basic observations. Since these workers have also not been able to draw the attention of molecular biologists to the topic of K$^+$ transport in bacteria, progress in this field has declined rather than increased. The present situation is that low-affinity K$^+$ uptake by different species is supposed to occur via a variety of systems, including uniporters, symporters, and ATPases (Table 2), and that, except for system KtrII from *Enterococcus hirae,* nothing is known about the molecular properties of these systems (see Chapter IID).

Clearly, the subject of K$^+$ uptake by bacteria is complex. Bacteria other than *Escherichia coli* or *Enterococcus hirae* will almost certainly contain more than one system too. Nevertheless, for many species it is now possible to inactivate these systems one after the other and to complement the K$^+$ transport defect by reintroduction of the appropriate gene(s) via shuttle vectors or conjugative plasmids. In addition, strains of *Escherichia coli* can be constructed in which the effect of expression of K$^+$ transport genes from other bacteria can be tested by the feature of restoration of growth at low external K$^+$ concentrations. Hence, molecular biological studies on low-affinity K$^+$ uptake systems of species other than *E. coli* are now possible.

ACKNOWLEDGMENTS

I thank Wolfgang Epstein, Kenneth Rudd, and Baz Jackson for communicating unpublished results to me and for helpful discussion. The previous and present members of my group, Eva Limpinsel, Angela Hamann, Katharina Montag, Dirk Bossemeyer, Andreas Schlösser, and Manfred Schleyer, have contributed to the work described in Section II of this article. This research has been supported by the Deutsche Forschungs Gemeinschaft (SFB171, Project C1), the Fonds der Chemischen Industrie, and the European Community (contract SC-1-0334-C(A)).

REFERENCES

1. **Walderhaug, M. O., Dosch, D. C., and Epstein, W.,** Potassium transport in bacteria, in *Ion Transport in Prokaryotes,* Rosen, B. P. and Silver, S., Eds., Academic Press, New York, 1987, 84.
2. **Dosch, D. C., Helmer, G. L., Sutton, S. H., Salvacion, F. F., and Epstein, W.,** Genetic analysis of potassium transport loci in *Escherichia coli:* evidence for three constitutive systems mediating uptake of potassium, *J. Bacteriol.,* 173, 687, 1991.

3. **Schlösser, A., Hamann, A., Schleyer, M., and Bakker, E. P.,** The K$^+$-uptake systems TrkG and TrkH from *Escherichia coli:* a pair of unusual transport systems involved in osmoregulation, in *Molecular Mechanisms of Transport,* Quagliariello, E. and Palmieri, F., Eds., Elsevier, Amsterdam, 1992, 51.

4. **Epstein, W.,** Bacterial transport ATPases, in *Bacterial Energetics,* Vol. 12, The Bacteria: A Treatise on Structure and Function, Academic Press, Orlando, 1990, 87.

5. **Epstein, W. and Kim, B. S.,** Potassium transport loci in *Escherichia coli* K-12, *J. Bacteriol.,* 108, 639, 1971.

6. **Epstein, W. and Laimins, L.,** Potassium transport in *Escherichia coli:* diverse systems with common control by osmotic forces, *Trends Biochem. Sci.,* 5, 21, 1980.

7. **Helmer, G. L.,** A Genetic and Kinetic Analysis of the Trk Transport System of *Escherichia coli,* Ph.D. thesis, University of Chicago, Chicago, IL, 1982.

8. **Dosch, D. C.,** A Study of the Trk Transport System of *Escherichia coli,* Ph.D. thesis, University of Chicago, Chicago, IL, 1985.

9. **Rhoads, D. B., Waters, F. B., and Epstein, W.,** Cation transport in *Escherichia coli.* VIII. Potassium transport mutants, *J. Gen. Physiol.,* 67, 325, 1976.

10. **Helmer, G. L. and Epstein, W.,** Rate-limiting components of Trk-mediated potassium uptake in *Escherichia coli* K-12, *Biophys. J.,* 33, 61a, 1981.

11. **Bossemeyer, D., Schlösser, A., and Bakker, E. P.,** Specific cesium transport via the *Escherichia coli* Kup (TrkD) K$^+$ uptake system, *J. Bacteriol.,* 171, 2219, 1989.

12. **Bossemeyer, D., Borchard, A., Dosch, D. C., Helmer, G. L., Epstein, W., Booth, I. R., and Bakker, E. P.,** K$^+$ transport protein TrkA of *Escherichia coli* is a peripheral membrane protein that requires other *trk* gene products for attachment to the cytoplasmic membrane, *J. Biol. Chem.,* 264, 16403, 1989.

13. **Epstein, W. and Rudd, K.,** personal communication, 1991.

14. **Schlösser, A., Kluttig, S., Hamann, A., and Bakker, E. P.,** Subcloning, nucleotide sequence and expression of *trkG,* a gene that encodes an integral membrane protein involved in potassium uptake via the Trk system of *Escherichia coli, J. Bacteriol.,* 173, 3170, 1991.

15. **Schlösser, A.,** unpublished data, 1991.

16. **Rhoads, D. B. and Epstein, W.,** Energy coupling to net K$^+$ transport in *Escherichia coli* K-12, *J. Biol. Chem.,* 252, 1394, 1977.

17. **Rhoads, D. B. and Epstein, W.,** Cation transport in *Escherichia coli.* IX. Regulation of K transport, *J. Gen. Physiol.,* 72, 283, 1978.

18. **Epstein, W., Whitelaw, V., and Hesse, J.,** A K$^+$ transport ATPase in *Escherichia coli, J. Biol. Chem.,* 253, 6666, 1978.

19. **Bakker, E. P. and Harold, F. M.,** Energy coupling to potassium transport in *Streptococcus faecalis.* Interplay of ATP and the protonmotive force, *J. Biol. Chem.,* 255, 433, 1980.

20. **Stewart, L. M. D., Bakker, E. P., and Booth, I. R.,** Energy coupling to K$^+$ uptake via the Trk system in *Escherichia coli:* the role of ATP, *J. Gen. Microbiol.,* 131, 77, 1985.

21. **Silver, S.,** Transport of cations and anions, in *Bacterial Transport,* Rosen, B. P., Ed., Marcel Dekker, New York, 1978, 222.

22. **Wagner, E. F., Ponta, H., and Schweiger, M.,** Development of *Escherichia coli* virus T1. The role of the protonmotive force, *J. Biol. Chem.,* 255, 534, 1980.

23. **Bakker, E. P.,** The role of ATP and the proton-motive force in potassium transport by the *Escherichia coli* TrkA system, in *Vectorial Reactions in Electron and Ion Transport,* Palmieri, F., Ed., Elsevier/North Holland, Amsterdam, 1981, 315.

24. **Michels, M. and Bakker, E. P.,** Low affinity potassium uptake system in *Bacillus acidocaldarius, J. Bacteriol.,* 169, 4341, 1987.

25. **West, I. C. and Mitchell, P.,** Stoicheiometry of lactose-proton symport across the plasma membrane of *Escherichia coli, Biochem. J.,* 132, 587, 1973.

26. **Zilberstein, D., Schuldiner, S., and Padan, E.**, Proton electrochemical gradient of *Escherichia coli* cells and its relation to active transport of glucose, *Biochemistry,* 18, 669, 1979.

27. **Booth, I. R., Mitchell, W. J., and Hamilton, W. A.**, Quantitative analysis of proton-linked transport systems. The lactose permease of *Escherichia coli, Biochem. J.,* 182, 687, 1979.

28. **Harold, F. M. and Papineau, D.**, Cation transport and electrogenesis by *Streptococcus faecalis.* I. The membrane potential, *J. Membr. Biol.,* 8, 27, 1972.

29. **Kashket, E. R. and Barker, S. L.**, Effects of potassium ions on the electrical and pH gradients across the membrane of *Streptococcus lactis* cells, *J. Bacteriol.,* 130, 1017, 1977.

30. **Bakker, E. P. and Mangerich, W. E.**, Interconversion of components of the bacterial proton motive force by electrogenic potassium transport, *J. Bacteriol.,* 147, 820, 1981.

31. **Kroll, R. G. and Booth, I. R.**, The role of potassium transport in the generation of a pH gradient in *Escherichia coli, Biochem. J.,* 198, 691, 1981.

32. **Benyoucef, M., Rigaud, J.-L., and Leblanc, G.**, The electrochemical proton gradient in *Mycoplasma* cells, *Eur. J. Biochem.,* 113, 498, 1981.

33. **Abee, T., Hellingwerf, K. J., and Konings, W. N.**, Effect of potassium ions on proton motive force in *Rhodobacter sphaeroides, J. Bacteriol.,* 170, 5647, 1988.

34. **McCarthy, J. E. G., Ferguson, S. J., and Kell, D. B.**, Estimation with an ion-selective electrode of the membrane potential in cells of *Paracoccus denitrificans* from the uptake of the butyltriphenylphosphonium cation during aerobic and anaerobic respiration, *Biochem. J.,* 196, 311, 1981.

35. **Golby, P., Carver, M., and Jackson, J. B.**, Membrane ionic currents in *Rhodobacter capsulatus.* Evidence for electrophoretic transport of K⁺, Rb⁺ and NH₄⁺, *Eur. J. Biochem.,* 187, 589, 1990.

36. **Booth, I. R.**, Regulation of cytoplasmic pH in bacteria, *Microbiol. Rev.,* 49, 359, 1985.

37. **Meury, J. and Kepes, A.**, The regulation of potassium fluxes for the adjustment and maintenance of potassium levels in *Escherichia coli, Eur. J. Biochem.,* 119, 165, 1981.

38. **Meury, J., Robin, A., and Monnier-Champeix, P.**, Turgor controlled K⁺ fluxes and their pathways in *Escherichia coli, Eur. J. Biochem.,* 151, 613, 1985.

39. **Mulder, M. M.**, Energetic Aspects of Bacterial Growth: A Mosaic Non-Equilibrium Thermodynamic Approach, Ph.D. thesis, University of Amsterdam, 1988.

40. **Dinnbier, U., Limpinsel, E., Schmid, R., and Bakker, E. P.**, Transient accumulation of potassium glutamate and its replacement by trehalose during adaptation of growing cells of *Escherichia coli* K-12 to elevated sodium chloride concentrations, *Arch. Microbiol.,* 150, 348, 1988.

41. **Mulder, M. M., Teixeira de Mattos, M. J., Postma, P. W., and Van Dam, K.**, Energetic consequences of multiple K⁺ uptake systems in *Escherichia coli, Biochim. Biophys. Acta,* 851, 223, 1986.

42. **Buurman, E. T., Teixeira de Mattos, M. J., and Neijssel, O. M.**, Futile cycling of ammonium ions via the high affinity potassium uptake system (Kdp) of *Escherichia coli, Arch. Microbiol.,* 155, 391, 1991.

43. **Hamann, A.**, Kaliumtransport bei *Escherichia coli:* molekularbiologische Untersuchungen der Komponenten der niedrigaffinen Kaliumaufnahmesysteme, Ph.D. thesis, University of Osnabrück, Osnabrück, Germany, 1991.

44. **Bossemeyer, D.**, Kaliumtransport bei *Escherichia coli:* biochemische und genetische Untersuchungen der niedrigaffinen Kaliumaufnahmesysteme, Ph.D. thesis, University of Osnabrück, Osnabrück, Germany, 1989.

45. **Rossmann, M. G., Liljas, A., Brändén, C.-I., and Banaszak, L. J.**, Evolutionary and structural relationships among dehydrogenases, in *The Enzymes,* Vol. II, Boyer, P. D., Ed., Academic Press, New York, 1975, 61.

46. **Taylor, W. R. and Green, N. M.**, The predicted secondary structures of the nucleotide-binding sites of six cation-transporting ATPases lead to a probable tertiary fold, *Eur. J. Biochem.,* 179, 241, 1989.

47. **Kakahigashi, K. and Inokuchi, H.,** Nucleotide sequence between the *fadB* gene and the *rrnA* operon from *Escherichia coli, Nucleic Acids Res.,* 18, 6439, 1990.

48. **von Heijne, G.,** The distribution of positively charged residues in bacterial inner membrane proteins correlates with the transmembrane topology, *EMBO J.,* 5, 3021, 1986.

49. **Lengeler, J. W., Bockmann, J., Heuel, H., and Titgemeyer, F.,** The enzymesII of the PTS as carbohydrate transport systems: what evolutionary studies tell us on their structure and function, in *Molecular Mechanisms of Transport,* Quagliariello, E. and Palmieri, F., Eds., Elsevier, Amsterdam, 1992, 77.

50. **Walderhaug, M. O., Polarek, J. W., Voelkner, P., Daniel, J. M., Hesse, J. E., Altendorf, K., and Epstein, W.,** KdpD and KdpE, proteins that control expression of the *kdpABC* operon, are members of the two component sensor-effector class of regulators, *J. Bacteriol.,* 174, 2152, 1992.

51. **Meury, J. and Kohiyama, M.,** Role of heat shock protein DnaK in osmotic adaptation of *Escherichia coli, J. Bacteriol.,* 173, 4404, 1991.

52. **Meury, J. and Schlösser, A.,** unpublished data, 1992.

53. **Hemmingsen, S. M., Woodford, C., van der Vies, S. M., Tilly, K., Dennis, D. T., Georgeopoulos, C. P., Hendrix, R. W., and Ellis, R.,** Homologous plant and bacterial proteins chaperone oligomeric protein assembly, *Nature (London),* 333, 330, 1988.

54. **Bakker, E. P.,** Ionophore antibiotics, in *Antibiotics,* Vol. 5 (Part 1), Hahn, F. E., Ed., Springer-Verlag, Heidelberg, 1979, 67.

55. **Bachmann, B. J.,** Linkage map of *Escherichia coli* K-12, edition 8, *Microbiol. Rev.,* 54, 130, 1990.

56. **Montag, K., Schleyer, M., and Bakker, E. P.,** unpublished data, 1990–1992.

57. **Médigue, C., Bouché, J. P., Hénaut, A., and Danchin, A.,** Mapping of sequenced genes (700 kbp) in the restriction map of the *Escherichia coli* chromosome, *Mol. Microbiol.,* 4, 169, 1443.

58. **Bakker, E. P.,** pH-dependent transport of rubidium by the constitutive potassium uptake system TrkA of *Escherichia coli* K-12, *FEMS Microbiol. Lett.,* 16, 229, 1983.

59. **Roth, W. G., Porter, S. E., Leckie, M. P., Porter, B. E., and Dietzler, D. N.,** Restoration of cell volume and the reversal of carbohydrate transport and growth inhibition of osmotically upshocked *Escherichia coli, Biochem. Biophys. Res. Commun.,* 126, 442, 1985.

60. **Epstein, W.,** personal communication, 1991.

61. **Walderhaug, M. O., Litwack, E. D., and Epstein, W.,** Wide distribution of homologs of *Escherichia coli* Kdp K$^+$-ATPase among Gram-negative bacteria, *J. Bacteriol.,* 171, 1192, 1989.

62. **Hafer, J., Siebers, A., and Bakker, E. P.,** The high-affinity K$^+$-translocating ATPase complex from *Bacillus acidocaldarius* consists of three subunits, *Mol. Microbiol.,* 3, 487, 1989.

63. **Jasper, P.,** Potassium transport system of *Rhodopseudomonas capsulata, J. Bacteriol.,* 133, 1314, 1978.

64. **Golby, P., Cotton, N. P. J., Carver, M., and Jackson, J. B.,** Evidence that the low-affinity K$^+$ transport system of *Rhodobacter capsulatus* is a uniporter. The effects of ammonium on the transporter, *Arch. Microbiol.,* 157, 125, 1992.

65. **Hellingwerf, K. J., Friedberg, I., Lolkema, J., Michels, P. A. M., and Konings, W. N.,** Energy coupling of facilitated transport of inorganic ions in *Rhodopseudomonas sphaeroides, J. Bacteriol.,* 150, 1183, 1982.

66. **Davidson, V. L. and Knaff, D. B.,** ATP-dependent K$^+$ uptake by a photosynthetic purple sulfur bacterium, *Arch. Biochem. Biophys.,* 213, 358, 1988.

67. **Rhoads, D. B., Woo, A., and Epstein, W.,** Discrimination between Rb$^+$ and K$^+$ by *Escherichia coli, Biochim. Biophys. Acta,* 469, 45, 1977.

68. **Abee, T., Palmen, R., Hellingwerf, K. J., and Konings, W. N.,** Osmoregulation in *Rhodobacter sphaeroides, J. Bacteriol.,* 172, 149, 1990.

69. **Reed, R. H. and Walsby, A. E.**, Changes in cell turgor pressure in response to increases in external NaCl concentration in the gas-vacuolate cyanobacterium *Microcystis* sp., *Arch. Microbiol.*, 143, 290, 1985.

70. **Leblanc, G. and Le Grimellec, C.**, Active K^+ transport in *Mycoplasma mycoides* var. Capri. Net and unidirectional K^+ movements, *Biochim. Biophys. Acta*, 554, 156, 1979.

71. **Leblanc, G., and Le Grimellec, C.**, Active transport in *Mycoplasma mycoides* var. Capri. Relationships between K^+ distribution, electrical potential and ATPase activity, *Biochim. Biophys. Acta*, 554, 168, 1979.

72. **Benyoucef, M., Rigaud, J.-L., and Leblanc, G.**, Cation transport mechanisms in *Mycoplasma mycoides* var. Capri cells. Na^+-dependent K^+ accumulation, *Biochem. J.*, 208, 529, 1982.

73. **Benyoucef, M., Rigaud, J.-L., and LeBlanc, G.**, Cation transport in *Mycoplasma mycoides* var. Capri Cells. The nature of the link between K^+ and Na^+ transport, *Biochem. J.*, 208, 539, 1982.

74. **Eisenstadt, E.**, Potassium content during growth and sporulation in *Bacillus subtilis, J. Bacteriol.*, 112, 264, 1972.

75. **Whatmore, A. M., Chudek, J. A., and Reed, R. H.**, The effects of osmotic upshock on the intracellular solute pools of *Bacillus subtilis, J. Gen. Microbiol.*, 136, 2527, 1990.

76. **Whatmore, A. M. and Reed, R. H.**, Determination of turgor pressure in *Bacillus subtilis*: a possible role for K^+ in turgor regulation, *J. Gen. Microbiol.*, 136, 2521, 1990.

77. **Smith, L. T.**, Characterization of γ-glutamyl kinase from *Escherichia coli* that confers proline overproduction and osmotic tolerance, *J. Bacteriol.*, 164, 1088, 1985.

78. **Michels, M. and Bakker, E. P.**, Generation of a large, protonophore sensitive proton motive force and pH difference in the acidophilic bacteria *Thermoplasma acidophilum* and *Bacillus acidocaldarius, J. Bacteriol.*, 161, 231, 1985.

79. **Apell, H.-J. and Solioz, M.**, Electrogenic transport by the *Enterococcus hirae* ATPase, *Biochim. Biophys. Acta*, 1017, 221, 1990.

80. **Kakinuma, Y., Igarashi, K., Konishi, K., and Yamato, I.**, Primary structure of the alpha-subunit of vacuolar-type Na^+-ATPase in *Enterococcus hirae* — Amplification of a 1000-bp fragment by polymerase chain reaction, *FEBS Lett.*, 292, 64, 1991.

81. **Moll, R. and Schäfer, G.**, Chemiosmotic H^+ cycling across the membrane of the thermoacidophilic archaebacterium *Sulfolobus acidocaldarius, FEBS Lett.*, 232, 359, 1988.

82. **Epstein, W. and Davies, M.**, Potassium-dependent mutants of *Escherichia coli* K-12, *J. Bacteriol.*, 101, 836, 1970.

83. **Bakker, E. P., Booth, I. R., Dinnbier, U., Epstein, W., and Gajewska, A.**, Evidence for multiple K^+ export systems in *Escherichia coli, J. Bacteriol.*, 169, 3743, 1987.

84. **Brenner, D. J.**, Facultative anaerobic Gram-negative rods, in *Bergey's Manual of Systematic Bacteriology*, Vol. 1, Krieg, N. R. and Holt, J. G., Eds., Williams & Wilkins, Baltimore, 1984, 408.

85. **Damper, P. D., Epstein, W., Rosen, B. P., and Sorensen, E. N.**, Thallous ion is accumulated by potassium transport systems in *Escherichia coli, Biochemistry*, 18, 4165, 1979.

86. **Erecinska, M. C., Deutsch, C. J., and Davis, J. S.**, Energy coupling to K^+ transport in *Paracoccus denitrificans, J. Biol. Chem.*, 256, 278, 1981.

87. **Reed, R. H., Rowell, P., and Stewart, W. D. P.**, Uptake of potassium and rubidium ions by the cyanobacterium *Anabaena variabilis, FEMS Microbiol. Lett.*, 11, 233, 1981.

88. **Reed, R. H. and Stewart, W. D. P.**, Evidence for turgor sensitive K^+ influx in the cyanobacterium *Anabaena variabilis* ATCC29413 and *Synechocystis* PCC6714, *Biochim. Biophys. Acta*, 812, 155, 1985.

89. **Reed, R. H., Warr, S. R. C., Richardson, D. L., Moore, D., and Stewart, W. D. P.**, Multiphasic osmotic adjustment in a euryhaline cyanobacterium, *FEMS Microbiol. Lett.*, 28, 225, 1985.

90. **Avery, S. V., Codd, G. A., and Gadd, G. M.,** Caesium accumulation and interactions with other monvalent cations in the cyanobacterium *Synechocystis* PCC 6803, *J. Gen. Microbiol.,* 137, 405, 1991.

91. **Sprott, G. D., Shaw, K. M., and Jarrell, K. F.,** Methanogenesis and the K$^+$ transport system are activated by divalent cations in ammonium-treated cells of *Methanospirillum hungatei, J. Biol. Chem.,* 260, 9244, 1985.

92. **Kanner, B. I. and Racker, E.,** Light-dependent proton and rubidium translocation in membrane vesicles from *Halobacterium halobium, Biochem. Biophys. Res. Commun.,* 64, 1054, 1975.

93. **Wagner, G., Hartmann, R., and Oesterhelt, D.,** Potassium uniport and ATP synthesis in *Halobacterium halobium, Eur. J. Biochem.,* 89, 169, 1978.

94. **Garty, H. and Caplan, S. R.,** Light-dependent rubidium transport in intact *Halobacterium halobium* cells, *Biochim. Biophys. Acta,* 459, 532, 1977.

Chapter IID

K$^+$ TRANSPORT IN *ENTEROCOCCUS HIRAE**

Yoshimi Kakinuma

TABLE OF CONTENTS

---

* Most of the work in this review has been done with *E. hirae* ATCC 9790, formerly called
   Streptococcus faecalis (faecium) ATCC 9790, and mutants derived from it.

0-8493-6982-7/93/$0.00 + $.50

I. INTRODUCTION

The Gram-positive bacterium *Enterococcus hirae,* which is found in the intestine of higher animals, has several virtues as an experimental organism for membrane bioenergetics: first, it lacks cytochrome and can generate a proton motive force only from the hydrolysis of ATP by the F_0F_1 H^+-ATPase (Figure 1). Second, its simple metabolic pathways allow for the precise calculation of ATP yields from the few compounds it can metabolize, such as glucose and arginine. Third, it is easy to deplete of energy because it does not make energy store compounds. Finally, like other Gram-positive organisms, it is sensitive to ionophores and inhibitors that act on the cell membrane. One limitation of this organism has been the absence of a convenient system for genetic manipulation, although this is now being studied intensively. Taking advantages of these characteristics, Harold and his colleagues contributed most of the available information about cation transport in *E. hirae.*[1] One of the major findings with *E. hirae* is that ATP plays a much more important role in transmembrane transport than it does in nonfermentative organism, probably arising from the inability of this organism to generate a large proton potential because of lack of the respiratory chain.

Two K^+ uptake systems, designated KtrI and KtrII,[3] and one extrusion system[4] have been reported in this organism. KtrI looks similar to the Trk system of *Escherichia coli,* but the other two systems show unique features as ATP-drive cation pump. These transport systems are summarized in this chapter, and the physiological significance of the K^+ circulation in *Enterococcus hirae* is discussed.

II. K^+ UPTAKE SYSTEMS

A. KtrI

This system is the major potassium uptake system under most conditions, has a pH optimum near 7, an apparent K_m of about 0.2 mM, and attains a maximal rate of about 70 nmol K^+ min^{-1} mg^{-1} cells. The latter is comparable to the rate of glycolysis (100 to 150 nmol lactate per minute per milligram cells), and under certain conditions cells do take up K^+ at a rate that approaches their overall metabolic rate.[5] KtrI also transports Rb^+ with kinetics similar to K^+, and it appears to be constitutive; an early observation[6] that the rates of K^+ and Rb^+ uptake via KtrI increased in *Enterococcus hirae* grown on K^+-deficient medium possibly reflects an increased activity of the inducible F_0F_1 H^+-ATPase, whose expression is probably regulated by the cytoplasmic pH, in such a medium.[7,8] There is also no evidence indicating an activation of KtrI activity by a change of turgor pressure.

A potassium concentration gradient, $[K^+]_{in}/[K^+]_{out}$, of 5×10^4 or even 10^5 is established by this system. The membrane potential is at most -150 mV and often much less,[5,9] judging from a series of studies that pioneered

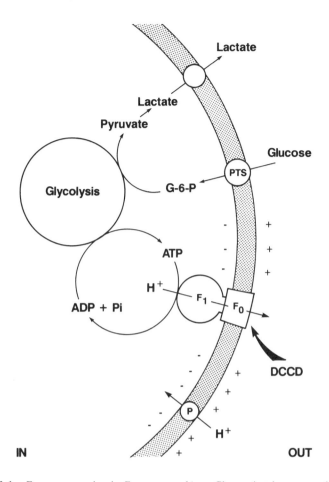

FIGURE 1. Energy conversion in *Enterococcus hirae*. Glucose is taken up as glucose-6-phosphate (G6P) via the phosphotransferase system (PTS) and converted to pyruvate in *glycolysis*. ATP produced by substrate-level phosphorylation is hydrolyzed by the F_0F_1-ATPase with the accompanying translocation of protons (H^+), and the resulting electrochemical proton gradient ($\Delta\bar{\mu}_{H^+}$) powers various proton-coupled membrane reactions. Dicyclohexylcarbodiimide (DCCD) covalently binds the specific subunit of the F_0 portion and blocks proton translocation. The $\Delta\bar{\mu}_{H^+}$ is short-circuited by the action of the protonophore (P) in the membrane. In other streptococci it has been proposed that an efflux of protons mediated by lactate efflux can generate the $\Delta\bar{\mu}_{H^+}$.[2] However, this hypothesis is not addressed here because its physiological importance is still unknown.

the use of lipid-soluble cations (tetraphenylphosphonium ion, TPP^+) to measure the membrane potential in bacteria. It follows that the cellular K^+ pool cannot simply be in electrochemical equilibrium with the membrane potential, so electrogenic uniport is ruled out for KtrI.[5]

The energy requirement of KtrI was inferred from experiments where a membrane potential was imposed artificially by shifting the cells equilibrated at pH 6 to a buffer at pH 8 in the presence of a protonophore. N,N'-dicyclohexylcarbodiimide (DCCD) was present in all experiments to block the H^+-ATPase and prevent generation of the proton potential by ATP hydrolysis. When a membrane potential alone was imposed, K^+ uptake was insignificant. There was somewhat greater K^+ uptake when glucose was present and a protonophore was omitted. A high rate of K^+ uptake was induced where both glucose and a protonophore were present. The requirement for glucose shows that some metabolic product, possibly ATP, is required for KtrI activity. Thus, K^+ uptake by KtrI requires the cells to generate both ATP and an electrical potential. Potassium uptake depolarizes the cells, suggesting that it is accompanied by the influx of a positive charge.[5] From the mechanistic viewpoint, KtrI might be either a pump or a porter. Which mechanism fits this system? For a while, Solioz and his colleagues[10] have insisted that the vanadate-sensitive ATPase, which they purified from the cell membrane of *E. hirae,* is KtrI itself, but they have now withdrawn this claim, as will be described later. Considering the V_{max} of KtrI activity, K^+-stimulated ATPase activity of membrane vesicles would be detectable easily if the H^+-ATPase was blocked. Repeated efforts to find a K^+-stimulated ATPase with the properties of KtrI have not been fruitful. We should tentatively favor the hypothesis that KtrI is a secondary porter that mediates symport of K^+ with proton(s) and is regulated by phosphorylation (or by some other ATP-dependent covalent modification).[5]

By the trick of coupling the uptake of K^+ to that of a proton, the driving force would be greatly increased, and a potassium concentration gradient of 10^4 can be compatible with a membrane potential of -120 mV.[5] Regulation of KtrI by ATP would serve a physiological role by inhibiting rapid K^+ loss in energy-depleted cells. The energetics of KtrI are similar to those of the Trk system of *Escherichia coli* as described in Chapter IIC.

A unique mutant with an alteration in this system is strain 576B, in which the competitive inhibition of K^+ uptake by Na^+ is markedly increased.[11] This mutation appears to reduce the specificity of the cation binding site, especially at pH 7.5, and as a result the high concentration of Na^+ in the medium inhibits K^+ uptake and growth.

B. KtrII

A second uptake system for K^+ in *Enterococcus hirae,* one not dependent on the proton motive force, was discovered by Kobayashi[12] during studies of mutant strain AS25, which is defective in the H^+-ATPase and generation of the proton motive force. The low magnitude of the proton potential in this mutant would not support high activity of KtrI, yet this strain can accumulate K^+ 5000-fold, and in the presence of DCCD K^+ is accumulated about 800-fold. This uncoupler-resistant uptake system, KtrII, has a pH optimum around 9, does not transport Rb^+ very well, and has a K_m for K^+ of 0.5 mM and a V_{max} of about 20 nmol K^+ per minute per milligram cells. Induction of KtrII activity does not respond to K$^+$ deprivation, nor is it repressed by excess K$^+$; it is synthesized in response to the cell's need to expel Na$^+$ because its activity is induced at a high level in mutant AS25 when grown in high-Na$^+$ media. High levels of synthesis of KtrII in the wild-type strain require growth in high-Na$^+$ media containing a protonophore. Thus, the behavior of KtrII appears to be quite unlike Kdp, the K$^+$-translocating ATPase of *Escherichia coli* (see Chapter IIB). KtrII cannot scavenge trace amounts of K$^+$ from the medium, but it does permit the cells to grow under conditions that render KtrI inoperative.[13,14]

Enterococcus hirae contains two Na$^+$ extrusion systems: the Na$^+$/H$^+$ antiporter,[13,15] which is probably constitutive, and a Na$^+$-ATPase, which is induced by an increase in the cytoplasmic Na$^+$ level.[14] The counterion for Na$^+$ efflux via the Na$^+$-ATPase has not been defined clearly. Harold and I proposed a hypothesis that KtrII is a manifestation of the Na$^+$-ATPase and that this enzyme transports K$^+$ by exchange for Na$^+$. First, potassium uptake via KtrII required Na$^+$ to be presented in the cytoplasm; overall, KtrII mediated an equimolar exchange of K$^+$ for Na$^+$. Second, the uptake of K$^+$ required the cells to generate ATP, but did not depend on the electrochemical gradients of either Na$^+$ or H$^+$; the Na$^+$-ATPase generated little or no membrane potential under the experimental conditions, and the K$^+$ concentration gradient, $[K^+]_{in}/[K^+]_{out}$, of 800 established by KtrII was dissipated by addition of valinomycin, excluding the uniport model of KtrII proposed by Skulachev.[14,16] Third, when the proton potential was dissipated, extrusion of Na$^+$ from intact cells was markedly stimulated by external K$^+$, but not by Rb$^+$ or TPP$^+$. These results suggest that the movements of both K$^+$ and Na$^+$ are coupled to ATP hydrolysis.

Furthermore, growth conditions that lead to the expression of KtrII consistently induced enhanced levels of the sodium-stimulated ATPase identified earlier in studies of Na$^+$ efflux (Table 1). Moreover, mutant strain Nak1, which is defective in the Na$^+$-ATPase, lacked KtrII, and the revertant Nak1R that recovered the former activity also possessed the latter.[17] KtrII activity was also absent in strain 7683, another Na$^+$-ATPase mutant, and was recovered in its revertant.[3,18] There is also a qualitative correlation between the initial rates of K$^+$ uptake and the levels of Na$^+$-ATPase (Table 1). These results, indicating that K$^+$ extrusion via KtrII is strictly coupled to Na$^+$ extrusion via the Na$^+$-ATPase, can be explained by two possibilities: (1) a direct exchange of Na$^+$ for K$^+$ ions by the Na$^+$-ATPase; and (2) the combined effect of two separate ATPases, K$^+$ uptake by a putative electrogenic K$^+$-ATPase, KtrII, and Na$^+$ efflux by the Na$^+$-ATPase; in DCCD-treated, Na$^+$-loaded cells, in which Na$^+$ extrusion by the Na$^+$/H$^+$ antiporter is blocked, net K$^+$ uptake by the assumed constitutive KtrII would be totally limited by Na$^+$ efflux by an inducible Na$^+$-ATPase in an indirect manner.[19,20]

Uptake activities of K$^+$ and Rb$^+$ were examined in Na$^+$-loaded cells of strain 9790 grown either on KTY (high K$^+$, low Na$^+$) medium or on KTY supplemented with 0.5 *M* NaCl in order to induce the Na$^+$-ATPase (Figure

TABLE 1
Correlation between KtrII Activity and the Na$^+$-Stimulated ATPase

| Strain[a] | Medium[b] | KtrII activity[c] | | Na$^+$-ATPase activity[c] (nmol/min/mg protein) |
		Initial rate of K$^+$ uptake (nmol/min/mg cells)	$[K^+]_{in}/[K^+]_{out}$ in steady state	
9790	KTY	0.0	1	0.5
9790	KNaTY	1.6	180	10
9790	KNaTY plus CCCP	5.3	490	42
Nak1	KNaTY plus CCCP	0.0	1	0.6
Nak1R	KNaTY plus CCCP	4.6	460	38
AS25	NaTY	13	800	36

Note: Cells were grown on the media listed below. The KtrII activity, in the presence of DCCD to block KtrI, and the Na$^+$-stimulated ATPase activity of the membrane vesicles were assayed as described elsewhere.[3]

[a] 9790, parent strain; Nak1, lack of the Na$^+$-ATPase; Nak1R, revertant of Nak1; AS25, lack of the H$^+$-ATPase.

[b] All the media contained tryptone, yeast extract, and glucose. K stands for 60 mM K$_2$HPO$_4$, Na for 60 mM Na$_2$HPO$_4$. CCCP, carbonylcyanide m-chlorophenylhydrazone (20 μM).

[c] These activities were assayed as described elsewhere.[3]

From Kakinuma, Y. and Harold, F. M., *J. Biol Chem.*, 260, 2086, 1985. With permission.

2); DCCD is absent in the assay buffer. Based on the cation specificity of KtrI and KtrII as above described, we regard Rb$^+$ uptake as KtrI and K$^+$-specific uptake as KtrII, respectively. The Na$^+$/H$^+$ antiporter can operate under the experimental conditions, and Na$^+$ extrusion via the Na$^+$-ATPase should not be rate limiting for the accumulation by both systems. Rb$^+$ uptake by KtrI was unaffected by the growth conditions, but K$^+$-specific uptake activity by KtrII obviously increased in cells grown in KTY containing NaCl (Figures 2A and 2B), indicating that KtrII is actually induced in response to the high Na$^+$ content of the medium. In Nak1 cells grown on KTY medium supplemented with NaCl, Rb$^+$ uptake was normally observed, but K$^+$-specific uptake was not.[21] The results are circumstantial and not strong enough to completely exclude the proposal by Walderhaug and co-workers,[19,20] but I favor a direct exchange model of K$^+$ for Na$^+$ by the Na$^+$-ATPase as the simplest explanation.

If KtrII is the ATPase itself, whichever mechanism fits it, one should expect the membrane-bound ATPase activity that is stimulated by K$^+$ ions. However, we have observed little or no stimulation of the ATPase activity by K$^+$ ions, even together with Na$^+$ ions. Whatever the *in vitro* activity of the ATPase of membrane vesicles is representative of that in the intact cells remains open to question, but this must be counted again as an objection to the "ATPase" hypothesis. We now understand that the Na$^+$-ATPase of *E. hirae* is not a P-type ATPase. This is inferred from (1) its resistance to vanadate; and (2) its molecular properties, which are now being uncovered:

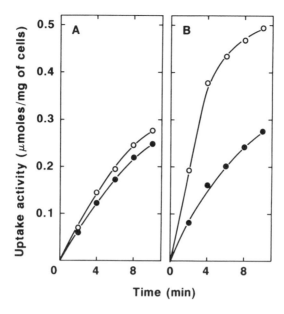

FIGURE 2. Induction of KtrII system in Na⁺-rich medium. The wild-type strain 9790 was grown on KTY medium (A) or on KTY supplemented with 0.5 M NaCl (B), loaded with Na⁺, and resuspended in 50 mM Na⁺-N-tris(hydroxymethyl)methylglycine buffer, pH 8.2. The suspension was supplemented with 10 mM glucose (at -10 min), and uptake was initiated by the addition of 1 mM KCl or RbCl (at 0 min), respectively. ○, KCl; ●, RbCl. (From Kakinuma, Y. and Igarashi, K., *J. Biol. Chem.*, 263, 14166, 1988. With permission.)

the enzyme consists of a peripheral headpiece as the catalytic portion and a membrane-embedded portion.[22] The headpiece resembles that of V-type ATPases,[23] but the nature of the membrane portion is still unknown. Thus, the mechanism of ion coupling by the Na⁺-ATPase of this organism is quite unlike that of the Na⁺/K⁺-ATPase in vertebrates. Cloning and sequencing of the Na⁺-ATPase gene is now in progress, and I expect that it will give hints on the mechanism of KtrII activity in this ATPase.

III. K⁺ EXTRUSION SYSTEM

In several organisms, K⁺ extrusion has been ascribed to secondary antiport of K⁺ for protons, as described in Chapter IIE. The nature of K⁺ efflux and its relation to K⁺ uptake in *Enterococcus hirae* have not been fully studied so far. Passive K⁺ efflux observed in the presence of uncouplers requires ATP (or some other product of metabolism), but whether such efflux is through KtrI (putative ATP-regulated K⁺/H⁺ symporter) or some other pathways is not known.[5] Mutant strain 325B, isolated as requiring elevated concentrations of K⁺ for growth, is defective in K⁺ retention. The normal kinetics of K⁺ flux in this mutant suggest that it affects a separate system, probably one that mediates K⁺ efflux.[24]

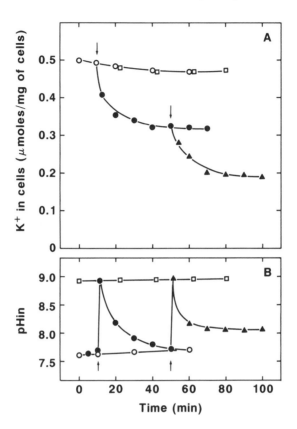

FIGURE 3.   Effect of diethanolamine on uphill K$^+$ extrusion and the internal pH. Cells were grown on KTY medium and suspended in buffer *a* (50 mM 2-(cyclohexylamino)ethanesulfonic acid (CHES)-KOH, pH 9.0, containing 0.2 M KCl and 50 mM diethanolamine) or buffer *b* (50 mM CHES-KOH, pH 9.0, containing 0.2 M KCl and 10 mM glucose). Diethanolamine (pH 9.0) was added to buffer *b* at 10 min and 50 min, respectively. The cellular K$^+$ (A) and the intracellular pH (B) were determined as described elsewhere.[4] The cytoplasmic water space is taken to be 2 μl/mg of cells.[35] Buffer *a:* □, no addition; Buffer *b:* ○, no addition; ●, 20 mM diethanolamine at 10 min.; ▲, 50 mM diethanolamine at 50 min. (From Kakinuma, Y. and Igarashi, K., *J. Biol. Chem.,* 263, 14166, 1988. With permission.)

A unique K$^+$ extrusion system was recently discovered in this organism: K$^+$ extrusion occurred only when the cytoplasmic pH was alkaline, and it required the generation of ATP.[4] As shown in Figure 3, glycolyzing cells suspended in an alkaline medium extruded K$^+$, even against a K$^+$ concentration gradient, provided the medium contained a weak permeant base (e.g., diethanolamine or methylamine). The amine renders the cytoplasmic pH alkaline (see Figure 1 in Chapter IF); when conditions were arranged to keep the cytoplasm neutral, no K$^+$ extrusion was seen. K$^+$ extrusion required the presence of either glucose or arginine and was unaffected by protonophores or by inhibition of the H^+-ATPase. When the medium contained radioactive methylamine, the cells accumulated the base to an extent stoichiometrically equivalent to the K^+ lost. Concomitantly, the cytoplasmic pH fell from 8.8 to 7.6, at which point K^+ extrusion ceased: this system is strictly regulated by cytoplasmic pH. It is hard to distinguish K^+/ammonium (methylammonium) antiport, which will be described in Chapter IIIB, from K^+/H^+ antiport. However, the K^+/H^+ antiport mechanism would be applicable to this system since (1) K^+/amine (ammonium) exchange occurs only at alkaline pH, as expected if the unprotonated form of amine was the species crossing the membrane; (2) a high concentration of amine is required, suggesting that the amine serves to dissipate the pH gradient (inside acidic) at alkaline pH; and (3) most importantly, radioactive methylamine uptake coupled to K^+ extrusion does not display saturation kinetics even at extracellular concentrations as high as 50 mM. Thus, an ATP-driven transport system that expels K^+ by exchange for H^+ functions at an alkaline cytoplasmic pH. The physiological role of this system will be discussed in Section V.

IV. VANADATE-SENSITIVE ATPase

The issue of potassium transport in *Enterococcus hirae* has been confused by reports of a vanadate-sensitive ATPase that has been solubilized by Triton® X-100 and purified to homogeneity from cell membranes by Solioz and his group.[25] This ATPase consists of a single 78-kDa polypeptide and is 50% inhibited by 3 μM vanadate. The enzyme forms a phosphorylated intermediate on an aspartyl residue.[26] Although the ATPase activity was activated only by valinomycin in the presence of K^+ ions after reconstituting the purified enzyme into proteoliposomes with asolectin, this enzyme was regarded as a K^+-translocating ATPase. As one most important item of evidence for this speculation, K^+ movements were observed in the reconstituted system. K^+-loaded proteoliposomes showed ATP-dependent downhill K^+ efflux, although it occurred at a rather low rate and generated only small gradients of K^+. K^+ efflux was blocked by an antibody raised against purified enzyme.[10] Subsequent screening of a DNA expression library of *E. hirae* genomic DNA with a polyclonal antibody to the purified ATPase led to the isolation of a clone encoding a 63-kDa protein whose sequence is homologous to P-type ATPases, and the sequence homology to that of the KdpB protein of *Escherichia coli* was emphasized.[27] Although no K^+ dependence of ATPase activity was seen in native membranes or in any detergent-solubilized preparations, people easily accepted this ATPase as the first purified, reconstitutively active, bacterial P-type K^+-ATPase.[28] Since this ATPase activity was observed in cells in which the Na^+-ATPase, KtrII, is not induced, Solioz and his colleagues[10] claimed this enzyme to be the KtrI system.

All of these conclusions have recently been called into serious question. Recent work by Solioz and others has shown that the protein encoded by the cloned gene does not encode the ATPase that was purified because disruption of the gene on the chromosome does not lead to loss of the ATPase.[20,29] The homology of this ATPase with other ATPases is actually only moderate, there being only 24% identity with KdpB and less with other P-type ATPases.[20] In recent biochemical studies they suggest that it is a proton ATPase;[30] in proteoliposomes which were prepared by basically the same procedure as that in which they reported K^+ efflux, they observed the generation of a membrane potential that was intravesicular positive, independent of the ionic composition of the assay medium. Although ATP-driven H^+ movement was not demonstrated in reconstituted proteoliposomes, they conclude that the membrane potential is generated by an electrogenic influx of protons. Thus, K^+ transport activity in the reconstituted preparation could be due to secondary movements of K^+ through leak pathways, driven by the membrane potential generated by proton movement. Although it is another doubtful matter that this ATPase physiologically functions as an electrogenic proton pump like F_0F_1-ATPase,[30] we appreciate their withdrawal of the idea that this enzyme is a K^+ transport ATPase and the KtrI system.

V. PHYSIOLOGICAL SIGNIFICANCE OF K^+ CIRCULATION IN *ENTEROCOCCUS HIRAE*

Although ion circulation is certainly important for cell physiology,[31] an elegant work by Harold and Van Brunt[32] showed that cation circulation is not obligatory for the growth of *E. hirae;* this organism can grow quite well on a rich medium even when the proton potential and other cation gradients have been dissipated by the ionophore gramicidin D, which renders the membrane permeable to H^+, Na^+, and K^+, provided the medium pH is kept near 8.0 and the concentration of external K^+ is high. In other words, the maintenance of a high cytoplasmic K^+ level and of a slightly alkaline cytoplasmic pH would be the essence of bacterial growth; the key functions of these are not particularly clear.

The diversity of K^+ transport systems seen in *E. hirae* is closely related to ion homeostasis in its natural habitat. *E. hirae* survives in K^+-deficient media (100 μM) and over a broad range of environmental pH (from 5 to 11).[33] At acid pH, the proton motive force generated by the H^+-ATPase, although smaller than that of aerobes, suffices to drive KtrI (K^+/H^+ symporter) and other proton-linked transport systems (Figure 4A). The cytoplasmic pH must be alkalinized to within 7.5 to 8.0 for optimal growth by an interplay of electrogenic H^+ expulsion via the H^+-ATPase and KtrI,[34] but as the cytoplasmic pH rises above the pH optimum of the H^+-ATPase (around 6.5) it will diminish. It is unknown whether the K^+ efflux pathway as seen in strain 325B[24] participates in the K^+ circulation at acid pH.

The proton motive force is drastically decreased at external pH above 8. At pH 9 to 10 *E. hirae* still grows well, and the cytoplasmic pH is acidified

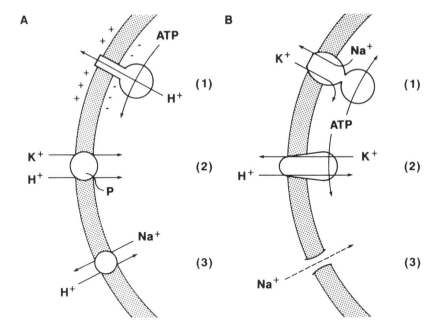

FIGURE 4. Model of K^+ circulation in *Enterococcus hirae*. (A) Activity in media poor in Na^+ and at low pH: (1) the F_0F_1 H^+-translocating ATPase; (2) KtrI, a regulated K^+/H^+ symporter; (3) Na^+/H^+ antiporter. (B) Activity in media rich in Na^+ and at high pH (F_0F_1 H^+-ATPase and KtrI are omitted because of their smaller contributions): (1) KtrII, Na^+/K^+-ATPase; (2) the K^+/H^+ antiport pump (K^+/H^+-ATPase?); (3) the nature of this Na^+ pathway is unknown (Na^+/H^+ antiporter?).

to 8 to 8.5; growth of *E. hirae* at alkaline pH requires acidification of the cytoplasmic pH since nigericin, an electroneutral K^+/H^+ antiport ionophore, completely blocks growth at pH 9.5, but not at pH 7.5, in K^+-rich medium.[35] As a result the proton potential across the cell membrane goes to zero at alkaline pH, and in some cases turns positive by one unit or more. Thus, neither the KtrI system nor the Na^+/H^+ antiporter function at alkaline pH. We imagine that KtrII functions for K^+ accumulation under alkaline pH condition at which KtrI does not operate (Figure 4B). Mutant Nak1, defective in both KtrII and the Na^+-ATPase, was selected by alkaline sensitivity in high Na^+, low K^+ media.[17] Amplification of both KtrII and the Na^+-ATPase was remarkable at alkaline pH.[36] In addition, *E. hirae* cannot grow well in alkaline media poor in both K^+ and Na^+, but the inclusion of Na^+ into media restores growth,[37] suggesting that the Na^+-dependent system, KtrII, is probably the only means by which the cells can accumulate K^+ in alkaline media poor in K^+.

In several bacteria, acidification of the cytoplasmic pH has been attributed to secondary porters of Na^+/H^+ or K^+/H^+.[38] This cannot be true for *E. hirae*. Even in the medium containing 150 to 200 mM K^+, this organism acidifies the cytoplasm and grows well at alkaline pH. At this condition, the Na^+/H^+ antiporter cannot acidify the cytoplasmic pH because the membrane potential is too small to drive it. The H^+ influx via the secondary K^+/H^+ antiporter, if it was present, cannot account for the magnitude of the established pH gradient.[35] The ATP-driven K^+/H^+ antiport pump probably plays an important role in pH regulation at alkaline pH (Figure 4B); a mutant defective in this pump, isolated as one of the alkaline-sensitive mutants in high-K^+ media, failed to acidify the cytoplasm at alkaline pH.[39] Thus, the existence of diverse K^+ transport systems allows the organism to cope with an environment subject to fluctuations in both ionic composition and pH.

VI. CONCLUSION

There should be a relationship between the kinds of transport systems found in any given organism and its habitat or metabolic economy. In this context the characteristics of the K^+ transport systems described here may be relevant to the fact that growing *Enterococcus hirae* maintain a significantly lower proton potential than do growing aerobic bacteria, e.g., *Escherichia coli*, under some conditions. This may be one reason why this organism has evolved various ATP-drive cation pumps, here KtrII and the K^+/H^+ antiport system functioning at alkaline pH.

ACKNOWLEDGMENTS

I thank Dr. F. M. Harold for invaluable support at the beginning of this work and critical reading of the manuscript.

Work from my laboratory described in this review was supported in part by a grant-in-aid for scientific research from the Ministry of Education, Science and Culture, Japan.

REFERENCES

1. **Heefner, D. L.**, Transport of H^+, K^+, Na^+ and Ca^{2+} in *Streptococcus, Mol. Cell. Biochem.*, 44, 81, 1982.
2. **Otto, R., Sonnenberg, A. S. M., Veldkamp, H., and Konings, W. N.**, Generation of an electrochemical proton gradient in *Streptococcus cremoris* by lactate efflux, *Proc. Natl. Acad. Sci. U.S.A.*, 77, 5502, 1980.
3. **Kakinuma, Y. and Harold, F. M.**, ATP-driven exchange of Na^+ and K^+ ions by *Streptococcus faecalis*, *J. Biol. Chem.*, 260, 2086, 1985.
4. **Kakinuma, Y. and Igarashi, K.**, Active potassium extrusion regulated by intracellular pH in *Streptococcus faecalis*, *J. Biol. Chem.*, 263, 14166, 1988.
5. **Bakker, E. P. and Harold, F. M.**, Energy coupling of potassium transport in *Streptococcus faecalis*, *J. Biol. Chem.*, 255, 433, 1980.

6. **Abrams, A. and Smith, J. B.,** Increased membrane ATPase and K⁺ transport rates in *Streptococcus faecalis* induced by K⁺ restriction during growth, *Biochem. Biophys. Res. Commun.,* 44, 1488, 1971.

7. **Abrams, A. and Jensen, C.,** Altered expression of the H⁺-ATPase in *Streptococcus faecalis* membranes, *Biochem. Biophys. Res. Commun.,* 122, 151, 1984.

8. **Kobayashi, H., Suzuki, T., Kinoshita, N., and Unemoto, T.,** Amplification of the *Streptococcus faecalis* proton-translocating ATPase by a decrease in cytoplasmic pH, *J. Bacteriol.,* 158, 1157, 1984.

9. **Harold, F. M. and Papineau, D.,** Cation transport and electrogenesis by *Streptococcus faecalis.* I. The membrane potential, *J. Membr. Biol.,* 8, 27, 1972.

10. **Fürst, D. and Solioz, M.,** The vanadate-sensitive ATPase of *Streptococcus faecalis* pumps potassium in a reconstituted system, *J. Biol. Chem.,* 261, 4302, 1986.

11. **Harold, F. M. and Baarda, J. R.,** Inhibition of potassium transport by sodium in a mutant of *Streptococcus faecalis, Biochemistry,* 6, 3107, 1967.

12. **Kobayashi, H.,** Second system for potassium transport in *Streptococcus faecalis, J. Bacteriol.,* 150, 506, 1982.

13. **Harold, F. M. and Kakinuma, Y.,** Primary and secondary transport of cations in bacteria, *Ann. N.Y. Acad. Sci.,* 456, 375, 1985.

14. **Kakinuma, Y. and Igarashi, K.,** Sodium-translocating adenosine triphosphatase in *Streptococcus faecalis, J. Bioenerg. Biomembr.,* 21, 679, 1989.

15. **Kakinuma, Y.,** Sodium/proton antiporter in *Streptococcus faecalis, J. Bacteriol.,* 169, 3886, 1987.

16. **Skulachev, V. P.,** Membrane-linked energy transductions. Bioenergetic functions of sodium: H⁺ is not unique as a coupling ion, *Eur. J. Biochem.,* 151, 199, 1985.

17. **Kakinuma, Y. and Igarashi, K.,** Mutants of *Streptococcus faecalis* sensitive to alkaline pH lack Na⁺-ATPase, *J. Bacteriol.,* 172, 1732, 1990.

18. **Harold, F. M., Baarda, J. R., and Pavlasova, E.,** Extrusion of sodium and hydrogen ions as the primary process in potassium ion accumulation by *Streptococcus faecalis, J. Bacteriol.,* 101, 152, 1970.

19. **Walderhaug, M. O., Dosch, D. C., and Epstein, W.,** Potassium transport in bacteria, in *Ion Transport in Prokaryotes,* Rosen, B. P. and Silver, S., Eds., Academic Press, San Diego, 1987, 85.

20. **Epstein, W.,** Bacterial transport ATPases, in *The Bacteria,* Vol. 12, Bacterial Energetics, Krulwich, T. A., Ed., Academic Press, New York, 1990, 87.

21. **Kakinuma, Y.,** unpublished results, 1990.

22. **Kakinuma, Y. and Igarashi, K.,** Release of the component of *Streptococcus faecalis* Na⁺-ATPase from the membranes, *FEBS Lett.,* 271, 102, 1990.

23. **Kakinuma, Y. and Igarashi, K.,** Some features of the *Streptococcus faecalis* Na⁺-ATPase resemble those of the vacuolar-type ATPases, *FEBS Lett.,* 271, 97, 1990.

24. **Harold, F. M., Harold, R. L., Baarda, J. R., and Abrams, A.,** A genetic defect in retention of potassium by *Streptococcus faecalis, Biochemistry,* 6, 1777, 1967.

25. **Hugentobler, G., Heid, I., and Solioz, M.,** Purification of a putative K⁺-ATPase from *Streptococcus faecalis, J. Biol. Chem.,* 258, 7611, 1983.

26. **Fürst, P. and Solioz, M.,** Formation of a β-aspartyl phosphate intermediate by the vanadate-sensitive ATPase of *Streptococcus faecalis, J. Biol. Chem.,* 260, 50, 1985.

27. **Solioz, M., Mathews, S., and Fürst, P.,** Cloning of the K⁺-ATPase of *Streptococcus faecalis, J. Biol. Chem.,* 262, 7358, 1987.

28. **Gennis, R. B.,** *Biomembranes,* Springer-Verlag, New York, 1989, 311.

29. **Solioz, M. and Bienz, D.,** personal communication in Reference 19.

30. **Apell, H.-J. and Solioz, M.,** Electrogenic transport by the *Enterococcus hirae* ATPase, *Biochim. Biophys. Acta,* 1017, 221, 1990.

31. **Harold, F. M.,** *The Vital Force: A Study of Bioenergetics,* W. H. Freeman, New York, 1986.

32. **Harold, F. M. and Van Brunt, J.**, Circulation of H$^+$ and K$^+$ across the plasma membrane is not obligatory for bacterial growth, *Science,* 197, 372, 1977.

33. **Deibel, R. H. and Seeley, H. W., Jr.**, Genus I. Streptococcus, in *Bergey's Manual of Determinative Bacteriology,* 8th ed., Buchanan, R. E. and Gibbons, N. E., Eds., Williams & Wilkins, Baltimore, 1974, 490.

34. **Bakker, E. P. and Mangerich, W. E.**, Interconversion of components of the bacterial proton motive force by electrogenic potassium transport, *J. Bacteriol.,* 147, 820, 1981.

35. **Kakinuma, Y.**, Lowering of cytoplasmic pH is essential for growth of *Streptococcus faecalis* at high pH, *J. Bacteriol.,* 169, 4403, 1987.

36. **Kakinuma, Y. and Igarashi, K.**, Amplification of the Na$^+$-ATPase of *Streptococcus faecalis* at alkaline pH, *FEBS Lett.,* 261, 135, 1990.

37. **Kakinuma, Y.**, unpublished results, 1988.

38. **Booth, I. R.**, Regulation of cytoplasmic pH in bacteria, *Microbiol. Rev.,* 49, 359, 1985.

39. **Kakinuma, Y. and Igarashi, K.**, in *Abstr. of 14th Annu. Meet. Japan Bioenergetics Group,* Nagoya, 1988, 100.

Chapter IIE

K$^+$ EFFLUX SYSTEMS

Ian R. Booth, Roseileen M. Douglas, Gail P. Ferguson, Andrew J. Lamb, Andrew W. Munro, and Graeme Y. Ritchie

TABLE OF CONTENTS

I. INTRODUCTION

Potassium efflux systems appear to be ubiquitous in Gram-negative bacteria, but there is less certainty about their distribution among the Gram-positive genera. At the present time there is only limited evidence relating to the number of potassium efflux systems, their mechanisms, and their functions within the cell. Only in *Escherichia coli* has there been any systematic evaluation of the number of such transport systems, and even here there is doubt about the number of efflux systems. Consequently, this chapter is constructed around the proposed functions of potassium efflux systems, the current understanding of the mechanisms involved in potassium efflux, and the unanswered questions. This chapter will not seek to discuss ion-conducting channels recently reported for a number of organisms; this will be the subject of another chapter (Chapter IIH).

A. FUNCTIONS OF POTASSIUM EFFLUX SYSTEMS
1. Control over Cytoplasmic pH and Cell Turgor

Potassium efflux systems have been proposed to have two major functions in bacterial cells: regulation of cytoplasmic pH and adjustment of cell turgor. Rosen and his colleagues[12] first suggested that a potassium proton antiport, acting to exchange internal potassium for external protons, could bring about the acidification of the cytoplasm. Such a process would be essential for cells that are able to grow at alkaline pH with their cytoplasmic pH at a more acidic value than that of their environs. Further, the potassium proton antiport would provide an alternative acidification mechanism to the sodium-proton antiport (see Chapters IA and IB).

It has been proposed that bacterial cells require a relatively constant cell turgor to provide a stretch factor for the growth of the bacterial cell wall. Turgor regulation in bacterial cells requires that the organism be able to respond to changes in the external or internal osmolarity by adjusting the osmolarity of the cytoplasm.[1] It is now well established that the accumulation of potassium provides the major mechanism for the adjustment of cytoplasmic osmolarity in the majority of bacterial genera that possess rigid cell walls. This implies that the functioning of potassium transport systems, both for uptake and for efflux, is subjected to regulation of their activity such that they contribute to the controlled accumulation of potassium. Coupled with the control over potassium is the balancing of the composition of the osmotically active components of the cytoplasm.[2,3] The accumulation of appropriate compatible solutes (trehalose, proline, α-aminobutyrate, glycine betaine) contributes to the turgor pressure of the cell and allows the cell to achieve turgor regulation at lower cytoplasmic potassium levels.[2] It is in the reduction of cytoplasmic osmotic pressure through the release of potassium to the medium that potassium efflux systems are proposed to have a significant role.

An ammonium/potassium antiport has also been characterized in *E. coli* cells[4] (see also Chapter IIIB). Addition of methylammonium to cells provoked a release of $^{86}Rb^+$ in approximately equimolar ratio to the methylammonium taken up by the cells. The exchange was not inhibited by FCCP, an uncoupler of oxidative phosphorylation, and has been suggested to be energy independent. Whether this exchange is a manifestation of the osmotic effect of the accumulation of methylammonium provoking potassium loss is unclear (see below). However, the presence of a specific ammonium/potassium exchange system cannot be excluded.

2. Nonphysiological Activation of Potassium Efflux

In addition to the two physiologically important roles for potassium efflux systems described above, it has been shown that at least two other mechanisms exist that will elicit potassium efflux: the addition of certain reagents that form adducts with glutathione[5,6] and the addition of dinitrophenol and related compounds.[7] An important distinction is that the physiological importance of these two "activators" of potassium efflux is unknown.

As will be seen below, we now have a good understanding of the mechanism of gating of potassium efflux by glutathione metabolites, but the dinitrophenol (DNP)-elicited efflux has been studied very little and remains a mystery.

3. Assays of Potassium Efflux Systems
a. *Direct Measurement of Potassium Loss from Cells*

This simple assay relies on the measurement of the potassium content of cells by flame photometry or atomic absorption spectroscopy. For accurate estimation of potassium loss one must be able to separate cells from medium quickly, and both centrifugation and filtration have proved effective in achieving this.[8,9] Potassium loss elicited by alkalinization of the cytoplasm, by the addition of *N*-ethymaleimide (NEM) and DNP, and in response to high turgor due to the accumulation of glycine betaine have been studied in this way.

A more sophisticated version of this assay uses either $^{42}K^+$ or $^{86}Rb^+$ and assays the loss of radioactivity from cells by scintillation counting after separating cells from medium by either filtration or centrifugation.

b. *Assay in Inverted Membrane Vesicles*

In inverted membrane vesicles the cytosolic face of the membrane is exposed to the experimenter; thus, flux into the vesicle represents efflux from the cell. Assays of potassium movement can be monitored with a radioactive ion (e.g., $^{204}Tl^+$)[12] or from the quenching of the fluorescence of weak bases such as 9-amino acridine or quinacrine. In the latter system one can obtain a direct readout of activity from the spectrophotometer, and this can offer distinct advantages to the experimenter. When such vesicles are energized either with ATP or with a respiratory substrate (e.g., lactate), a pH gradient (inside acid) and a membrane potential (inside positive) are generated due to active proton transfer into the lumen of the vesicle. The pH gradient can be maximized by addition of chloride ions, which are relatively freely permeant and, thus, cross the membrane and partially dissipate the membrane potential.[10] Weak bases, which will accumulate on the acid side of the membrane, are transferred to the lumen of the energized vesicle. Quenching of the fluorescence of the base arises from stacking, which only occurs at high concentrations of the base. Any ion which is translocated across the membrane by exchange for a proton reverses the quenching by reducing the magnitude of the transmembrane pH gradient and, thus, also of the intravesicular concentration of the dye. Consequently, potassium-proton antiport activity which will result in alkalinization of the lumen will be observed as a reversal of fluorescence quenching. It is important to note, however, that the magnitude of fluorescence quenching is not linearly dependent on the magnitude of the transmembrane pH gradient. The following relationship between fluorescence quenching and the pH gradient has been proposed:

$$\log(Q/1 - Q) \sim \Delta pH^{12a}$$

Thus, small changes in the pH gradient are usually seen as large changes in fluorescence.

4. Regulation of Cytoplasmic pH

The proposal by Rosen and his colleagues that the potassium-proton antiport could play a role in the regulation of cytoplasmic pH was based on analysis of the properties of the antiporter. Thus, the antiport was found to have an alkaline pH optimum (ca. pH 7.8 to 8.2[11]), consistent with a role in adjusting the cytoplasmic pH whenever it became too high (the "normal" value of the cytoplasmic pH is ca. pH_i 7.4 to 7.6[8]). In contrast with the Na^+ and Ca^{2+} antiports, the K^+-H^+ antiport was not inhibited by thiocyanate, and it was concluded that the system was electroneutral. Consequently, the activity of the system is dictated by the potassium gradient, the pH gradient, and the cytoplasmic pH. The efflux system exhibited sigmoidal kinetics with respect to potassium, but at concentrations that were well below those normally found in cells. However, the data do suggest the possibility that the antiport may be multimeric.[11] The effect of the pH gradient on activity was not investigated.

The above analysis suffers from two minor limitations. The first is that we now know that the potassium transport activity measured may be the sum of several independent systems. Second, the changes in fluorescence observed in these membrane preparations are quite small; therefore, the quantitation requires great care.

Subsequent to the above biochemical analysis, Rosen and his colleagues[12] isolated a pH-sensitive mutant that had lost K^+-H^+ antiport activity. The mutant KHA1 was isolated after mutagenesis with ethylmethylsulfate followed by penicillin amplification at pH 8.5 in media containing high potassium levels. This growth phenotype has been confirmed in our own laboratory. However, the phenotype may be the result of more than one mutation or may be due to a deletion since we have not been able to either complement the lesion from plasmid gene banks, transduce the mutation out by bacteriophage P1, or obtain revertants that have recovered the ability to grow at alkaline pH. In the vesicle assay system both thallium uptake and potassium-dependent reversal of quinacrine fluorescence were reduced, indicating that all of the systems responsible for these activities had been eliminated by the mutation(s). Unfortunately, as is indicated above, it has not proved possible to take the genetics analysis of this mutant further.

Studies on alkalinization of the cytoplasm have implicated potassium-proton exchangers in control over cytoplasmic pH in *Vibrio alginolyticus* (see Chapter IIA). Thus, addition of diethanolamine to cells at pH 9.0 elicited a transient alkalinization of the cytoplasm.[13] The acidification phase of the transient was accompanied by the release of potassium ions, presumably by a K^+-H^+ antiport. Quantitatively, the loss of potassium appeared to be regulated by the cytoplasmic pH; thus, the proposal was made that the antiport function regulated pH_i.[13] There is one major caveat to this experiment. The alkalinization of the cytoplasm is caused by the entry of diethanolamine down its concentration gradient, and it is the association of the weak base with a proton that causes the rise in pH_i. The influx of diethanolamine thus increases the turgor pressure of the cytoplasm, which may in itself activate potassium efflux (see below). Extrapolating from the published data, there appears to be an equimolar exchange of diethanolamine for potassium, which would be consistent with turgor balancing (see also Chapter IF). Consequently, the antiport may be responding to high turgor rather than pH_i or indeed to both, or there may be separate antiports that allow K^+-H^+ exchange in response to the separate signals.

In a further series of experiments clearer evidence for the role of a K^+-H^+ antiport in pH regulation was obtained.[13] Cells were resuspended at different pH values and with different transmembrane potassium gradients, and the final steady-state value of pH_i was determined. The experimental result was that the final pH_i of the cells was independent of the transmembrane potassium gradient as long as that gradient exceeded the value needed to establish the "normal" value of pH_i.[13] This is strong evidence that the activity of at least one potassium-proton exchanger is regulated by the cytoplasmic pH.

Finally, a similar series of experiments has shown that alkalinization of the cytoplasm of *Streptococcus faecalis* provokes potassium efflux in a manner similar to that observed in *Vibrio*[14] (see also Chapter ID). The potassium efflux system required active metabolism to generate ATP, but did not require the activity of the proton-translocating ATPase. Although this specific information is not available for the *Vibrio* experiments, the suggestion here that the antiport may be energized directly by ATP is unusual, although not without precedent in the streptococci.[15]

5. Regulation of Cell Turgor

Epstein and his colleagues[16] have established that for *E. coli* the principal determinant of the cytoplasmic potassium content is the osmotic pressure of the growth medium. Subsequently, it has been established that potassium uptake is regulated by the turgor pressure.[17,18] Studies on the efflux of potassium have been more limited, but have established that potassium efflux is elicited by the accumulation of glycine betaine in cells that are subjected to osmotic stress.[19] Thus, when cells are subjected to an upshock (increase in the external osmotic pressure), potassium is accumulated to a new level that is usually equivalent to one half the increase in the osmotic pressure.[20] This accumulation may be transient if the cells can synthesize the compatible solute trehalose.[21] The accumulated potassium is released when the cells are subsequently exposed to glycine betaine or proline.[19] Thus, the antiport serves to relieve the high turgor of the cell by exchanging potassium for protons.

B. IDENTIFYING SEPARATE FUNCTIONS

The problem in describing potassium efflux phenomena is that in no organism is it clear how many potassium efflux systems are present since the genetic analysis of these systems is very much in its infancy. Thus, it is impossible to allocate the different phenomena to the functioning of a specific system. For reasons that will become obvious, this article will concentrate on *E. coli,* in which there has been the most detailed analysis of potassium efflux systems to date.

II. POTASSIUM EFFLUX SYSTEMS IN *ESCHERICHIA COLI*

Genetic analysis of *E. coli* has revealed that there are at least three potassium efflux systems, two of which are encoded by the *kefB* and *kefC* genes (Figure 1). Unpublished work by Epstein and his colleagues[40] has identified genes that may encode at least one, and possibly two, other potassium efflux systems. Recent work has shown that the KefB and KefC systems are responsible for the glutathione-regulated potassium efflux, but probably do not contribute to efflux elicited by alkaline pH, dinitrophenol, or high turgor. Whether the genes recently identified by Epstein encode efflux systems and whether these systems are responsible for turgor-regulated efflux remain to be established.

A. KefB AND KefC

In the early 1970s Epstein and colleagues[9] initiated the genetic analysis of potassium transport and retention in *E. coli*. Among the many mutant loci

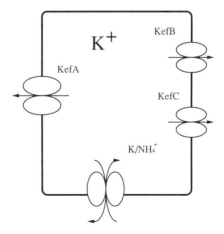

FIGURE 1. Potassium efflux systems in *E. coli*. The major system involved in turgor regulation, and possibly pH regulation, is believed to be KefA.[40] KefB and KefC are the two systems regulated by glutathione metabolites (see text). The fourth system is a potassium/ammonium exchange system described in Reference 4.

identified were *trkB* and *trkC;* mutations at the former locus caused only a minor increase in the potassium requirement of cells, whereas *trkC* mutants possessed a severe impairment of growth when the potassium content of the growth medium was reduced below 0.1 mM. Mutations at either the *trkC* or the *trkB* locus caused a failure of potassium retention. The rate of potassium loss from these mutants was accelerated dramatically compared with the parent strain.[22] It thus seemed likely that these genes encoded protein(s) involved in potassium efflux systems.

A striking observation was that *trkC* mutants were unstable, yielding "revertants" at a rate of 10^{-6}. This property was used to isolate transposon insertions in the *trkB* and *trkC* genes,[7] selecting for cells that recovered the ability to grow on low-potassium media after mutagenesis with a range of transposable genetic elements. All Tn*10* insertions that generated "revertants" of the *trkC* mutant strain TK121 to growth on low-potassium medium lay in the *trkC* gene. Similarly, with *trkB* all of the insertions mapped to the *trkB* region of the genome. Using these transposon insertions it was established that *trkB* and *trkC* encoded independent systems. Once it was established that these genetic loci affected the efflux of potassium the nomenclature was changed to *kefB (trkB)* and *kefC (trkC)*.[19]

The role of these two systems in potassium efflux phenomena was established using transposon insertion mutants that lacked both KefB and KefC activity. From these studies it emerged that the *kefB* and *kefC* loci encode two separate potassium efflux systems that are inhibited by glutathione and are activated by specific glutathione metabolites.[6,19] The KefC system has been characterized in some detail at both the physiological and the molecular level (Figure 2).[6,23,24]

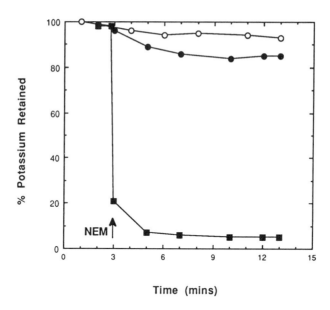

Time (mins)

FIGURE 2. Potassium efflux from *Escherichia coli* cells induced by the addition of *N*-ethyl-maleimide. Cells were prepared as described in Reference 6 and were incubated at 37°C in medium lacking potassium. *N*-Ethylmaleimide (NEM; 0.5 m*M*) was added after 3 min. Symbols: (○,●) strain MJF276 (KefB⁻, KefC::Tn10; (■) strain MJF276 carrying plasmid pkC11. Filled symbols: cells treated with NEM; open symbols: untreated control cells. (From Munro, A. W. et al., *Mol. Microbiol.*, 5, 607, 1991. With permission.)

1. The Role of Glutathione in Potassium Retention

A role for glutathione in the control of potassium retention was proposed by Kepes and co-workers[5,25] after they observed and analyzed the effects of sulfhydryl reagents such as NEM on potassium retention by *E. coli* cells. The pool of cytoplasmic potassium was recovered by cells after the addition of an excess of a reducing agent.[5,26] (A slow recovery also may occur after the removal of NEM by washing cells without the requirement for the addition of a reducing agent.[41]) It was proposed that glutathione, the major soluble thiol of the cell, was essential for potassium retention and that its depletion through reaction with NEM caused the potassium efflux.[25] Subsequently, mutants deficient in the synthesis of glutathione were shown to leak potassium, and such mutants failed to exhibit NEM-stimulated potassium efflux. However, the mutants used in this study were isolated after chemical mutagenesis,[27] and it has been suggested that they also possessed defects in potassium uptake.[40] When the NEM phenomenon was reinvestigated using well-defined mutants in glutathione synthesis (*gshA*::Tn*10kan*), potassium leakage provoked by the absence of glutathione was less significant than in the earlier study.[6] Supplying the mutant with glutathione restored both the potassium pool and the NEM-elicited efflux in both types of mutant.[6,25] Kepes and his colleagues[25,28] have shown the efflux system can also be regulated by a non-thiol-containing analogue of glutathione, ophthalmic acid.[25,28]

With Frag56 (a well-defined glutathione deficient strain) it was possible to demonstrate that the potassium efflux observed with NEM was due to activation of the KefB and KefC systems. In this strain rapid potassium efflux was not observed when cells were resuspended in potassium-free medium,[6] suggesting that the absence of glutathione was insufficient to activate KefC as had been proposed.[25] In Frag56 NEM did not elicit rapid potassium efflux.[6] Further, in normal strains some sulfhydryl reagents that completely modify the glutathione pool do not elicit potassium efflux. Thus, iodoacetate and iodoacetamide reduce the free glutathione pool to close to zero, but do not elicit significant potassium efflux.[6] More significantly, iodoacetate and iodacetamide block the action of NEM. Such cells exhibit normal NEM-elicited potassium efflux if they are reincubated with glutathione prior to NEM treatment. These studies can be explained only if specific glutathione metabolites, e.g., those derived from the reaction of NEM with glutathione, can act as activators of the efflux system.

The synthesis of derivatives of glutathione is well established.[29] In NEM- and iodoacetate-treated cells the glutathione pool was found to be converted to *N*-ethylsuccinimido-*S*-glutathione (ESG) and *S*-carboxymethyl-glutathione (CMG), respectively.[6] Each of these metabolites could be broken down by cells during incubation with reducing agents, regenerating free glutathione.[6] In addition, the reducing agents release glutathione from mixed dithiols in the cells. It is the breakdown of ESG and the release of glutathione that leads to the reversal of NEM-elicited activation of KefC, preventing further potassium loss. The recapture of potassium is assumed to take place through the uptake systems since it has been shown that reversal does not take place when the uptake systems are inactivated by appropriate genetic lesions.[26] Since the Trk systems are inhibited by NEM in a nonreversible manner[6] it is assumed that it is the residual activity of these systems that functions to recover the potassium lost through the efflux systems.

Potassium efflux can be activated by other sulfhydryl reagents such as 1-chloro-2,4-dinitrobenzene (CDNB), *p*-chloromercuribenzoate, *p*-chloro-mercuribenzene sulfonate, phenylmaleimide, or diamide. CDNB and diamide offer further insights into the efflux system and its regulation. The conjugate of glutathione and CDNB (2,4-dinitrophenyl-*S*-glutathione; DNG) is very stable and cannot be broken down readily by the cell.[41] Consequently, there is no reducing agent-provoked reversal of the potassium efflux elicited by CDNB. Clearly, the nature of the C-S bond in DNG is quite different from that in ESG. Despite the dinitrophenyl group, DNG is not permeant through the membrane to a significant extent; consequently, DNG cannot be used as an externally added activator of the efflux process.

Diamide oxidizes glutathione to the disulfide form via a chemical intermediate,[30] and efflux provoked by this compound may reflect redox control

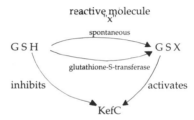

FIGURE 3.   Control over KefC activity. Glutathione is an inhibitor of the KefC system. When glutathione reacts, either spontaneously or via glutathione-S-transferase, with a reactive compound to form "GSX", the KefC potassium efflux system is activated. Note that the inhibitory action of glutathione is independent of the SH group of the peptide,[5] but for activation the presence of the SH group is essential. Detoxification of the glutathione metabolites is achieved either by the glyoxalase I and II enzymes or via another route yet to be identified.

over KefC, as has been suggested.[31] However, high-pressure liquid chromatography (HPLC) analysis of the products formed between diamide and cellular glutathione has revealed some glutathione adducts.[41] Thus, the activation may reflect the adduct rather than the oxidized form of glutathione. Further, other oxidizing agents (hydrogen peroxide, duroquinone) do not provoke potassium loss through KefC. This would suggest that the oxidation of glutathione does not play a significant regulatory role in KefC activity.

2. Is There a Role for Glutathione in KefC Regulation?

The effect of NEM on potassium retention by *E. coli* cells is explained by the formation of a glutathione adduct that activates the KefC system. The original experiments by Kepes' group suggested that glutathione is a negative regulator of potassium efflux.[25] The cloning of the structural gene for KefC allowed this proposal to test more rigorously. A cell lacking glutathione but with multiple copies of the *kefC* gene should exhibit a major leak of potassium; however, this was not found to be the case.[41] Only when a strain was constructed that lacked both potassium uptake systems and the capacity to synthesize glutathione was the multicopy plasmid bearing the *kefC* gene effective in reducing the capacity of the cell to accumulate potassium. The growth deficiency at low potassium concentration could be prevented by incubation with glutathione. Consequently, it is apparent that although glutathione does act negatively on the efflux system the activity of the system in the absence of glutathione is quite low.

The control over KefC activity is depicted in Figure 3.

A high-affinity binding site for glutathione has been reported in the cytoplasm of *E. coli* and has been suggested to be implicated in the regulation of potassium retention.[28] The involvement in potassium retention is based solely on the binding specificity of the protein, which was similar to the relative effectiveness of several glutathione analogues at overcoming the effects of glutathione deficiency. Thus, ophthalmic acid (glutamylaminobutyr-

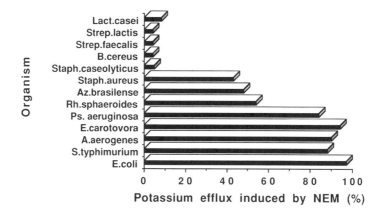

FIGURE 4. KefC distribution in bacteria. The distribution of KefC activity is based on potassium efflux elicited with N-ethylmaleimide (NEM), the formation of N-ethylsuccinimido-S-glutathione (ESG) in cells, and in specific instances by reversal of efflux by dithiothreitol (DTT). See Reference 24 for details. (Data taken from Douglas, R. M. et al., *J. Gen. Microbiol.*, 137, 1999, 1991. With permission.)

ylglycine) and glutamylcystinylbisglycine can substitute for glutathione in restoring potassium tightness to glutathione-deficient mutants. Similarly, both compete with glutathione for a protein binding site in the cytoplasm. However, the protein described is a soluble protein, whereas all the evidence at present suggests that KefC activity is regulated by binding of glutathione to a membrane protein. While the observation of glutathione binding by soluble proteins is interesting, it is unlikely to be a component of the regulation of KefC activity.

3. The Role of Glutathione Gating in Cell Physiology

In analyzing the role of the KefC system in cell physiology there are three principal questions to be addressed:

1. Is the most significant factor the inactivation of the system by glutathione or the activation by glutathione metabolites?
2. What are the properties of mutants that lack the KefC system?
3. Is the KefC system limited to the enteric bacteria?

Of the three questions, a definitive answer is available only to number 3. In a recent survey KefC activity was detected in all the Gram-negative bacterial species examined, but was absent from Gram-positive species with the exception of *Staphylococcus aureus*[24] (Figure 4). KefC activity was defined as NEM-elicited potassium efflux that could be prevented by iodoacetate in cells that possessed glutathione and could synthesize ESG and CMG. As with *E. coli*, the NEM-elicited potassium efflux could be reversed by incubation with reducing agents, and iodoacetate-treated cells exhibited NEM-elicited potassium efflux after incubation with glutathione.[24] Considerable variation in KefC activity has been observed in the different organisms, although the significance of this observation is unclear. A notable factor is that there was no correlation between the magnitude of the glutathione pool and KefC activity revealed by NEM activation. Indeed, some Gram-positive bacteria are able to accumulate significant glutathione pools, but have no detectable KefC activity. KefC activity has also been reported to occur in *V. alginolyticus.*[32]

Homology at the DNA level to the *E. coli kefC* gene was very poor and indicated that even within the enteric bacteria there was little sequence conservation between the genes for KefC.[24] It was also evident that the DNA sequences from the 5' end of the gene were more highly conserved than those from the 3' end.[24]

The widespread distribution of KefC activity among Gram-negative bacteria argues for a function for the transport system that has led to its retention by a range of divergent species. Clearly the answers to questions 1 and 2 above may shed some light on this function. Mutants lacking both KefC and KefB activity were isolated with ease, suggesting that under the conditions of their construction these strains suffered no growth disadvantage.[7] However, this may simply indicate that other compensating changes can occur by adjusting the balance of gene expression for other systems with similar functions. Under the growth regimes investigated no advantage or disadvantage has been observed for mutants lacking KefB and KefC. It may be that the construction of strains lacking other potassium efflux systems may shed some light on the value of KefC to *E. coli.*

Glutathione is utilized by cells as part of the mechanism for detoxifying methylglyoxal (MG), a natural metabolite produced whenever there is an imbalance between flux in the upper (from hexoses to triose phosphates) and lower (from triose phosphates to pyruvate)[33] parts of the glycolytic sequence. Methylglyoxal is a potent inhibitor of bacterial growth, and MG-resistant mutants excrete glutathione.[34] Externally supplied methylglyoxal does activate KefC, indicating that there could be a link between the production of this toxic metabolite and the role of glutathione-regulated potassium efflux.[41] In addition to its role in the detoxification of methylglyoxal, glutathione is also involved in the metabolism of a number of toxic compounds, such as CDNB (see above), and reduces the toxicity of this and similar compounds to the cell.[35] The primary route for CDNB metabolism is via glutathione-S-transferase, but there may be a second route for detoxification which operates in the absence of glutathione.[41] Thus, glutathione may occupy a significant niche in the resistance of the cell to certain toxic metabolites. Conceivably, the ability of such metabolites to activate KefC and lead to spontaneous potassium loss may be an important function for the transport system. Transient potassium loss during CDNB-induced stress may reduce the damage to the cell arising from the potent mutagenicity of CDNB; i.e., potassium loss would serve as a general inhibitor of cell growth to prevent damage by toxic compounds.

Glutathione levels within the cell have been observed to be elevated under conditions of osmotic stress.[6,20,36] Whether this is an adaptation that leads to tighter control over KefC activity is not clear. However, an equally plausible explanation is that glutathione synthesis increases as a mass action consequence of the elevated glutamate pools that are created during osmoadaptation. Alternatively, the increased salt concentration of the cytoplasm in osmoadapting cells may affect the feedback control over glutathione synthesis. Thus, there may be either an important or a prosaic explanation of the high levels of glutathione in osmoadapting cells. The former is favored by the observation of reduced growth rates for glutathione-deficient cells at high osmolarity — a phenomenon that is independent of KefC activity.[20]

B. STRUCTURE OF THE KefC SYSTEM

The structural gene for the *E. coli* KefC system has been cloned recently and the DNA sequence determined.[23] The cloned fragment encodes a putative polypeptide of 620 amino acids. On SDS-polyacrylamide gels a membrane protein of approximately 60 kDa has been detected, which is somewhat smaller than the predicted molecular mass of 79 kDa. Gene fusions of *kefC* to the *lacZ* truncated gene[23] form hybrid β-galactosidase proteins that partition to the cytoplasmic membrane. Analysis of the amino acid sequence of the putative KefC protein has revealed that it is highly asymmetric in the distribution of hydrophobic residues — indeed, the pattern more closely resembles proteins of the multiple drug resistance (MDR) class rather than the "12 helix" pattern of many proton-linked transporters. The carboxy-terminal one third of the protein is essential for KefC activity,[23] but is predominantly hydrophilic in nature and is unlikely to be inserted into the membrane unless in a β-sheet conformation. The amino terminus region is very hydrophobic and could fold across the membrane many times. There are no well-defined turns in the sequence; consequently, without more detailed genetic analysis it is not possible to define the probable number of transmembrane spans.

There is no strong amino acid homology between KefC and other proteins in the data bank, but there are sequence similarities with a number of glutathione-utilizing enzymes such as dehalogenase[37] and glyoxalase.[38] Other enzymes that utilize glutathione and glutathione-like molecules may also show some similarity to KefC,[42] but these again are weak and often involve different regions of KefC. The similarities to dehalogenase and to glyoxalase start within the hydrophobic domain (ca. amino acid 180) and extend into the "hydrophilic domain" close to the carboxy terminus. Both dehalogenase and glyoxalase are soluble proteins and are known to possess hydrophobic pockets.[39] The sequence similarities lend credence to a model of KefC in which only the region close to the amino acid terminus (the first 200 amino acids) is involved in the transport of potassium, and the rest of the protein forms a regulatory domain. Glutathione and glutathione adducts bind to the latter region to close and to activate the system, respectively (Figure 5).

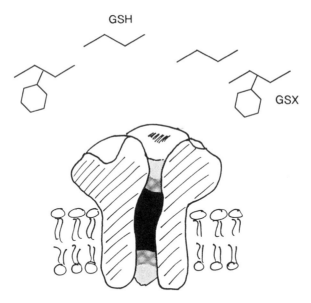

FIGURE 5. Artist's impression of the KefC system. The system is proposed to be oligomeric (here shown as a trimer although there is no specific evidence favoring this particular oligomeric state) based on genetic data. The system is closed when glutathione (GSH) is bound, but is active in the presence of GSX, where X is a moderately hydrophobic grouping such as that derived from N-ethylmaleimide or chloro-2,4-dinitrobenzene. Other less hydrophobic groupings such as the carboxymethyl derivative are only weak activators.

1. The Mechanism of KefC

Two lines of evidence have led to the suggestion that KefC is a regulated potassium-proton antiport.[19,26] First, NEM addition to *E. coli* cells was observed to result in acidification of the cytoplasm concomitant with potassium efflux, while the membrane potential was observed to increase.[19] Activation of KefC by NEM also provoked the accumulation of sodium ions; this was said to be due to the secondary action of the Na^+/H^+ antiport exchanging internal protons for external sodium rather than Na^+ transfer via KefC.[19] The absence of sodium ions from the incubation limited potassium efflux in the presence of NEM.

Second, in *kef*C missense mutants it was observed that the effect of the amino acid substitution was to provoke a potassium leak that led a lower potassium gradient than could be sustained by the membrane potential[26] (i.e., the potassium gradient did not equilibrate with the membrane potential via the mutated KefC protein). The same phenomena were observed with NEM-treated cells.[26] These data provide circumstantial support for a potassium-proton antiport as the mechanism for KefC.

It is possible that KefC forms a glutathione-gated potassium channel. While the evidence for this is not very strong, some properties of the system would be consistent with channel formation. First, the protein is present at very low levels in *E. coli* cells, possibly as few as ten molecules per cell. The rate of potassium efflux would correspond to around 10^5 to 10^6 potassium ions per second per KefC unit.[23] This activity would place the system at the lower edge of the range associated with channels and at the upper end of the exchanger type of system. Second, the system is probably multimeric with cooperativity between the subunits.[41] Thus, the cloned *kefC* gene can suppress the leakage of potassium associated with the missense mutations affecting the chromosomal *kefC* locus. The rate of NEM-induced potassium loss from such a mutant carrying the wild-type gene on a plasmid is extremely rapid. The degree to which this effect is seen is determined by the level of expression of the cloned gene.[41] These data can be explained if the mutant and the wild-type KefC proteins interact in the membrane to form a complex that has properties intermediate between those of the parent and the mutant. Potassium leakage via the mutant protein would be suppressed by the wild-type subunit, and NEM-elicited release via the wild-type subunit would be stimulated by the mutant subunit. Preliminary data obtained with cross-linking reagents have shown that KefC can be cross-linked to give a high-molecular-mass protein. Oligomeric structure would be consistent with channel formation, although it would not exclude antiport function.

III. CONCLUSION

The KefC system is the best understood of the potassium efflux systems in bacteria. Clearly there is still substantial progress to be made, not just on the specifics of this system, but also on the genetics, physiological function, and molecular mechanism of other potassium efflux systems in bacteria. These systems are intrinsically difficult to work with, but they perform essential functions in the cell and we cannot afford to ignore them. The last 20 years has seen many substantial advances in the understanding of the regulation of gene expression phenomena. However, it remains true that cells have remarkable capacities to adapt without recourse to changes in gene expression. It is through the analysis of the systems responsible for such adaptation that further understanding of the physiology of bacterial cells will come, and the information derived from such studies will complement and complete our understanding of bacterial cells.

ACKNOWLEDGMENTS

The authors are indebted to many scientists who have cooperated to further their research. However, outstanding amongst those is Wolf Epstein, who has set the stage for the understanding of potassium transport in bacterial cells. We are also grateful to Evert Bakker, Etana Padan, Shimon Schuldiner, and Chris Higgins for many stimulating discussions on cation transport and cell physiology. Funding for the Aberdeen research group has been provided by the Science and Engineering Research Council, NATO, The Nuffield Foundation,and the Natural Environmental Research Council.

REFERENCES

1. **Epstein, W.,** Osmoregulation by potassium transport in *Escherichia coli, FEMS Microbiol. Rev.,* 39, 73, 1986.
2. **Booth, I. R., Cairney, J., Sutherland, L., and Higgins, C. F.,** Enteric bacteria and osmotic stress: an integrated homeostatic system, *J. Appl. Bacteriol. Symp. Suppl.,* 35, 1988.
3. **Csonka, L. N.,** Physiological and genetic responses of bacteria to osmotic stress, *Microbiol. Rev.,* 53, 121, 1989.
4. **Jayakamur, A., Epstein, W., and Barnes, E. M., Jr.,** Characterisation of ammonium (methylammonium)/potassium antiport in *Escherichia coli, J. Biol. Chem.,* 260, 7528, 1985.
5. **Meury, J., Lebail, S., and Kepes, A.,** Opening of potassium channels in *Escherichia coli* by thiol reagents and recovery of potassium tightness, *Eur. J. Biochem.,* 113, 33, 1980.
6. **Elmore, M. J., Lamb, A. J., Ritchie, G. Y., Douglas, R. M., Munro, A., Gajewska, A., and Booth, I. R.,** Activation of potassium efflux from *Escherichia coli* by glutathione metabolites, *Mol. Microbiol.,* 4, 405, 1990.
7. **Booth, I. R., Epstein, W., Giffard, P. M., and Rowland, G. C.,** Roles of the *trkB* and *trkC* gene products of *Escherichia coli* in potassium transport, *Biochimie,* 67, 83, 1985.
8. **Booth, I. R.,** Regulation of cytoplasmic pH in bacteria, *Microbiol. Rev.,* 49, 359, 1985.
9. **Epstein, W. and Kim, B. S.,** Potassium transport loci in *Escherichia coli, J. Bacteriol.,* 108, 639, 1971.
10. **Reenstra, W. W., Patel, L., Rottenberg, H., and Kaback, H. R.,** Electrochemical proton gradient in inverted membrane vesicles from *Escherichia coli, Biochemistry,* 19, 1, 1980.
11. **Brey, R. N., Rosen, B. P., and Sorensen, E. N.,** Cation/proton antiport systems in *Escherichia coli.* Properties of the potassium/proton antiport, *J. Biol. Chem.,* 255, 39, 1980.
12. **Plack, R. H. and Rosen, B. P.,** Cation/proton antiport systems in *Escherichia coli.* Absence of potassium/proton antiporter activity in a pH-sensitive mutant, *J. Biol. Chem.,* 255, 3824, 1980.
13. **Nakamura, T., Tokuda, H., and Unemoto, T.,** K^+/H^+ antiporter functions as a regulator cytoplasmic pH in a marine bacterium *Vibrio alginolyticus, Biochim. Biophys. Acta,* 776, 330, 1984.
14. **Kakinuma, Y. and Igarashi, K.,** Active potassium extrusion regulated by intracellular pH in *Streptococcus faecalis, J. Biol. Chem.,* 263, 14,166, 1988.
15. **Heefner, D. L. and Harold, F. M.,** ATP-driven sodium pump in *Streptococcus faecalis, Proc. Natl. Acad. Sci. U.S.A.,* 79, 2798, 1982.
16. **Epstein, W. and Scultz, S. G.,** Cation transport in *Escherichia coli.* V. Regulation of cation content, *J. Gen. Physiol.,* 49, 224, 1965.
17. **Rhoads, D. B. and Epstein, W.,** Cation transport in *Escherichia coli.* IX. Regulation of K^+ transport, *J. Gen. Physiol.,* 72, 283, 1978.

18. **Meury, J., Robin, A., and Monnier-Champeix, P.,** Turgor controlled fluxes and their pathways in *Escherichia coli, Eur. J. Biochem.,* 151, 613, 1985.

19. **Bakker, E. P., Booth, I. R., Dinnbier, U., Epstein, W., and Gajewska, A.,** Evidence for multiple potassium export systems in *Escherichia coli, J. Bacteriol.,* 169, 3743, 1987.

20. **McLaggan, D., Logan, T. M., Lynn, D. G., and Epstein, W.,** Involvement of γ-glutamyl peptides in osmoadaptation of *Escherichia coli, J. Bacteriol.,* 172, 3611, 1990.

21. **Dinnbier, U., Limpinsel, E., Schmid, R., and Bakker, E. P.,** Transient accumulation of potassium glutamate and its replacement by trehalose during adaptation of growing cells of *Escherichia coli* K-12 to elevated sodium chloride concentrations, *Arch. Microbiol.,* 150, 348, 1988.

22. **Rhoads, D. B., Waters, F. B., and Epstein, W.,** Cation transport in *Escherichia coli.* VIII. Potassium transport mutants, *J. Gen. Physiol.,* 67, 325, 1976.

23. **Munro, A. W., Ritchie, G. Y., Lamb, A. J., Douglas, R. M., and Booth, I. R.,** The cloning and DNA sequence of the gene for the glutathione-regulated potassium efflux system KefC of *Escherichia coli, Mol. Microbiol.,* 5, 607, 1991.

24. **Douglas, R. M., Roberts, J. A., Munro, A. W., Ritchie, G. Y., Lamb, A. J., and Booth, I. R.,** The distribution of homologues of the *Escherichia coli* KefC K⁺ efflux system in other bacterial species, *J. Gen. Microbiol.,* 137, 1999, 1991.

25. **Meury, J. and Kepes, A.,** Glutathione and the gated potassium channels of *Escherichia coli, EMBO J.,* 1, 339, 1982.

26. **Bakker, E. P. and Mangerich, W. E.,** *N*-Ethylmaleimide induces K⁺-H⁺ antiport activity in *Escherichia coli* K-12, *FEBS Lett.,* 140, 177, 1982.

27. **Apontoweil, A. and Berends, W.,** Isolation and characterisation of glutathione-deficient mutants of *Escherichia coli, Biochim. Biophys. Acta,* 399, 10, 1975.

28. **Meury, J. and Robin, A.,** A high-affinity site for glutathione in the cytoplasm of *Escherichia coli* and its possible role in potassium retention, *Eur. J. Biochem.,* 148, 113, 1985.

29. **Kermack, W. O. and Matheson, N. A.,** The synthesis of some analogues of glutathione, *Biochem. J.,* 65, 45, 1957.

30. **Hibberd, K. A., Berget, P. B., Warner, H. R., and Fuchs, J. A.,** Role of glutathione in reversing the deleterious effects of a thiol-oxidising agent in *Escherichia coli, J. Bacteriol.,* 133, 1150, 1978.

31. **Meury, J. and Robin, A.,** Glutathione-gated K⁺ channels of *Escherichia coli* carry out K⁺ efflux controlled by the redox state of the cell, *Arch. Microbiol.,* 154, 475, 1990.

32. **Nakamura, T., Tokuda, H., and Unemoto, T.,** *N*-Ethylmaleimide desensitises the pH-dependent K⁺-H⁺ antiporter in the marine bacterium *Vibrio alginolyticus, Biochem. Biophys. Res. Commun.,* 136, 1030, 1986.

33. **Ackerman, R. S., Cozzarelli, N. R., and Epstein, W.,** Accumulation of toxic concentrations of methylglyoxal by wild-type *Escherichia coli* K-12, *J. Bacteriol.,* 119, 357, 1974.

34. **Murata, K., Tani, K., Kato, J., and Chibata, I.,** Excretion of glutathione by methylglyoxal-resistant mutants of *Escherichia coli, J. Gen. Microbiol.,* 120, 545, 1980.

35. **Summer, K.-H. and Goggelmann, W.,** Mutagenicity of 1-fluoro-2,4-dinitrobenzene is affected by bacterial glutathione, *Mut. Res.,* 70, 173, 1980.

36. **Munro, G. F., Hercules, K., Morgan, J., and Sauerbier, W.,** Dependence of the putrescine content of *Escherichia coli* on the osmotic strength of the medium, *J. Biol. Chem.,* 247, 1272, 1972.

37. **La Roche, S. D. and Leisinger, T.,** Sequence analysis and expression of the bacterial dichloromethane dehalogenase structural gene, a member of the glutathione-S-transferase supergene family, *J. Bacteriol.,* 172, 164, 1990.

38. **Murata, K., Inoue, Y., Rhee, H.-I., and Kimura, A.,** 2-Oxoaldehyde metabolism in microorganisms, *Can. J. Microbiol.,* 35, 423, 1989.

39. **Douglas, K. T. and Al-Timara, A.,** Recognition determinants of glutathione-binding sites: ribonucleotide reductase system and other enzymes, in *Thioredoxin and Glutaredoxin Systems: Structure and Function,* Holmgren, A., Ed., Raven Press, New York, 1986, 155.
40. **Epstein, W.,** personal communication.
41. **Booth, I. R., Douglas, R. M., Ferguson, G. P., Lamb, A. J., Munro, A. W., and Ritchie, G.Y.,** unpublished data.
42. **Douglas, K. T.,** personal communication.

Chapter IIF

REGULATION OF GENE EXPRESSION DURING OSMOREGULATION: THE ROLE OF POTASSIUM GLUTAMATE AS A SECONDARY SIGNAL OF OSMOTIC STRESS

Ian R. Booth

TABLE OF CONTENTS

0-8493-6982-7/93/$0.00 + $.50

I. INTRODUCTION

The initial events occurring during osmotic upshock of enteric bacteria are now well established. The rapid loss of water, cell shrinkage, and activation of potassium transport leading to swelling of the cell are well documented and represent the short-term adaptation of the cell to osmotic stress.[1,2] Recent attention has focused on changes in gene expression that occur secondarily to the short-term adaptation of the bacterial cell.[3] Several genetic loci have been identified which are specifically transcribed following a rapid increase in the external osmotic pressure (Table 1). Much of the interest has focused on the mechanisms by which regulation of expression of these genes is affected by changes in the external osmolarity. A central theme has been the means by which an external signal is perceived and transmitted to the cytoplasmically located genome.[3] These studies have revealed that there are several mechanisms by which gene expression is controlled by osmotic pressure rather than a single global regulatory protein. However, underlying these mechanisms is the change in cytoplasmic potassium glutamate and the consequent increase in the ionic strength of the cytoplasm.[1-3] It is in the transmission of these changes into a specific signal for expression of selected genes that has provoked some disagreement.[1-8] These studies have highlighted the role of changes in the negative supercoiling of the DNA, the role of the TopA gene product, possible salt regulation of RNA polymerase specificity, and a possible involvement of histone-like protein H1 in regulation of gene expression. Indeed, the relative significance of these factors is the basis of disagreements between groups studying the same genes. The objective of this chapter will be to try to establish that which is accepted as common ground by all groups and then to explore the nature of the controversy and the roots of the disagreement.

A. SHORT-TERM ADAPTATION

Bacterial cells maintain an outwardly directed turgor pressure that is required in Gram-negative bacteria for the extension of cell walls during growth.[1,2] It is widely held to be a truism that the turgor pressure is maintained relatively constant and in adapted cells is independent of the osmotic pressure of the environment. However, in the absence of methods for the direct measurement of turgor in most bacterial cells this belief cannot be verified, and there are some indications that enteric bacteria can adjust their turgor when growth is limited by factors other than the osmolarity of the environment, a condition frequently encountered in natural situations.[9] The turgor pressure derives from the accumulation of solutes in the cytoplasm both via their synthesis and through their selective transport across the membrane.[10-12] The major osmotically active species within the cell are potassium and glutamate ions,[1] and in the enteric bacteria it is the accumulation of these ions that constitutes the short-term adaptation to osmotic stress.

TABLE 1
Osmoregulated Genes of Known Function in Enteric Bacteria

Function	Gene	Induction/repression	OsmZ[a]
Outer membrane permeability	*ompF*	Repression	Yes
	ompC	Induction	Yes
Synthesis of MDO[b]	*mdoG,H*	Repression	?
Transport of potassium	*kdpA,B,C*	Induction	?
Betaine transport	*proP*	Induction	?
	proU	Induction	Yes
Betaine synthesis	*betA,B*	Induction	Yes
Choline transport	*betT*	Induction	Yes
Trehalose synthesis	*otsA,B*	Induction	?
Trehalose breakdown	*treA*	Induction	Yes

Note: The above genes have been shown to be induced or repressed in response to an elevation of the osmotic stress experienced by the cells. Other genes have been reported to be induced by osmotic stress,[66] but their functions are unknown and they have been excluded from the above analysis.

[a] OsmZ: Yes — indicates that the expression of the gene is observed at low osmolarity in the presence of a mutation in the *osmZ* gene; ? — indicates that this property appears not to have been investigated.

[b] MDO, membrane-derived oligosaccharide.[67]

Upon an increase in the osmolarity of the medium, water flows out of the cell and shrinkage occurs. By definition there is an immediate loss of cell turgor that stimulates the activity of the major potassium transport system Trk (consisting of two systems, TrkG and TrkH; see Chapter IIC). Potassium accumulation continues until turgor is restored. If the activity of these transport systems is insufficient to restore turgor, the high-affinity Kdp transport system is induced.[13] As a consequence of the activation of potassium uptake by low turgor, the internal potassium content of enteric bacteria is a function of medium osmolarity.[2] In the absence of significant compatible solute synthesis there is rise in internal potassium equivalent to one half the rise in external osmolarity.

The counterion to potassium has been a subject of much experimentation and conjecture. Variously, it has been suggested that the increase in cytoplasmic glutamate is either sufficient[14] or insufficient to match the increase in the cellular potassium.[12,14,15] Recent careful measurements suggest that osmotic upshock produces an almost quantitatively equivalent increase in glutamate.[14] There are two important caveats to this generalization. First, the cytoplasmic concentrations of other anionic species, glutathione and γ-glutamylglutamine, increase after osmotic upshock.[14] However, in quantitative terms they are not significant. More significant, particularly in the shift from very low osmolarity to moderate osmolarity, there is an exchange of external potassium for organic cations such as putrescine.[16] This latter exchange does mean that there is less requirement for glutamate synthesis under these conditions, although undoubtedly some synthesis still occurs. These two changes may both be important for regulation of cell physiology and gene expression.[17] Thus, mutants unable to make glutathione have a growth disadvantage compared to the parent in a medium of high osmolarity. This effect is independent of the role of glutathione in controlling potassium permeability through the KefC system[14] (see Chapter IIE).

The accumulation of potassium is followed by its loss from cells if compatible solutes are available or can be synthesized. Compatible solutes are low-molecular-weight, neutral compounds that can be accumulated by cells to high intracellular concentrations without impairing enzyme function. Indeed, most compatible solutes will also protect enzymes from the deleterious effects of high ionic strength solutions.[18] When compatible solutes are accumulated by cells of most bacterial cells there is an iso-osmotic displacement of potassium glutamate from the cell.[12] Consequently, the protective effect of the solute lies both in the direct protection of enzymes and in the lowering of the ionic strength of the cytoplasm that is a result of their accumulation by the cell[1] (this description applies most certainly to potassium ions, but the fate of the accumulated glutamate is not clear at the present time). In the enteric bacteria the major compatible solutes are trehalose, betaine, and proline. Trehalose can be synthesized from glucose, but the accumulation of betaine and proline is usually dependent on their transport across the membrane (although some enteric bacteria can synthesize betaine from choline, which is also transported from the environment).[1,3,19] The synthesis of trehalose takes place in most strains of *Escherichia coli* utilized for the study of osmoregulation.[20-22] Consequently, the short-term adaptation is followed by a gradual change in the osmotic balance of the cytoplasm away from the accumulation of potassium glutamate and in favor of trehalose.[12] Trehalose synthesis in *E. coli* appears to be somewhat limited quantitatively; therefore, at high osmolarity trehalose never completely replaces potassium glutamate. In some strains the synthesis of trehalose requires the induction of the *tre* genes, but in other strains the expression is semiconstitutive due to the presence of amber-suppressor mutations — a phenomenon that is still incompletely understood.[21,22] Clearly, however, the change in the cytoplasm and the specific properties of the strain must be borne in mind when conducting physiological experiments.

The accumulation of proline and betaine is activated by osmotic stress and requires the presence of these solutes in the growth medium.[10,11] In cells rapidly shifted from low osmolarity to high osmolarity the accumulation of proline and betaine occurs through the activation of a semiconstitutive transport system, ProP.[23] Thus, if these compatible solutes are available in the growth medium their accumulation will take place within seconds of imposition of osmotic upshock. Relatively low concentrations of betaine or proline are required to give maximum accumulation of the compatible solute. In the growth medium at conventional bacterial densities (i.e., 0.03 to 0.5 mg dry weight cells per milliliter) as little as 100 μM betaine is sufficient to satisfy *Salmonella typhimurium* and cause maximum growth stimulation at osmotic pressures up to 1 Osm.[69] The accumulation of proline or betaine is sufficient to prevent trehalose synthesis and to reduce the expression of many genes that are activated by osmotic stress (Table 1). Furthermore, as indicated above, the accumulation of proline or betaine activates the loss of potassium from the cell due to turgor activation of potassium export systems[12] (see Chapter IIE). Complex media such as L broth contain both proline and betaine derived from yeast extract. It is desirable that studies on osmoregulation are conducted in minimal medium lacking these two compatible solutes. When this precaution is not taken it is likely that the effects of osmotic shifts on gene expression are scaled down, and interpretation of the observations may prove difficult.

The effect of compatible solute transport and synthesis is to reduce the accumulation of potassium glutamate. High concentrations of potassium glutamate in the cytoplasm impair enzyme function and thus growth. This inhibition can be relieved by the accumulation of compatible solutes, although at high osmotic pressures the maximum growth rates that can be established fall well below 50% of the rates observed at low osmolarity. Almost certainly this reflects the inhibition by low water activity of enzyme activities located at the cell surface which are not protected by compatible solutes. Thus, the accumulation of compatible solutes reduces potassium glutamate accumulation but does not completely relieve osmotic stress except at very low osmotic pressures (<0.4 Osm for *S. typhimurium*).

II. GENE EXPRESSION DURING OSMOADAPTATION

Osmotic upshock causes a transient inhibition of expression of genes, the severity of which is determined by the degree of osmotic stress.[24,25] The basis of this transient shutdown of gene expression is not clear, but it may involve elements of the stringent response[26] linked to changes in the pattern of carbon metabolism as carbon flow to glutamate is increased at the expense of the synthesis of other monomers. Equally important may be the transient inhibition of transport of nutrients associated with the loss of turgor, which will lead to short-lived starvation for carbon.[27] Once transcription and translation recommence, several genes are transcribed at enhanced rates (Table 1) while others are now repressed. The expression of some of these genes has a positive effect on growth (e.g., *proU* and *kdp* when the concentration of betaine or potassium, respectively, in the environment is very low, and the *tre* genes for trehalose synthesis), while the absence of others appears not to affect the growth rate and may be purely fortuitous (e.g., *ompC*). Not all of the regulatory systems have been investigated in detail, but several appear to be grouped with the mechanism first described for the *proU* locus. The initial signal for the expression of this locus has been shown to be the increase in potassium and, possibly, also in glutamate.

A. THE ROLE OF POTASSIUM IN REGULATION OF GENE EXPRESSION

The possible role of potassium glutamate in regulating gene expression during osmoadaptation became evident from a comparison of the effects of osmotic pressure on the expression of two loci, *kdp* and *proU*.[28]

The *kdpA* through *kdpE* genes encode the three structural components of the high-affinity Kdp potassium transport system (*kdpA* through *kdpC*) and the two regulatory genes (*kdpD* and *kdpE*). The *kdpD* and *kdpE* genes encode a histidine protein kinase/response regulator couple[69] that acts as a positive regulator of expression of the *kdpA* through *kdpC* genes.[13] In a careful analysis of the regulation of the expression of these genes, Epstein and his colleagues determined that the locus was regulated by the turgor pressure of the cell and proposed that the KdpD protein was a turgor sensor attached to the outer membrane and presumably passing into the cytoplasm across the inner membrane.[13] Whenever the cells experienced a loss of turgor the *kdpA* through *kdpC* genes were expressed. Loss of turgor occurs whenever cells are subjected to an osmotic upshock; the loss of turgor is transient and, similarly, the increased expression of the *kdp* genes is short-lived, returning to the basal level within minutes. Loss of turgor is also experienced by the cells whenever the potassium content of the environment is too low for turgor to be maintained. In this instance, low turgor persists and expression of the *kdp* locus continues. Cells will grow at a reduced growth rate in media with a potassium concentration low enough for sustained *kdp* gene expression to occur. It is not clear whether these cells divide once turgor has reached the desired value (i.e., through the slow accumulation of potassium) or whether they are growing with a reduced steady-state turgor value. Clearly, the expression of the *kdp* genes indicates that at some time during the cell cycle the turgor is low, but more detailed experimentation would be required to ascertain whether expression is continuous or discontinuous.

The *proU* locus encodes a high-affinity transport system for betaine[29] which is induced whenever *E. coli* or *S. typhimurium* is subjected to osmotic upshock.[30] No *proU*-specific regulatory protein has been detected to date despite extensive searches conducted via a variety of genetic stratagems in several laboratories.[31-33] Using *in vitro* experiments and antibodies to sigma-70, the *proU* promoter was shown to be dependent on this sigma factor *in vitro*; thus, it is unlikely that there is a specific sigma factor involved in regulating this locus.[7] The *proU* promoter has been reported to be independent of integration host factor (IHF), although its activity may be modulated by the histone-like protein H1 (see below). An S-30 extract from cells grown at high osmolarity was reported to be more effective at activating the *proU* promoter *in vitro;* since this was a dialyzed extract, it is assumed that this factor is a protein.[7] No further data have been published on this mystery component. Coupled with the absence of specific DNA sequences close to the promoter that resemble regulatory elements, it remains a strong probability that there is no specific regulatory protein. In conclusion, there is little evidence for a specific regulatory protein for the *proU* locus.

In an attempt to understand the physiological switch for the ProU operon a comparison with the turgor-regulated *kdp* operon was undertaken.[28]

1. Comparison of *kdp* and *proU* Expression Signals

A comparison of the *kdp* and *proU* gene systems reveals that although both can be classed as osmoregulated genes they are controlled in quite different ways; this led to the proposal that the *proU* genes appear to be switched on by the increased potassium glutamate accumulation in the cytoplasm. Consider the following observations:[28]

1. Expression of ProU is restricted to media of high osmolarity, while that of Kdp can be induced by growth of cells in media of low osmolarity by the simple expedient of lowering the external potassium concentration.

2. Induction of Kdp in media of high osmolarity and moderate potassium is transient, while that of ProU is sustained.

3. Addition of betaine or proline to media at high osmolarity reduces the expression of ProU, but has no effect on expression of Kdp.

4. High potassium concentrations repress the Kdp system, whereas they are essential for the expression of the ProU system. This last effect was shown to be specific to the ProU locus since the *kdp* genes and other gene systems employed as controls were expressed when induced in cells limited for potassium.

These observations laid the foundation for the hypothesis that the expression of the ProU locus was dependent on the accumulation of potassium glutamate in the cytoplasm. Subsequently, it was proposed that potassium glutamate, either itself or as increased ionic strength, was a "second messenger" transmitting the external osmolarity to a variety of enzyme and regulatory systems located in the cytoplasm.[1-3] In general, this hypothesis has been supported by other experimental observations on several systems and is widely accepted as part of the mechanism by which osmoregulation is coordinated.

2. How is Potassium Glutamate Sensed?

The controversy that surrounds the regulation of the *proU* locus is mainly concerned with this specific question. The common ground is that the osmotically induced changes in potassium glutamate are usually on the order of 100 mM or greater, with a basal level at low osmolarity (ca. 70 mOsm growth media) or around 180 to 220 mM. It is difficult to envisage proteins with specific binding sites with sufficient discrimination to detect the difference between 200 mM and 300 mM potassium glutamate. On the other hand, the conformation of many proteins is affected by the ionic strength of the incubation medium, and for this reason it has been proposed that proteins are likely to sense the change in cytoplasmic ionic strength. This idea has been criticized because in *in vitro* experiments salts other than potassium glutamate have been found to be inhibitory rather than stimulatory.[5] This is suggested to mean that it is potassium glutamate per se that is the regulatory molecule, rather than ionic strength. It must be borne in mind, however, that approximately 3 million years of evolution have led to the almost universal adoption of potassium as the intracellular cation and glutamate as the major anion. Proteins have evolved to function in this environment, and it should come as no surprise that they exhibit optimum activity with these ions. This does not rule out the possibility that changes in potassium glutamate are sensed as changes in ionic strength; rather, it emphasizes that proteins are best adapted to that which they experience most frequently.

Originally, two mechanisms were envisaged that would explain the mechanism of activation of gene expression by potassium glutamate.[3] The first was that a regulatory protein binds to a site adjacent to the promoter of the osmoregulated gene and effects either positive or negative control over expression. High potassium glutamate would either enhance (positive regulation) or inhibit (negative regulation) expression through altered conformations of the DNA or the regulatory protein. The second model involved no intermediary protein, but envisaged a change in promoter-RNA polymerase interaction as a result of increased potassium glutamate in the cytoplasm. Further analysis of the regulation of the *proU* promoter has revealed that neither is likely to be correct, but that the degree of negative supercoiling of the DNA, the activity of histone-like proteins, and the sensitivity of RNA polymerase to salt are all important facets of the regulatory mechanism[5-7,32,34-37]

3. Activation of RNA Polymerase by Salt

Data from *in vitro* analysis of expression of the *proU* locus are consistent with salt activation of the RNA polymerase such that the *proU* genes are transcribed selectively.[5-7]

Studies from two groups have demonstrated the expression from the *proU* promoter in S-30 extracts from *E. coli* cells[5,7] and with purified RNA polymerase.[6] Potassium glutamate was found to stimulate the activity of the *proU* promoter differentially when compared with other promoters. Although initial results suggested that the concentration range within which potassium glutamate was effective was below the physiological range,[1,5,7,12] subsequent experimentation[6] suggested an optimum concentration of 450 mM, which is at the lower end of the range observed in *E. coli* cells subjected to an up-shock.[1,12] Other ion combinations were less effective at activating transcription,[5,7] leading to the proposal that the effect of potassium glutamate was not an ionic strength activation, as had originally been proposed,[2,3,28] but was due to allosteric modulation of RNA polymerase by the salt.[6] However, it can be argued that, when cells have evolved to function in a cytoplasm that is essentially devoid of other ions, the enzyme systems work best with potassium glutamate. In support of this, the other salt combination to which cells might have been exposed frequently during evolution, potassium acetate, was found to stimulate expression, though not as effectively as the glutamate salt.[5,7] It is worth noting that many ion effects on transcription that have been observed in S-30 extracts have proved difficult to reproduce *in vivo;*[38] it is the preexisting demonstration of the effects of potassium accumulation in cells that give credibility to the *in vitro* data. Interestingly, the *proU* promoter was reported to be reduced in its activity when the pH of the incubation was reduced.[6] Similar data have been obtained recently with whole cells; this could indicate that the transient alkalinization of the cytoplasm observed immediately after upshock[12] could aid expression of the *proU* locus.[70]

Comments on two curious observations need to be made. First, when transcription from linear templates was attempted it was reported that these systems were inactive.[5,7] Such data would clearly support a model in which DNA supercoiling played a role in the regulation of gene expression. The authors suggested that the lack of activity of the linear template was to be expected since this was not the normal state of the DNA. However, linear templates have been used very successfully *in vitro* by other workers, and such an explanation cannot be acceptable. If one extended this logic one could argue that the high optimum salt concentration for the activity of the *proU* promoter is expected since these are the conditions under which it was designed to act. Second, it was reported that betaine did not reduce the transcription of the *proU* promoter when potassium glutamate was present.[6] This observation would not be consistent with ionic strength as the activator of the RNA polymerase since it is well attested that salt-dependent changes in enzyme activity are usually reversed by betaine. Further, it was observed that potassium glutamate-induced inhibition of transcription of the *bla* gene was greatly reduced with identical concentrations of glycine betaine. Possibly, this is evidence for a more specific interaction of the RNA polymerase with potassium glutamate or that the primary activation event is the change in DNA supercoiling, which would not be affected by addition of betaine to the *in vitro* assay mix.

Thus, the data obtained with *in vitro* incubations are consistent with the *in vivo* measurements, but they do not prove that the mechanism of activation of the *proU* (and other osmoregulated promoters) is solely or predominantly through salt activation of RNA polymerase.

4. DNA Supercoiling

The circular chromosome of the enteric bacteria is believed to be organized into approximately 50 independently supercoiled regions. The DNA is negatively supercoiled and is in a dynamic state with respect to this parameter due to the actions of DNA gyrase and DNA topoisomerase I.[39-41] DNA gyrase is the enzyme that introduces negative supercoils into the DNA and is one of the sites of action of the antibiotics novobiocin and coumermycin. Topoisomerase I relaxes the DNA, and in *E. coli* mutations that lead to loss of this enzyme are lethal except under aerobic conditions in media of low osmolarity.[42] There is indirect evidence from the measurement of the linking number of reporter plasmids that the superhelical tension is regulated homeostatically. It is believed that the expression of the structural genes for DNA gyrase and topoisomerase I are themselves regulated by the degree of negative supercoiling.[40] The role of gyrase and topoisomerase I in DNA replication, recombination, and transcription has been reviewed recently.[40]

Transcription of genes can be either enhanced or reduced by changes in negative supercoiling. Increased negative supercoiling can promote DNA strand separation and, thus, the initiation of transcription. With the *lac*p[s] promoter, a weak catabolite repressor protein (CRP)-independent promoter in the *lac* operon, low to moderate levels of DNA supercoiling stimulated strand separation 40-fold, but further increases caused a reduction from the maximum rate of open complex formation.[43] It is commonly observed that promoters have supercoiling optima, suggesting that DNA strand separation is not the only parameter affected by the degree of negative supercoiling. Borowiec and Gralla[43] investigated the influence of supercoiling on the open complex formation by *lac*p[s] mutants that were affected in the -35 region, spacer length, and -10 region. The authors found that the stronger promoters, arising from promoter-up point mutations, had lower optima for negative supercoiling than did the native promoter. Consistent with this observation, a mutation at the -35 region that reduced promoter strength was stimulated by high levels of supercoiling that were inhibitory to either the native promoter or the promoters with mutations that increased their activity. Thus, from this example there appears to be a correlation between promoter strength and sensitivity to supercoiling. Strong promoters would have lower optimum values for supercoiling, while weak promoters would require greater changes in supercoiling for their promoter activity to be seen. It is important that essentially similar trends are seen whether the promoter strength determinant derives from the -35 region, the -10 region, or the spacing of these two elements. It should be noted that this analysis can be extended to promoters that are controlled by the binding of regulatory proteins. The negative supercoiling can stimulate the interaction of regulatory proteins with the DNA. Where a regulon exists it is often observed that the DNA binding sites for the regulatory protein show different degrees of similarity to the consensus binding site. Such differences give rise to a hierarchy in the order with which these sites become saturated with the regulatory protein as the protein concentration is varied (e.g., by phosphorylation, by variation of the concentration of a coregulatory molecule, or by synthesis or degradation).[44] It is easy to envisage DNA supercoiling affecting the hierarchy by altering DNA conformation, leading to the sensitivity of some genes in a regulon to changes in supercoiling. Furthermore, models of DNA supercoiling change as a mechanism of regulation of gene expression are often criticized because large regions of the chromosome are affected, whereas environmental signals affect the expression of a small number of genes. Clearly, as is indicated by the work of Borowiec and Gralla,[43] the structure of the promoter and operator will influence the sensitivity of the expression of the gene to changes in the degree of negative supercoiling.

Further evidence for the potential of DNA supercoiling to alter promoter activity comes from the analysis of the *leu500* mutation in *S. typhimurium*.[45,46] The *leu500* mutation is an A to G transition in the −10 region of the *leu* promoter (which creates a GCC sequence at the center of the −10 Region)[45] and inactivates the promoter of the *leu* operon, leading to auxotrophy for this amino acid. Suppressor mutations that conferred leucine prototrophy *(supX)* have been shown to be mutations that eliminate the topoisomerase I activity. Reporter plasmids isolated from *supX* strains are oversupercoiled.[46] The effects of the *supX* mutation can be overcome by low concentrations of coumermycin, which inhibits DNA gyrase and lowers the level of negative supercoiling. Recently, deletions of the *topA* gene encoding topoisomerase I have been constructed *in vitro;* these confer the same phenotype as *supX* mutations.[46] Thus, it is very likely that the *leu500* promoter mutation can be overcome by elevation of the level of negative supercoiling.

From the above it is clear that promoter activity can be influenced by the degree of negative supercoiling. However, the evidence cited above is based on mutations introduced into genes rather than natural systems. To place DNA supercoiling in the panoply of control mechanisms affecting bacterial adaptation, one needs two types of observations: first, that environmental change affects alterations of the level of negative DNA supercoiling; second, that natural promoters respond to these changes in DNA supercoiling. In general, these two requirements have now been met.

Both anaerobics and high osmolarity have been found to increase the degree of negative supercoiling of DNA.[4,32,36,40,47] In addition, supercoiling changes have also been implicated in growth transitions (diauxie), changes in growth temperature, growth phase, and pH[39,40] (elevated supercoiling is found in plasmids isolated from cells grown at neutral and slightly alkaline pH[68]). A role for DNA supercoiling in adaptation to anaerobic conditions was inferred from the slow adaptation to anaerobic conditions of specific *gyrB* mutants.[48] Subsequently, it has been shown that reporter plasmids and the chromosome are more negatively supercoiled when isolated from anaerobically grown cells.[36,47] Consistent with the increased state of negative supercoiling, it was observed that the amount of DNA gyrase bound to the chromosome was increased in anaerobic cells.[40] One important point here is that the degree of supercoiling observed in isolated DNA is possibly an underestimate due to the removal, by phenol treatment, of proteins that stabilize DNA supercoils. Similarly, it has been shown that elevation of the osmolarity of the growth environment provokes increases in the negative supercoiling of reporter plasmids.[8,17,32,36] The increase in negative supercoiling was reversed by the addition of betaine or proline to cells prior to the isolation of the DNA, indicating that the DNA conformation responds similarly to many of the genes regulated by osmotic pressure.[17,20,28,29,32,49] Consistent with these observations is the finding that *topA* mutations that eliminate activity of topoisomerase I in *E. coli* are only lethal under anaerobic conditions and at high osmolarity.[41] This suggests that these growth conditions are associated with elevated DNA supercoiling which, if not reduced by the activity of the topoisomerase, prevents DNA replication and possibly transcription.

There has been at least one report casting doubt on the significance of DNA supercoiling;[50] the authors failed to observe the effects reported in several other studies conducted by different groups. Given the weight of evidence from several other groups, it seems likely that this report is in part incorrect. The assays of DNA supercoiling can be very difficult and certainly require very careful experimentation. It seems likely that this apparent contradiction could be resolved by a short visit of the dissenting authors to one of the other laboratories.

Various strands of evidence point to the role of negative supercoiling in the regulation of gene expression. First, the expression of *proU, tppB* (anaerobically induced tripeptide permease), and *ompC* (the gene for the outer membrane porin synthesized in cells grown at high osmolarity) is prevented by DNA gyrase inhibitors that would relax the DNA.[17,32,36] In contrast, *gyrB* and *tonB* (which encodes a component of the iron siderophore uptake system) are induced by treatment of cells with novobiocin or coumermycin.[40,51] The *tonB* locus is repressed under anaerobic conditions, whereas *tppB* is induced. Indeed, the latter was originally recognized as an anaerobically induced gene and was subsequently shown to be induced also by high osmolarity.[36] The phenomenon of cross-induction of genes by other environmental signals that affect supercoiling is typical of many genes that are sensitive to supercoiling. Thus, for example, induction of *proU* by high osmolarity is enhanced by anaerobic conditions and by slightly alkaline pH, all conditions that increase the negative supercoiling of the DNA.[36,68] However, cross-induction should not be taken to be an indication that genes are regulated by supercoiling. For example, the expression of the *ompC* gene, although affected by DNA supercoiling, is controlled by many environmental signals and involves regulation via both supercoiling and the EnvZ/OmpR two-component regulatory system plus other factors that may not act directly through these mechanisms.[17,52]

Genetic studies also cement a link between the expression of the above genes and the topology of the DNA. Strains carrying missense mutations in the *topA* gene of *S. typhimurium* exhibit increased supercoiling of reporter plasmids and express the *proU* gene when grown at low osmolarity.[32] In contrast, *topA* deletion strains do not express ProU even when grown at high osmolarity (see below). When suppressor mutants *(tos)* of the *topA* missense mutation were isolated, they were found to map in the *gyr* genes and reduced DNA supercoiling (even in the absence of the original *topA* mutant).[32] The expression of ProU at high osmolarity was reduced in the *tos* mutants.[32] The *tos* mutations did not suppress the effects of the *topA* deletion;[32] this has undermined the importance of supercoiling as a regulatory mechanism for some researchers.[5,6,50] However, a similar result was obtained more recently with the *leu500* promoter;[46] a *topA* deletion strain was able to express the *leu* genes even in the presence of *tos* mutations that lowered the mean superhelical density of reporter plasmids. The leucine phenotype was found to correlate more closely with the presence or absence of the TopA protein, as was observed for the *proU* locus.[32] (Note that whereas TopA was essential for expression of *proU* its absence was essential for the expression of *leu500*). Coumermycin, however, did prevent expression of *leu500* in the *topA* deletion strain, suggesting that oversupercoiling may be the root of the effects of *topA* deletions.[46] In this context, it is worth noting that of the genes studied to date *proU* and *leu500* are the most sensitive to supercoiling. Finally, it is clear that while reporter plasmids do provide an indication of the degree of negative supercoiling they are not completely reliable and, hence, should be used in conjunction with other experimental approaches.[34,46]

The search for a regulatory protein for the ProU locus led to the identification of the *osmZ* gene, which has been shown to encode the histone-like protein H1.[34,35,37,53] Mutations of the *osmZ* locus were identified by their ability to lower the threshold osmotic pressure at which the expression of the ProU locus occurs.[32] It should be noted that osmotic regulation of the *proU* locus was not affected, merely the induction curve was shifted to lower values of osmolarity, giving expression at low osmolarity. The *osmZ* locus was found to be the site of mutations that affect the expression of many different gene systems in a range of enteric bacteria (Table 1). Some of these gene systems have been shown to be osmoregulated; others are induced by high temperature (off at 28°C, on at 37°C) and are involved in pathogenic responses.[35] A third group of genes, the *bgl* operon, appears to have no specific regulation and is usually activated by DNA rearrangements.[35,54] The *osmZ* mutations have been shown to increase the frequency of gene switching due to site-specific inversion events.[54] Thus, mutations at the *osmZ* locus exert many pleiotropic effects.

Plasmid DNA isolated from *osmZ* mutants is usually more negatively supercoiled than that from the wild type; this would be consistent with its effects on *proU* expression.[32] However, more recently it has been reported that a *Tn10* insertion in the promoter region of the *osmZ* gene that has been observed to lower the supercoiling of reporter plasmids[35] increases the expression of the ProU locus.[35] A similar observation was made with the analogous gene in *Shigella flexnerii*. Other *Tn10* insertions that are within the *osmZ* gene, and which appear to eliminate the gene product entirely, increase the expression of the ProU locus, and reporter plasmids exhibit increased negative

DNA supercoiling.[37] Thus, as with the *topA* deletion described above, there is significant variance in the outcome of measurements of the reporter plasmid supercoiling. It has been suggested that the different phenotypes of the *osmZ* alleles may result from localization of the mutations in the 3' end of the gene encoding the C-terminus of the gene.[35] DNA binding activity is thought to lie in the amino-terminal region; thus, truncated proteins could have effects that are different from either a complete deletion or the presence of the unaltered protein.

The mechanism of action of protein H1, the product of the *osmZ* locus, is far from clear. The H1 protein is 15- to 16-kDa neutral, DNA-binding protein present in the cell at around 20,000 copies.[55] The protein has a very high content of acid and basic residues (ca. 37% of the residues have charged side chains) that are organized into two regions that are strongly basic, one in which the charged residues are predominantly acidic, and two other regions where the amino acid distribution consists of approximately equal numbers of acidic and basic residues.[35] Synthesis of the H1 protein is not subject to osmotic regulation,[35] but the protein does accumulate in stationary phase cells;[56] thus, its level may rise in cells in which the growth rate is severely restricted by high osmotic pressure. The cell appears to be very sensitive to the amount of the H1 protein since it cannot be cloned, in an expression-competent state, on a multicopy plasmid.[35,37] Consistent with this observation, multicopy plasmids carrying the N-terminal region caused slow growth, and the plasmids acquired mutations readily or were lost from the cells at a high rate.[35] The wild-type *osmZ* gene suppressed the effect of *osmZ* mutations and was found to reduce the basal expression of the *proU* locus at the lower end of the osmolarity spectrum.[37]

It has been shown *in vitro* that addition of pure H1 protein inhibits transcription; thus, a role for this protein as a transcriptional silencer has been proposed.[37] However, alteration of the H1 protein does not always lead to increased transcription, and the effects of H1 can be overcome by high osmotic pressure, anaerobiosis, or high temperature. It could be that H1 stabilizes loosely supercoiled DNA by forming a compaction mechanism and that its removal is compensated for by an increase in supercoiling to restore the DNA to a more compact state. Although H1 does not change the supercoiling of DNA when pure protein and plasmids are incubated *in vitro,* it does cause the compaction of the DNA as revealed by an increase in the sedimentation velocity of the H1-DNA complex.[56] Further, interaction of pure H1 with plasmids carrying the *lacUV5* promoter inhibits open complex formation, an observation consistent with a role for H1 in altering the topology of the DNA.[56] If H1 activity was impaired by high potassium glutamate this could lead to a loosening of the DNA coiling and consequent increased supercoiling to recover compaction. In this view the role of H1 in *proU* expression would be purely incidental, with mutations simply provoking increased supercoiling. An alternative view is that the supercoiling is secondary and that the role of inactivation of H1, either by mutation or by potassium glutamate, is to release the *proU* from its silenced state. No data are available on this point at the present time.

5. Osmoregulated Promoters
a. ProU

Several studies have mapped the *proU* promoter.[33,57-59] Both S1 nuclease studies and primer extension have been used to define the start point of transcription in both *E. coli* and *Salmonella typhimurium*. While the two methods frequently give subtly different results — indicating different numbers of transcription start points — the emerging consensus appears to be that in *S. typhimurium* there is a single major transcription start site approximately 60 nucleotides 5′ to the translation start site of the *proV* gene. (*proV* is the first structural gene of the *proU* operon.) In *E. coli* there also may be a minor promoter located 5′ to the major promoter described in *S. typhimurium*. Recent work in *E. coli*[8] in which the roles of both promoters have been analyzed by deletion analysis has led to the proposal that both are regulated by osmotic stress and that they contribute to similar extents to osmotic regulation of the *proU* operon. This suggestion is at odds with the measurements of the levels of mRNA derived from the two start points, which indicate that the upstream promoter is a much less significant activity.[57] Further, the analysis of promoter mutants also indicates that the downstream promoter is the major activity (see below). Perhaps more significantly, it was observed that removal of sequences downstream of the promoter and lying within the coding region for *proV* led to increased basal expression of the *proU* promoter.[8] It has been suggested that there is a region of DNA that acts negatively in *cis* on the activity of the *proU* promoter and that this sequence is neutralized by changes in DNA supercoiling. Plasmids carrying this region and the region upstream as far as the *proU* promoter have been shown to adopt unusual levels of supercoiling when cloned on low-copy-number vectors.[8] This observation would be consistent with this region affecting the topology of the DNA.

Analysis of the major promoter reveals that potential -35 and -10 regions are moderately good with respect to the consensus, possessing the highly conserved TTGxxx (-35) and TAxxxT (-10) with a 16-base spacing.[33,57-59] The isolation of promoter-up and promoter-down mutations in these two sequences[60] suggests that these are components of the major promoter. To date there is little to indicate why the promoters should be particularly sensitive to changes in either DNA supercoiling or potassium glutamate. The promoter has a GGG motif as part of the -10 region (and notably the *treA* promoter has GG in the -10 region and is less sensitive to supercoiling than *proU*[61] and the *leu500* mutation has GCC in the -10 region[45]), which could indicate an extra energy requirement for melting of the DNA that could be provided by increased DNA supercoiling. The upstream region is rich in polyA and has been shown to adopt an altered conformation in gel retardation assays.[8] However, while this region has been shown to be important for the level of *proU* promoter activity, it does not affect induction by osmotic pressure.[60] Further, the region including the -35 and upstream sequences has been replaced with new sequences and does not affect osmotic induction.[60]

b. Other Promoters

Among the other promoters that are regulated by osmotic pressure, the sequences for the *treA, ompC,* and *betT* genes are available.[19,61,62] The only one of these which shares a feature with *proU* is the *treA* promoter; both have G moieties in the -10 region. All three promoters do, however, respond in the same way to the *osmZ* mutation (i.e., increased expression at low osmolarity) and glycine betaine (i.e., reduced expression).[17,20,32,49] Further, the *bet* genes have been reported to be induced by other conditions that are known to influence DNA supercoiling, e.g., anaerobiosis and high temperature.[63] Curiously, the *treA* promoter was reported not to be sensitive to trehalose synthesis.[61] One would have expected that the synthesis of this compatible solute would reduce expression of the *treA* gene. The failure to observe such repression may relate to the low rate of synthesis of this compatible solute, but this aspect deserves further analysis.

It is not surprising that there are few common elements to the promoters for different osmoregulated promoters since their common activator is accepted to be a salt and there are different ways in which this could act. Perhaps the best illustration of this is the recent demonstration of the role of potassium in the activation of OmpC expression at high osmolarity.[64,65] Mizuno and his colleagues[64] have recently shown that the phosphorylated state of the OmpR protein, the positive activator of the *ompC* promoter, is regulated by potassium in the range that is physiologically relevant. The EnvZ/OmpR couple are members of the family of histidine protein kinase/response regulators in which EnvZ is the membrane-located sensor protein that autophosphorylates at the expense of ATP.[52] OmpR can be phosphorylated and OmpR-P can be dephosphorylated by EnvZ. The cytoplasmic concentration of OmpR-P determines the balance of *ompF* and *ompC* promoter activity through binding to high-affinity sites *(ompF)* and low-affinity sites *(ompC)*.[52] The balance of the kinase and phosphatase activities of the EnvZ protein determines the concentration of OmpR-P[64,65] (Figure 1). The phosphatase activity has recently been shown to be inhibited by high potassium concentrations.[64] Thus, high osmotic pressure, leading to accumulation of potassium glutamate, leads to expression of the OmpC porin through the inactivation of the phosphatase activity of the EnvZ protein. Since the same effects can be shown with a solubilized EnvZ fragment that lacks the N-terminal sensory domain[65] that would normally be located in the periplasm, it is clear that a major aspect of the regulation of porin synthesis is the sensing of cytoplasmic potassium levels. Consistent with this is the observation that glycine betaine reverses

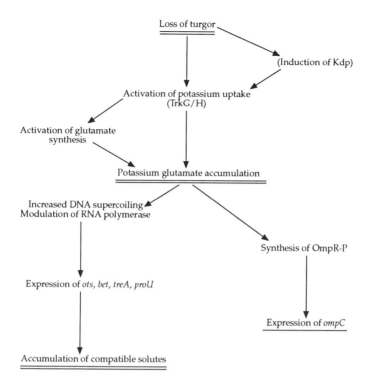

FIGURE 1. Summary of main features of osmoregulation of gene expression in the enteric bacteria (see text for details).

the osmotic induction of the porins, leading to expression of the OmpF porin at high osmolarity.[49]

III. SUPERCOILING VS. ACTIVATION OF RNA POLYMERASE

At this stage of our understanding of the regulation of *proU* and other osmoregulated genes it seems to be unnecessary to choose between the two models described above for osmoregulation of gene expression. There are sufficient good data to be explained that neither hypothesis can claim exclusive accuracy as a model. There are some inconsistencies in the supercoiling data: thus, for example, it is known that the *proU* locus can be osmoregulated normally when inserted at the *attB* site, which implies that the essential information for regulation is carried within the *proU* DNA. If the DNA domains are supercoiled independently this would require these two regions (i.e., the normal *ProU* locus and *attB*) to respond identically with regard to osmotically induced changes in DNA supercoiling. However, it seems unlikely that the independent domains are differentially supercoiled since it is known that the level of supercoiling is regulated homeostatically (via the supercoiling dependence of the *topA* and *gyr* promoters), and if the different domains were differentially coiled then this could lead to a breakdown in the balance of gyrase and topoisomerase. A further difficulty to be resolved is that if the synthesis of gyrase and topoisomerase are regulated by supercoiling it is difficult to envisage how any change in DNA supercoiling can be other than short lived since an increase in DNA supercoiling should promote the synthesis of TopA, leading to relaxation. This could be overcome if the activities of the two enzymes controlling supercoiling were differentially affected by potassium glutamate. Clearly, the expression of the *proU* locus is at a high level for several generations after the imposition of osmotic upshock; thus, the signal for expression must be sustained. Thus, either the cell must be able to stabilize high levels of supercoiling, or the expression of *proU* cannot be regulated by supercoiling. The former appears most likely given the range of functions that appear to be regulated by DNA topology,[35,37,53] as revealed by isolation of H1 mutants in a variety of organisms. In this context, the mechanism of transmission of the signal is also unclear: is the H1 protein the transducer of osmotic pressure within the cytoplasm? It is clear that DNA supercoiling is regulated by factors other than the potassium glutamate level in the cell; how are the signals from growth rate, temperature, pH, and anaerobiosis integrated with those from osmotic stress? I have already referred to the effects of temperature and anaerobiosis, which do not significantly affect the potassium glutamate levels. Further, glycine betaine does not completely repress the expression of any of the osmoregulated genes,[28,29] although it is effective in lowering the potassium glutamate to a level close to that observed in cells at low osmolarity.[12] It is likely that the activities of the topoisomerase and the gyrase are affected by other regulatory molecules. Indeed, it has been suggested that ATP/ADP ratios may play a significant role,[47] although AMP (the changes in the concentration of this molecule with the energy status of the cell are much greater than those for ADP and ATP, making the task of sensing change much easier) and ppGpp are more likely effectors.

To date, the *in vitro* data are insufficiently discriminating to explain completely the osmoregulation of the *proU* locus (i.e., the difference between other systems and *proU* is not sufficiently marked). None of the studies has been able to reproduce the very characteristic induction pattern of the *proU* operon that is observed as a function of osmotic pressure *in vivo*. The evidence for the lack of a role of supercoiling in *proU* expression in *E. coli* lacks conviction when compared with other studies in the same organism.

IV. CONCLUSIONS

It is striking that there are elements in both analyses which look convincing, and it may be that since supercoiled templates are necessary for the *in vitro* assay and potassium glutamate accumulation is essential for expression *in vivo* both are forces directing the osmoregulation of *proU* and other osmoregulated genes.

There are a large number of questions that remain unanswered despite much effort and ingenuity of experimental design. What is clear is that potassium glutamate is the secondary signal for osmoregulation of gene expression and that it elicits changes in both DNA topology and gene expression. Inhibition of these changes in DNA supercoiling with appropriate inhibitors prevents the changes in gene expression and, thus, establishes a causal link between the two phenomena. Similarly, preventing potassium accumulation or provoking its release by the addition of glycine betaine reduces the expression of the osmoregulated genes and lowers the supercoiling of reporter plasmids. The classical genetic approach of the isolation of mutations that change the pattern of gene expression leads to the isolation of mutations affecting the enzymes involved in regulating the level of DNA supercoiling and the proteins involved in DNA compaction. *In vitro* experiments point to the *proU* promoter having a high salt requirement for activation, but also a requirement for supercoiled DNA templates. The balance of the data points to a major link between the state of supercoiling of the DNA and osmotic regulation of gene regulation. Thus, the task of signaling osmotic stress must be added to the important role of turgor regulation as the major roles of cytoplasmic potassium.

ACKNOWLEDGMENTS

The author wishes to thank his colleagues, especially Chris Higgins, with whom he has collaborated on investigations of osmoregulation in the enteric bacteria. Particular thanks are also due to Wolf Epstein, Gordon Stewart, and Evert Bakker for many useful discussions. The work of the Aberdeen group has been supported by the Science and Engineering Research Council and the Agricultural and Food Research Council.

REFERENCES

1. **Epstein, W.,** Osmoregulation by potassium transport in *Escherichia coli, FEMS Microbiol. Rev.,* 39, 73, 1986.
2. **Booth, I. R., Cairney, J., Sutherland, L., and Higgins, C. F.,** Enteric bacteria and osmotic stress: an integrated homeostatic system, *J. Appl. Bacteriol. Symp. Suppl.,* 35, 1988.
3. **Booth, I. R. and Higgins, C. F.,** Enteric bacteria and osmotic stress: intracellular potassium glutamate as a secondary signal of osmotic stress?, *FEMS Microbiol. Rev.,* 75, 239, 1990.

4. **Higgins, C. F., Dorman, C. J., and Ni Bhriain, N.**, Environmental influences on DNA supercoiling: a novel mechanism for the regulation of gene expression, in *The Bacterial Chromosome*, Drlica, K., Ed., American Society for Microbiology, Washington, D.C., 1990, 421.

5. **Ramirez, R., Prince, W. S., Bremer, E., and Villarejo, M.**, *In vitro* reconstitution of osmoregulated expression of *proU* of *Escherichia coli*, *Proc. Natl. Acad. Sci. U.S.A.*, 86, 1153, 1989.

6. **Prince, W. S. and Villarejo, M. R.**, Osmotic control of *proU* transcription is mediated through direct action of potassium glutamate on the transcription complex, *J. Biol. Chem.*, 265, 17637, 1990.

7. **Jovanovich, S. B., Record, M. T., and Burgess, R. R.**, In an *Escherichia coli* coupled transcription-translation system, expression of the osmoregulated gene *proU* is stimulated at elevated potassium concentrations and by an extract from cells grown at high osmolality, *J. Biol. Chem.*, 264, 7821, 1989.

8. **Dattananda, C. S., Rajkumari, K., and Gowrishankar, J.**, Multiple mechanisms contribute to osmotic inducibility of *proU* operon expression in *Escherichia coli:* demonstration of two osmoresponsive promoters and of a negative regulatory element within the first structural gene, *J. Bacteriol.*, 173, 7481, 1991.

9. **Welsh, D. T., Reed, R. H., and Herbert, R. A.**, The role of trehalose in the osmoadaptation of *Escherichia coli* NCIB 9484; interaction of trehalose, K$^+$, and glutamate during osmoadaptation in continuous culture, *J. Gen. Microbiol.*, 137, 745, 1991.

10. **Perroud, B. and Le Rudulier, D.**, Glycine betaine transport in *Escherichia coli:* osmotic modulation, *J. Gen. Microbiol.*, 161, 393, 1985.

11. **Strom, A. R., Falkenberg, P., and Landfald, B.**, Genetics of osmoregulation in *Escherichia coli:* uptake and biosynthesis of organic solutes, *FEMS Microbiol. Rev.*, 39, 79, 1986.

12. **Dinnbier, U., Limpinsel, E., Schmid, R., and Bakker, E. P.**, Transient accumulation of potassium glutamate and its replacement by trehalose during adaptation of growing cells of *Escherichia coli* K-12 to elevated sodium chloride concentrations, *Arch. Microbiol.*, 150, 348, 1988.

13. **Laimins, L. A., Rhoads, D. B., and Epstein, W.**, Osmotic control of *kdp* operon expression in *Escherichia coli*, *Proc. Natl. Acad. Sci. U.S.A.*, 78, 464, 1981.

14. **McLaggan, D., Logan, T. M., Lynn, D. G., and Epstein, W.**, Involvement of γ-glutamyl peptides in osmoadaptation of *Escherichia coli*, *J. Bacteriol.*, 127, 3631, 1990.

15. **Richey, B., Cayley, D. S., Mossing, M. C., Kolka, C., Anderson, C. F., Farrar, T. C., and Record, M. T., Jr.**, Variability of the intracellular ionic environment of *Escherichia coli*, *J. Biol. Chem.*, 262, 7157, 1987.

16. **Munro, G. F., Hercules, K., Morgan, J., and Sauerbier, W.**, Dependence of the putrescine content of *Escherichia coli* on the osmotic strength of the medium, *J. Biol. Chem.*, 247, 1272, 1972.

17. **Graeme-Cook, K. A., May, G., Bremer, E., and Higgins, C. F.**, Osmotic regulation of porin expression: a role for DNA supercoiling, *Mol. Microbiol.*, 3, 1287, 1989.

18. **Pollard, A. and Wyn Jones, R. G.**, Enzyme activities in concentrated solutions of glycine betaine and other solutes, *Planta*, 144, 291, 1979.

19. **Lamark, T., Kaasen, I., Eshoo, M. W., Falkenberg, P., McDougall, J., and Strom, A. R.**, DNA sequence and analysis of the *bet* genes encoding the osmoregulatory choline-glycine betaine pathway of *Escherichia coli*, *Mol. Microbiol.*, 5, 1049, 1991.

20. **Glaever, H. M., Styrvold, O. B., Kaasen, I., and Strom, A. R.**, Biochemical and genetic characterization of osmoregulatory trehalose synthesis in *Escherichia coli*, *J. Bacteriol.*, 170, 2841, 1988.

21. **Styrvold, O. B. and Strom, A. R.**, Synthesis, accumulation and excretion of trehalose in osmotically stressed *Escherichia coli* K-12 strains: influence of amber suppressors and function of the periplasmic trehalase, *J. Bacteriol.*, 173, 1187, 1991.

22. **Rod, M. L., Alam, K. Y., Cunningham, P. R., and Clark, D. P.,** Accumulation of trehalose by *Escherichia coli* K-12 at high osmotic pressure depends on the presence of amber suppressors, *J. Bacteriol.,* 170, 3601, 1988.

23. **Cairney, J., Booth, I. R., and Higgins, C. F.,** *Salmonella typhimurium proP* gene encodes a transport system for the osmoprotectant glycine betaine, *J. Bacteriol.,* 164, 1218, 1985.

24. **Jovanovich, S. B., Martinell, M., Record, M. T., Jr., and Burgess, R. R.,** Rapid response to osmotic upshift by osmoregulated genes in *Escherichia coli* and *Salmonella typhimurium, J. Bacteriol.,* 170, 534, 1988.

25. **Park, S. F., Stirling, D. A., Hulton, C. S. J., Booth, I. R., Higgins, C. F., and Stewart, G. S. A. B.,** A novel, non-invasive promoter probe vector: cloning of the osmoregulated *proU* promoter of *Escherichia coli* K12, *Mol. Microbiol.,* 3, 1011, 1989.

26. **Harshman, R. B. and Yamasaki, H.,** MSI accumulation induced by sodium chloride, *Biochemistry,* 11, 615, 1972.

27. **Roth, W. G., Leckie, M., and Dietzler, D. N.,** Osmotic stress drastically inhibits active transport of carbohydrates by *Escherichia coli, Biochem. Biophys. Res. Commun.,* 126, 434, 1985.

28. **Sutherland, L., Cairney, J., Elmore, M. J., Booth, I. R., and Higgins, C. F.,** Osmotic regulation of transcription: induction of the *proU* betaine transport gene is determined by the accumulation of intracellular potassium, *J. Bacteriol.,* 168, 805, 1986.

29. **Cairney, J., Booth, I. R., and Higgins, C. E.,** Osmoregulation of gene expression in *Salmonella typhimurium proU* encodes an osmotically-induced betaine transport system, *J. Bacteriol.,* 164, 1224, 1985.

30. **Csonka, L. N.,** A third L-proline permease in *Salmonella typhimurium* which functions in media of elevated osmotic strength, *J. Bacteriol.,* 151, 1433, 1983.

31. **Druger-Liotta, J., Prange, V. J., Overdier, D. G., and Csonka, L. N.,** Selection of mutants that alter the osmotic control of transcription of *Salmonella typhimurium proU* operon, *J. Bacteriol.,* 169, 2449, 1987.

32. **Higgins, C. F., Dorman, C. J., Stirling, D. A., Waddell, L., Booth, I. R., May, G., and Bremer, E.,** A physiological role for DNA supercoiling in the osmotic regulation of gene expression in *S. typhimurium* and *E. coli, Cell,* 52, 569, 1988.

33. **May, G., Faatz, E., Lucht, J. M., Haardt, M., Bolliger, M., and Bremer, E.,** Characterisation of the osmoregulated *Escherichia coli proU* promoter and identification of ProV as a membrane-associated protein, *Mol. Microbiol.,* 3, 1521, 1989.

34. **Hulton, C. S., Seirafi, A., Hinton, J. C. D., Sidebotham, J. M., Waddell, L., Pavitt, G. D., Owen-Hughes, T., Spassky, A., Buc, H., and Higgins, F.,** Histone-like protein H1 (H-NS), DNA supercoiling and gene expression in bacteria, *Cell,* 63, 631, 1990.

35. **Higgins, C. F., Hinton, J. C. D., Hulton, C. S. J., Owen-Hughes, T., Pavitt, G. D., and Seirafi, A.,** Protein H1: a role for chromatin structure in the regulation of bacterial gene expression and virulence?, *Mol. Microbiol.,* 4, 2007, 1990.

36. **Ni Bhriain, N., Dorman, C. J., and Higgins, C. F.,** An overlap between osmotic and anaerobic stress responses: a potential role for DNA supercoiling in the coordinate regulation of gene expression, *Mol. Microbiol.,* 3, 933, 1989.

37. **May, G., Dersch, P., Haardt, M., Middendorf, A., and Bremer, E.,** The *osmZ (bglY)* gene encodes the DNA-binding protein H-NS (H1a) component of the *Escherichia coli* K12, *Mol. Gen. Genet.,* 224, 81, 1990.

38. **Richey, B., Cayley, D. S., Mossing, M. C., Kolka, C., Anderson, C. F., Farra, T. C., and Record, M. T.,** Variability of the intracellular ionic environment of *E. coli:* differences between *in vitro* and *in vivo* effects of ion concentrations on protein-DNA interactions and gene expression, *J. Biol. Chem.,* 262, 7157, 1987.

39. **Pruss, G. J. and Drilica, K.,** DNA supercoiling and prokaryotic transcription, *Cell,* 56, 521, 1989.

40. **Drilica, K., Pruss, G. J., Burger, R. M., Franco, R. J., Hsieh, L.-S., and Berger, B. A.,** Roles of DNA topoisomerases in bacterial chromosome structure and function, in *The Bacterial Chromosome,* Drilica, K., Ed., American Society for Microbiology, Washington, D.C., 1990, 195.

41. **Richardson, S. M. H., Higgins, C. F., and Lilley, D. M. J.,** The genetic control of DNA supercoiling in *Salmonella typhimurium, EMBO J.,* 3, 1745, 1984.

42. **Dorman, C. J., Lynch, A. S., Ni Bhriain, N., and Higgins, C. F.,** DNA supercoiling in *Escherichia coli: topA* mutations can be suppressed by DNA amplifications involving the *tolC* locus, *Mol. Microbiol.,* 3, 531, 1989.

43. **Borowiec, J. A. and Gralla, J. D.,** All three elements of the *lac* p^s promoter mediate its transcriptional response to DNA supercoiling, *J. Mol. Biol.,* 195, 89, 1987.

44. **de Crombrugghe, B., Busby, S., and Buc, H.,** Activation of transcription by the cyclic AMP receptor protein, in *Biological Regulation and Development,* Goldberger, R. F. and Yamamoto, K. R., Eds., Plenum Press, New York, 1984, 129.

45. **Gemmill, R. M., Tripp, M., Friedman, S. B., and Calvo, J. M.,** Promoter mutation causing catabolite repression of the *Salmonella typhimurium* leucine operon, *J. Bacteriol.,* 158, 948, 1984.

46. **Lilley, D. M. J. and Higgins, C. F.,** Local DNA topology and gene expression: the case of the *leu-*500 promoter, *Mol. Microbiol.,* 5, 779, 1991.

47. **Hsieh, L.-S., Burger, R. M., and Drilica, K.,** Bacterial DNA supercoiling and [ATP]/ [ADP] changes associated with a transition to anaerobic growth, *J. Mol. Biol.,* 219, 443, 1991.

48. **Yamamoto, N. and Droffner, M. L.,** Mechanisms determining aerobic or anaerobic growth in the facultative anaerobe *Salmonella typhimurium, Proc. Natl. Acad. Sci. U.S.A.,* 82, 2077, 1985.

49. **Barron, A., May, G., Bremer, E., and Villarejo, M.,** Regulation of envelope protein composition during adaptation to osmotic stress in *Escherichia coli, J. Bacteriol.,* 167, 433, 1986.

50. **Ramirez, R. M. and Villarejo, M.,** Osmotic signal transduction to *proU* is independent of DNA supercoiling in *Escherichia coli, J. Bacteriol.,* 173, 879, 1991.

51. **Dorman, C. J., Barr, G. C., Ni Bhriain, N., and Higgins, C. F.,** DNA supercoiling and the anaerobic and growth phase regulation of *tonB* gene expression, *J. Bacteriol.,* 170, 2816, 1988.

52. **Mizuno, T. and Mizushima, S.,** Signal transduction and gene regulation through the phosphorylation of two regulatory components: the molecular basis for the osmotic regulation of the porin genes, *Mol. Microbiol.,* 4, 1077, 1990.

53. **Goransson, M., Sonden, B., Nilssen, P., Dagberg, B., Forsman, K., Emanuelsson, K., and Uhlin, B. E,** Transcriptional silencing and thermoregulation of gene expression in *Escherichia coli, Nature (London),* 344, 682, 1990.

54. **Lejeune, P. and Danchin, A.,** Mutations in the *bglY* gene increase the frequency of spontaneous delections in *Escherichia coli* K-12, *Proc. Natl. Acad. Sci. U.S.A.,* 87, 360, 1990.

55. **Drlica, K. and Rouviere-Yaniv, J.,** Histone-like proteins of bacteria, *Microbiol. Rev.,* 51, 301, 1987.

56. **Spassky, A., Rimsky, S, Garreau, H., and Buc, H.,** H1a, an *Escherichia coli* DNA-binding protein which accumulates in stationary phase, strongly compacts DNA *in vitro, Nucleic Acids Res.,* 12, 5321, 1984.

57. **Stirling, D. A., Hulton, C. S. J., Waddell, L., Park, S. F., Stewart, G. S. A. B., Booth, I. R., and Higgins, C. F.,** Molecular characterisation of the proU loci of *Salmonella typhimurium* and *Escherichia coli* encoding osmoregulated glycine betaine transport systems, *Mol. Microbiol.,* 3, 1025, 1989.

58. **Gowrishankar, J.,** Nucleotide sequence of the osmoregulatory *proU* operon of *Escherichia coli, J. Bacteriol.,* 171, 1923, 1989.

59. **Overdier, D. G., Olson, E. R., Erickson, B. D., Ederer, M. M., and Csonka, L. N.,** Nucleotide sequence of the transcriptional control regulated *proU* operon of *Salmonella typhimurium* and identification of the 5' endpoint of the *proU* mRNA, *J. Bacteriol.,* 171, 4694, 1989.
60. **Lucht, J. M. and Bremer, E.,** Characterisation of mutations affecting the osmoregulated *proU* promoter of *Escherichia coli* and identification of 5' sequences required for high-level expression, *J. Bacteriol.,* 173, 801, 1991.
61. **Repoila, F. and Cutierrez, C.,** Osmotic induction of the periplasmic trehalase in *Escherichia coli* K12: characterization of the *treA* gene promoter, *Mol. Microbiol.,* 5(3), 747, 1991.
62. **Mizuno, T., Chou, M. Y., and Inouye, M.,** A comparative study of the genes for the three porins of the *Escherichia coli* outer membrane: DNA sequence of the osmoregulated *ompC* gene, *J. Biol. Chem.,* 258, 6932, 1983.
63. **Eshoo, M. W.,** *lac* fusion analysis of the bet genes of *Escherichia coli:* regulation by osmolarity, temperature, oxygen, choline and glycine betaine, *J. Bacteriol.,* 170, 5208, 1988.
64. **Tokishita, S., Yamada, H., Aiba, H., and Mizuno, T.,** Transmembrane signal transduction and osmoregulation in *Escherichia coli:* II. The osmotic sensor, EnvZ, located in the isolated cytoplasmic membrane displays its phosphorylation and dephosphorylation abilities as to the activator protein, OmpR, *J. Biochem.,* 108, 488, 1990.
65. **Aiba, H., Mizuno, T., and Mizushima, S.,** Transfer of phosphoryl group between two regulatory proteins involved in osmoregulatory expression of the *ompF* and *ompC* genes in *Escherichia coli, J. Biol. Chem.,* 264, 8563, 1989.
66. **Guitierrez, C., Barondess, J., Manoil, C., and Beckwith, J.,** The use of Tn*phoA* to detect genes for the cell envelope proteins subject to a common regulatory stimulus, *J. Mol. Biol.,* 195, 289, 1987.
67. **Kennedy, E. P. and Rumley, M. K.,** Osmotic regulation of membrane-derived oligosaccharides in *Escherichia coli, J. Bacteriol.,* 170, 2457, 1988.
68. **Booth, I. R.,** unpublished data.
69. **Epstein, W.,** personal communication.
70. **Thomas, A. and Booth, I. R.,** unpublished data.

Chapter IIG

CELL K$^+$, INTERNAL pH, AND ACTIVE TRANSPORT IN *RHODOBACTER SPHAEROIDES*

Tjakko Abee and Wil N. Konings

TABLE OF CONTENTS

0-8493-6982-7/93/$0.00 + $.50

I. INTRODUCTION

The cell envelope of Gram-negative bacteria consists of a cytoplasmic membrane and an outer membrane which are separated by the peptidoglycan-containing periplasm.[1] The protein machinery involved in extracting solutes from the environment is located in this cell envelope.

The structure and functions of the Gram-negative cell envelope have been studied most extensively in *Escherichia coli* and other Enterobacteriaceae. The peptidoglycan layer together with the outer membrane give the cell its rigidity and shape. The outer membrane functions as a molecular sieve. Hydrophilic solutes with molecular masses smaller than 600 to 1000 Da can freely enter the periplasmic space through pore-forming proteins (porins) in this outer membrane. In addition, these cells can synthesize porins with a special function such as the uptake of maltodextrins or vitamin B_{12}.[1] Recently, the three-dimensional structure of a *Rhodobacter capsulatus* porin was analyzed.[2]

In the periplasmic space the solutes meet the cytoplasmic membrane that serves as the main selective barrier of the cell. Specific uptake of nutrients (and release of metabolites) is mediated by carrier proteins embedded in the phospholipid bilayer of the cytoplasmic membrane. Like in all bacteria, this membrane also plays a major role in the various modes of energy-transducing processes of the Gram-negative, purple nonsulfur, photoheterotrophic bacterium *R. sphaeroides*.[3-6]

II. ELECTRON TRANSFER CHAINS IN RHODOSPIRILLACEAE

A. AEROBIC RESPIRATORY CHAINS

R. sphaeroides, like most members of the Rhodospirillaceae, can grow aerobically in the dark.[5] When grown in the presence of high oxygen concentrations (25% O_2 atmosphere), *R. sphaeroides* has a typical Gram-negative cell envelope, and no photosynthetic intracytoplasmic membranes (ICMs) are present.[6] Energy for growth is then acquired from electron transfer through a linear respiratory chain.[5] A scheme of this chain is shown in Figure 1. Succinate dehydrogenase, NADH dehydrogenase, and hydrogenase can catalyze the input of electrons from reducing equivalents into the quinone pool. The electrons are subsequently transferred from ubiquinone into branch I. This branch is very similar to the mitochondrial respiratory chain and contains two proton-translocating sites. The other branch (branch II) contains cytochrome *o* oxidase, in which electron transfer proceeds without coupled proton translocation.[7] Electron transfer from the ubiquinone cytochrome b-c_1 oxidoreductase (bc_1 complex) to the terminal aa_3-type cytochrome c oxidase is mediated by the soluble cytochrome c_2, which is located in the periplasmic space.

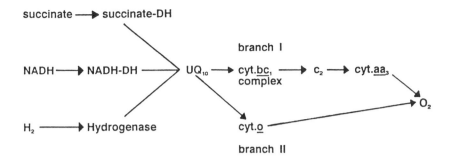

FIGURE 1. Scheme of the aerobic branched electron transfer chain in *Rhodobacter sphaeroides*. See text for details.

B. ANAEROBIC RESPIRATORY CHAINS

Some members of the Rhodospirillaceae are also able to grow anaerobically in the dark if provided with a suitable carbon source and electron acceptors. Anaerobic respiration has been studied most extensively in *R. capsulatus*.[3] Three distinct pathways of anaerobic respiration are known, and all three involve terminal oxidoreductases located in the periplasmic space. These enzymes are the nitrate reductase, the nitrous oxide reductase, and the dimethylsulfoxide/trimethylamine-*N*-oxide (DMSO/TMAO) reductase. Electron transport via these anaerobic pathways has been shown to be coupled to proton translocation across the cytoplasmic membrane.[3,8,9]

C. PHOTOSYNTHETIC ELECTRON TRANSFER CHAIN

Conversion of light energy into useful forms of cellular energy requires the presence of membrane-bound photosystems. When after aerobic growth oxygen is removed from the culture, a series of events is triggered which, through a process of membrane invagination, results in the differentiation of the cytoplasmic membrane into specialized domains and the formation of the photosynthetic ICM.[4,6] The structure of the ICM is illustrated in Figure 2. Three different pigment-protein complexes are predominant in mature ICM of *R. sphaeroides:* two types of light-harvesting complexes, designated B800—850 and B875 on the basis of their near-infrared absorption maxima, and the reaction center (RC) complex.[4]

When photons are absorbed by the B800—850 complexes, the excitation energy is transferred to the photochemical RC complex via the B875 complexes, where it oxidizes the special bacteriochlorophyll dimer (P870). The liberated electron is transferred sequentially via a molecule of bacteriopheophytin and the primary quinone (Q_A) to the secondary quinone (Q_B), which are both bound to the RC. Q_A and Q_B are subsequently oxidized by the quinone pool, from which electrons migrate via the bc_1 complex to cytochrome c_2, which rereduces P870. The activity of both the RC and the cytochrome bc_1 complex contribute to the generation of the proton motive force (Δp) via electron and electron plus proton translocation, respectively.[4,6]

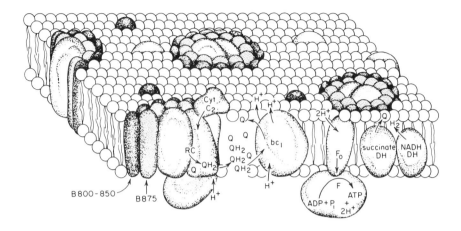

FIGURE 2. Illustration of a portion of the intracytoplasmic membrane (ICM) bilayer depicting the relative orientation of specific ICM components. DH, dehydrogenase; Q, ubiquinone. See text for details. (From Kiley, P. J. and Kaplan, S., *Microbiol. Rev.*, 52, 50, 1988. With permission.)

III. ENERGY COUPLING AND pH HOMEOSTASIS

Circulation of protons across the cytoplasmic membrane plays a primary role in energy transduction of bacterial cells.[10] Primary proton pumps such as the respiratory and photosynthetic electron transport chains can mediate electrogenic extrusion of protons across the cytoplasmic membrane and generate an electrochemical gradient of protons ($\Delta\bar{\mu}_{H^+}$/F mV). This electrochemical proton gradient exerts an inwardly directed force on the protons, the proton motive force (Δp), which equals $\Delta\bar{\mu}_{H^+}$/F and is composed of an electrical potential ($\Delta\Psi$) and a chemical gradient of protons or pH gradient across the cytoplasmic membrane (ΔpH). The ΔpH equals the internal pH (pH_{in}) minus the pH of the bacterial environment (pH_{out}):

$$\Delta\bar{\mu}_{H^+}/F = \Delta p = \Delta\Psi - Z\Delta pH \text{ (mV)} \qquad (1)$$

where $Z = 2.3$ (RT/F) and R is the gas constant, T the absolute temperature, and F the Faraday constant. Usually, Δp, $\Delta\Psi$, and $-Z\Delta pH$ have negative values.

The Δp can be used to drive energy-requiring processes such as ATP synthesis, transhydrogenation of pyridine nucleotides, flagellar rotation, solute transport, and protein translocation.[11,12]

There is now substantial evidence that in a variety of bacteria the cytoplasmic pH is maintained within a relatively narrow range. The cytoplasmic pH of acidophilic bacteria ranges between 6.0 to 7.0; of neutrophilic bacteria, between 7.5 to 8.0; and of alkalophilic bacteria, between 8.5 to 9.0.[13,14] Efficient pH homeostatic mechanisms are needed to prevent fluctuations in the cytoplasmic pH upon changes in the external pH. It is generally assumed that the cytoplasmic pH is regulated by the activity of various cation transport systems. K^+ transport plays an important role in internal pH regulation in *Escherichia coli, E. faecalis,* and *L. lactis.*[13-20]

The internal pH can affect a large number of cellular processes. It has long been recognized that cell metabolism, i.e., enzyme activity, is affected by the intracellular pH. Solute transport systems also appear to be regulated by the internal pH. Bacterial solute transport systems can be classified in three groups according to the mode of energy coupling:[21,22]

1. Primary transport systems which utilize adenosine 5′-triphosphate (ATP) or another form of chemical energy to drive transport;[23] binding protein-dependent transport systems belong to this class[24,25]
2. Secondary transport systems which utilize electrochemical energy
3. Group translocation systems which couple the translocation to a chemical modification of the solute

These solute transport systems are located in the cytoplasmic membrane, and domains will be exposed to the aqueous phases at both sides of the membrane which can interact with protons. Several transport systems in lactic acid bacteria are activated by increasing the internal pH above neutrality. These effects must be allosteric in the case of phosphate bond-driven transport systems, but can be catalytic or allosteric or both in the case of proton-coupled solute transport systems.[26]

IV. REGULATION OF CYTOPLASMIC pH IN *RHODOBACTER SPHAEROIDES*

We have studied the regulation of the internal pH in the purple nonsulfur bacterium *R. sphaeroides,* which is capable of growth energized by aerobic respiration, anaerobic respiration, or anoxygenic photosynthesis.[3,5,6] In the absence of electron transfer *R. sphaeroides* maintains a significant $\Delta\Psi$. This anaerobic-dark potential is most likely generated by a K^+ diffusion potential since it is not affected by ATPase inhibitors, and increasing the external K^+ concentration results in a decrease of the $\Delta\Psi$, which becomes zero at high external K^+ concentrations.[27]

The Δp in *R. sphaeroides* has been determined under different conditions. Anaerobically in the dark and at alkaline pH values of the medium, the intracellular pH is more acidic. This results in a reversed ΔpH, while the $\Delta\Psi$ has the normal polarity (inside negative vs. outside).[27] The $\Delta\Psi$ of around -100 mV is thus partially compensated by a reversed ΔpH, resulting in a Δp of around -60 mV.

Under anaerobic-dark conditions *R. sphaeroides* maintains a dark potential $(\Delta\Psi)$ of -85 mV in 50 mM K-phosphate of pH 7.0. Upon illumination the $\Delta\Psi$ increases transiently and then decreases to -65 mV, but this decrease of the $\Delta\Psi$ is accompanied by an increase of the internal pH ($-Z\Delta pH = -62$ mV). The decrease of the $\Delta\Psi$ was not observed in the absence of K^+ or in the presence of 100 mM K-acetate, used to collapse any ΔpH formed. The net decrease of the $\Delta\Psi$ in *R. sphaeroides* upon illumination thus appears to be the result of electrogenic uptake of K^+ and is compensated by an increase of the ΔpH. The magnitude of the ΔpH and $\Delta\Psi$ are determined by the external K^+ concentration and the activity of the low-affinity K^+ uptake system. This was demonstrated by studying the effect of increasing K^+ concentrations on $\Delta\Psi$ and ΔpH generation in cells incubated at pH 6.0. Anaerobically in the light the addition of K^+ results in a decrease of the $\Delta\Psi$ and an increase of the ΔpH, which saturates at about 1.2 mM K^+. This value is close to that of the K_m for uptake via the constitutive K^+ uptake system of *R. sphaeroides*. This system also transports Rb^+, although with a lower affinity than for K^+.[28] Depolarization and increase of pH_{in} is also accomplished by the addition of Rb^+, which saturates at about 4 mM. In all these experiments $\Delta\Psi$ and $Z\Delta pH$ interconverted in such a manner that the Δp remained constant at about -132 mV.[27] These results suggest that at low external pH *R. sphaeroides* can only generate a large ΔpH in the presence of K^+. Saturation at 1.2 mM K^+ is most likely caused by a saturation of the K^+ transport system. In *Escherichia coli* and *E. faecalis* depolarization of the $\Delta\Psi$ by K^+ saturates at about 1 to 2 mM, which is close to the K_m values for K^+ uptake in these bacteria when grown in media with high concentrations of K^+.[17,19,20]

The effect of K^+ on the internal pH was followed as a function of the external pH in media with or without K^+ (Figure 3). Under aerobic conditions in the dark, cells in 50 mM Na-phosphate generate a ΔpH (inside alkaline) at an external pH lower than 7.3 and a reversed ΔpH (inside acid) at higher external pH values. In the presence of KCl (5 mM) a ΔpH (inside alkaline) can be generated over the whole pH range from pH 6 to 8. At all external pH values, in the presence of K^+, the internal pH is higher and varies from 7.5 to 8.3 with increasing external pH (Figure 3).[27] These results are in agreement with reports on intracellular pH measurements with ^{31}P-NMR in *R. sphaeroides*.[29] *R. sphaeroides* WS82 and *R. sphaeroides* forma spec. *denitrificans* have also been found to be unable to generate a ΔpH (inside alkaline) in the absence of K^+ at external pH values above 7.0.[30,31] At all pH values K^+ also reduced $\Delta\Psi$. However, the total Δp was relatively constant at about -130 mV. This value is very similar to that reported for Δp in *R.sphaeroides* WS8.[32]

V. A Kdp-LIKE ATPase IN *R. SPHAEROIDES*

In addition to its role in pH homeostasis, K^+ plays an important role in the maintenance of cell turgor of prokaryotes[33,34] (see also Chapters IIA and IIE). During growth of *Escherichia coli* at low external K^+, the cell turgor

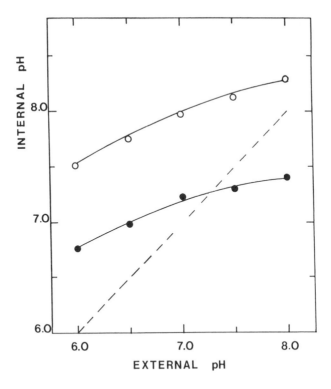

FIGURE 3. Effect of K⁺ on the internal pH of *Rhodobacter sphaeroides* cells at various external pH values. Cells were grown aerobically in the dark and resuspended in 50 mM Na-phosphate + 5 mM MgSO$_4$ in the absence (●) or presence (○) of 5 mM KCl, respectively. Intracellular pH was determined from the distribution of ¹⁴C-methylamine or ¹⁴C-benzoate after separation of the cells from the medium by silicon oil centrifugation. (From Abee, T., Helling-werf, K. J., and Konings, W. N., *J. Bacteriol.*, 170, 5647, 1988. With permission.)

diminishes because of K⁺ limitation in the medium. Under these conditions this organism synthesizes a K⁺-ATPase (Kdp). This enzyme transports K⁺ with high affinity, which allows the cell to scavenge the medium for K⁺ to restore cell turgor.[35,36] The Kdp system is composed of three subunits, KdpA, B, and C, and belongs to the class of P-type ATPases[37-40] (see also Chapter IIB). Recently, a Kdp-like high-affinity K⁺-translocating ATPase was iden-tified in the thermoacidophilic bacterium *Bacillus acidocaldarius*.[41,42]

The previous section focused on the regulation of the internal pH in *R. sphaeroides* grown at high K⁺ concentrations. *R. sphaeroides* can also grow in a minimal succinate medium at low K⁺ concentrations. K⁺ transport studies in these cells revealed the presence of a second, high-affinity K⁺ uptake system ($K_m < 10$ μM).[43] In contrast to the results reported in Reference 44, Southern blotting experiments with chromosomal DNA from various photo-trophic bacteria showed that *R. sphaeroides* and *R. capsulatus* DNA hybridizes with the *kdp*ABC genes of *E. coli*.[45] Subsequent immunological studies with

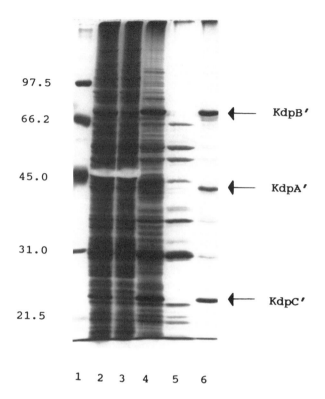

FIGURE 4. Partial purification of the K$^+$-ATPase. The proteins present in the different fractions were separated by sodium-dodecylsulfate polyacrylamide gel electrophoresis. Lane 1: marker proteins; lane 2: membranes of cells grown at low K$^+$ concentrations; lane 3: fraction that was not solubilized by decyl-β-D-maltoside; lane 4: fraction solubilized by this detergent; lane 5: fraction that was not bound to a Fractogel TSK AF-Red-sepharose column; lane 6: fraction eluted from the Fractogel column.

antibodies against the catalytic KdpB protein or the whole KpdABC complex showed cross-reactivity with a 70.0-kDa *R. sphaeroides* protein that was expressed only in cells grown in media with low K$^+$ concentrations.[46] Subsequently, the K$^+$-stimulated ATPase was partially purified from membranes of these cells by solubilization with decyl-β-D-maltoside followed by AF-Red affinity chromatography (Figure 4). The K$^+$-stimulated ATPase from *R. sphaeroides* is composed of three subunits with molecular weights of 435, 70.0, and 23.5 kDa which are almost identical to those of the *E. coli* KdpA, B, and C proteins.[39]

The role of this high-affinity K$^+$-translocating ATPase in the regulation of the internal pH in *R. sphaeroides* was studied. Cells were grown at high and low K$^+$ concentrations. After harvesting, washing, and resuspending in 50 mM Na-phosphate (pH 6.0) containing 5 mM MgSO$_4$ the ΔpH was determined as a function of the external K$^+$ concentration. Remarkably, both

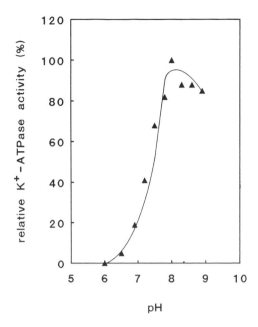

FIGURE 5. pH dependence of K⁺-ATPase activity in everted membrane vesicles from *Rhodobacter sphaeroides*. ATPase activity (100% value is 1.95 mmol ATP per minute per gram protein) was determined in 50 mM Tris-Hepes containing 2 mM MgSO$_4$ and 1 mM KCl at 37°C.

types of cells require high concentrations (2 mM) of K⁺ for the generation of a large ΔpH (inside alkaline).[47] In the absence of K⁺, under energized conditions, *R. sphaeroides* can only build up a pH gradient ($-Z$ΔpH) of -40 mV (pH$_{in}$ = 6.8). The K⁺-ATPase activity was measured in everted membranes as a function of the pH. The activity decreased rapidly at acid pH values, but remained high at alkaline pH (Figure 5). At pH 6.8 the K⁺-ATPase activity was only 10% of the activity found at pH 8.0. It appears that under these conditions the activity of the K⁺-stimulated ATPase of *R. sphaeroides* is too low to result in a decrease of the ΔΨ and an increase of the ΔpH.

VI. REGULATION OF SOLUTE TRANSPORT BY THE INTERNAL pH

The internal pH of bacteria (and to some extent of membrane vesicles) is largely determined by the ionic composition of the medium (see Section V). Effects of ions on transport therefore may be secondary, i.e., due to changes in the internal pH, and conclusions about the mechanism of energy coupling to transport can be drawn only when the effects of the internal pH, either directly or indirectly, on solute transport are known.

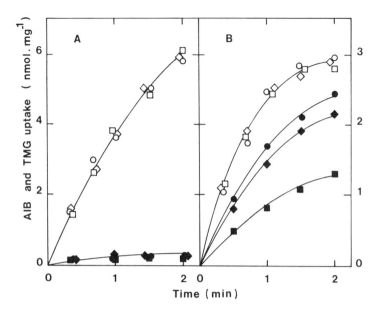

FIGURE 6.   Effect of medium composition on AIB (A) and TMG (B) transport in *Rhodobacter sphaeroides* 4P1. Cells were washed and resuspended in 50 mM K-phosphate (open symbols) or 50 mM Na-phosphate (closed symbols), pH 6 (■), 7 (●), and 8 (◆), containing a 5 mM MgSO$_4$. Uptake of ^{14}C-AIB and ^{14}C-TMG was measured aerobically in the dark. (From Abee, T. et al., *J. Bacteriol.*, 171, 5148, 1989. With permission.)

Stimulation of solute transport by K$^+$ has been observed in a number of bacteria. Extensive studies in *R. sphaeroides* have indicated that solute transport systems are only active when electron transfer in the respiratory or cyclic electron transfer chain takes place. Furthermore, uptake of L-alanine was found to require the presence of K$^+$.[48] The properties of an L-alanine uptake system have been studied and compared with those of H$^+$/lactose symport in *R. sphaeroides* 4P1, a strain in which the lactose carrier of *E. coli* has been cloned and functionally expressed.[49] Lactose transport was studied in membrane vesicles from this strain. The generation of a Δp via electron transfer by the linear electron transfer chain (Figure 1) leads to lactose uptake.[7] Under these conditions no uptake of alanine is observed, suggesting that the transport mechanism of alanine differs from that of lactose. Additional studies showed that transport of L-alanine is mediated by a binding protein-dependent transport system and is driven by phosphate bond energy.[50]

Further transport studies were performed with the nonmetabolizable analogues for alanine and lactose, 2-α-amino-isobutyric acid (AIB) and methyl-β-D-thiogalactopyranoside (TMG), respectively. Intact cells of *R. sphaeroides* 4P1 were washed and resuspended in 50 mM K-phosphate or Na-phosphate at pH values between 6 to 8. Aerobically in the dark, the initial uptake rate of AIB is about 14-fold higher in K-phosphate (4.2 nmol/min/mg protein) than in Na-phosphate (0.3 nmol/min/mg protein) (Figure 6). The differences

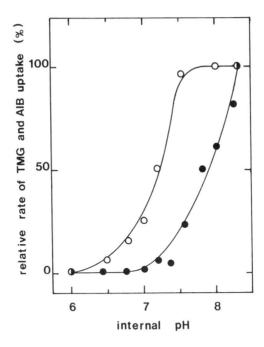

FIGURE 7. Regulation of AIB and TMG uptake in *Rhodobacter sphaeroides* by the internal pH. The internal pH was varied by washing and resuspending the cells in 50 m*M* K-phosphate of the desired pH (pH range 6.0 to 8.3) containing 100 m*M* K-acetate and 5 m*M* MgSO$_4$. The initial uptake rates of ^{14}C-AIB (●) and ^{14}C-TMG (○) were measured between 5 and 20 s under aerobic-dark conditions. The maximal rates of AIB and TMG uptake (100% values) at an internal pH of 8.3 were 5.9 and 4.4 nmol/min/mg protein, respectively. (From Abee, T. et al., *J. Bacteriol.*, 171, 5148, 1989. With permission.)

in the rates of TMG uptake in K-phosphate and Na-phosphate are less pronounced. In K-phosphate the initial rate of TMG uptake is 4.0 nmol/min/mg protein in the pH range 6 to 8, while in Na-phosphate the rate of TMG uptake is 4, 1.8, and 2.2 times lower at pH 6, 7, and 8, respectively. The uptake rates of AIB and TMG in Na-phosphate buffer are restored to the level in K-phosphate buffer by the addition of 2 m*M* KCl or 5 m*M* RbCl. These results show that AIB and TMG uptake in *R. sphaeroides* is stimulated by K⁺.

In the previous section it was shown that K⁺ plays a central role in the regulation of the internal pH in *R. sphaeroides*. Therefore, we investigated the effect of the internal pH on solute transport in *R. sphaeroides* 4P1 (Figure 7). The internal pH was varied by washing and resuspending the cells in 50 m*M* K-phosphate adjusted to different pH values in the presence of 100 m*M* K-acetate. Acetate dissipates the pH gradient and consequently equilibrates the internal pH with the external pH. Under these conditions the Δp consists solely of a ΔΨ of − 132 mV which remains constant in the pH range 6 to 8.3, while the intracellular ATP concentration remains at about 2.2 m*M*. AIB and TMG uptake increases with increasing internal pH under conditions of constant Δp. From the dependence of TMG and AIB transport on the internal pH, pK_a values of 7.2 and 7.8 are estimated, respectively.[50]

The pK_a of 7.2 for lactose transport in *R. sphaeroides* is 0.9 pH units higher than the value of 6.3 which has been determined in *E. coli* cells and in reconstituted liposomes.[51,52] The reason for this discrepancy is unclear; one factor might be the difference in phospholipid composition of the cytoplasmic membranes of *E. coli* and *R. sphaeroides*.

The stimulation of solute transport by K^+ in *R. sphaeroides* thus appears to be due to an increase of the intracellular pH. The dependence of solute transport in *R. sphaeroides* on both K^+ and the activity of an electron transfer chain can now be explained by an effect of the internal pH, which subsequently influences the activities of the lactose and the binding protein-dependent L-alanine transport system.[50] Recently, evidence was presented that binding protein-dependent transport of C4-dicarboxylates in *R. capsulatus* also is regulated by the internal pH. A significant activation of malate uptake occurred at intracellular pH values greater than 7.[53,54]

VII. CONCLUDING REMARKS

Most bacteria function optimally when the internal pH is maintained slightly alkaline. Under those conditions the balance between the metabolic processes is optimal and the organism can achieve the highest growth rate.

In order to maintain the internal pH slightly alkaline, the coordinated action of primary and secondary ion transport systems is required. Especially K^+ and Na^+ transport systems have been found to play important roles in pH homeostasis. It is the internal pH itself that controls the activity of these transport systems.

When the environmental conditions vary and the organism is not capable of maintaining the optimum internal pH, the metabolism will become unbalanced and the rate of cell biosynthesis will decline. The rate at which metabolic energy becomes available for the cell will also decline. A decrease of the energy requirements for uptake of precursors for biosynthesis will be advantageous to the organism.

The mechanism of regulating solute transport systems by the internal pH can be important for survival of organisms which encounter transient states of energy supply. This is most clearly visualized for the purple nonsulfur bacterium *R. sphaeroides*. These phototrophic bacteria are widely distributed in nature and are commonly found in flat, stagnant water bodies with rapid production and decomposition of organic matter. In their natural environment these bacteria are constantly subjected to fluctuations in energy supply. The simultaneous presence of enzymes involved in aerobic and anaerobic electron transfer together with photosynthetic enzymes allows these bacteria to respond immediately to changes in the energy supply from the environment.

Under energized conditions the organism maintains an internal pH of about 8.0 and the activities of the solute transport systems are optimal. When the energy supply becomes limited, i.e., low O_2 concentrations and/or low light intensities, the internal pH decreases and at a pH_{in} of 7.0 solute transport is inhibited completely. The organism thus saves energy by decreasing its transport activity under conditions of limited energy supply.

Under energized conditions, i.e., in the presence of high O_2 concentrations or high light intensities, the increase in electron transfer rates results in an increase of the internal pH. The solute transport systems are reactivated and the organism can resume its metabolic activities. The important role of K^+ in optimalizing cell metabolic activities is stressed by the observation that *R. sphaeroides* shows strong chemotactic responses to K^+.[55]

It is concluded that regulation of solute transport by the internal pH can be an effective mechanism for the organism to survive energetically unfavorable conditions.

ACKNOWLEDGMENTS

Purification of the phototrophic Kdp-ATPase was performed in excellent collaboration with Annette Siebers at the Department of Microbiology, University of Osnabrück, Germany, and we wish to thank Karlheinz Altendorf for his continuous interest in this project. This work was supported by The Netherlands Foundation for Chemical Research, which is subsidized by the Netherlands Organization for the Advancement of Pure Scientific Research.

REFERENCES

1. **Nikaido, H. and Vaara, M.,** Molecular basis of bacterial outer membrane permeability, *Microbiol. Rev.,* 49, 1, 1985.
2. **Weiss, M. S., Wacker, T., Weckesser, J., Welte, W., and Schultz, G. E.,** The three-dimensional structure of porin from *Rhodobacter capsulatus* at 3 Å resolution, *FEBS Lett.,* 267, 268, 1990.
3. **Ferguson, S. J., Jackson, J. B., and McEwan, A. G.,** Anaerobic respiration in the Rhodospirillaceae: characterization of pathways and evaluation of roles in redox balancing during photosynthesis, *FEMS Microbiol. Rev.,* 46, 117, 1987.
4. **Chory, J., Donohue, T. J., Varga, A. R., Staehelin, L. A., and Kaplan, S.,** Induction of the photosynthetic membranes of *Rhodopseudomonas sphaeroides:* biochemical and morphological studies, *J. Bacteriol.,* 159, 540, 1984.
5. **Smith, L. and Pinder, D. B.,** Aerobic respiratory chains in Rhodospirillaceae, in *The Photosynthetic Bacteria,* Clayton, R. K. and Sistrom, W. R., Eds., Plenum Press, New York, 1978, 641.
6. **Kiley, P. J. and Kaplan, S.,** Molecular genetics of photosynthetic membrane biosynthesis in *Rhodobacter sphaeroides, Microbiol. Rev.,* 52, 50, 1988.
7. **Abee, T., Hellingwerf, K. J., and Konings, W. N.,** The relation between electron transfer, proton motive force and lactose transport in membrane vesicles from aerobically grown *Rhodobacter sphaeroides, Biochim. Biophys. Acta,* 932, 250, 1988.

8. **Richardson, D. J., King, G. F., Kelly, D. J., McEwan, A. G., Ferguson, S. J., and Jackson, J. B.,** The role of auxiliary oxidants in maintaining the redox balance during phototrophic growth of *Rhodobacter capsulatus* on propionate or butyrate, *Arch. Microbiol.,* 150, 131, 1988.

9. **McEwan, A. G., Cotton, N. P. J., Ferguson, S. J., and Jackson, J. B.,** The role of auxiliary oxidants in the maintenance of a balanced redox poise for photosynthetic bacteria, *Biochim. Biophys. Acta,* 810, 140, 1985.

10. **Mitchell, P.,** Chemiosmotic coupling in oxidative phosphorylation, *Biol. Rev.,* 41, 445, 1966.

11. **Hellingwerf, K. J. and Konings, W. N.,** Energy flow in bacteria: the main free energy intermediates and their regulatory role, *Adv. Microbiol. Physiol.,* 26, 125, 1985.

12. **Schiebel, E., Driessen, A. J. M., Hartl, F. U., and Wickner, W.,** $\Delta\bar{\mu}_{H^+}$ and ATP function at different steps of the catalytic cycle of pre-protein translocase, *Cell,* 64, 927, 1991.

13. **Bakker, E. P.,** The role of alkali-cation transport in energy coupling of neutrophilic and acidophilic bacteria: an assessment of methods and concepts, *FEMS Microbiol. Rev.,* 75, 319, 1990.

14. **Booth, I. R.,** Regulation of cytoplasmic pH in bacteria, *Microbiol. Rev.,* 49, 359, 1985.

15. **Kroll, R. G. and Booth, I. R.,** The role of potassium transport in the generation of a pH gradient in *Escherichia coli, Biochem. J.,* 198, 691, 1981.

16. **Kroll, R. G. and Booth, I. R.,** The relationship between intracellular pH, the pH gradient and potassium transport in *Escherichia coli, Biochem. J.,* 216, 709, 1983.

17. **Bakker, E. P. and Mangerich, W. E.,** Interconversion of components of the bacterial proton motive force by electrogenic potassium transport, *J. Bacteriol.,* 147, 820, 1983.

18. **Poolman, B., Hellingwerf, K. J., and Konings, W. N.,** Regulation of the glutamate/glutamine transport system by the intracellular pH in *Streptococcus lactis, J. Bacteriol.,* 169, 2272, 1987.

19. **Kashket, E. R.,** Effects of K^+ and Na^+ on the proton motive force of respiring *Escherichia coli* at alkaline pH, *J. Bacteriol.,* 163, 423, 1985.

20. **Kashket, E. R. and Barker, S. L.,** Effects of potassium ions on the electrical and pH gradients across the membrane of *Streptococcus lactis* cells, *J. Bacteriol.,* 130, 1017, 1977.

21. **Konings, W. N., de Vrij, W., Driessen, A. J. M., and Poolman, B.,** Primary and secondary transport systems in Gram positive bacteria, in *Sugar Transport and Metabolism in Gram-Positive Bacteria,* Reizer, J. and Peterkofsky, A., Eds., Ellis Horwood, Chichester, England, 1986, 270.

22. **Higgens, C. F., Gallagher, M. P., Hyde, S. C., Mimmack, M. L., and Pearce, S. R.,** Periplasmic binding protein-dependent transport systems: the membrane associated components, *Philos. Trans. R. Soc. London Ser. B,* 326, 353, 1990.

23. **Rosen, B. P.,** ATP-coupled solute transport systems, in *Escherichia coli and Salmonella typhimurium: Cellular and Molecular Biology,* Neidhardt, F. C., Ed., American Society for Microbiology, Washington, D.C., 1987, 760.

24. **Furlong, C. E.,** Osmotic-shock-sensitive transport systems, in *Escherichia coli and Salmonella typhimurium: Cellular and Molecular Biology,* Neidhardt, F. C., Ed., American Society for Microbiology, Washington, D.C., 1987, 768.

25. **Ames, G. F.-L.,** Structure and mechanism of bacterial periplasmic transport systems, *J. Bioenerg. Biomembr.,* 20, 1, 1988.

26. **Poolman, B., Driessen, A. J. M., and Konings, W. N.,** Regulation of solute transport in Streptococci by external and internal pH values, *Microbiol. Rev.,* 51, 498, 1987.

27. **Abee, T., Hellingwerf, K. J., and Konings, W. N.,** Effects of potassium ions on proton motive force in *Rhodobacter sphaeroides, J. Bacteriol.,* 170, 5647, 1988.

28. **Hellingwerf, K. J., Friedberg, I., Lolkema, J. S., Michels, P. A. M., and Konings, W. N.,** Energy coupling of facilitated transport of inorganic ions in *Rhodospeudomonas sphaeroides, J. Bacteriol.,* 150, 1183, 1982.

29. **Nicolay, K., Lolkema, J. S., Hellingwerf, K. J., Kaptein, R., and Konings, W. N.,** Quantitative agreement between the values for the light induced ΔpH in *Rhodopseudomonas sphaeroides* measured with automated flow dialysis and ^{31}P-NMR, *FEBS Lett.,* 123, 319, 1981.

30. **Armitage, J. P. and Evans, M. C. N.,** Control of the proton motive force in *Rhodopseudomonas sphaeroides* in the light and dark and its effects on the initiation of flagellar rotation, *Biochim. Biophys. Acta,* 806, 42, 1985.

31. **Kundu, B. and Nicholas, D. J. D.,** Proton motive force in a photodenitrifier, *Rhodopseudomonas sphaeroides* forma specialis *denitrificans, Biochem. Cell Biol.,* 64, 328, 1986.

32. **Elferink, M. G. L., Hellingwerf, K. J., and Konings, W. N.,** The role of the proton motive force and electron flow in solute transport in *Escherichia coli, Eur. J. Biochem.,* 153, 161, 1985.

33. **Higgens, C. F., Cairney, D., Stirling, A., Sutherland, L., and Booth, I. R.,** Osmotic regulation of gene expression: ionic strength as an intracellular signal?, *Trends Biochem. Sci.,* 12, 339, 1987.

34. **Epstein, W.,** Osmoregulation by potassium transport in *Escherichia coli, FEMS Microbiol. Rev.,* 39, 73, 1986.

35. **Laimins, L. A., Rhoads, D. B., and Epstein, W.,** Osmotic control of *kdp* operon expression in *Escherichia coli, Proc. Natl. Acad. Sci. U.S.A.,* 78, 464, 1981.

36. **Walderhaug, M. O., Dosch, D. C., and Epstein, W.,** Potassium transport in bacteria, in *Ion Transport in Prokaryotes,* Rosen, B. P., Ed., Academic Press, New York, 1987, 85.

37. **Epstein, W.,** The Kdp system: a bacterial K⁺ transport ATPase, *Curr. Top. Membr. Transp.,* 23, 153, 1985.

38. **Siebers, A., Wieczorek, K., and Altendorf, K.,** K⁺-ATPase from *Escherichia coli:* isolation and characterization, *Methods Enzymol.,* 157, 668, 1988.

39. **Siebers, A. and Altendorf, K.,** The K⁺-translocating Kdp-ATPase from *Escherichia coli:* purification, enzymatic properties and production of complex- and subunit-specific antisera, *Eur. J. Biochem.,* 178, 131, 1988.

40. **Epstein, W., Walderhaug, M. O., Polarek, J. W., Hesse, J. E., Dorus, E., and Daniel, J. M.,** The bacterial Kdp K⁺-ATPase and its relation to other transport ATPases, such as the Na⁺/K⁺- and Ca²⁺-ATPases in higher organisms, *Philos. Trans. R. Soc. London Ser. B,* 326, 479, 1990.

41. **Bakker, E. P., Borchard, A., Michels, M., Altendorf, K., and Siebers, A.,** High-affinity potassium uptake system in *Bacillus acidocaldarius* showing immunological cross-reactivity with the Kdp system from *Escherichia coli, J. Bacteriol.,* 169, 4342, 1987.

42. **Hafer, J., Siebers, A., and Bakker, E. P.,** The high-affinity K⁺-translocating ATPase complex from *Bacillus acidicaldarius* consists of three subunits, *Mol. Microbiol.,* 3, 487, 1989.

43. **Abee, T. and Hellingwerf, K. J.,** unpublished results, 1990.

44. **Walderhaug, M. O., Litwack, E. D., and Epstein, W.,** Wide distribution of homologs of *Escherichia coli* Kdp K⁺-ATPase among Gram-negative bacteria, *J. Bacteriol.,* 171, 1192, 1989.

45. **Abee, T. and Bakker, E. P.,** unpublished results, 1989.

46. **Knol, J., Siebers, A., and Abee, T.,** unpublished results, 1990–1991.

47. **Abee, T.,** unpublished results, 1990–1991.

48. **Elferink, M. G. L., Hellingwerf, K. J., and Konings, W. N.,** The relation between electron transfer, proton motive force and energy consuming processes in cells of *Rhodopseudomonas sphaeroides, Biochim. Biophys. Acta,* 730, 379, 1987.

49. **Elferink, M. G. L., Hellingwerf, K. J., Nano, F. E., Kaplan, S., and Konings, W. N.,** The lactose carrier of *Escherichia coli* functionally incorporated in *Rhodopseudomonas sphaeroides* obeys the regulatory conditions of the phototrophic bacterium, *FEBS Lett.,* 164, 185, 1983.

50. **Abee, T., van der Wal, F. J., Hellingwerf, K. J., and Konings, W. N.,** Binding protein-dependent alanine transport in *Rhodobacter sphaeroides* is regulated by the internal pH, *J. Bacteriol.,* 171, 5148, 1989.

51. **Page, M. G. P.,** The role of protons in the mechanism of galactoside transport via the lactose permease of *Escherichia coli, Biochim. Biophys. Acta,* 897, 112, 1987.

52. **Page, M. G. P., Rosenbusch, J. P., and Yamato, I.,** The effects of internal pH on proton sugar symport activity of the lactose permease purified from *Escherichia coli, J. Biol. Chem.,* 263, 15897, 1988.

53. **Hamblin, M. J., Shaw, J. G., Curson, J. P., and Kelly, D. J.,** Mutagenesis, cloning and complementation analysis of C4-dicarboxylate transport genes from *Rhodobacter sphaeroides, Mol. Microbiol.,* 4, 1567, 1990.

54. **Shaw, J. G. and Kelly, D. J.,** Binding protein dependent transport of C4-dicarboxylates in *Rhodobacter capsulatus, Arch. Microbiol.,* 155, 466, 1991.

55. **Poole, P. S., Brown, S., and Armitage, J. P.,** Potassium chemotaxis in *Rhodobacter sphaeroides, FEBS Lett.,* 260, 88, 1990.

Chapter IIH

STRETCH-ACTIVATED CHANNELS IN PROKARYOTES

Mario Zoratti and Alexandre Ghazi

TABLE OF CONTENTS

I. INTRODUCTION

Net unidirectional ion transport generates, by definition, an electrical current. In biological systems ranging from mammalian neurons to monocellular algae the tools of electrophysiology have been instrumental in the progress of our understanding of ion transport. In the late sixties the electrophysiology of bacteria had a promising start with a series of investigations concerning an ''excitability inducing material'' (excreted toxins or porins) isolated from Gram-negative bacteria and capable of inducing stepwise changes in the conductance of a lipid bilayer.[1,2] Voltage-dependent channels that showed multiple conductance substates were described at the single-channel level.[3,4] Subsequently, the porins present in the outer membrane of bacteria or mitochondria were isolated and reconstituted in lipid bilayers, and their electrical properties were characterized in detail in this system.[5-9] Since porins are not involved in ion transport across the cell membrane they are not discussed here. By contrast, there has been very little electrophysiological analysis of the cytoplasmic membrane of either Gram-negative or Gram-positive organisms, or even of outer membranes in their native state. Until recent years only one study had been reported, in which a microelectrode was inserted into giant *Escherichia coli* cells.[10] This lack of initiative has been due in part to the overwhelming influence of the chemiosmotic dogma, which requires that the inner membrane is essentially impermeable to ions. In addition, almost all bacteria are much too small for obtaining reliable electrophysiological data with microelectrodes. Recently this situation has drastically changed with the successful application of the patch-clamp technique[11,12] to bacteria (see Figure 1). Results have now been reported on giant cells[13] or spheroplasts derived from *E. coli*,[13-15] on protoplasts obtained from Gram-positive organisms,[16,17] and on proteoliposomes containing partially purified membrane fractions.[18-20] These studies revealed the presence of mechanosensitive, stretch-activated channels in both Gram-negative[14] and Gram-positive[16] bacteria. Mechanosensitive channels are pores whose state (open or closed) is controlled by mechanical forces (i.e., the hydrostatic pressure applied on the membrane). Nonporin, nonmechanosensitive channels are also present in bacteria,[15,18] but investigations of their properties are only beginning. This chapter deals, therefore, only with stretch-activated channels and their relevance to bacterial ion transport and osmoadaptation.

The discovery of bacterial stretch-activated channels added another interesting item to the lengthening list of membranes in which mechanosensitive channels have been found since 1984,[21] which includes examples from all taxa (for reviews see the articles by Sachs[22] and Morris[23]). At present much interest is devoted to this class of channels. Nonbacterial stretch-activated channels include cation-, K^+-, anion-, and nonselective channels. The conductances of these channels are generally in the range of tens of picosiemens. The bacterial stretch-activated channels, however, are a notable exception in

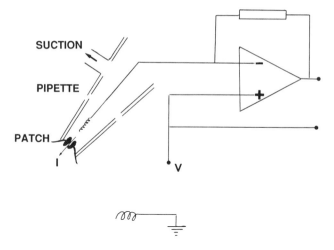

FIGURE 1. An illustration of the patch-clamp technique and its application to the study of stretch-activated channels. A glass pipette is filled with an electrolyte solution bathing a wire electrode connected to a low-noise electronic device. The tip of the pipette is pressed onto the surface of the target cell (or proteoliposome). Suction is applied to the inside of the pipette, resulting in the formation of a high-resistance seal between the membrane and the glass. Suction is then released. The current I can then flow between the pipette electrode and the bath (grounded) electrode only through any open channels that may be present in the membrane patch obstructing the pipette mouth. The patch can be excised, thus exposing the inner surface of the membrane to the bath solution. The electronic device allows the voltage V to be clamped at the desired value and the current I to be amplified and recorded. The opening or closing of a single channel produces measurable discrete jumps in the current level. Channel activity can be controlled by voltage, ligand binding, or mechanical stimuli, depending on the type of channel. In the last case, which is of interest here, the stimulus is exerted by applying controlled suction to the inside of the pipette via a side arm of the holder.

this respect (see below). The channels respond to tension (T), i.e., stress in the plane of the membrane,[23,24] which may be calculated by Laplace's law (T = PR/2) from the applied pressure (P) provided that the radius of curvature of the membrane patch (R) is known. Since the radius is difficult to determine, the usual practice is to relate the probability that the channel is open (Po) to the applied pressure. In most cases investigated so far, membrane stretch was found to affect mainly the kinetics of channel opening, while the channel closing rate was stretch independent.[21,23,25]

Depending on the biological systems, stretch-activated channels have been speculated to play a role in cell division, the sensing of gravity or other forces, motility, touch, smooth muscle contraction, chemotactic and avoidance reactions, and osmoregulation. Some of these effects might be mediated by the influx of Ca^{++} through the stretch-activated channels. While much progress is being made in the electrophysiological characterization of mechanosensitive channels, no protein or gene has yet been identified.

II. EXPERIMENTAL PROCEDURES

Since planar bilayers cannot be easily subjected to stretch, the electrical properties of stretch-activated channels can be properly studied only in patch-clamp experiments (Figure 1). Normal bacterial cells cannot be used in these experiments because they are too small as targets for the patch-clamp electrode. One approach to circumvent this problem is to patch-clamp giant proteoliposomes obtained from the fusion of membrane vesicles with liposomes. The dehydration-rehydration procedure of Criado and Keller[26] has been applied to bacterial membrane fractions to produce proteoliposomes of suitable size.[18,19] The technique allows the study of purified membranes, but the system is rather remote from native conditions, and the channels might not all have the same orientation in the membrane. However, it has been verified experimentally that stretch sensitivity survives the elaborate procedure required for preparing proteoliposomes.[18,19]

A second approach uses giant proto- or spheroplasts. In the case of Gram-positive bacteria, the procedure consists simply of allowing the cells to grow in the presence of lysozyme, which destroys the cell wall and prevents cell division. Protoplasts with diameters up to 4 μm can thereby be obtained from *Streptococcus faecalis* and *Bacillus subtilis*.[16,17] In the case of Gram-negative bacteria, cell wall synthesis can be blocked with penicillin, resulting in spheroplasts with diameters up to 15 μm.[15] Alternatively, septation can be inhibited by cephalexin. This results in the formation of very long, snake-like cells which, upon treatment with EDTA and lysozyme, yield suitable spheroplasts.[14] Buechner et al.[13] have obtained targets of suitable size also by UV or Mg^{++} treatment of the cells or by using mutants.

III. PROPERTIES OF THE BACTERIAL STRETCH-ACTIVATED CHANNELS

A large stretch-activated channel was discovered and first characterized by Martinac et al.[14] in *E. coli* spheroplasts. The channel is reversibly activated by application of pressure and the opening probability changes with pressure, following a Boltzmann distribution, in which the free energy of gating is linearly dependent on pressure. The channel is also voltage dependent: as the voltage increases, less pressure is needed to activate the channel. It has an unusually large conductance (from 630 to 950 pS, depending on the polarity of the voltage) and is slightly selective for anions.

Subsequent studies by Berrier et al.,[18] performed on isolated membranes of *E. coli* fused into giant liposomes, showed that different stretch-activated channels coexist in *E. coli* membranes. Although, due to their number, the full characterization of these channels in terms of voltage dependence, pressure dependence, and selectivity has not been fully completed, the channels are easily distinguished on the basis of their different conductances and kinetics.

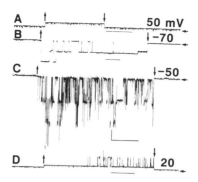

FIGURE 2. Stretch-activated channels of *E. coli*. Recordings of unitary currents (in 100 m*M* KCl) from four different patches excised from giant proteoliposomes. The giant proteoliposomes are obtained by subjecting a mixture of liposomes and *E. coli* purified inner membranes to a cycle of dehydration-rehydration. The closed levels are indicated by horizontal arrows. Application of suction (upward arrow) results in the opening of the channels present in the patch, which close upon release of suction (downward arrow). The membrane potential is indicated on the right of each trace; vertical bar: 20 pA; horizontal bar: 0.5 s. (From Berrier, C., Coulombe, A., Houssin, C., and Ghazi, A., *FEBS Lett., 259*, 27, 1989. With permission.)

Up to now, conductances of approximately 100, 150, 300, 500, 1000, 1500, and 2000 pS (in 100 m*M* KCl) have been recorded in this type of preparation. The recordings in Figure 2, obtained from different patches excised from *E. coli* giant proteoliposomes, show how the application of suction leads to the opening of channels of various sizes and kinetics, which close upon release of suction.

A study of penicillin-induced *E. coli* spheroplasts by Szabo' et al.[15] revealed the complex substate structure and the cooperative behavior of the channels and suggested that the presence of several different conductances may be due to the cooperative gating of only a limited number of channels.[15] The stretch-activated activity of Gram-positive organisms closely resembles that of *E. coli*. The application of stretch to the membrane of *S. faecalis* or *B. subtilis* protoplasts elicited single-step events of various sizes, with conductances ranging up to 6 nS in 350 m*M* KCl.[16,17,27] The abundance of different huge conductances, the substate structure, and the often-observed decay, upon repeated stretch cycles or aging, to give lower conductance channels derived from the larger ones support the idea that also in these species the events may be due to cooperative gating in multicomponent arrays. The centerpiece of this system may be a channel with a conductance of 1 to 1.2 nS in 350 m*M* KCl. It is itself endowed with substates, and it may well simply represent the most frequent assembly of smaller "building blocks".

The abundance of different conductances and their tendency to degenerate with time complicate the investigation of the properties of the channels. The following generalizations concerning the channels of Gram-positive bacteria can, however, be made already:[27] membrane stretch affects both the opening and the closing transition rates; i.e., an increase in applied pressure leads to more frequent as well as longer-lasting events. The channels are also modulated by voltage: the probability of the open state increases with depolarization, and both opening and closing rates are affected.

It is still a matter of debate where the stretch-activated channels of Gram-negative bacteria are located. Giant spheroplasts, prepared by growth of the bacteria in the presence of cephalexin followed by a lysozyme-EDTA treatment, may have preserved their outer membrane, and the seal may thus be formed on this membrane or on the inner membrane. The Madison group favors a localization of stretch-activated channels in the outer membrane.[13] This proposal is based on patch-clamp experiments performed on a mutant which spontaneously forms giant cells. If the cell envelopes are intact in these bacteria, the seal has to be formed on the outer membrane. In addition it was observed that perfusion of lysozyme in the patch-clamp bath results in an increase in the activity of stretch-activated channels in excised patches. This observation, which indicates that the cell wall is accessible to lysozyme in the bath, also indicates that a seal formed on the outer membrane. Nevertheless, it should be noted that the formation of high-resistance seals on the outer membrane implies that only a minute fraction of the porins present in this membrane are open (less than 1 out of 10^4). If this was to be the case, our current vision of outer membrane functioning would have to be revised.

Evidence in favor of stretch-activated channels being located in the inner membrane came from experiments in which different membrane fractions of *E. coli* (outer membranes, contact zones, and inner membranes) were purified on sucrose gradients and separately reconstituted in giant proteoliposomes.[18] Stretch-activated channels were observed in 50% of the seals performed on liposomes reconstituted with inner membranes and only in 6% of those performed on liposomes reconstituted with outer membrane under similar conditions (lipid/protein ratio) of reconstitution. Although the efficiency of reconstitution of the different fractions is unknown, these results are strong arguments in favor of the presence of stretch-activated channels in the inner membrane. This would be consistent with their presence in the plasma membrane of Gram-positive bacteria. It is not yet clear how these different observations can be reconciled, and it might be that there are stretch-activate channels in both the outer and the inner membrane, where they might have different functions.

The molecular mechanism of gating of the stretch-activated channels is unknown. Two hypotheses, which are not necessarily exclusive, have been proposed. In the first, mechanical stimulation is transmitted by some kind of cytoskeletal elements.[28] In the second, channel opening is induced by changes in the lipid surrounding the channel. This latter idea is strongly supported by the finding by Martinac et al.[29] that a variety of amphipathic molecules (including phenotiazines and local anesthetics) can activate bacterial stretch-activated channels directly. This effect is explained according to the bilayer couple hypothesis of Sheetz and Singer[30] which predicts that, given the asymmetry in electric charge between the two leaflets of the membrane, differential insertion of these molecules in the two leaflets would result in a stress in the membrane. The fact that stretch sensitivity is maintained in reconstituted systems also argues in favor of a direct role of the lipid bilayer. Nonetheless, a role for the cytoskeleton (or the exoskeleton, i.e., the cell wall in the case of bacteria) in regulating channel activity is probable, as evidenced from the reported effect of lysozyme.[13]

IV. PUTATIVE PHYSIOLOGICAL ROLES

The first response of *E. coli* to a decrease in the osmolarity of the external medium is a loss of K^+,[31,32] which is responsible for the maintenance of turgor under normal conditions. Upon a strong osmotic downshock, cells also lose metabolites,[33,34] in particular amino acids, without lysis of the cells. We recently measured the rates of these processes[35] and found that K^+ is lost at a rate as high as 800 nmol per milligram dry weight per second, which is two orders of magnitude higher than that of the known K^+ transport systems.[32,36-38] The rate of metabolite efflux could not be measured accurately, but efflux is complete in less than 5 s, suggesting extremely high rates for the release of these solutes as well. These high rates and the nonspecificity of the efflux suggest the involvement of pores activated by the osmotic shock. The hypothesis is supported by the observation[35] that gadolinium (Gd^{3+}), previously reported to inhibit eukaryotic stretch-activated channels,[39] inhibited the efflux of metabolites such as lactose and ATP from shocked cells as well as stretch-induced activity in patch-clamp experiments with both Gram-positive and Gram-negative bacteria, at submillimolar concentrations. K^+ efflux induced by osmotic shock was not blocked by Gd^{3+},[35,40] but only slowed,[35] suggesting that only a part of it takes place via the channels mediating the efflux of large molecules. A relevant observation from electrophysiological experiments is that the sensitivity of the various conductances to Gd^{3+} is not the same.[27,35,41] While detailed characterization has not yet been carried out, it appears that the smaller conductances require higher concentrations to be inhibited or in some cases may not be inhibited at all.

Taken as a whole, these results supply further evidence in favor of the channel location being in the inner membrane of Gram-negative bacteria. Furthermore, they suggest that these channels play a role in osmoregulation. The size of the bacterial stretch-activated channels is unusual as compared to their eukaryotic counterparts. Some of the bacterial stretch-activated conductances are the highest observed among channels in general. For a comparison, OmpC and OmpF porin trimers show a conductance of about 200 pS in 100 mM KCl.[8,9] Also, the large number of different conductances is puzzling. Perhaps an explanation can be found in the necessity for bacteria to adapt to media of very different osmolarities. Upon an osmotic upshock bacteria accumulate potassium[31,42] and concomitantly synthesize glutamate. Later, potassium and glutamate are partially substituted with trehalose.[43,44] Bacteria also can accumulate the osmoprotectants proline or glycine betaine if they are present in the medium.[45,46] Cells, adapted to a medium with high osmolarity, which are shifted back to a normal medium or to a medium of low osmolarity will have to excrete these species very rapidly. This could be possible through stretch-activated channels large enough to accommodate molecules of the size of a disaccharide or an amino acid. When the cells are subjected to only small fluctuations in the osmolarity of the medium, excretion of small amounts of potassium is enough to maintain an optimal turgor. Under these conditions loss of metabolites should be avoided. Indeed, *E. coli* cells subjected to slow dilution with water lose only K^+ (and not larger metabolites).[47] Consistently, we observed that K^+, lactose, and ATP exit from *E. coli* cells at different thresholds of osmotic downshock intensity.[41] A parallel observation, made in patch-clamp experiments, is that small *E. coli* stretch-activated channels are activated at lower suction than larger stretch-activated channels.[41] Thus, a multiplicity of bacterial stretch-activated channels with different sizes and different thresholds of activation by pressure may represent an advantage for a species confronted with very different media. A role in other aspects of bacterial physiology is not excluded. In particular, it has been suggested that, because of their sensitivity to membrane deformation, these channels may be involved in the control of osmotaxis,[14,48] i.e., the avoidance of regions of nonoptimal osmolarity.

V. CONCLUSION

Stretch-activated channels constitute a new element of the ion transport machinery of bacterial cells, characterized by low selectivity and high capacity, which is expected to function mainly in the efflux mode. Together with active transport systems triggered by a release in membrane tension, they presumably play a role in the maintenance of turgor within defined limits, by acting as a pressure-activated valve. Future studies should aim at identifying more selective high-affinity inhibitors of the channels, capable of leading to the eventual identification of the protein(s) forming the pores and of their gene(s).

REFERENCES

1. **Mueller, P. and Rudin, D. O.**, Action potential phenomena in experimental bimolecular lipid membranes, *Nature (London)*, 213, 603, 1967.
2. **Mueller, P. and Rudin, D. O.**, Resting and action potentials in experimental bimolecular lipid membranes, *J. Theor. Biol.*, 18, 222, 1968.
3. **Bean, R. C., Shepherd, W. C., Chan, H., and Eicher, J.**, Discrete conductance fluctuations in lipid bilayer protein membranes, *J. Gen. Physiol.*, 53, 741, 1969.

4. **Ehrenstein, G., Lecar, H., and Nossal, R.,** The nature of the negative resistance in bimolecular lipid membranes containing excitability-inducing material, *J. Gen. Physiol.,* 55, 119, 1970.

5. **Schindler, H. and Rosenbusch, J. P.,** Matrix protein from *Escherichia coli* outer membrane forms voltage-controlled channels in lipid bilayers, *Proc. Natl. Acad. Sci. U.S.A.,* 75, 3751, 1978.

6. **Benz, R., Janko, K., Boos, W., and Laüger, P.,** Formation of large, ion-permeable membrane channels by the matrix protein (porin) of *Escherichia coli, Biochim. Biophys. Acta,* 511, 305, 1978.

7. **Benz, R.,** Porin from bacterial and mitochondrial outer membranes, *Crit. Rev. Biochem.,* 19, 145, 1985.

8. **Benz, R.,** Structure and function of porins from Gram-negative bacteria, *Annu. Rev. Microbiol.,* 42, 359, 1988.

9. **Jap, B. K. and Walian, P. J.,** Biophysics of the structure and function of porins, *Q. Rev. Biophys.,* 23, 367, 1990.

10. **Felle, H., Porter, J. S., Slayman, C. L., and Kaback, H. R.,** Quantitative measurement of membrane potential in *Escherichia coli, Biochemistry,* 19, 3585, 1980.

11. **Hamill, O. P., Marty, A., Neher, E., Sakmann, B., and Sigworth, F. J.,** Improved patch-clamp techniques for high-resolution current recordings from cells and cell-free membrane patches, *Pflügers Arch.,* 391, 85, 1981.

12. **Sakmann, B. and Neher, E., Eds.,** *Single-Channel Recording,* Plenum Press, New York, 1983.

13. **Buechner, M., Delcour, A. H., Martinac, B., Adler, J., and Kung, C.,** Ion channel activities in *Escherichia coli* outer membrane, *Biochim. Biophys. Acta,* 1024, 111, 1990.

14. **Martinac, B., Buechner, M., Delcour, A. H., Adler, J., and Kung, C.,** Pressure-sensitive ion channel in *Escherichia coli, Proc. Natl. Acad. Sci. U.S.A.,* 84, 2297, 1987.

15. **Szabo', I., Petronilli, V., Guerra, L., and Zoratti, M.,** Cooperative mechanosensitive ion channels in *Escherichia coli, Biochem. Biophys. Res. Commun.,* 171, 280, 1990.

16. **Zoratti, M. and Petronilli, V.,** Ion-conducting channels in a Gram-positive bacterium, *FEBS Lett.,* 240, 105, 1988.

17. **Zoratti, M., Petronilli, V., and Szabo', I.,** Stretch-activated composite ion channels in *Bacillus subtilis, Biochem. Biophys. Res. Commun.,* 168, 443, 1990.

18. **Berrier, C., Coulombe, A., Houssin, C., and Ghazi, A.,** A patch-clamp study of inner and outer membranes and of contact zones of *E. coli,* fused into giant liposomes. Pressure-activated channels are localized in the inner membrane, *FEBS Lett.,* 259, 27, 1989.

19. **Delcour, A. H., Martinac, B., Adler, J., and Kung, C.,** A modified reconstitution method used in patch-clamp studies of *Escherichia coli* ion channels, *Biophys. J.,* 56, 631, 1989.

20. **Delcour, A. H., Martinac, B., Adler, J., and Kung, C.,** Voltage-sensitive ion channel of *Escherichia coli, J. Membr. Biol.,* 112, 267, 1989.

21. **Guharay, F. and Sachs, F.,** Stretch-activated single ion channel currents in tissue-cultured embryonic chick skeletal muscle, *J. Physiol. (London),* 352, 685, 1984.

22. **Sachs, F.,** Mechanical transduction in biological systems, *Crit. Rev. Biomed. Eng.,* 16, 141, 1988.

23. **Morris, C.,** Mechanosensitive ion channels, *J. Membr. Biol.,* 113, 93, 1990.

24. **Gustin, M. C., Zhou, X.-L., Martinac, B., and Kung, C.,** A mechanosensitive ion channel in the yeast plasma membrane, *Science,* 242, 762, 1988.

25. **Cooper, K. E., Tang, J. M., Rae, J. L., and Eisenberg, R. S.,** A cation channel in frog lens epithelia responsive to pressure and calcium, *J. Membr. Biol.,* 93, 259, 1986.

26. **Criado, M. and Keller, B.,** A membrane fusion strategy for single-channel recordings of membranes usually non-accessible to patch-clamp pipette electrodes, *FEBS Lett.,* 224, 172, 1987.

27. **Szabo', I., Petronilli, V., and Zoratti, M.,** unpublished data, 1991.

28. **Sokabe, M., Sachs, F., and Jing, Z.,** Quantitative video microscopy of patch clamped membranes. Stress, strain, capacitance, and stretch channel activation, *Biophys. J.,* 59, 722, 1991.

29. **Martinac, B., Adler, J., and Kung, C.,** Mechanosensitive ion channels of *E. coli* activated by amphipaths, *Nature (London),* 348, 261, 1990.

30. **Sheetz, M. and Singer, S. J.,** Biological membranes as bilayer couples. A molecular mechanism of drug-erythrocyte interactions, *Proc. Natl. Acad. Sci. U.S.A.,* 71, 4457, 1974.

31. **Epstein, W. and Schultz, S. G.,** Cation transport in *Escherichia coli.* V. Regulation of cation content, *J. Gen. Physiol.,* 49, 221, 1965.

32. **Booth, I.,** Potassium efflux systems, in *Alkali Cation Transport Systems in Prokaryotes,* Bakker, E. P., Ed., CRC Press, Boca Raton, FL, 1992, chapter IIE.

33. **Britten, R. J. and McClure, F. T.,** The amino acid pool of *Escherichia coli, Bacteriol. Rev.,* 26, 292, 1962.

34. **Tsapis, A. and Kepes, A.,** Transient breakdown of the permeability barrier of the membrane of *Escherichia coli* upon hypoosmotic shock, *Biochim. Biophys. Acta,* 469, 1, 1977.

35. **Berrier, C., Coulombe, A., Szabo', I., Zoratti, M., and Ghazi, A.,** *Eur. J. Biochem.,* 206, 559, 1992.

36. **Rhoads, D. B. and Epstein, W.,** Cation transport in *E. coli.* IX. Regulation of K^+ transport, *J. Gen Physiol.,* 72, 283, 1978.

37. **Siebers, A.,** The Kdp-ATPase, in *Alkali Cation Transport Systems in Prokaryotes,* Bakker, E. P., Ed., CRC Press, Boca Raton, FL, 1992, chapter IIB.

38. **Bakker, E.,** Constitutive K^+ uptake systems, in *Alkali Cation Transport Systems in Prokaryotes,* Bakker, E. P., Ed., CRC Press, Boca Raton, FL, 1992, chapter IIC.

39. **Yang, X.-C. and Sachs, F.,** Block of stretch-activated ion channels in *Xenopus* oocytes by gadolinium and calcium ions, *Science,* 243, 1068, 1989.

40. **Schleyer, M. and Bakker, E.,** personal communication, 1991.

41. **Berrier, C., Coulombe, A., and Ghazi, A.,** unpublished results, 1991.

42. **Epstein, W.,** Osmoregulation by potassium transport in *Escherichia coli, FEMS Microbiol. Lett.,* 39, 73, 1986.

43. **Larsen, P., Sydness, L. K., Lanfald, B., and Strom, A. R.,** Osmoregulation in *Escherichia coli* by accumulation of organic osmolytes: betaines, glutamic acid, and trehalose, *Arch. Microbiol.,* 147, 1, 1987.

44. **Dinnbier, U., Limpinsel, E., Schmid, R., and Bakker, E. P.,** Transient accumulation of potassium glutamate in its replacement by trehalose during adaptation of growing cells of *Escherichia coli* K-12 to elevated sodium chloride concentrations, *Arch. Microbiol.,* 150, 348, 1988.

45. **Perroud, B. and Le Rudelier, D.,** Glycine betaine transport in *Escherichia coli:* osmotic modulation, *J. Bacteriol.,* 161, 393, 1985.

46. **Cairney, J., Booth, I. R., and Higgins, C. F.,** Osmoregulation of gene expression in *Salmonella typhimurium: proU* encodes an osmotically induced betaine transport system, *J. Bacteriol.,* 164, 1224, 1985.

47. **Meury, J., Robin, A., and Monnier-Champeix, P.,** Turgor-controlled K^+ fluxes and their pathways in *Escherichia coli, Eur. J. Biochem.,* 151, 613, 1985.

48. **Li, C., Boileau, A. J., Kung, C., and Adler, J.,** Osmotaxis in *Escherichia coli, Proc. Natl. Acad. Sci. U.S.A.,* 85, 9451, 1988.

Chapter IIJ

BACTERIOCIN AND BACTERIOPHAGE CHANNELS IN PROKARYOTES

Lucienne Letellier

TABLE OF CONTENTS

I. INTRODUCTION

Bacteriocins are toxins produced by a wide variety of Enterobacteriaceae.[1] Genes encoding these proteins are carried on plasmids that confer to the host bacterium immunity to the produced toxin. Bacteriocins kill bacteria of the same or a related family as that of the producing cell. Bacteriocins have different targets and modes of action: those acting in the cytoplasm cleave 16S ribosomal RNA (colicin E3 and cloacin DF13) or DNA (colicin E2).[2] The cellular target of colicin M appears to be the enzymes involved in peptidoglycan formation.[3] The analysis of these activities are beyond the scope of this review. The cytoplasmic membrane is the target of the so-called pore or channel-forming bacteriocins. The understanding of the mode of action of colicins (i.e., the bacteriocins produced by *Escherichia coli*) is more advanced that of other bacteriocins. This chapter will therefore focus mainly on the properties of the channel-forming colicins A, B, N, E1, Ia, Ib, and K.

These colicins are proteins of high molecular weight (40 to 70 kDa).[4] They are synthesized in large quantities, are generally released in the culture medium,[5] and are easily purified. Their genes have been cloned, and since 1982 their nucleotide sequence has been determined, which makes their genetic and chemical manipulation possible.[6,7] Although they are water-soluble proteins, they insert spontaneously into lipid bilayers where they form ion channels.[4,8,9] To exert their lethal activity, these colicins first bind specifically to an outer membrane receptor. These receptors, whose physiological function is to transport solutes, have also been parasitized by bacteriophages.[10] Following binding, colicins are translocated through the envelope. This translocation involves the participation of the genes products of *tolQ, R, A, B* (for colicins A, E1, and K) or of *tonB* (for colicins B, Ia, and Ib).[11] The TonB and Tol pathways are also used by bacteriophages T1 and Φ80 and the filamentous bacteriophages, respectively.[8,12] The observations that colicins form ion channels *in vitro*, that they induce a collapse of the electrochemical gradient of protons ($\Delta\bar{\mu}_{H^+}$), and that they increase the permeability of the cell envelope to K^+ and several solutes *in vivo* have led to the concept that their lethal activity is the result of the formation of ion channels in the cytoplasmic membrane.

E. coli cells can be infected by a wide range of bacteriophages. Despite great variations in the size and nature of their genetic material and in their morphology,[13,14] the first steps of the infectious process of most phages share common properties. After random collision between the virion and the bacteria, the phage irreversibly attaches to a specific receptor located on the outer membrane. The phage genome is then ejected from the capsid, transferred through the envelope, and injected in the cytoplasm. The transfer of the phage nucleic acid (only one or a few copies per bacterium) is rapid: the double-stranded DNA of phage T4, whose length is 50 times that of the bacteria, is injected in less than 1 min. The study of DNA transport is therefore rather difficult. Starting in the sixties several groups observed that bacteriophages, like colicins, increased the permeability of the cell envelope to several solutes and cellular K$^+$. This has led to the general concept that bacteriophage infection causes the formation of "holes or localized lesions" in the envelope.[15,16] Only a few groups attempted to analyze these ion fluxes in greater details and to correlate them with DNA penetration. The recent development of a selective K$^+$ electrode[17] has allowed the quantitative analysis of these K$^+$ fluxes without the need of separation methods (filtration or centrifugation) and, thus, under the physiological conditions of infection. The results obtained strongly suggest that ion channels are involved in the transfer of phage DNA through the envelope. A similar approach could be used to gather new informations on the *in vivo* properties of the channel-forming colicins. This chapter deals essentially with the experiments relevant to these channels. Several reviews and books give greater coverage of the properties of bacteriocins and bacteriophages.[2,4,18-20]

II. COLICIN CHANNELS

A. EARLY *IN VIVO* OBSERVATIONS: COLICINS E1, K, AND Ia AFFECT THE SOLUTE CONTENT OF BACTERIA

Early experiments on uptake and efflux of various solutes in colicin E1-, K-, and Ia-treated cells led to the conclusion that the primary target of these colicins is the cytoplasmic membrane.[4,18] Colicins E1 and K prevented the accumulation of amino acids and thiomethyl-β-D-galactosides and caused their efflux from preloaded cells.[18,21] However, there was no general permeabilization of the cell membrane since the transport of α-methyl glucosides (which occurs via the phosphoenolpyruvate-dependent phosphotransferase system) was not inhibited.[21] Colicin K and colicin A inhibit the active transport of K$^+$ and cause the loss of intracellular K$^+$.[22,23] Hirata et al.[24] also observed that colicin K induced a K$^+$ efflux; since no concomitant macromolecular anion leakage could be detected, they concluded that the efflux of K$^+$ was accompanied by the entry of cations or efflux of small anions. However, K$^+$ efflux and Na$^+$ influx are not always coupled since the omission of Na$^+$ from the medium did not affect the kinetics of K$^+$ efflux (this is a situation very similar to that observed in the case of the channels induced by phage

T1 and T4; see Sections III.B and C). Wendt[25] established a quantitative relationship between the killing activity of colicin K and the extent of K^+ efflux: a single lethal hit was sufficient to empty a cell of K^+.

By the end of the sixties, Fields and Luria[26,27] suggested that the action of these colicins on energy metabolism can be explained within the framework of the chemiosmotic theory of Mitchell:[28] "Colicins would act if they were uncouplers by promoting leakage of protons through the membrane and dissipating the energy needed for ATP synthesis."

B. COLICINS DO NOT ACT AS PROTONOPHORES

Feingold[29] tested Fields and Luria's hypothesis. Using the elegant approach of Mitchell and Moyle,[30] he examined the effect of colicin E1 treatment on the response of *E. coli* to an imposed acid load; in contrast to the protonophore carbonyl cyanide *m*-chlorophenyl hydrazone (CCCP), colicin E1 did not increase the permeability of the membrane to H^+. However, it is possible that the harsh treatment to which the colicin-treated cells were submitted (they were incubated for 30 min with colicin E1, filtered, scraped from the filter, and resuspended in the medium before being submitted to the acid pulse) may have impaired their channel activity (see below). The experiments of Lusk and Nelson[31] confirmed the difference between uncouplers and colicins: colicin K and E1 inhibited the energy-dependent exchange of intracellular Mg^{2+} with extracellular Mg^{2+} and caused a net loss of $^{28}Mg^{2+}$, whereas the internal pool of Mg^{2+} was unaffected by uncouplers. They also observed that the net $^{28}Mg^{2+}$ efflux depended on the relative concentration of colicins, a property shared by K^+ efflux.

C. COLICINS DISSIPATE THE TRANSMEMBRANE ELECTRICAL POTENTIAL

Gould and Cramer[32] gave the first convincing proof that colicin E1 and K dissipate the transmembrane electrical potential ($\Delta\Psi$) by analyzing the H^+ permeability of an anaerobic suspension of *E. coli*. Colicin E1 caused an increase in the rate and extent of H^+ efflux after an oxygen pulse, in a manner similar to mobile charge-compensating counterions (e.g., SCN^-, K^+ plus valinomycin).[33] The increase in the rate and extent of H^+ efflux was larger in a medium containing K^+, Li^+, or Na^+ than in choline chloride MOPS-Tris medium. The H^+/O ratio in the presence of colicin increased from 0.59 to 2.33 when the external K^+ concentration was increased from 0.3 to 7.8 mM. The authors compared the rate of colicin-induced H^+ influx to the rate of K^+ efflux; the kinetics of both fluxes were qualitatively similar. However, the H^+/K^+ ratio never exceeded 0.1, suggesting that the influx of H^+ was not sufficient to compensate for the K^+ efflux (this situation is also found in the case of the channel induced by phage T4 — see Section III.C).[17] Clearly the net K^+ efflux must be accompanied by an efflux of anions or/and an influx of cations to maintain macroscopic electroneutrality. The authors concluded that the increase in ionic conductance was either due to an ion channel created by the colicin itself or induced in the membrane through structural changes; whatever the origin of this ion channel, the resulting permeability should cause the dissipation of the membrane potential.

Weiss and Luria[34] demonstrated that *E. coli* cells were effectively depolarized upon addition of colicin K. They measured the membrane potential ($\Delta\Psi$) from the accumulation of the lipophilic cation triphenylmethylphosphonium (TPMP⁺). Addition of colicin appeared to depolarize the cells completely since the addition of CCCP did not induce a further loss of TPMP⁺ (the authors pointed out the difficulties of determining correct $\Delta\Psi$ at low values). The pH gradient (ΔpH) which was measured from the accumulation of the weak butyric acid was not abolished by colicin K. Tokuda and Konisky[35] published a quite similar observation on colicin Ia: *E. coli* cells pretreated with colicin A could not generate a membrane potential, but retained a ΔpH. Cytoplasmic membrane vesicles were prepared from control and colicin Ia-treated cells: only those from the control were able to generate a membrane potential.

D. COLICINS FORM CHANNELS *IN VITRO*
1. Lipid Bilayer Experiments

In 1973 Luria proposed that the alterations caused *in vivo* by the colicins could be due to the creation by the colicins of specific "channels."[18] The subsequent report that colicin K forms channels *in vitro* appeared simultaneously with the observations that colicins depolarize *E. coli* cells.[34,35] In a series of elegant experiments, Schein, Kagan, and Finkelstein[36] demonstrated that colicins inserted spontaneously into planar lipid bilayers where they formed ion channels. The formation of the channels required an acidic pH. The single-channel conductance in 100 mM KCl was about 1.5 to 3 pS; the channels were almost always open at large negative voltage (transcompartment), the "switching voltage" being -31 mV. These results were confirmed for colicin E1, colicin Ia and Ib, A, B, and N (reviewed in References 8 and 9). All of these colicins form channels with small conductances (15 to 20 pS in 1 M NaCl at neutral pH) and poor selectivity between anions and cations. Both the results of Schein et al.[36] and those of Slatin[37] suggested that the colicin channels are formed by a monomer. The C-terminal peptide fragment of colicins which was isolated by proteolytic cleavage was shown to carry channel activity similar to that of the colicins.[38-40]

2. Liposome Experiments

The properties of colicins as both membrane and soluble proteins have turned out to be more complicated than initially suspected. The spontaneous insertion of colicins into liposomes is only possible at acidic pH (pH < 5), conditions in which they also promote the fusion of vesicles.[41,42] Tokuda and Konisky[43] showed that colicin E1 and Ia, but not E2 and E3 (which are both "enzymatic" colicins), incorporated into small liposomes abolished a

K$^+$-valinomycin-induced diffusion potential. Rb$^+$, Na$^+$, phosphate, and choline leak out the vesicles, but dextran and inulin do not. However, this permeabilization does not require a membrane potential as it does in lipid bilayers. Davidson et al.[44] claimed that the dependence on membrane potential could be recovered if large vesicles were used. The main difficulty with these liposome experiments is that one cannot discriminate between the ion flux originating from the perturbation of the lipid bilayer created by the colicins and a true movement of ions through the channel. Voltage gating of the colicin E1 channel was recently documented by Cramer et al.[45] by labeling colicin E1 with ^{125}I lipophilic photoaffinity probes. It was suggested that the membrane potential drives the channel peptide into the lipid bilayer.

The size of the channel was estimated from the ability of liposomes to retain entrapped low-molecular-weight nonelectrolyte solutes in the presence of colicins.[46] Solutes of molecular weight equal to or greater than 92 (glycerol) were retained in the liposomes. The hydration radius of glycerol being 0.31 nm, this would suggest that the radius of the colicin channel is smaller than that of glycerol. These results do not agree with those obtained by Raymond et al.[47] using lipid bilayers. The measurements of reversal potentials, under conditions of asymmetric media, indicated that the colicin channel could conduct large ions as glucosamine or NAD; they concluded that the average diameter of lumen of the colicin E1 channel is probably ca. 1.2 to 1.6 nm. These results are difficult to reconcile with the observation that the diameter of the gramicidin channel, whose single channel conductance is higher than that of the colicin channel, is only 0.4 nm.[47]

3. Colicins Depolarize Isolated Cytoplasmic Membrane Vesicles

Although the main target of colicins is the *E. coli* envelope, they incorporate into vesicles isolated from the *E. coli* inner membrane only after a freeze and thaw cycle. Under these conditions, colicin Ia depolarizes vesicles prepared from *E. coli* strains which are either resistant (defective in their capacity to bind colicin), tolerant (mutated in the translocation machinery), or sensitive to colicins.[48] Colicin Ib added to membrane vesicles after a freeze and thaw cycle also inhibits the generation of a membrane potential.[49]

E. INDIRECT EVIDENCE SUGGESTING THAT COLICIN A FORMS CHANNELS *IN VIVO*

We have recently analyzed the kinetics of K$^+$ efflux induced by colicin A using a valinomycin-K$^+$ electrode.[50] Addition of colicin A to *E. coli* cells caused, after a lag, a net efflux of K$^+$. The initial rate of efflux increased linearly with increasing colicin concentration up to 50 molecules per cell. The rate of this K$^+$ efflux was larger than the rate of efflux via the known K$^+$ transport systems,[51] which suggested that colicin A causes a new route for K$^+$ efflux to open. This new efflux route is likely to be channel mediated since a single colicin molecule added to one bacteria was able to induce an efflux of 3×10^5 K$^+$ ions per second, which is fast compared to the turnover rate of mobile carriers. The linear relationship between the initial rate of efflux and the amount of colicin suggests a lack of cooperation between colicin molecules. Presumably, each colicin forms a single channel.

The existence of a threshold of membrane potential for channel opening and closing *in vivo* was inferred from the following observations:

1. The cells were depolarized to the same plateau value (85 mV negative inside at pH 6.8) independent of the amount of colicins added; these $\Delta\Psi$ values were also those required to trigger the K$^+$ efflux.
2. A decrease of $\Delta\Psi$ below 85 mV after the efflux had started, by addition of the protonophore 3,3′,4′-tetrachlorosalicylanilide (TCS), resulted in an immediate arrest of the efflux; a reincrease of $\Delta\Psi$ above 85 mV by addition of bovine serum albumin (which desorbs amphiphilic molecules from membranes) reinitiated the efflux.

The fact that the colicin channel closes below a threshold membrane potential raises the question of how colicins kill bacteria. The decrease of $\Delta\Psi$ alone cannot explain the death of the bacteria since they can remain viable with a reduced $\Delta\bar{\mu}_{H^+}$.[52] It is probably that the large decrease in internal K$^+$ is responsible in part for the arrest of protein synthesis. Channel-forming colicins also have the unusual property of inducing a depletion (but not a leakage) of internal ATP concomitant with the K$^+$ efflux (80% of the ATP may be loosed in less than 2 min).[24,27] These results were recently confirmed in the case of colicin A: there is a tight correlation between the rate and extent of K$^+$ efflux and the rate and extent of the decrease of internal ATP. However, ATP is not hydrolyzed by the F$_1$F$_0$-ATPase to regenerate $\Delta\bar{\mu}_{H^+}$ as previously thought since the internal ATP concentration decreased similarly in a strain deleted for the F$_1$F$_0$-ATPase. Rather, we think that the ATP decrease, which is presumably the primary event causing cell death, is mainly the consequence of the colicin-mediated efflux of inorganic phosphate.[53]

E. coli cells are protected against colicin K and E1 by strict anaerobiosis.[18] Similarly, pretreatment of the bacteria with respiratory inhibitors such as cyanide or with protonophores prior to the addition of colicin K or E1 also prevents the occurrence of irreversible cellular damage.[54,55] Since all these treatments are known to decrease or abolish $\Delta\Psi$, it is likely that they prevent the insertion and/or opening of the channel in the inner membrane. Colicins which do not form channels appear to be also "sensitive to the energized state" of the cells.[55] This suggests that $\Delta\Psi$ does not specifically affect the channel activity, but rather allows the correct insertion of one of the colicin domains traversing the inner membrane. However, these results should be interpreted with caution since it was reported that colicin E2 preparations may be contaminated by the pore-forming colicin Ia.[36]

The K$^+$ efflux caused by colicin A is preceded by a lag time. During this period, the colicin binds to its receptor and is translocated through the envelope. From the comparison of the lag time required to initiate K^+ efflux under energized and deenergized conditions it appears that one of the steps following binding occurs in deenergized cells. However, it is not clear whether or not colicin A is translocated and inserted into the deenergized membrane and $\Delta\Psi$ is necessary for channel opening or if $\Delta\Psi$ is required for insertion and channel opening.[50]

F. COLICIN A UNFOLDS DURING ITS TRANSLOCATION AND SPANS THE WHOLE CELL ENVELOPE WHEN ITS CHANNEL IS FORMED[56]

At high colicin concentrations the measurement of the lag time before K^+ efflux is a good approximation of the time needed for translocation. The translocation time of colicin A denatured in 8 M urea prior to being added to cells was one half that of native colicin A. This suggests that the rate-limiting step in translocation may be the partial unfolding of the polypeptide chain. Unfolding had no effect on the characteristics of the channel. Import of polypeptides in *E. coli* thus appears to be reminiscent of import processes across the mitochondrial membranes.[57]

Addition of trypsin to the external medium caused an immediate closing of colicin A and B channels (as judged from the arrest of the K^+ efflux), although colicins A and B share neither the same receptor nor the same translocation machinery.[56] The protease does not gain access to either the periplasmic space or the cytoplasmic membrane. These colicins are therefore accessible from the external medium when forming the pore in the inner membrane. Although colicins are elongated molecules (their axial ratio is high — 8 to 10 — and electron microscopy of crystals of colicin Ia show that the molecule is at least 7 nm long), they are not long enough to traverse the entire envelope.[8,11] This suggests that colicins remain unfolded once the pore has been formed in the inner membrane.

These trypsin experiments yield new insights into the so-called "trypsin rescue state". Early experiments led to the conclusion that the action of trypsin on the pore-forming colicins K and Ib is divided into stages 1 and 2 depending on whether or not cell viability can be restored by adding trypsin.[58,59] It is likely that stage 1 is that at which the pore can be closed and the main transport activities required for the vital activity of the cells can be restored; stage 2 may take place when this closing occurs too long after the addition of colicin, i.e., when the levels of internal K^+ and of internal ATP are too low for the cell to recover.[60]

What triggers the unfolding of colicins? The polypeptide chain of colicins is organized linearly into three domains which carry different functions: the C-terminal domain forms the channel, while the central domain binds to the receptor. There is good indirect evidence that the N-terminal domain is involved in the translocation step.[8] It is likely that interactions of the colicins with their receptors trigger the first step of unfolding in the outer membrane.

Further unfolding into the periplasmic space may be favored by the interaction of colicins with their translocation machinery. It has recently been shown that the N-terminal domain of colicin A interacts *in vitro* with the C-terminal domain of Tol A, a protein associated with the inner membrane and protruding into the periplasmic space.[11] This interaction may be relevant to the *in vivo* translocation mechanism.

Colicin A-treated *E. coli* envelopes were recently fractionated. Colicin A was identified by monoclonal antibody labeling and found essentially in a fraction corresponding to the contact sites between the inner and outer membrane.[61] It is thus likely that the translocation of colicins, like the translocation of phage DNA,[62] takes place at particular sites in the envelope. Colicin import thus appears to share some common features with the import mechanism of most precursors into the inner mitochondrial membrane, which probably also occurs at contact sites between the two membranes.[63]

G. COLICIN CHANNELS *IN VIVO* AND *IN VITRO*: ARE WE LOOKING AT THE SAME CHANNEL?

There has been considerable progress in the elucidation of the structure of the channel-forming colicins in solution in recent years. The C-terminal domain of colicin A has been crystallized and its structure has been solved to 2.5 Å resolution. A model for its insertion and channel formation has been proposed. To account for the size of the channel lumen and other constraints it was proposed that the channel is formed by an oligomer of colicin (reviewed in Reference 8). However, most of the *in vitro* and *in vivo* data argue for a monomeric channel.[1,8,45,50] The recent findings suggesting that colicins remain in an extended configuration while forming the channels *in vivo*[56] are also difficult to reconcile with *in vitro* studies which give no evidence for unfolding.

Recent data suggest that colicin A interacts with the Tol A protein *in vitro*.[64] It may be that the interaction between these proteins in the inner membrane is sufficient either to create a local perturbation which mimics the behavior of channels or even to create a composite channel. This type of model would explain discrepancies between *in vitro* and *in vivo* data. Recently, it has become possible to overproduce and to characterize the proteins involved in the translocation of colicins. Their study is of major importance to the understanding of *in vivo* functioning of colicins.

III. BACTERIOPHAGE CHANNELS

A. MODIFICATION OF HOST MEMBRANE PERMEABILITY TO K⁺ AND OTHER SOLUTES INDUCED BY PHAGE T4 AND THE RELATED T-EVEN PHAGES

T-even phages are very large phages containing a double-stranded DNA molecule of approximately 160 kilobase pairs (kbp). When the phage attaches

to the receptor, the tail contracts, the tip of the internal tube of the phage tail
sheet is brought close to the cytoplasmic membrane, and the DNA is injected
through the cytoplasmic membrane.[13] Almost 40 years ago Puck and Lee[65]
observed a transient leakage of ^{32}P and ^{35}S lasting about 5 min after the
beginning of the infection by phage T2. Silver et al.[66] showed that the infection
by phage T2 led to a rapid and transient leak of cytosolic Mg^{2+} and K^+. The
extent of ion loss increased with the number of added phages. Leakage did
not occur at 4°C (i.e., under conditions where the phage attaches to its
receptor, but does not inject its DNA). They concluded that the changes in
permeability correspond to the formation of "localized holes or lesions" in
the membrane which are repaired a few minutes after infection by a "sealing"
reaction. This "sealing" is not dependent on the synthesis of a phage protein
since it occurs even in the presence of chloramphenicol. These permeability
changes do not depend on the presence of the phage DNA since T4 ghosts
(phages depleted of their DNA and internal proteins by osmotic shock) also
cause the release of several metabolites and ions.[67]

Shapira et al.[68] analyzed the influx and efflux of ^{42}K caused by the T4
phage and ghosts. Cells infected by phage T4 were able to reaccumulate K^+,
whereas ghost-infected cells lost this faculty. The authors also concluded that
there is a "repair mechanism" which operates after infection by phage, but
not by ghosts.

B. PERMEABILITY CHANGES INDUCED BY T-ODD PHAGES

T-odd phages also contain a double-stranded DNA, but in contrast to the
T-even phages they do not possess a contractile tail. Phages T3 and T7 are
related to each other, but T1 and T5 are different from each other and from
the other T phages.[20]

Phage T1 has a double-stranded DNA about 16 μm long. The changes
in permeability caused by this phage have been documented by two groups.
Wagner et al.[69,70] have shown that T1 inhibits host gene expression very early
and independently of viral protein synthesis. This inhibition was associated
with several membrane alterations leading to a decrease in the internal K^+
concentration. K^+ efflux was accompanied by $^{45}Ca^{2+}$ influx, with kinetics
very similar to the K^+ efflux. Their results, however, provided arguments
against a generalized "hole hypothesis" since neither β-galactosidase nor *o*-
nitrophenyl β-D-galactopyranoside could cross the membrane and the membrane remained energized, although at a lower level. These authors favored
the idea that the naked DNA molecule is actively transported in symport with
protons.[69,70] This attractive model agrees with the hypothesis formulated by
Grinius: $\Delta\bar{\mu}_{H^+}$ would be used as the driving force for phage nucleic acid
transport.[71] However, such a model does not fit in with data showing that
the DNA of phage T5 and λ[72,73] can be injected into a cell deprived of ATP
and in the absence of $\Delta\bar{\mu}_{H^+}$.

Keweloh and Bakker[74,75] performed the first detailed analysis of the ion
fluxes caused by phages: addition of T1 resulted in a rapid loss of internal

K$^+$ which was accompanied by the uptake of Na$^+$; Li$^+$ could replace Na$^+$ efficiently, and the loss of K$^+$ also occurred when the external medium contained either Tris, bis-Tris propane, or choline. T1 also induced Mg^{2+} loss, an increase of ΔpH, a decrease of $\Delta\Psi$ and of the concentration of internal ATP. Interestingly, the cation gradients could be restored to levels close to that of control cells by incubating the cells in high Mg^{2+} (5 mM), but not in low Mg^{2+} (0.4 mM). Although the authors did not exclude the possibility that the permeability changes were due to the formation of channels, they favored a mechanism in which the function of the permeability changes would be to depolarize the cells completely so that upon entry the phage DNA polyanion should not have to overcome the membrane potential.

Phage T5 possesses a noncontractile tail and a double-stranded DNA of 113 kbp. The process of DNA injection is very unusual: first, 8% of the DNA is injected (first step transfer [FST] of DNA); then there is a pause of about 4 min during which proteins encoded by this fragment are synthesized. Two of these proteins, A1 and A2, are required for the transfer of the remaining 92% of the DNA or the second step transfer (SST) of DNA.[76] Infection by this phage is also accompanied by a permeabilization of the envelope to K$^+$ and Mg^{2+}, the inhibition of the $\Delta\bar{\mu}_{H^+}$-dependent transport of proline, and a stimulation of the phosphoenolpyruvate phosphotransferase-mediated uptake of α-methyl glucoside.[77]

Phages belonging to the T7 group have a double-stranded DNA with a length of 12 μm. A drastic decrease of the K$^+$ content (as judged from ^{86}Rb efflux measurements) was observed by Ponta et al.[78] upon infection by phage T7 and T3. These authors proposed that the drop in ionic strength of the infected cells is a regulatory step producing an optimal environment for the activity of the phage-induced enzymes. However, this hypothesis is not supported by the results of Kuhn and Kellenberger,[79] who observed that phage production by T3 and T7 was also correct if there was no cation leakage.

To our knowledge there has been no direct demonstration of K$^+$ leakage upon infection by the temperate phage λ. However, it was observed that 5 min after infection there was a 95% decrease of the $\Delta\bar{\mu}_{Na^+}$-dependent accumulation of proline and that the rate of transport returned to normal 20 min after infection. Interestingly, phage proteins are required for the release of transport inhibition.[80]

C. THE PHAGE-INDUCED TRANSIENT PERMEABILIZATION OF THE CELL ENVELOPE IS PROBABLY DUE TO THE FORMATION OF ION CHANNELS

We have analyzed the kinetics of the K$^+$ efflux caused by T4 phages and ghosts and by phage T5.[16,17,81] Phage T5 caused a two-step K$^+$ efflux, the timing of which correlated to the timing of DNA penetration. T4 caused a one-step efflux the rate of which was six times higher for ghosts than for phage. A single phage and ghost induced the release of 10^6 to 10^7 ions per second and per bacteria, respectively, which strongly indicates that the efflux is channel mediated (see Section I.E.1). The initial rates of efflux increased linearly with the number of phages or ghosts added per cell, which suggests that one phage or ghost induces the formation of one channel.

The similarity of their characteristics (same K_m of efflux, similar dependence on $\Delta\Psi$ and temperature) suggests that the channels formed by T4 phage and ghost originate from the same protein. However, the phage channel was only transiently open, but the ghost channel remained open and efflux continued until the complete depletion of K^+. This suggests that the phage DNA is not required for the opening of the channels, but is required for their closing. The existence of a closing device associated with the phage DNA is also suggested in the case of phage T5, the channel of which closes both after the first and second steps of DNA transfer.

The channels formed by T4 phages and ghosts are poorly selective for cations. Li^+ and Na^+ were interchangeable in the external medium without affecting the behavior of the channels. H^+ and Na^+ competed for entry into the channel. The stoichiometry of H^+ influx and K^+ efflux decreased from 1 to 0.1 when the concentration of Na^+ increased from 20 to 100 mM. Furthermore, for a given concentration of Na^+ in the external medium (100 mM), the stoichiometry between H^+ influx and K^+ efflux varied from 0.1 to pH 6.9 to 0.5 at pH 7.8. These results suggest that protons alone cannot ensure the electrical compensation of the K^+ efflux. Simultaneously with these ion fluxes, the cytoplasmic membrane is depolarized.[82] This depolarization and the ion fluxes ceased simultaneously and the membrane became partially repolarized when the DNA was injected. The depolarization caused by ghosts was irreversible; this observation is compatible with the K^+ efflux data. They both explain the killing activity of the ghosts.[67]

The conformation of the phage T4 channel depends on $\Delta\Psi$: a decrease of $\Delta\Psi$ below 85 mV by addition of the protonophore TCS strongly decreased the K^+ efflux and even prevented it when the cells were totally depolarized. On the other hand, the conductance of K^+ through the phage T5 channel was not affected by changes in $\Delta\Psi$. These results suggest that the two types of channels have different characteristics and presumably, therefore, a different origin.

D. IS THE FORMATION OF THESE CHANNELS RELATED TO DNA PENETRATION?

There is no direct means of following the transfer of DNA through the envelope. Indirect approaches have been used, an example being that described by Labedan and Goldberg,[83] who determined the kinetics of penetration of a T4 DNA mutant by measuring the extent of the degradation of its DNA by a cytoplasmic exonuclease. Thus, only indirect correlations can be made between DNA penetration and channel formation. These correlations are summarized in Table 1. From these correlations it is tempting to propose

TABLE 1
Relationship Between Phage-Induced K⁺ Efflux and DNA Penetration

Phage	K⁺ efflux	Phage DNA penetration
T4	Efflux in one step lasting about 1 min at 37°C	Injection in one step and in less than 2 min
	Break in the Arrhenius plot at 29°C	Break in the Arrhenius plot of DNA degradation at 29°C
	Closing of the channel when $\Delta\Psi$ is below 80 mV	No DNA penetration when $\Delta\Psi$ is below 80 mV
	Superinfecting phages do not induce K⁺ efflux	DNA of superinfecting phage is not injected
T5	Efflux in one step in the presence of chloramphenicol	Only the FST DNA is injected in the presence of chloramphenicol
	Efflux in two steps	Injection in two steps
	Similar timing of the two fluxes and two steps of DNA injection	
	FST efflux even when $\Delta\Psi = 0$	FST-DNA injected when $\Delta\Psi = 0$

Note: FST = first step transfer.

that the main function of these phage-induced channels is to mediate the transfer of the double-stranded phage DNA.

E. ARE THESE CHANNELS CONSTITUTED BY PHAGE OR HOST PROTEINS?

The differences in the characteristics of the T4 and T5 channels make it unlikely that the same host protein catalyzes the transfer of both DNA. The involvement of phage proteins in the penetration of phage T5 DNA was suggested by recent fractionation experiments.[62] To identify putative phage protein(s) involved in the formation of these channels, *E. coli* cells were infected with radioactively labeled phage; after injection of the FST DNA fragment, the phage capsids were removed, and the cells were broken in a French pressure cell and fractionated according to the protocol described by Ishidate et al.[84] This method not only permits the separation of the inner membrane from the outer membrane, but also allows the isolation of contact sites between the two membranes. After a flotation gradient, proteins belonging to the phage tail were recovered in both the fractions containing the contact sites and the outer membrane. The electrophoretic banding pattern of the phage proteins indicated that the contact sites were enriched in the protein pb2. Moreover, infected cells were significantly enriched in contact sites as compared to intact cells. When the cells were infected at 4°C (i.e., under conditions where the phage does not inject its DNA), there was no enrichment of contact sites and very little radioactivity was found in this fraction and in the outer membrane. These results suggest that both contact sites and pb2 are involved in the translocation of phage T5 DNA. pb2 is the only protein of

T5 straight fiber; it is a multimer the length of which would be sufficient to span the whole envelope and, thus, to gain access to the inner membrane. This model is supported by the recent results of Feucht et al.[85] They showed that isolated pb2 could form pores when reconstituted into lipid bilayers. From the conductance data they calculated that the diameter of the pore would be 2 nm; this would be large enough to allow the passage of the double-stranded phage DNA.

F. IS CHANNELING A GENERAL MECHANISM FOR PHAGE DNA PENETRATION?

It is probable that at least the tailed phages containing a double-stranded DNA share a common mode of DNA penetration. The transient permeabilization of the cell envelope following infection by these phages probably results from the formation of channels.

Although much effort has been devoted to the study of the small single-stranded DNA phages, the mechanism by which their DNA traverses the cell envelope is unknown. The closely related F-specific filamentous (Ff) phages fd, f1, and M13 infect *E. coli* strains which harbor the conjugative F plasmid. Infection is initiated by the binding of one end of the phage to the tip of the F pilus. This binding is mediated by the N-terminus of the gene 3-encoded viral protein (g3p). The gene 3-encoded protein has been purified. The oligomeric form of g3p forms large pores when reconstituted into lipid bilayers; the estimated diameter of the pores suggests that they are large enough to allow the passage of the single-stranded DNA. Although there is only indirect evidence that the oligomers exist *in vivo,* this may be an encouraging result.[86]

IV. FEATURES COMMON TO BACTERIOPHAGE AND BACTERIOCIN CHANNELS

The idea that phages and colicins share a common evolutionary origin is not new.[87] It has been proposed that bacteriocins would behave as defective phages.[88] In several bacteria, at least one of the virulence factors (diphtheria, scarlet fever, etc.) is an exotoxin whose structural gene has been shown to be carried by a bacteriophage.

If the hypothesis that the bacteriophage channels are formed by phage proteins is correct, then the same questions may be asked for colicin and bacteriophage channels: how are these soluble proteins translocated through the envelope, and how do they integrate into the hydrophobic inner membrane? The fact that some phages and colicins share common translocation pathways[10] and that the putative T5 channel protein and colicin A are found in the contact sites between the two membranes suggests that they may use similar strategies of translocation and insertion. When reaching their final target they both cause permeability changes. Although the characteristics of the channels are different, the efflux of K^+ caused by phage and colicins are kinetically and quantitatively correlated with a decrease of internal ATP which occurred independently of the functioning of the F_1F_0-ATPase.[53] It is thus likely that both phages and colicins cause similar damage to the cell envelope.

Luria, reviewing the "status of the colicin problem" in 1973[18] wrote about the analogies between phages and colicins and concluded that "after all, it would be surprising if several agents that damage the membrane without disrupting it did not have some effect in common".

REFERENCES

1. **Jacob, F., Siminovitch, L., and Wollman, L.,** Sur la biosynthèse d'une colicine et sur son mode d'action, *Ann. Inst. Pasteur* Paris, 83, 295, 1952.
2. **Konisky, J.,** Colicins and other bacteriocins with established modes of action, *Annu. Rev. Microbiol.,* 36, 125, 1982.
3. **Harkness, R. E. and Braun, V.,** Colicin M inhibits peptidoglycan biosynthesis by interferring with lipid carrier recycling, *J. Biol. Chem.,* 264, 6177, 1989.
4. **Konisky, J.,** The bacteriocins, in *The Bacteria,* Vol. 6, Ornston, L. and Sokatch, J. R., Eds., Academic Press, New York, 1978, 71.
5. **Lazdunski, C., Baty, D., Geli, V., Cavard, D., Morlon, J., Lloubès, R., Howard, S. P., Knibiehler, M., Chartier, M., Varenne, S., Frenette, M., Dasseux, J. L., and Pattus, F.,** The membrane channel-forming colicin A: synthesis, secretion, structure, action and immunity, *Biochim. Biophys. Acta,* 947, 445, 1988.
6. **Yamada, M., Ebina, Y., Miyata, T., Nakazawa, T., and Nakazawa, A.,** Nucleotide sequence of the structural gene for colicin E1 and predicted structure of the protein, *Proc. Natl. Acad. Sci. U.S.A.,* 79, 2827, 1982.
7. **Morlon, J., Lloubès, R., Varenne, S., Chartier, M., and Lazdunski, C.,** Complete nucleotide sequence of the structural gene for colicin A, a gene translated in a non-uniform rate, *J. Mol. Biol.,* 170, 271, 1983.
8. **Pattus, F., Massotte, D., Wilmsen, H. U., Lakey, J., Tsernoglou, D., Tucker, A., and Parker, M. W.,** Colicins: prokaryotic killer-pores, *Experientia,* 46, 180, 1990.
9. **Parker, M. W., Pattus, F., Tucker, A. D., and Tsernoglou, D.,** Structure of the membrane-pore-forming fragment of colicin A, *Nature (London),* 337, 93, 1989.
10. **Nikaido, H. and Vaara, M.,** Outer membrane, in *Escherichia coli and Salmonella typhimurium,* Vol. 1, Neidhardt, F. C., Ed., American Society for Microbiology, Washington, D.C., 1987, 7.
11. **Webster, R. E.,** The *tol* gene products and the import of macromolecules into *E. coli, Mol. Microbiol.,* 5, 1005, 1991.
12. **Cramer, W. A., Dankert, J. R., and Uratani, Y.,** The membrane channel-forming bacteriocidal protein, colicin E1, *Biochim. Biophys. Acta,* 737, 173, 1983.
13. **Goldberg, E. B.,** in *Virus Receptors,* Randall, L. and Philipson, L., Eds., Chapman and Hall, London, 1980, 115.
14. **Matthews, C. K., Kutter, E. M., Mosig, G., and Berget, P. B., Eds.,** *Bacteriophage T4,* American Society for Microbiology, Washington, D.C., 1983.
15. **Silver, S., Levine, E., and Spielman, P. M.,** Cation fluxes and permeability changes accompanying bacteriophage infection of *E. coli, J. Virol.,* 2, 763, 1968.
16. **Letellier, L. and Boulanger, P.,** Involvement of ion channels in the transport of phage DNA through the cytoplasmic membrane of *E. coli, Biochimie,* 71, 167, 1989.

17. **Boulanger, P. and Letellier, L.,** Characterization of ion channels involved in the penetration of phage T4 DNA into *E. coli* cells, *J. Biol. Chem.,* 263, 9767, 1988.

18. **Luria, S.,** Colicins, in *Bacterial Membranes and Walls,* Leive, L., Ed., Vol. 1, Marcel Dekker, New York, 1973, 293.

19. **Konisky, J.,** Specific transport systems and receptors for colicins and phages, in *Bacterial Outer Membranes, Biogenesis and Functions,* Inouye, M., Ed., John Wiley & Sons, New York, 1979, 320.

20. **Calendar, R., Ed.,** *The Bacteriophages,* Vol. 1 and 2, Plenum Press, New York, 1988.

21. **Konisky, J., Gilchrist, J., Nieva Gomez, D., and Gennis, R.,** Effects of colicin Ia on membrane functions of *E. coli,* in *Molecular Aspects of Membrane Phenomena,* Kaback, H. R. et al., Eds., Springer-Verlag, New York, 1975, 193.

22. **Nomura, M. and Maeda, A.,** Mechanisms of action of colicins, *Zentralbl. Backteriol. Parasitenkd. Infektionskr. Hyg. Abt. 1 Orig.,* 196, 216, 1965.

23. **Dandeu, J. P., Billault, A., and Barbu, E.,** Action des colicines sur le vitesse de sortie du potassium intracellulaire, *C. R. Acad. Sci.,* 269, 2044, 1969.

24. **Hirata, H., Fukui, S., and Ishikawa, S.,** Initial events caused by colicin K infection: cation movement and depletion of ATP pool, *J. Biochem.,* 65, 843, 1969.

25. **Wendt, L.,** Mechanism of colicin action: early events, *J. Bacteriol.,* 104, 1236, 1970.

26. **Fields, K. L. and Luria, S.,** Effects of colicins E1 and K on transport systems, *J. Bacteriol.,* 97, 57, 1969.

27. **Fields, K. L. and Luria, S.,** Effects of colicins E1 and K on cellular metabolism, *J. Bacteriol.,* 97, 64, 1969.

28. **Mitchell, P.,** Chemiosmotic coupling in oxidative and photosynthetic phosphorylation, *Biol. Rev. Cambridge Philos. Soc.,* 41, 445, 1966.

29. **Feingold, D.,** The mechanism of colicin E1 action, *J. Membr. Biol.,* 3, 372, 1970.

30. **Mitchell, P. and Moyle, J.,** Acid-base titration across the membrane system of rat-liver mitochondria: catalysis by uncouplers, *Biochem. J.,* 104, 588, 1967.

31. **Lusk, J. and Nelson, D.,** Effects of colicins E1 and K on permeability to magnesium and cobaltous ions, *J. Bacteriol.,* 112, 148, 1972.

32. **Gould, J. and Cramer, W.,** Studies on the depolarization of the *E. coli* cell membrane by colicin E1, *J. Biol. Chem.,* 252, 5491, 1977.

33. **Scholes, P. and Mitchell, P.,** Respiration drive proton translocation in *Micrococcus denitrificans, J. Bioenerg.,* 1, 309, 1970.

34. **Weiss, M. and Luria, S.,** Reduction of membrane potential, an immediate effect of colicin K, *Proc. Natl. Acad. Sci. U.S.A.,* 75, 2483, 1978.

35. **Tokuda, H. and Konisky, J.,** Mode of action of colicin Ia: effect of colicin on the *E. coli* proton electrochemical gradient, *Proc. Natl. Acad. Sci. U.S.A.,* 75, 2579, 1978.

36. **Schein, S., Kagan, B., and Finkelstein, A.,** Colicin K acts by forming voltage-dependent channels in phospholipid bilayer membranes, *Nature (London),* 276, 159, 1978.

37. **Slatin, S.,** Channels formed by colicin E1 in planar lipid bilayers are monomers, *Biophys. J.,* 53, 155a, 1988.

38. **Bullock, J., Cohen, F., Dankert, J., and Cramer, W.,** Comparison of the macroscopic and single channel conductance properties of colicin E1 and its COOH-terminal tryptic peptide, *J. Biol. Chem.,* 258, 9908, 1983.

39. **Dankert, J., Uratani, Y., Grabau, C., Cramer, W., and Hermodson, M.,** On a domain structure of colicin E1, *J. Biol. Chem.,* 257, 3857, 1982.

40. **Ohno-Iwashita, Y. and Imahori, K.,** Assignment of the functional loci in the colicin E1 molecule by characterization of its proteolytic fragment, *J. Biol. Chem.,* 257, 6446, 1982.

41. **Pattus, F., Cavard, D., Crozel, V., Baty, D., Adrian, M., and Lazdunski, C.,** pH-dependent membrane fusion is promoted by various colicins, *EMBO J.,* 4, 2469, 1985.

42. **Massotte, D. and Pattus, F.,** Colicin N and its thermolytic fragment induce phospholipid vesicle fusion, *FEBS Lett.,* 257, 447, 1989.

43. **Tokuda, H. and Konisky, J.**, Effect of colicins Ia and E1 on ion permeability of liposomes, *Proc. Natl. Acad. Sci. U.S.A.*, 76, 6167, 1979.

44. **Davidson, V., Cramer, W., Bishop, L., and Brunden, K.**, Dependence of the activity of colicin E1 in artificial membrane vesicles on pH, membrane potential and vesicle size, *J. Biol. Chem.*, 259, 594, 1984.

45. **Cramer, W., Cohen, F. S., Merill, A., and Song, H.**, Structure and dynamics of the colicin E1 channel, *Mol. Microbiol.*, 4, 519, 1990.

46. **Uratani, Y. and Cramer, W. A.**, Reconstitution of colicin E1 into dimyristoyl-phosphatidylcholine vesicles, *J. Biol. Chem.*, 256, 4017, 1981.

47. **Raymond, L., Slatin, S., and Finkelstein, A.**, Channels formed by colicin E1 in planar lipid bilayers are large and exhibit pH-dependent ion selectivity, *J. Membr. Biol.*, 84, 173, 1985.

48. **Tokuda, H. and Konisky, J.**, *In vitro* depolarization of *E. coli* membrane vesicles by colicin Ia, *J. Biol. Chem.*, 253, 7731, 1978.

49. **Weaver, C., Kagan, B., Finkelstein, A., and Konisky, J.**, Mode of action of colicin Ib: formation of ion-permeable membrane channels, *Biochim. Biophys. Acta*, 645, 137, 1981.

50. **Bourdineaud, J. P., Boulanger, P., Lazdunski, C., and Letellier, L.**, *In vivo* properties of colicin A: channel activity is voltage dependent but translocation may be voltage independent, *Proc. Natl. Acad. Sci. U.S.A.*, 87, 1037, 1990.

51. **Booth, I.**, K$^+$ efflux systems, in *Alkali Cation Transport Systems in Prokaryotes*, Bakker, E. P., Ed., CRC Press, Boca Raton, FL, 1992, chapter IIE.

52. **Kinoshita, N., Unemoto, T., and Kobayashi, H.**, Protonmotive force is not obligatory for growth of *E. coli*, *J. Bacteriol.*, 160, 1074, 1984.

53. **Guihard, G., Benedetti, H., and Letellier, L.**, submitted.

54. **Okamoto, K.**, Requirement of heat and metabolic energy for the expression of inhibitory action of colicin K, *Biochim. Biophys. Acta*, 389, 370, 1975.

55. **Jetten, A. M. and Jetten, M.**, Energy requirement for the initiation of colicin action in *E. coli*, *Biochim. Biophys. Acta*, 387, 12, 1975.

56. **Bénédetti, H., Lloubès, R., Lazdunski, C., and Letellier, L.**, Colicin A unfolds during its translocation and spans the whole cell envelope when the pore is formed, *EMBO J.*, 11, 441, 1992.

57. **Neupert, W., Hartl, F.-U., Craig, E., and Pfanner, N.**, How do polypeptides cross the mitochondrial membrane?, *Cell*, 63, 447, 1990.

58. **Plate, C. and Luria, S.**, Stage in colicin K action, as revealed by the action of trypsin, *Proc. Natl. Acad. Sci. U.S.A.*, 69, 2030, 1972.

59. **Dankert, J., Hammond, S., and Cramer, W.**, Reversal by trypsin of the inhibition of active transport by colicin E1, *J. Bacteriol.*, 143, 594, 1980.

60. **Letellier, L., Lazdunski, K., Bénédetti, H., Bourdineaud, J. P., and Boulanger, P.**, *In vivo* properties of colicin A: channel activity and translocation, NATO-ASI ser., Springer-Verlag, in press.

61. **Bénédetti, H., Boulanger, P., Guihard, G., Lazdunski, C., and Letellier, L.**, manuscript in preparation.

62. **Guihard, G., Boulanger, P., and Letellier, L.**, Involvement of phage T5 tail proteins and contact sites between the inner and outer membrane of *E. coli* in phage T5 DNA injection, *J. Biol. Chem.*, 267, 3173, 1992.

63. **Schleyer, M. and Neupert, W.**, Transport of proteins into mitochondria: translocational intermediates spanning contact sites between outer and inner membranes, *Cell*, 43, 339, 1985.

64. **Bénédetti, H., Lazdunski, C., and Lloubès, R.**, Colicins A and E1 interact with TolA protein, a component of their translocation system, *EMBO J.*, 8, p. 1982, 1991.

65. **Puck, T. and Lee, H.**, Mechanism of cell wall penetration by viruses. Demonstration of cyclic permeability change accompanying virus infection of *E. coli* B, *J. Exp. Med.*, 101, 151, 1955.

66. **Silver, S., Levine, E., and Spielman, P.**, Cation fluxes and permeability changes accompanying bacteriophage infection of *E. coli*, *J. Virol.*, 2, 763, 1968.

67. **Duckworth, D. and Winkler, H.**, Metabolism of T4 bacteriophage ghost-infected cells, *J. Virol.*, 9, 917, 1972.

68. **Shapira, A., Giberman, E., and Kohn, A.**, Recoverable K^+ fluxes variations following adsorption of T4 phage and their ghosts on *E. coli* B, *J. Gen. Virol.*, 23, 159, 1974.

69. **Wagner, E., Ponta, H., and Schweiger, M.**, Development of *E. coli* virus T1: the role of protonmotive force, *J. Biol. Chem.*, 255, 534, 1980.

70. **Wagner, E., Ponta, H., and Schweiger, M.**, Development of *E. coli* virus T1: ATP-mediated discrimination of gene expression, *J. Biol. Chem.*, 255, 540, 1980.

71. **Kalasauskaite, E. and Grinius, L.**, The role of energy yielding ATPase and respiratory chain at early stages of bacteriophage T4 infection, *FEBS Lett.*, 99, 287, 1979.

72. **Filali Maltouf, A. and Labedan, B.**, Host cell metabolic energy is not required for injection of bacteriophage T5 DNA, *J. Bacteriol.*, 153, 124, 1983.

73. **Filali Maltouf, K. and Labedan, B.**, The energetics of the injection process of phage λ DNA and the role of the ptsM/pel-encoded protein, *Biochem. Biophys. Res. Commun.*, 130, 1093, 1985.

74. **Keweloh, H. and Bakker, E.**, Permeability changes in the cytoplasmic membrane of *E. coli* K12 early after infection with bacteriophage T1, *J. Bacteriol.*, 160, 347, 1984.

75. **Keweloh, H. and Bakker, E.**, Increased permeability and subsequent resealing of the host cell membrane early after infection with bacteriophage T1, *J. Bacteriol.*, 160, 354, 1984.

76. **Lanni, Y.**, First step transfer DNA of bacteriophage T5, *Bacteriol. Rev.*, 32, 227, 1968.

77. **Glenn, J. and Duckworth, D.**, Ion fluxes during T5 bacteriophage infection of *E. coli*, *Arch. Biochem. Biophys.*, 201, 576, 1980.

78. **Ponta, H., Altendorf, K. H., Schweiger, M., Hirsch Kaufmann, M., Pfennig Yeh, M., and Herrlich, P.**, *E. coli* membranes become permeable to ions following T7 virus infection, *Mol. Gen. Genet.*, 149, 145, 1976.

79. **Kuhn, A. and Kellenberger, E.**, Productive phage infection in *E. coli* with reduced internal levels of the major cations, *J. Bacteriol.*, 163, 906, 1985.

80. **Wilson, H. and Okabe, A.**, Second function of the S gene of bacteriophage λ, *J. Bacteriol.*, 152, 1091, 1982.

81. **Boulanger, P. and Letellier, L.**, Ion channels are likely to be involved in the two steps of phage T5 penetration into *E. coli* cells, *J. Biol. Chem.*, 267, 3168, 1992.

82. **Labedan, B. and Letellier, L.**, Membrane potential changes during the first steps of coliphage infection, *Proc. Natl. Acad. Sci. U.S.A.*, 78, 215, 1981.

83. **Labedan, B. and Goldberg, E.**, Requirement for membrane potential in injection of phage T4 DNA, *Proc. Natl. Acad. Sci. U.S.A.*, 76, 4669, 1979.

84. **Ishidate, K., Creeger, E. S., Zrike, J., Deb, S., Glauner, B., Mc Allister, T. J., and Rothfield, L. I.**, Isolation of differenciated membrane domains from *E. coli* and *S. typhimurium*, including a fraction containing attachment sites between the inner and outer membranes and the murein skeleton of the cell envelope, *J. Biol. Chem.*, 261, 428, 1986.

85. **Feucht, A., Schmid, A., Benz, R., Schwarz, H., and Heller, K. J.**, Pore formation associated with the tail-tip protein pb2 of phage T5, *J. Biol. Chem.*, 265, 18561, 1990.

86. **Wuttke, G., Keppner, J., and Rashed, I.**, Pore forming properties of the absorption protein of filamentous phage fd, *Biochim. Biophys. Acta*, 985, 239, 1989.

87. **Bradley, D.**, Ultrastructure of bacteriophages and bacteriocins, *Bacteriol. Rev.*, 31, 230, 1967.

88. **Reeves, P.**, The bacteriocins, in *Molecular Biology: Biochemistry and Biophysics*, Part II, Springer-Verlag, Berlin, 1972, 17 and 83.

Section III: NH$_4^+$ Transport Systems

Chapter IIIA

NH$_4^+$ TRANSPORT SYSTEMS

Diethelm Kleiner

TABLE OF CONTENTS

0-8493-6982-7/93/$0.00 + $.50

I. INTRODUCTION

The microbial community is famous for its high capacity to degrade a vast variety of nitrogenous substrates, ranging from macromolecules (e.g., proteins, nucleic acids, murein) to simple inorganic compounds like N_2. Some of these conversions are carried out for energy gain (e.g., ammonia and nitrite oxidation, dentifrication, respiratory oxidation or fermentation of organic molecules), most of them for assimilation of nitrogen and — to a lesser extent — carbon or phosphorus. Generally, microorganisms encounter a variety of nitrogenous compounds in the environment. Among them ammonia* occupies a central position, being ubiquitous and being the preferred nitrogen source of many, if not most, microbes. This preference is reflected by the phenomenon of "catabolite repression": the presence of ammonia suppresses the assimilation of other nitrogenous substrates, notably in the well-studied enterobacteria.[1,2] Another outstanding characteristic of ammonia is its position at the junction of microbial nitrogen catabolism and anabolism: almost all nitrogen sources are intracellularly degraded to ammonia and/or glutamate, which are the starting points for the synthesis of amino acids, nucleic acids, and cofactors. Both positions — as preferred nitrogen source or as junction in nitrogen metabolism — are associated with the phenomenon of its transport across the cytoplasmic membrane, as will be outlined below.

II. GENERAL REMARKS ON TRANSPORT

Transport of a molecule across a hydrophobic barrier like the cytoplasmic membrane occurs either by nonspecific diffusion through the lipid bilayer or by specific permeation (only specific transport proteins serve as gates). The rate of nonspecific diffusion at a given moment obeys Fick's first law:[3]

$$dn/dt = -P \times A \times (c_1 - c_2) \tag{1}$$

where dn/dt is the number (mole) of substrate molecules passing through a membrane with area A per unit time, c_1 and c_2 are the concentrations of the diffusing species on both sides of the membrane, and P is the permeability coefficient. P is a complex term which depends on both membrane specific parameters (thickness, composition) and characteristics of the diffusing molecule (molecular weight, solubility in aqueous phase and membrane) and which generally is not accessible by theoretical calculations.

For our purposes it is sufficient to consider the bacterial cytoplasmic membrane as rather similar among the prokaryotes. Then differences in P mainly depend on the molecular parameters of the permeant compound. The

---

* The terms "ammonia" and "methylamine" are used to cover both the protonated and the unprotonated forms; definite protonated states will be indicated by the chemical symbols (e.g., NH_4^+).

overriding parameter is the charge: with few exceptions, ions are very soluble in the aqueous phase and rather insoluble in the lipid phase. Accordingly, ions (like NH$_4$$^+$) are expected to diffuse very slowly across lipid membranes. In contrast, uncharged molecules (like NH$_3$) are supposed to pass through membranes much faster. Less important characteristics influencing P are polarity and molecular weight.

In the ammonia system, rapid protonation and deprotonation according to the equation

$$NH_3 + H^+ \leftrightarrows NH_4^+; pK_a = 9.25 \tag{2}$$

always results in coexistence of NH$_3$ and NH$_4$$^+$ in a given system. Thus, we must consider both species when dealing with transport, even when NH$_4$$^+$ comprises more than 99% of the total sum, as is the case at pH values below 7.25.

As discussed above, a relatively rapid nonspecific diffusion is expected for NH$_3$, while the ion NH$_4$$^+$ should require a specific transport system to ensure biologically significant permeation. In contrast to nonspecific diffusion, which at most leads to equilibration of the concentrations on both sides of the membrane, specific transport may result in accumulation if transport is coupled to an energy-yielding reaction. The energy required (ΔG) depends on the size of the concentration gradient c_1/c_2:

$$\Delta G = RT \times \ln(c_1/c_2) \tag{3}$$

where T is the temperature and R is the gas constant. Both diffusion of NH$_3$ and active (energy-dependent) transport of NH$_4$$^+$ will be dealt with below.

III. DIFFUSION OF NH$_3$

The first evidence for rapid diffusion of NH$_3$ across some biomembranes, especially of plants, was obtained about 100 years ago.[4] Since then, fast diffusion of NH$_3$ across any lipid membrane has been widely taken for granted.[4-10] Quantitative determinations of P(NH$_3$), however, were lacking until the mid-1970s, when the permeability coefficient P(NH$_3$) = 9 $\times$ 10^{-6} m/s was determined for the alga *Nitella clavata*.[11] About 10 years later P(NH$_3$) values were measured for the membranes of the bacteria *Synechococcus* R-2 (P = 6.4 $\times$ 10^{-6} m/s),[12] *Rhodobacter sphaeroides* (P = 2.6 $\times$ 10^{-6} m/s),[13] and *Klebsiella pneumoniae* (P = 18 $\times$ 10^{-6} m/s).[14] The P(NH$_3$) values for plant[11,15] and bacterial membranes center around a value of 7 $\times$ 10^{-6} m/s. In view of the wide variations obtained for permeation of various molecules through artificial bilayer lipid membranes[16] (e.g., between 5 $\times$ 10^{-6} and 120 $\times$ 10^{-6} m/s for H$_2$O), the differences in bacterial P(NH$_3$) values are acceptable. It is not surprising that P(H$_2$O) and P(NH$_3$) do not differ much since the properties of the two molecules are also very similar (molecular weights of 18 and 17 and electric dipole moments of 1.87 and 1.47,[17] respectively). These values are about 100 to 1000 times higher than the permeability coefficients for hydrophilic organic compounds through bilayer lipid membranes.[16]

In conclusion, ample experimental evidence supports the notion that NH_3 is permeating fast (on same order as H_2O) through plant and prokaryotic cytoplasmic membranes. This implies that its diffusion across these membranes can be considered negligible only when the concentration gradients are very small or when superimposing active transport processes are much faster.

IV. ACTIVE TRANSPORT OF NH_4^+

A. DISTRIBUTION OF CARRIERS

The notion of fast, nonspecific NH_3 diffusion has suppressed the concept of active ammonia transport for some time. Therefore, the first report on active ammonia uptake by *Penicillium chrysogenum*[18] went largely unnoticed. The last decade, however, has witnessed the discovery of about 50 bacterial ammonia carriers (Table 1). Apart from the organisms listed in Table 1, ammonia transport was reported for the following strains (see also previous reviews[58,59,59a]): *Nitrosococcus oceanus,*[60] *Azotobacter paspali,*[61] *Azomonas macrocytogenes,*[61] *Azomonas agilis,*[61] *Beijerinckia mobilis,*[61] *Derxia gummosa,*[61] *Alcaligenes latus,*[61] *Anabaena azollae,*[62] *Anabaena variabilis,*[62] *Anabaena doliolum,*[63,65] *Anabaena cylindrica,*[64] and *Nostoc* CAN.[64] Table 1 and the preceding list show that ammonia carriers exist in many groups of eubacteria. To the best of my knowledge, so far no ammonia transport system has been discovered in an archaebacterium.

B. CRITICAL ASSESSMENT OF EXPERIMENTAL TECHNIQUES

Since the following considerations have been discussed extensively in a previous review,[58] only the main arguments will be repeated here.

1. Determination of NH_4^+ Gradients

The first indications for active ammonia transport came from determinations of large ammonia gradients across many bacterial membranes (Table 1), the formation of which could not be attributed to a trapping mechanism (i.e., diffusion of NH_3 into the cell and subsequent protonation due to a low internal pH):[58,59] in neutral environments bacteria generally maintain a ΔpH (internally alkaline) which would promote NH_3 efflux. Therefore, these observations strongly indicate active transport of either NH_3 or NH_4^+, with the energy requirement of Equation 3.

A critical point is the determination of the intracellular NH_4^+ pool. For reliable measurements, fast extraction methods must be employed to avoid changes in pool size due to rapid NH_4^+ turnover or decomposition of labile compounds. Errors should be largely minimized by rapid centrifugation of the cells through silicone oil layers into an extraction buffer[66] followed by microdiffusion (expulsion of gaseous NH$_3$ into a drop of acid after alkalinization of the sample) and NH$_3$ determination. This procedure does not result in decomposition of glutamine or carbamyl phosphate.[21] With similar methods about 100-fold concentration gradients were found in *Anacystic nidulans (Synechococcus) R-2,*[12] *Rhodobacter sphaeroides,*[13] *Azotobacter chroococcum,*[25] and *Klebsiella pneumoniae.*[14] Conflicting reports on intracellular NH$_4$$^+$ pool sizes may be due to differing growth conditions or different extraction methods. For *Anabaena cylindrica,* for example, pools between 0[67,68] and 1.8 mM[69] have been reported.

2. NH$_4$$^+$ Uptake Measurements

Frequently, NH$_4$$^+$ depletion of the medium has been employed to study NH$_4$$^+$ transport. Since the kinetics observed also can be explained by nonspecific NH$_3$ diffusion drawn by fast assimilation, such experiments must be complemented by demonstration of uphill gradients. Differences of K$_m$, pH profile, temperature profile, etc. between uptake and assimilating enzyme (usually by glutamine synthetase) can be used as supporting, but not conclusive, evidence for carrier-mediated uptake.

3. Transport of [^{14}C]-Methylammonium

The most versatile technique used to study transport involves intracellular accumulation of a radioactive substrate. Since no suitable nitrogen isotope is generally available, [^{14}C]-methylamine was introduced as an ammonium analogue.[18] The method is now widely accepted (see Table 1) and can be regarded as an indicator of ammonia transport if some of the following points are considered:[58,59a]

1. If the organism can use methylamine as a carbon or nitrogen source, specific methylamine carriers may be expressed which do not transport ammonia.
2. Competitive inhibition of [^{14}C]-methylamine transport by ammonia proves that there is binding to the same carrier site (for this simple first-order reaction). If the affinity of this site is higher for NH$_4$$^+$ than for methylamine (i.e., if the K$_i$ for NH$_4$$^+$ is smaller than the K$_m$ for methylamine; see Table 1), NH$_4$$^+$ can be inferred to be the natural substrate (especially if the organisms cannot assimilate methylamine [see point 1]).
3. Growing evidence indicates the existence of NH$_4$$^+$ carriers which do not accept methylamine as substrate.[42,50,54,59a] In some cases, methylamine-independent NH$_4$$^+$ transport may be catalyzed by K$^+$ carriers[70] (see Chapter IIIC). In contrast, the ammonia/methylamine carriers are only slightly inhibited by K$^+$ ions[30,32] (methylamine uptake [Table 1] was mainly assayed in K$^+$-rich buffers).

TABLE 1
Distribution and Some Properties of Prokaryotic NH_4^+ Carriers

Species	NH_4^+ gradient (fold)	K_m for NH_4^+ (μM)	K_m for Methylamine (μM)	Repression by NH_4^+	Ref.
Gram-Positive Eubacteria					
Bacillus acidocaldarius	—	8[a]	500	Yes	18a
Clostridium aceticum	—	—	—	Yes	19
Clostridium formicoaceticum	—	—	—	Yes	20
Clostridium pasteurianum	60	9[a]	150	—	21
Frankia CpI1	—	<2[a]	2	Yes	22
Gram-Negative Eubacteria					
Alcaligenes eutrophus	—	<0.1[a]	35–111	Yes	23
Azospirillum brasilense	—	5–10[a]	50–200	Yes	24
Azospirillum lipoferum	—	—	—	Yes	24
Azotobacter chroococcum	110	10[a]	140	Yes	25
Azotobacter vinelandii	120	1;3[a]	25;61	Yes	26–28
Bradyrhizobium japonicum	—	—	2	Yes	29
Bradyrhizobium 32H1 (cowpea)	—	—	—	Yes	36
Escherichia coli	—	0.5	36	Yes	30
Klebsiella pneumoniae	40–100	7[a]	140	Yes	31, 32
Paracoccus denitrificans	—	<50[a]	70	Yes	33
Pseudomonas aeruginosa	—	—	40–100	Yes	34
Rhizobium ANU289	—	0.4[a]	6.6	Yes	35
Rhizobium MNF 2030 (cowpea)	—	—	—	Yes	37
Rhizobium leguminosarum	60	7;15	117	Yes	38,39

					Ref.
Rhizobium meliloti	—	1.2[a]	2	Yes	40
Rhizobium trifolii	150	2	—	Partial	37
Rhodobacter capsulatus	—	1[a],4[a]	22;50	Yes	41, 42
	—	1.7;11.1	—	—	43
	—	2000	—	—	44
Rhodobacter palustris	—	18	130	Yes	41
Rhodobacter sphaeroides	—	16[a];4	113;16	Yes	41, 45
	100–500	<1	—	—	13, 54
Rhodomicrobium vanielii	—	10	65	Yes	41
Rhodospirillum rubrum	—	7[a]	110	Yes	41
Rhodospirillum tenue	—	—	—	Yes	41
Xanthobacter autotrophicus	—	—	230	Yes	46
Cyanobacteria					
Anabaena 7120	—	8;80 2.5;70	—	Yes	47
Anabaena flos-aquae	—	—	7.1	—	48
Anacystis nidulans (Synechococcus) R-2	150;3000	<1[a]	7.2	Yes	12,49–51
Anacystis nidulans IU625	—	50;357	100;833	—	52
Anabaena cycadeae	—	—	—	Yes	53
Nostoc calcicola	16–75	10;200	—	—	55
Nostoc muscorum	10–50	11;66	—	No	56
Synechococcus elongatus	10	37	—	—	57

[a]   K$_i$ of ammonia for methylamine uptake.

4. Accumulation of [^{14}C]-methylamine may be demonstrated by quantitative analysis of intracellular pools[21,50] or by chasing out[71] the unmetabolized [^{14}C]-methylamine with inhibitors, unlabeled methylamine, or NH_4^+. Recently, however, it was shown, that — at least under certain conditions — ammonia also can expel some of the metabolized methylamine (as γ-methyl glutamine) from the cells.[34]

C. ENERGY COUPLING

Apart from gradient formation, an energy requirement for ammonia transport has been inferred from dependence on an energy source or from inhibition of uptake by inhibitors of energy metabolism.[58] It should be kept in mind, however, that in most bacteria active ammonia uptake is followed by its ATP-dependent binding to glutamate, yielding glutamine:

$$\text{glutamate} + NH_4^+ + \text{ATP} \rightarrow \text{glutamine} + \text{ADP} + P_i \qquad (4)$$

which is catalyzed by the enzyme glutamine synthetase (GS). Nonavailability of an energy source or inhibition of energy production may decrease ATP content and cause a metabolic jam, which in turn could block uptake. Thus, selective inhibition of transport should be regarded only as evidence for an energy requirement when ATP pools remain largely unaffected. For some strains, these obstacles were removed by the construction of mutants lacking GS, but still exhibiting energy-dependent methylamine accumulation without its metabolism.[72,73]

A critical evaluation of experimental evidence shows that ammonia transport by many bacteria requires energy in the form of the membrane potential $\Delta\Psi$.[58,59a] For *Escherichia coli*, however, the K^+ potential has been proposed as the driving force of an NH_4^+/K^+ antiport system[30] (see Chapter IIIB). In this respect, the NH_4^+/K^+ exchange in some methanogenic bacteria[74] and *Vibrio alginolyticus*[75] is of interest. Those authors do not propose a direct coupling, but rather passive diffusion of NH_3 and subsequent stimulation of a K^+/H^+ antiporter by internal alkalinization.[74] Under those experimental conditions, however, (i.e., high extracellular NH_4^+), any putative NH_4^+ carrier is expected to be repressed (see Section IV.D).

Both $\Delta\Psi$-driven uniport and K^+-driven antiport imply that NH_4^+ is the molecular species transported, not the uncharged NH_3.

D. REGULATION

Catabolic reaction sequences generally start with transport of the substrate into the cell. If carrier mediated, this step is a prime target for regulation on both the metabolic and the genetic level (i.e., regulation of activity and of synthesis).

1. Regulation of Activity

NH$_4^+$ transport is markedly inhibited by glutamine in *E. coli*,[76,77] *R. sphaeroides*,[78] and *Rhizobium leguminosarum*.[38] Support for this regulation in other prokaryotes is derived from inhibition by the glutamine analogue methionine sulfoximine (MSX).[58] The results are controversial because MSX is also a powerful inhibitor of GS, and inhibition of the latter presumably would cause intracellular NH$_4^+$ accumulation and decrease transport (Equation 4). On the basis of inhibition kinetics, however, several investigators were able to discriminate between the effects of MSX on GS and NH$_4^+$ carrier in *Anabaena doliolum*,[63] *Anacystic nidulans*,[52] and *Anabaena* 7120.[47] Furthermore, one transport system (the so-called "*MSX*-sensitive" one, see Section IV.F.1) in a GS$^-$ mutant of *Anabaena cycadeae* was strongly inhibited by MSX.[53,72]

Taken together, the results summarized suggest that the activity of the NH$_4^+$ carrier in many prokaryotes is regulated by the intracellular glutamine pool. This control mechanism is biologically significant since the intracellular glutamine pool responds quickly and strongly to the extracellular ammonia level.[31,79] As will be outlined in Section IV.D.2, glutamine has multiple regulatory functions in nitrogen catabolism.

2. Regulation of Transcription

As summarized in Table 1, synthesis of almost all NH$_4^+$ carriers is repressed when the organisms are cultivated with abundant NH$_4^+$ as the nitrogen source. In enterobacteria this effect is paralleled by repression of numerous operons coding for enzymes of nitrogen catabolism[1,2] (e.g., nitrogenase, assimilatory nitrate reductase, urease, enzymes for amino acid degradation and transport). This regulation, which reflects the previously mentioned preference of enterobacteria for NH$_4^+$ as nitrogen source, is mainly mediated by the "nitrogen control" (Ntr) system. Ntr is comprised of the genes *ntrA* (= *glnF*), *ntrB* (= *glnL*), and *ntrC* (= *glnG*). The autoregulated *ntrA* gene codes for a special sigma factor which recognizes specific promoter sequences and, thus, in complex with the RNA polymerase initiates transcription of a distinct class of enzymes, among them — but not exclusively — the Ntr-controlled proteins.[1,2,80] transcription of the latter ones in addition must be activated (directly or indirectly) by the phosphorylated *ntrC* protein (= NtrC-P); (Figure 1). Unphosphorylated NtrC is not effective. Phosphorylation and dephosphorylation of NtrC are catalyzed by the NtrB protein, which via a complicated signal processing cascade responds to the intracellular levels of 2-oxoglutarate (2-OG) and glutamine (Gln) in such a way that a high 2-OG/Gln ratio promotes phosphorylation, whereas a low 2-OG/Gln ratio promotes dephosphorylation of NtrC[1,2] (Figure 1). As a consequence, the Ntr system coordinates carbon and nitrogen catabolic sequences: if the carbon supply and metabolism are faster than nitrogen supply and metabolism,

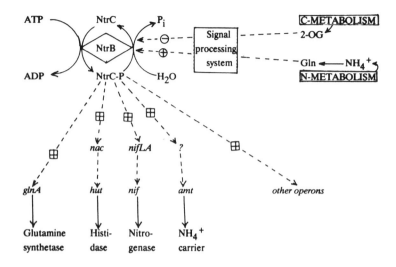

FIGURE 1. The nitrogen control *(ntr)* system in enterobacteria. For further explanations see text.

the ensuing high 2-OG/Gln ratio promotes NtrC phosphorylation and, thus, activation (directly or via subsystems like *nifLA* or *nac* in Figure 1) of a number of operons which in turn may provide additional NH_4^+ *(nif, hut),* and at the same time leads to synthesis of GS, which efficiently metabolizes low intracellular NH_4^+ levels. Under carbon limitation, however, nitrogenous organic compounds must serve as carbon sources. Surplus nitrogen is excreted as NH_3, which signals dephosphorylation of NtrC and, thus, repression of Ntr-controlled operons.

Early experiments with *K. pneumoniae*[32] and more recent studies with *E. coli*[73] (see Chapter IIIB for a detailed presentation) strongly indicated that the NH_4^+ carriers — at least in enterobacteria — are Ntr regulated. The physiological significance will be outlined in Section IV.E.

3. Mutants Deficient in NH_4^+ Transport (Amt⁻)

In addition to certain mutants defective in the *ntr* system, a number of pleiotropic (mostly regulatory) mutants exhibit the Amt⁻ (NH_4^+ transport negative) phenotype (Table 2). The control circuits involved await detailed description. It is evident that NH_4^+ transport is affected by numerous mutations.

In addition, three Amt⁻ strains have been described[52,87,88] which may contain a mutation in the structural gene of the carrier. Only the *E. coli* mutant[88] has been analyzed in detail (see Chapter IIIB).

E. PHYSIOLOGICAL FUNCTION

As has been discussed above, many bacteria under nitrogen limitation derepress a number of enzymes which may enhance nitrogen supply. The

TABLE 2
Pleiotropic Mutants Deficient in Ammonium Transport (Amt⁻)

Species	Symbol or gene	Other phenotypes (apart from Amt⁻)	Ref.
Klebsiella pneumoniae	GlnR(= Ntr⁻)	GlnA⁻Nif⁻Hut⁻	32, 81
	AmtR	Poor growth on amino acids	82
Escherichia coli	ntrC		73
	gltF	Ntr⁻	76, 83
Rhodobacter sphaeroides	adgA	Nif⁻Nar⁻Aut⁻Put⁻	84
Rhodobacter capsulatus	—	Gln⁻	42, 43
Azospirillum brasilense	Mea^R	Methylamine resistance	85
Nostoc muscorum	—	Het⁻Nif⁻	86

Note: GlnR, regulator for glutamine synthetase; GlnA, glutamine synthetase; Nif, nitrogen fixation; Hut, histidine utilization; AmtR, ammonium transport regulator; Ntr, nitrogen control; *glt,* glutamate synthase (GOGAT); *adg,* ammonium-dependent growth; Nar, nitrate reductase; Aut, arginine utilization; Put, proline utilization; Het, heterocyst formation.

NH_4^+ provided is generally assimilated by GS, which has a high affinity for NH_4^+ ($K_m \approx 0.1$ mM). This, nevertheless, implies that the cells maintain low internal NH_4^+ and NH_3 levels which are higher than the extracellular concentrations. Thus, an outward NH_3 gradient will be formed, which — in view of the good permeability of NH_3 through biomembranes (Section III) — must lead to NH_3 leakage to the outside, where it will be protonated. Further migration might be retarded by temporary binding to acidic groups of the cell wall (membrane-derived oligosaccharides, lipopolysaccharides, teichoic and teichuronic acids). Finally — since under nitrogen limitation the NH_4^+ carriers are derepressed (Table 1) — a large portion should be recaptured as NH_4^+. For each imported NH_4^+ molecule, one H^+ must be ejected by an energy-dependent H^+ pump to maintain membrane energization. This futile cycle was termed "NH_3/NH_4^+ cyclic retention".[58]

If the parameters on the right side of Equation 1 can be determined, the diffusion rate can be calculated. For a nitrogen-fixing anaerobic continuous culture of *K. pneumoniae* a leakage of 0.6×10^{-18} mol NH_3 per second per single bacterium was estimated.[14] Glutamine formation (the main route of NH_3 assimilation) was measured as 0.1×10^{-18} mol/s in the same culture. Thus, NH_3 leakage was six times faster than assimilation, meaning that on the average each NH_3/NH_4^+ molecule passed through the cycle six times before being metabolized. The energy requirement for this process under these conditions was estimated as two ATPs per molecule. This considerably increases the cost of N_2 fixation.

According to the cyclic retention hypothesis, Amt⁻ mutants should constantly excrete small amounts of NH_3. This was observed for the AmtR mutant of *K. pneumoniae*[32] (Table 2), but not for an Amt⁻ mutant of *Anabaena variabilis.*[87] Possibly the latter mutant expresses more than one NH_4^+ carrier like other phototrophs[42,50,54,59a] (see Section IV.B.3).

Cyclic retention is in no way unique for the ammonia system. Retention of escaping metabolites by substrate-repressible high-affinity carriers was also demonstrated for a range of amino acids.[89] Enterobacteria, for example, frequently express two classes of carriers for certain amino acids: high- and low-affinity systems. The low-affinity system is only expressed when a certain amino acid (e.g., leucine) serves as the nitrogen source. The high-affinity system is only expressed under nitrogen-limiting conditions and probably is involved in cyclic retention.[89] Indeed, strains lacking the respective carriers continuously excrete the cognate amino acids.[89-91] Thus, cyclic retention is expected to be of biological significance for many metabolites, where P and $(c_1 - c_2)$ in Equation 1 are not negligibly small.

Returning to the ammonia system, the rationale of the Ntr system now can be summarized as follows. Under nitrogen limitation the organism switches on (1) operons which may provide NH_4^+ (e.g., *nif, hut*); (2) GS synthesis for efficient NH_4^+ assimilation; and (3) NH_4^+ carrier synthesis to keep the scarce NH_4^+ inside the cell.

F. TWO SPECIAL CASES
1. Cyanobacteria
The kinetics of NH_4^+ or methylamine uptake by cyanobacteria are usually biphasic: an initial rapid phase (lasting about 1 min) is followed by a slower second phase (lasting for more than 1 h).[47,48,50,53,59a,62-64,92] Generally, both phases do not occur after growth with NH_4^+ (see Table 1). Both phases also can be distinguished by the effect of MSX, which only inhibits the second (slow) phase of uptake. In several organisms each phase reveals complex kinetics with two different K_m values.[47,52,55,56]

These and other results have led to two controversial explanations of this complicated picture:

1. Only the initial phase represents transport, while the slow phase reflects metabolization by GS. The main arguments in favor of this hypothesis are:
   - Selective inhibition by MSX, a potent inhibitor of GS (see Section IV.D.1)
   - Chase removes only a quantity of [^{14}C]-methylamine approximately equal to that taken up during the rapid phase, even after prolonged incubation[53,62-64]
   - GS-deficient mutants of *Anabaena variabilis* exhibit reduced second-phase uptake[87]
2. Both phases represent transport into different compartments (cytoplasm and thylakoids). The intrathylakoid compartment represents about 10 to 40% of the cell volume. Since under illumination a pH gradient

around 2 was measured across a thylakoid membrane,[94] the intrathylakoid space functions as a considerable sink for NH_4^+ by trapping. According to this hypothesis, one phase reflects active transport into the cytoplasm and the other reflects diffusion (or transport) of NH_3 into the thylakoids. The main arguments supporting this hypothesis are:

- A GS-deficient mutant of *Anabaena cycadeae* still exhibits both phases of transport.[53,72]
- *In vivo* the second-phase methylamine transport is inhibited by MSX much faster than GS.[47,63]

In this connection it is remarkable that freshly isolated cyanobionts from the root nodules of *Cycas revoluta* lack the second uptake phase despite normal GS levels.[95] Interestingly, the cyanobiont was unable to assimilate ammonia shortly after isolation, in contrast to its free-living isolate, *Anabaena cycadeae*.[95]

Although both hypotheses have been discussed extensively, the present data do not yet allow a definite conclusion.

2. Rhizobia

Rhizobia actively accumulate NH_4^+ or methylamine in the free-living state (Table 1). Numerous experiments, however, have demonstrated absence of the NH_4^+ carrier during symbiosis in legume root nodules.[29,37-39,96] Consequently, in the symbiotic state N_2 fixed by the bacteroides should diffuse across bacteroid and peribacteroid membranes into the plant cytoplasm, where it is trapped by a lower pH[97] and is bound to glutamate (Equation 4). The mechanism of repression of rhizobial NH_4^+ carriers (and GS) is unknown. Putative roles are assigned to K^+, which regulates a variety of bacteroid-associated functions,[98] to microaerophilic conditions, or to asparagine,[40] which represses NH_4^+ carrier synthesis in *Rhizobium meliloti* and which is present in high amounts in the cytoplasm of some nodules.[99]

Repression of the NH_4^+ carrier in nodules is biologically important: in this way the energy-demanding cyclic retention of NH_3/NH_4^+ is avoided, and NH_3 diffusion into the plant cell is not obstructed.

REFERENCES

1. **Merrick, M. J.**, Regulation of nitrogen assimilation by bacteria, in *The Nitrogen and Sulphur Cycles*, Cole, J. A. and Ferguson, S. J., Eds., Cambridge University Press, London, 1988, 331.
2. **Stock, J. B., Ninfa, A. J., and Stock, A. M.**, Protein phosphorylation and regulation of adaptive responses in bacteria, *Microbiol. Rev.*, 53, 450, 1989.
3. **Tien, H. T.**, *Bilayer Lipid Membranes*, Marcel Dekker, New York, 1974, 165.
4. **Boron, W. F.**, Intracellular pH regulation, *Curr. Top. Membr. Transp.*, 13, 3, 1980.

5. **Henderson, P. J. F.,** Ion transport by energy-conserving biological membranes, *Annu. Rev. Microbiol.,* 25, 393, 1971.

6. **Jacobs, M. H.,** Some aspects of cell permeability to weak electrolytes, *Cold Spring Harbor Symp. Quant. Biol.,* 8, 30, 1940.

7. **Zarlengo, H. and Abrams, A.,** Selective uptake of ammonia and alkylamines into *Streptococcus faecalis* and their effect on glycolysis, *Biochim. Biophys. Acta,* 71, 65, 1962.

8. **Caldwell, P.,** Intracellular pH, *Int. Rev. Cytol.,* 5, 229, 1956.

9. **Lockwood, A. H., Finn, R. D., Campbell, J. A., and Richman, T. B.,** Factors that affect the uptake of ammonia by the brain: the blood-brain pH gradient, *Brain Res.,* 181, 259, 1980.

10. **LaNoue, K. F. and Schoolwerth, A. C.,** Metabolite transport in mitochondria, *Annu. Rev. Biochem.,* 48, 871, 1979.

11. **Ryan, T. E., Barr, C. E., and Zorn, J. P.,** Potassium transference in *Nitella, J. Gen. Physiol.,* 72, 203, 1978.

12. **Ritchie, R. J. and Gibson, J.,** Permeability of ammonia, methylamine and ethylamine in the cyanobacterium, *Synechococcus* R-2 *(Anacystic nidulans)* PCC 7942, *J. Membr. Biol.,* 95, 131, 1987.

13. **Ritchie, R. J. and Gibson, J.,** Permeability of ammonia and amines in *Rhodobacter sphaeroides* and *Bacillus firmus, Arch. Biochem. Biophys.,* 258, 332, 1987.

14. **Kleiner, D.,** Energy expenditure for cyclic retention of NH_3/NH_4^+ during N_2 fixation by *Klebsiella pneumoniae, FEBS Lett.,* 187, 237, 1985.

15. **Ritchie, R. J.,** The permeability of ammonia, methylamine and ethylamine in the charophyte, *Chara corallina* (C. australis), *J. Exp. Bot.,* 38, 67, 1987.

16. **Tien, H. T.,** *Bilayer Lipid Membranes,* Marcel Dekker, New York, 1974, chap. 6.

17. **Nelson, R. D., Lide, D. R., and Maryott, A. A.,** Selected values of electric dipole moments for molecules in the gas phase, in *Handbook of Chemistry and Physics,* 64th ed., Weast, R. C., Ed., CRC Press, Boca Raton, FL, 1984, table E-58.

18. **Hackette, S. L., Skye, G. E., Burton, C., and Segel, I. H.,** Characterization of an ammonium transport system in filamentous fungi with methylammonium-^{14}C as the substrate, *J. Biol. Chem.,* 245, 4241, 1970.

18a. **Michels, M. and Bakker, E. P.,** Low-affinity potassium uptake system in *Bacillus acidocaldarius, J. Bacteriol.,* 169, 4335, 1987.

19. **Bogdahn, M. and Kleiner, D.,** Inorganic nitrogen metabolism in two cellulose degrading clostridia, *Arch. Microbiol.,* 145, 159, 1986.

20. **Bogdahn, M., Andreesen, J. R., and Kleiner, D.,** Pathway and regulation of N_2, ammonium and glutamate assimilation by *Clostridium formicoaceticum, Arch. Microbiol.,* 134, 167, 1983.

21. **Kleiner, D. and Fitzke, E.,** Some properties of a new electrogenic transport system: the ammonium (methylammonium) carrier from *Clostridium pasteurianum, Biochim. Biophys. Acta,* 641, 138, 1981.

22. **Mazzucco, C. E. and Benson, D. R.,** (^{14}C) methylammonium transport by *Frankia* sp. strain CpI1, *J. Bacteriol.,* 160, 636, 1984.

23. **Jahns, T., Kaltwasser, H., and Kleiner, D.,** Ammonium (methylammonium) uptake by *Alcaligenes eutrophus* H16, *Arch. Microbiol.,* 145, 306, 1986.

24. **Hartmann, A. and Kleiner, D.,** Ammonium (methylammonium) transport by *Azospirillum* sp., *FEMS Microbiol. Lett.,* 15, 65, 1982.

25. **Narula, N. and Kleiner, D.,** Ammonium (methylammonium) transport by *Azotobacter chroococcum, FEMS Microbiol. Lett.,* 44, 193, 1987.

26. **Kleiner, D.,** Ammonium uptake by nitrogen-fixing bacteria. I. *Azotobacter vinelandii, Arch. Microbiol.,* 104, 163, 1975.

27. **Gordon, J. K. and Moore, R. A.,** Ammonium and methylammonium transport by the nitrogen-fixing bacterium *Azotobacter vinelandii, J. Bacteriol.,* 148, 435, 1981.

28. **Barnes, E. M. and Zimniak, P.,** Transport of ammonium and methylammonium ions by *Azotobacter vinelandii, J. Bacteriol.,* 146, 512, 1981.

29. **Udvardi, M. and Day, D. A.,** Ammonia (^{14}C-methylamine) transport across the bacteroid and peribacteroid membranes of soybean root nodules, *Plant Physiol.,* 94, 71, 1990.

30. **Jayakumar, A., Epstein, W., and Barnes, E. M.,** Characterization of ammonium (methylammonium)/potassium antiport in *Escherichia coli, J. Biol. Chem.,* 260, 7528, 1985.

31. **Kleiner, D.,** Ammonium uptake and metabolism by nitrogen-fixing bacteria. II. *Klebsiella pneumoniae, Arch. Microbiol.,* 111, 85, 1976.

32. **Kleiner, D.,** Ammonium (methylammonium) transport by *Klebsiella pneumoniae, Biochim. Biophys. Acta,* 688, 702, 1982.

33. **Holtel, A. and Kleiner, D.,** Regulation of ammonium and methylammonium transport in *Paracoccus denitrificans, Arch. Microbiol.,* 142, 285, 1985.

34. **Jahns, T. and Kaltwasser, H.,** Uptake and metabolism of methylammonium by *Pseudomonas aeruginosa, FEMS Microbiol. Lett.,* 72, 131, 1990.

35. **Howitt, S. M., Udvardi, M. K., Day, D. A., and Gresshoff, P. M.,** Ammonia transport in free-living and symbiotic *Rhizobium* sp. ANU289, *J. Gen. Microbiol.,* 132, 257, 1986.

36. **Gober, J. W. and Kashket, E. R.,** Methylammonium uptake by *Rhizobium* sp. strain 32H1, *J. Bacteriol.,* 153, 1196, 1983.

37. **Jin, H. N., Glenn, A. R., and Dilworth, M. J.,** Ammonium uptake by cowpea *Rhizobium* strain MNF 2030 and *Rhizobium trifolii* MNF 1001, *Arch. Microbiol.,* 149, 308, 1988.

38. **O'Hara, G. W., Riley, I. T., Glenn, A. R., and Dilworth, M. J.,** The ammonium permease in *Rhizobium leguminosarum* MNF 3841, *J. Gen. Microbiol.,* 131, 757, 1985.

39. **Glenn, A. R. and Dilworth, M. J.,** Methylamine and ammonium transport systems in *Rhizobium leguminosarum* MNF 3841, *J. Gen. Microbiol.,* 130, 1961, 1984.

40. **Pargent, W. and Kleiner, D.,** Characteristics and regulation of ammonium (methylammonium) transport in *Rhizobium meliloti, FEMS Microbiol. Lett.,* 30, 257, 1985.

41. **Alef, K. and Kleiner, D.,** Regulatory aspects of inorganic nitrogen metabolism in the *Rhodospirillaceae, Arch. Microbiol.,* 133, 239, 1982.

42. **Rapp, B., Landrum, D. C., and Wall, J. D.,** Methylammonium uptake by *Rhodobacter capsulatus, Arch. Microbiol.,* 146, 134, 1986.

43. **Sharak Genthner, B. R. and Wall, J. D.,** Ammonium uptake in *Rhodopseudomonas capsulata, Arch. Microbiol.,* 141, 219, 1985.

44. **Golby, P., Carver, M., and Jackson, J. B.,** Membrane ionic currents in *Rhodobacter capsulatus.* Evidence for electrophoretic transport of K$^+$, Rb$^+$ and NH$_4^+$, *Eur. J. Biochem.,* 187, 589, 1990.

45. **Ivanovsky, R. N., Vakulenko, V. P., and Kondratieva, E. N.,** Uptake of ammonium and methylammonium by *Rhodopseudomonas sphaeroides* 2R, in *Abstr. 4th Int. Symp. Photosynthetic Prokaryotes,* Bombannes, France, 1982, B31.

46. **Holtel, A.,** unpublished results, 1984.

47. **Rai, A. N. and Prakasham, R.,** Characteristics of the methylammonium/ammonium transport systems of the nitrogen-fixing cyanobacterium *Anabaena* 7120 (ATCC 27893), *Indian J. Biochem. Biophys.,* 26, 219, 1989.

48. **Turpin, D. H., Edie, S. A., and Canvin, D. T.,** *In vivo* nitrogenase regulation by ammonium and methylamine and the effect of MSX on ammonium transport in *Anabaena flos-aquae, Plant Physiol.,* 74, 701, 1984.

49. **Boussiba, S., Resch, C. M., and Gibson, J.,** Ammonia uptake and retention in some cyanobacteria, *Arch. Microbiol.,* 138, 287, 1984.

50. **Boussiba, S., Dilling, W., and Gibson, J.,** Methylammonium transport in *Anacystis nidulans* R-2, *J. Bacteriol.,* 160, 204, 1984.

51. **Boussiba, S. and Gibson, J.,** Regulation of methylammonium/ammonium transport in the unicellular cyanobacterium *Synechococcus* R-2 (PCC 7942), *FEMS Microbiol. Lett.,* 43, 289, 1987.

52. **Kashyap, A. K. and Singh, D. P.**, Ammonium transport in unicellular cyanobacterium *Anacystis nidulans, J. Plant Physiol.*, 121, 319, 1985.

53. **Singh, D. T., Ghosh, R., and Singh, H. N.**, Physiological characterization of the ammonium transport system in the free-living diazotrophic cyanobacterium *Anabaena cycadeae, J. Plant Physiol.*, 127, 231, 1987.

54. **Cordts, M. L. and Gibson, J.**, Ammonium and methylammonium transport in *Rhodobacter sphaeroides, J. Bacteriol.*, 169, 1632, 1987.

55. **Prasad, P. and Kashyap, A. K.**, Characteristics of ammonium transport in an alkalophilic diazotrophic cyanobacterium *Nostoc calcicola:* influence of temperature and methionine sulfoximine, *J. Plant Physiol.*, 136, 149, 1990.

56. **Kashyap, A. K. and Johar, G.**, Genetic control of ammonium transport in nitrogen fixing cyanobacterium *Nostoc muscorum, Mol. Gen. Genet.*, 197, 509, 1984.

57. **Prasad, P. and Kashyap, A. K.**, Ammonium transport in thermophilic cyanobacterium *Synechococcus elongatus, J. Gen. Appl. Microbiol.*, 36, 303, 1990.

58. **Kleiner, D.**, Bacterial ammonia transport, *FEMS Microbiol. Rev.*, 32, 87, 1985.

59. **Kleiner, D.**, The transport of NH_3 and NH_4^+ across biological membranes, *Biochim. Biophys. Acta*, 639, 41, 1981.

59a. **Boussiba, S. and Gibson, J.**, Ammonia translocation in cyanobacteria, *FEMS Microbiol. Rev.*, 88, 1, 1991.

60. **Glover, H. E.**, Methylamine, an inhibitor of ammonia oxidation in the marine nitrifying bacterium *Nitrosococcus oceanus, Arch. Microbiol.*, 132, 37, 1982.

61. **Wiegel, J. and Kleiner, D.**, Survey of ammonium (methylammonium) transport by aerobic N_2 fixing bacteria — the special case of *Rhizobium, FEMS Microbiol. Lett.*, 15, 61, 1982.

62. **Rai, A. N., Rowell, P., and Stewart, W. D. P.**, Evidence for an ammonium transport system in free-living and symbiotic cyanobacteria, *Arch. Microbiol.*, 137, 241, 1984.

63. **Singh, D. T., Modi, D. R., and Singh, H. N.**, Evidence for glutamine synthetase and ammonium (methylammonium) transport system as two distinct primary targets of methionine sulfoximine inhibitory action in the cyanobacterium *Anabaena doliolum, FEMS Microbiol. Lett.*, 37, 95, 1986.

64. **Kerby, N. W., Rowell, P., and Stewart, W. D. P.**, The uptake and metabolism of methylamine by N_2-fixing cyanobacteria, *Arch. Microbiol.*, 143, 353, 1986.

65. **Singh, S., Kashyap, A. K., and Singh, H. N.**, Developmental regulation of methylammonium (ammonium) transport activity in the cyanobacterium *Anabaena doliolum, FEMS Microbiol. Lett.*, 68, 163, 1990.

66. **Gerth, K.**, Microassay for ammonia by determination of ammonia nitrogen in a nitrogen analyzer, *Anal. Biochem.*, 144, 432, 1985.

67. **Ohmori, M. and Hattori, A.**, Effect of ammonia on nitrogen fixation by the blue-green alga *Anabaena cylindrica, Plant Cell Physiol.*, 15, 131, 1984.

68. **Meeks, J. C., Wycoff, K. L., Chapman, J. D., and Enderlin, C. S.**, Regulation of expression of nitrate and dinitrogen assimilation by *Anabaena* species, *Appl. Env. Microbiol.*, 45, 1351, 1983.

69. **Dharmawardene, M. W. N. and Stewart, W. D. P.**, Nitrogenase activity, amino acid pool patterns and assimilation in blue-green algae, *Planta*, 108, 133, 1972.

70. **Buurman, E. T., Teixeira de Mattos, M. J., and Neijssel, O. M.**, Futile cycling of ammonium ions via the high affinity potassium uptake system (Kdp) of *Escherichia coli, Arch. Microbiol.*, 155, 391, 1991.

71. **Kepes, A.**, The β-galactoside permease of *Escherichia coli, J. Membr. Biol.*, 4, 87, 1971.

72. **Singh, D. T., Rai, A. N., and Singh, H. N.**, Methylammonium (ammonium) uptake in a glutamine auxotroph of the cyanobacterium *Anabaena cycadeae, FEBS Lett.*, 186, 51, 1985.

73. **Jayakumar, A., Schulman, I., MacNeil, D., and Barnes, E. M.,** Role of the *Escherichia coli glnALG* operon in regulation of ammonium transport, *J. Bacteriol.,* 166, 281, 1986.

74. **Sprott, G. D., Shaw, K. M., and Jarrell, K. F.,** Ammonia/potassium exchange in methanogenic bacteria, *J. Biol. Chem.,* 259, 12602, 1984.

75. **Nakamura, T., Tokuda, H., and Unemoto, T.,** Effects of pH and movable cations on the potassium exit from the marine bacterium, *Vibrio alginolyticus,* and the manipulation of cellular cation contents, *Biochim. Biophys. Acta,* 692, 389, 1982.

76. **Servin-Gonzàlez, L. and Bastarrachea, F,** Nitrogen regulation of synthesis of the high affinity methylammonium transport system of *Escherichia coli, J. Gen. Microbiol.,* 130, 3071, 1984.

77. **Jayakumar, A., Hong, J.-S., and Barnes, E. M.,** Feedback inhibition of ammonium (methylammonium) ion transport in *Escherichia coli* by glutamine and glutamine analogs, *J. Bacteriol.,* 169, 553, 1987.

78. **Kleiner, D., Alef, K., and Hartmann, A.,** Uptake of methionine sulfoximine by some N$_2$ fixing bacteria, and its effect on ammonium transport, *FEBS Lett.,* 164, 121, 1983.

79. **Kleinschmidt, J. A. and Kleiner, D.,** Relationships between nitrogenase, glutamine synthetase, glutamine and energy charge in *Azotobacter vinelandii, Arch. Microbiol.,* 128, 412, 1981.

80. **Thöny, B. and Hennecke, H.,** The $-24/-12$ promoter comes of age, *FEMS Microbiol. Rev.,* 63, 341, 1989.

81. **Streicher, S. L., Shanmugam, K. T., Ausubel, F. M., Morandi, C., and Goldberg, R. B.,** Regulation of nitrogen fixation in *Klebsiella pneumoniae:* evidence for a role of glutamine synthetase as a regulator of nitrogenase synthesis, *J. Bacteriol.,* 120, 815, 1974.

82. **Castorph, H. and Kleiner, D.,** Some properties of a *Klebsiella pneumoniae* ammonium transport negative mutant (Amt$^-$), *Arch. Microbiol.,* 139, 245, 1984.

83. **Castaño, I., Bastarrachea, F., and Covarrubias, A. A.,** *gltBDF* operon of *Escherichia coli, J. Bacteriol.,* 170, 821, 1988.

84. **Zinchenko, V. V., Babykin, M. M., Shestakov, S., Allibert, P., Vignais, P. M., and Willison, J. C,** Ammonia dependent growth (Adg$^-$) mutants of *Rhodobacter capsulatus* and *Rhodobacter sphaeroides:* comparison of mutant phenotypes and cloning of the wild-type *(adgA)* genes, *J. Gen. Microbiol.,* 136, 2385, 1990.

85. **Turbanti, L., Bazzicalupo, M., Casalone, E., Fani, R., Gallori, E., and Polsinelli, M.,** Mutants of *Azospirillum brasilense* resistant to methylammonium, *Arch. Microbiol.,* 150, 421, 1988.

86. **Singh, A. K., Sailaja, M. V., and Singh, H. N.,** A class of glyphosate-selected mutants of the cyanobacterium *Nostoc muscorum* showing loss of ammonium transport activity (Amt$^-$), heterocyst formation (Het$^-$) and nitrogenase activity (Nif$^-$), *FEMS Microbiol. Lett.,* 60, 187, 1989.

87. **Reglinski, A., Rowell, P., Kerby, N. W., and Stewart, W. D. P.,** Characterization of methylammonium/ammonium transport in mutant strains of *Anabaena variabilis* resistant to ammonium analogues, *J. Gen. Microbiol.,* 135, 1441, 1989.

88. **Fabiny, J. M., Jayakumar, A., Chinault, A. C., and Barnes, E. M.,** Ammonium transport in *Escherichia coli:* localization and nucleotide sequence of the *amtA* gene, *J. Gen. Microbiol.,* 137, 983, 1991.

89. **Antonucci, T. K. and Oxender, D. L.,** The molecular biology of amino-acid transport in bacteria, *Adv. Microb. Physiol.,* 28, 145, 1986.

90. **Anderson, J. J. and Oxender, D. L.,** Genetic separation of high and low-affinity transport systems for branched-chain amino acids in *Escherichia coli* K-12, *J. Bacteriol.,* 136, 168, 1978.

91. **Szech, U., Braun, M., and Kleiner, D.,** Uptake and excretion of amino acids by saccharolytic clostridia, *FEMS Microbiol. Lett.,* 58, 11, 1989.

92. **Kerby, N. W., Rowell, P., and Stewart, W. D. P.,** Cyanobacterial ammonium transport, ammonium assimilation, and nitrogenase regulation, *N.Z. J. Mar. Freshwater Res.,* 21, 447, 1987.

93. **Rai, A. N., Singh, D. T., and Singh, H. N.,** Regulation of ammonium/methylammonium transport by ammonium in the cyanobacterium *Anabaena variabilis, Physiol. Plant.,* 68, 320, 1986.

94. **Falkner, G., Horner, F., Werdan, K., and Heldt, H. W.,** pH changes in the cytoplasm of the blue-green alga *Anacystis nidulans* caused by light-dependent proton flux into the thylakoid space, *Plant Physiol.,* 58, 717, 1976.

95. **Rai, A. N., Linblad, P., and Bergman, B.,** Absence of glutamine-synthetase-linked methylammonium (ammonium)-transport system in the cyanobiont of *Cycas* — cyanobacterial symbiosis, *Planta,* 169, 379, 1986.

96. **Laane, C., Krone, W., Konings, W., Haaker, H., and Veeger, C.,** Short-term effect of ammonium chloride on nitrogen fixation by *Azotobacter vinelandii* and bacteroids of *Rhizobium leguminosarum, Eur. J. Biochem.,* 103, 39, 1980.

97. **Kennedy, I. R.,** Primary products of symbiotic nitrogen fixation. I. Short-term exposure of *Serradella* nodules to $^{15}N_2$, *Biochim. Biophys. Acta,* 130, 285, 1966.

98. **Gober, J. W. and Kashket, E. R.,** K^+ regulates bacteroid-associated functions of *Bradyrhizobium, Proc. Natl. Acad. Sci. U.S.A.,* 84, 4650, 1987.

99. **Boland, M. J., Farnden, K. J. F., and Robertson, G. J.,** Ammonia assimilation in nitrogen fixing legume nodules, in *Nitrogen Fixation,* Vol. 2, Newton, W. E. and Orme-Johnson, W. H., Eds., University Park Press, Baltimore, 1980, 33.

Chapter IIIB

NH$_4^+$ TRANSPORT SYSTEMS IN *ESCHERICHIA COLI*

Eugene M. Barnes, Jr. and Arumugam Jayakumar

TABLE OF CONTENTS

I. ASSAY AND GENERAL PROPERTIES OF NH_4^+ TRANSPORT IN *E. COLI*

As holds true for other microbial ammonium transport systems, most of the information concerning the *Escherichia coli* system has been derived from studying the uptake of the [^{14}C]-methylammonium ion. Only cells grown on nitrogen sources other than ammonia are capable of significant accumulation of this tracer. Stevenson and Silver[1] originally described energy-dependent uptake of ^{14}C-methylammonium by *E. coli*. Accumulation of the radiolabel was inhibited by NH_4^+. Servin-Gonzalez and Bastarracha[2] also found an energy-linked process for ^{14}C-methylammonium incorporation which was repressed by growth of *E. coli* on excess ammonium. However, the assay method used in these earlier studies involved washing the cells following the ^{14}C-methylammonium uptake period. As a result, the highly permanent intracellular pool of chemically unaltered ^{14}C-methylammonium was washed out and the impermeant metabolite, [^{14}C]-γ-glutamylmethylamide, remained inside the cells. This led to very large underestimates of the actual rates of ^{14}C-methylammonium translocation. Furthermore, such procedures rely on the metabolic trap provided by glutamine synthetase. Our approach for determining rates of ^{14}C-methylammonium transport utilizes filtration of cells on porous polycarbonate membranes which need not be washed.[3] Alternatively, the cells may be separated from the medium by centrifugation through a layer of oil.[4] Using the filtration method with *E. coli*, accumulation of chemically unaltered ^{14}C-methylammonium against 100-fold concentration gradients was demonstrated.[5] If rapid pulses of ^{14}C-methylammonium were utilized, the kinetic parameters of the translocation process could be estimated ($K_m = 36$ μM methylammonium; $V_{max} = 4.0$ μmol $\cdot$ sec^{-1} $\cdot$ g protein^{-1}). Because of the very high rates of ^{14}C-methylammonium uptake and the small intracellular volume, this represents an underestimate of the true initial rates of transport.[5] The pH dependence of ^{14}C-methylammonium entry indicated that $CH_3NH_3^+$ is the permeant species. At micromolar levels, external NH_4^+ is removed from the assay medium even more rapidly than ^{14}C-methylammonium, so the study of NH_4^+ inhibition kinetics requires special procedures. If very dilute cell suspensions were used to avoid depletion of external NH_4^+ during the ^{14}C-methylammonium transport assay, a K_i value of 0.5 μM for competitive inhibition by NH_4^+ could be estimated.[5] Thus, the apparent affinity of this *E. coli* system is more than 70 times higher for NH_4^+ than for $CH_3NH_3^+$. The high affinity for NH_4^+ suggests that this transport system may enable cells to utilize very low concentrations of external ammonium efficiently. Glutamine synthetase, the primary ammonium-utilizing enzyme under nitrogen-limiting growth conditions for *E. coli*, has a relatively low affinity for NH_4^+ ($K_m = 1.8$ mM).[6] However, the cells grow well on agar containing less than 20 μM NH_4^+. Moreover, mutants with a defect in the high-affinity ammonium/^{14}C-methylammonium transport system are unable to grow on limiting NH$_4^+$.[7] At much higher (>10 mM) concentrations of ammonium, the high-affinity transport system is dispensable and the mutants grow normally on M9 minimal medium with NH$_4^+$ as the sole source of nitrogen. Under these conditions, NH$_4^+$ may enter *E. coli* via the Kdp system for K$^+$ transport[8] (see Chapter IIIC). Since *E. coli* strains with multiple defects in K$^+$ transport *(kdp trkA trkD)* can utilize 16 mM NH$_4^+$ as a nitrogen source,[9] unmediated diffusion of NH$_3$ also may be sufficient to support growth.

II. NH$_4^+$/K$^+$ ANTIPORT MECHANISM

Studies of K$^+$-depleted *E. coli* strains have shown that internal K$^+$ (K$_{in}^+$) is required for the accumulation of ^{14}C-methylammonium. When *E. coli* cells are subjected to hypoosmotic shock, the capacity to accumulate ^{14}C-methylammonium is lost. However, the activity can be regained by incubating cells with boiled shock fluid or with glucose plus K$^+$.[5] A similar reconstitution was also observed in K$^+$-depleted parental strains, but not in *E. coli* TK2205 *(kdp trkA trkD)*, demonstrating the requirement for K$_{in}^+$. It is, however, important to note that K$^+$-replete TK2205 cells accumulate ^{14}C-methylammonium normally. Thus, the transporters which are normally responsible for K$^+$ uptake by *E. coli* do not provide a significant pathway for ^{14}C-methylammonium entry. After reconstitution of K$^+$-depleted cells or with untreated, K$^+$-replete cells, rapid addition of external K$^+$ (K$_{out}^+$) produces a noncompetitive type of inhibition (k$_i \approx 1$ mM K$^+$) of ^{14}C-methylammonium uptake. The sensitivity of the inhibition by K$_{out}^+$ is affected markedly by the level of K$_{in}^+$. Cells with low K$_{in}^+$ are inhibited more potently by K$_{out}^+$ than are cells with high K$_{in}^+$. This suggests that the extent of ^{14}C-methylammonium entry is related to the magnitude of the transmembrane K$^+$ gradient. Furthermore, rapid increases in K$_{out}^+$ (but not Na$_{out}^+$) produced an exodus of previously accumulated ^{14}C-methylammonium. Similarly, addition of ammonium or unlabeled methylammonium caused a loss of ^{86}Rb$_{in}^+$. This ammonium- or methylammonium-dependent expulsion of ^{86}Rb$_{in}^+$ was not observed in ammonium-repressed cells. In derepressed *E. coli,* the molar ratio of ^{14}C-methylammonium entry to ^{86}Rb$^+$ exit was 1.12 ± 0.11, leading us to propose a ^{14}C-methylammonium/^{86}Rb$^+$ or NH$_4^+$/K$^+$ antiport mechanism for the transporter.[5] Under normal conditions, e.g., in cells maintaining substantial K$^+$ gradients, K$^+$ exodus via this antiporter could drive the uphill movement of NH$_4^+$ into the cell.

III. BIOENERGETICS OF ^{14}C-METHYLAMMONIUM ACCUMULATION IN *E. COLI*

The energy requirements for ^{14}C-methylammonium transport are complex; both ATP and a transmembrane electrochemical H$^+$ gradient ($\Delta\bar{\mu}_{H^+}$) are necessary. Washed *E. coli* cells show strain-specific differences in a requirement for added energy sources to drive ^{14}C-methylammonium accumulation. ML strains require the addition of glucose or pyruvate, while D-lactate is much less effective in supporting ^{14}C-methylammonium transport. This contrasts with $\Delta\bar{\mu}_{H^+}$-coupled systems (e.g., proline or lactose transporters), which are efficiently supported by D-lactate. Washed K12 strains (e.g., FRAG-1) do not seem to require an exogenous energy source for ^{14}C-methylammonium uptake unless they are energy depleted by prior treatment with uncoupler. Variations in the capacity to maintain K^+ gradients may underlie some of these strain differences.

The uptake of ^{14}C-methylammonium by all of the *E. coli* strains we have tested is quite sensitive to uncouplers (FCCP), as well as to inhibitors of respiration (CN^-) or ATP synthesis (arsenate). The effect of arsenate on the glucose-dependent uptake of ^{14}C-methylammonium shows a requirement for ATP which could be independent of $\Delta\bar{\mu}_{H^+}$. This was examined further using the *unc* (F_1F_0-ATPase$^-$) mutant WB1-1. WB1-1 cells are unable to generate ATP via $\Delta\bar{\mu}_{H^+}$, but can do so by means of glycolysis. The uptake of ^{14}C-methylammonium by the *unc* mutant was fully sensitive to FCCP, CN^-, or iodoacetate, showing a separate role for $\Delta\bar{\mu}_{H^+}$.[5] We were able to verify that the ATP-driven transport of ^{3}H-arginine was supported by glycolysis in WB1-1 cells exposed to CN^- or FCCP, while the $\Delta\bar{\mu}_{H^+}$-dependent uptake of ^{3}H-proline was blocked. On the other hand, the transport of ^{3}H-arginine or ^{14}C-methylammonium by wild-type cells was strongly inhibited by arsenate, but the accumulation of ^{3}H-proline was spared. This demonstrates that both ATP and $\Delta\bar{\mu}_{H^+}$ are necessary for ^{14}C-methylammonium transport.[5] It is not clear at present whether ATP and/or $\Delta\bar{\mu}_{H^+}$ are utilized directly to drive ^{14}C-methylammonium accumulation or indirectly to establish and maintain an electrochemical K^+ gradient ($\Delta\bar{\mu}_{K^+}$). It seems possible that $\Delta\bar{\mu}_{K^+}$ could provide the energy for uphill movement of ^{14}C-methylammonium via methylammonium/K^+ antiport. On the other hand, our attempts to stimulate ^{14}C-methylammonium uptake by an artificially imposed $\Delta\bar{\mu}_{K^+}$ in arsenate-poisoned cells or in membrane vesicles have not been successful. Thus, ATP or $\Delta\bar{\mu}_{H^+}$ may play a more direct role in ammonium transport.

IV. ROLE OF THE NITROGEN REGULON (Ntr) IN NH$_4^+$ TRANSPORT

The regulation of the high-affinity ammonium/methylammonium transport system of *E. coli* is multifaceted. In addition to the competitive inhibition of ^{14}C-methylammonium uptake exerted by exogenous ammonium, elevation of intracellular glutamine pools produces a noncompetitive feedback inhibition (see below). Potential artifacts introduced by these mechanisms must be avoided when cells are repressed by growth in ammonia-rich medium. This is usually accomplished by incubating washed cells in a nitrogen-free medium at 25°C prior to transport assays. As noted earlier, ^{14}C-methylammonium transport is strongly repressed in *E. coli* grown on excess NH$_4^+$.[2,5,10] Methylammonium cannot serve as a nitrogen source for *E. coli* and does not repress transport.[10] Cells grown on 20 m*M* arginine or glutamate as the sole nitrogen source have the highest transport activity. Incubation of these derepressed cells in media containing 10 m*M* NH$_4^+$ produces more than a 90% reduction in ^{14}C-methylammonium transport activity. The onset of this repression is relatively slow ($t_{1/2}$ = 90 to 110 min at 37°C) and can be completely prevented by addition of chloramphenicol to the medium. Likewise, derepression in an NH$_4^+$-free medium ($t_{1/2}$ = 50 to 60 min at 37°C) is also prevented by chloramphenicol.[11]

The expression of genes for nitrogen-utilizing enzymes in *E. coli* is controlled by the global nitrogen regulon.[12,13] Mutants carrying defects in this Ntr system have played an essential role in understanding the control of ammonium transport. The *glnALG* operon, which includes the structural gene for glutamine synthetase *(glnA)*, encodes the Ntr elements which direct gene expression. The molecular mechanism of Ntr, as originally delineated by Magasanik and co-workers,[12] is diagrammed in Figure 1. The *glnG (ntrC)* gene product, the NR$_I$ protein, binds Ntr-specific enhancers which selectively activate downstream Ntr promoters. Thus, *glnG* mutants are unable to derepress the high-affinity ^{14}C-methylammonium transport system. The *glnL (ntrB)* product, NR$_{II}$, is bifunctional, having a kinase which phosphorylates NR$_I$ and activates NR$_I$ binding to enhancers, and a phosphatase which inactivates enhancer binding. In concordance, *glnL* mutants lacking kinase activity failed to derepress ^{14}C-methylammonium transport, while mutants deficient only in phosphatase were constitutive for transport.[10] Ntr$^-$ (GlnL or GlnG) strains which are GlnA$^+$, but are unable to derepress histidase or ^{14}C-methylammonium transport, grow normally with excess NH$_4^+$ as sole nitrogen source, but will not grow on limiting (25 to 100 μ*M*) ammonium.[7] This shows that high-affinity utilization of ammonium and the transport of ^{14}C-methylammonium are regulated similarly in *E. coli*. Although it was originally found that *glnA* mutants failed to incorporate ^{14}C-methylammonium,[2] this was due to the loss of metabolic trapping and to polar effects of the mutations on *glnL* and *glnG*. It is now quite clear that *glnA* mutants with uncompromised downstream genes accumulate chemically unaltered ^{14}C-methylammonium normally.[10]

Involvement of the Ntr-specific σ^{54} RNA polymerase subunit has been shown using a *glnF (ntrA)* mutant (Table 1) which failed to express ^{14}C-methylammonium transport. On the other hand, a ΔglnB mutant was constitutive for transport. This is consistent with the role of P$_{II}$, the *glnB* product, as an activator of the phosphatase component of NR$_{II}$ (see Figure 1). The bifunctional product of the *glnD* gene acts as the intracellular sensor of NH$_4^+$.[12] The GlnD protein contains a uridylyltransferase which, upon activation by high cytoplasmic levels of 2-ketoglutarate (2-KG), adds a uridylyl group to P$_{II}$, rendering P$_{II}$ ineffective as a cofactor for NR$_{II}$. The other function of the GlnD protein is uridylyl cleavage which is stimulated by high intracellular

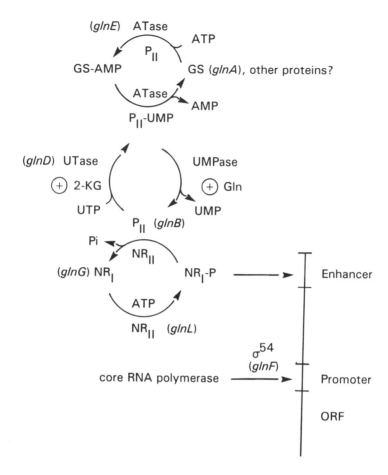

FIGURE 1. The nitrogen regulon of *E. coli*. The component mechanism is taken from Reference 12.

levels of glutamine, thus regenerating the P_{II} cofactor. Accordingly, high levels of intracellular ammonium, which elevate the glutamine pool, led to deuridylylation of P_{II}, dephosphorylation of NR_I, and inactivation of Ntr promoters. Ammonia limitation, which elevates the intracellular level of 2-KG and reduces that of glutamine, has the opposite effect. As expected, the mutant bearing a Tn*10* insertion into *glnD* will not express ^{14}C-methylammonium transport activity. Interestingly, a *gltB* (glutamate synthase) mutant also failed to derepress transport. Since glutamate synthase consumes substantial amounts of intracellular glutamine, a synthase deficiency leads to increased glutamine pools, which in turn shut off expression of Ntr-regulated genes.[13] Overall, the evidence for Ntr regulation of a gene or genes involved in the high-affinity ammonium/methylammonium transport system in *E. coli* is compelling, as was previously established for *glnA* and other genes involved in nitrogen utilization.

TABLE 1
**^{14}C-Methylammonium Transport by Nitrogen Regulatory
Mutants**

Strain	Genotype	Nitrogen Source[a]	^{14}C-methylammonium uptake[b] (nmol/mg · 5s)
RB9060	Parent	Glutamate	4.8
RB9010	Parent	Glutamine	3.0
RB9064	glnF208::Tn10	Glutamine	0.20
RB9040	glnD99::TN10	Glutamine	0.20
RB9060	ΔglnB2036	Glutamate	3.0
RB9060	ΔglnB2036	Ammonium	5.0
RB9017	glnE2313::Tn5	Glutamine	3.8
ET8050	gltB31::Tn5	Glutamate	0.35

[a] Cells were grown on the sole nitrogen source shown as described.[10] Where
    indicated, 20 mM ammonium was supplied as the acetate salt.
[b] Washed cells were assayed for transport according to Jayakumar et al.[10]

V. FEEDBACK INHIBITION OF NH$_4^+$ TRANSPORT BY GLUTAMINE

Existing high-affinity NH$_4^+$ transporters are also subject to control by
the glutamine pool.[2,11] We found that over 95% of the ^{14}C-methylammonium
transport activity in glutamate-grown, washed *E. coli* was blocked by incubation with 100 μM L-glutamine in the presence of chloramphenicol. The
rapid onset and rapid reversal of this phenomenon, as well as the insensitivity
to chloramphenicol, clearly distinguish feedback inhibition from repression.[11]
The inhibition of transport by external L-glutamine (K$_i$ = 18 μM) was noncompetitive with respect to the ^{14}C-methylammonium substrate. D-Glutamine
had no significant effect. The glutamine analogues γ-L-glutamyl hydroxamate
(K$_i$ = 360 μM) and γ-L-glutamyl hydrazide (K$_i$ = 800 μM) were also
noncompetitive inhibitors, suggesting that glutamine metabolism is not required. The role of the intracellular glutamine pool in the regulation of ^{14}C-
methylammonium transport was investigated using mutants carrying defects
in the operon of *glnP*, the gene for the L-glutamine transporter. The *glnP*
mutants had normal rates of ^{14}C-methylammonium uptake, but were refractory
to glutamine inhibition. Using ^{14}C-glutamine, we estimate that a 5 mM level
of internal glutamine would produce half-maximal inhibition of ^{14}C-methyl-
ammonium transport.[11] Since an 8 mM level of glutamine was found in *E.
coli* cells which had been briefly exposed to external NH$_4^+$,[14] the glutamine
concentrations necessary for transport inhibition appear to be physiologically
achievable. Glycylglycine, a noncompetitive inhibitor (K$_i$ = 43 μM), was
equipotent in blocking ^{14}C-methylammonium uptake in wild-type cells and
glnP mutants.[11] Peptide transport systems are likely to provide for glycylglycine accumulation.

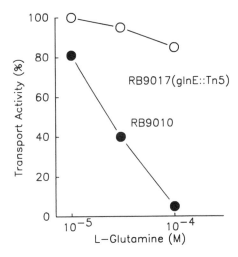

FIGURE 2.. Effect of L-glutamine on ^{14}C-methylammonium transport by *glnE* mutant and parental strains of *E. coli*. Washed cells were incubated with glutamine and then assayed for uptake of ^{14}C-methylammonium as described by Jayakumar et al.[11] Transport activity is expressed as a percentage of control values obtained in the absence of glutamine.

More recently, we have found evidence for a role of *glnE* in the inhibitory effects of glutamine on ^{14}C-methylammonium transport. This work was prompted by observations of Stadtman and co-workers[15] on the regulation of glutamine synthetase (GS). The *glnE* product (ATase) has two enzymatic functions, adenylyltransferase and adenylyl cleavage (Figure 1). The adenylyltransferase introduces one adenylyl group at Tyr-398 of the GS-subunit peptide, markedly reducing the biosynthetic activity of oligomeric GS. The adenylyltransferase activity is stimulated by P_{II}, while deadenylylation is enhanced by P_{II}-UMP. Accordingly (see Figure 1), elevation of intracellular glutamine produces an inhibition of glutamine synthetase. We have shown (Figure 2) that ^{14}C-methylammonium transport by a *glnE* mutant (RB9017) is refractory to glutamine inhibition, while the transport activity of the parental strain (RB9010) is fully sensitive to glutamine. While this suggests a role for protein adenylylation in the inhibition of ammonium/methylammonium transport by glutamine (see Figure 1), additional work is necessary to identify the target protein. Likely candidates would include the glutamine transporter and/or a component of the ammonium transport system.

VI. SELECTION OF STRUCTURAL GENE MUTANTS IN NH_4^+ TRANSPORT

As discussed above, defects in Ntr genes, as well as in other *E. coli* genes affecting glutamine pools, can prevent expression of the high-affinity ammonium/methylammonium transport system. Regulatory mutations are there-

TABLE 2
Nitrogen-Regulated Activities in Amt Mutants

Strain	Genotype	¹⁴C-methylammonium uptake[a] (nmol/mg · 5 s)		Gln synthetase[b] (μmol/min · mg)	
		Arg	NH₄⁺	Arg	NH₄⁺
W3110	F⁻ λ⁻ *thy*	5.41	0.56	1.58	0.25
AJ2652	*amtA*1::Tn*10*	0.22	0.23	1.48	0.23
AJ1577	*amtB*1::Tn*10*	0.20	0.43	1.18	0.28
RB9060	Δ*glnB*2306	6.07	4.31	1.25	1.17
AJ2654	*amtA glnB*	0.48	0.80	1.45	1.32
AJ1579	*amtB glnB*	0.24	0.42	1.28	1.35

[a] Cells were grown on the sole nitrogen source listed and assayed for transport as in Table 1.
[b] Cells were grown on the sole nitrogen source listed and extracts were prepared and assayed
 for γ-glutamyltransferase according to Jayakumar et al.[10]

fore far more common than mutations in the structural genes for ammonium transport. Parental strains of *E. coli* grow well on minimal agar with 25 to 100 μ*M* ammonium acetate as sole nitrogen source (Amt⁺), whereas Ntr⁻ GlnA⁺ mutants which are defective in ¹⁴C-methylammonium transport failed to grow, defining the Amt⁻ phenotype.[7] Since Ntr⁻ mutants are unable to utilize a variety of other nitrogen sources,[13] the growth of Amt⁻ mutants carrying defects in structural genes of the ammonium/methylammonium transport system should not be supported by 100 μ*M* ammonium acetate, but should be nearly normal with 20 m*M* ammonium and with arginine, proline, or glycine individually as the sole nitrogen source. Following Tn*10* transposon mutagenesis, four such Amt⁻ mutants with less than 10% of wild-type ¹⁴C-methylammonium transport activity were isolated independently. The properties of two of these mutants are shown in Table 2. Despite the Amt⁻ phenotype in AJ2652 and AJ1577, glutamine synthetase was expressed and repressed normally. As expected, the *glnB* deletion of RB9060 resulted in constitutive expression of glutamine synthetase and ¹⁴C-methylammonium transport. However, P1 transduction of the mutations of AJ2652 and AJ1577 into RB9060 gave Ntr-constitutive transductants (AJ2654 and AJ1579, respectively) which nevertheless failed to express transport. These and additional data[7] indicate that the ¹⁴C-methylammonium transport defects are not due to lesions affecting Ntr.

VII. MOLECULAR CLONING AND GENOMIC MAPPING OF *amtA*

The *amtA* gene has been cloned by complementation of the AJ2652 mutation using an *E. coli* genomic library. Two clones were obtained, and plasmid DNA from each was able to complement Ntr-controlled

[14]C-methylammonium transport activity in *amtA* mutants.[7] Using a restriction alignment program and hybridization to λ clones from the Kohara genomic library, we were able to map the position of *amtA* to 95.8 min on the *E. coli* chromosome.[16] Deletion analysis of both ends of the plasmid insert showed that the same 1.4-kb fragment was necessary for growth of cells on 100 μ*M* NH_4^+ as sole nitrogen source and for transport of ^{14}C-methylammonium. The sequence[17] from this region contains an open reading frame (ORF) encoding a 246-amino-acid (27-kDa) protein (Figure 3). This ORF contains the site of Tn*10* insertion in AJ2652. The upstream DNA sequence (Figure 4) has a ribosome binding site and a σ^{70}-dependent promoter. No sites for NR_I or σ^{54} binding were evident. The 5′ region also contains a portion of the divergently transcribed *cpdB* gene which encodes a periplasmic phosphodiesterase.[18] The *cpdB* region encoding the first 297 amino acids of the CpdB protein corresponds to nucleotides 891-1 of our sequence. Liu and Beacham[19] have identified a consensus sequence for cAMP receptor protein (CRP) binding which is centrally located between the *amtA* and *cpdB* promoters (Figure 4). Although cAMP and CRP appear to regulate expression of *cpdB*, additional studies are needed to explore a similar role in regulation of *amtA* transcription. The coding region of *E. coli amtA* shows high homology (82% by nucleotide and 88% by amino acid) to a fragmentary ORF from *Salmonella typhimurium*.[19] Since the *S. typhimurium* sequence also shows homologous promoter, CRP binding, and *cpdB* coding regions, this seems to represent a 5′ portion of the *Salmonella amtA* gene.

No other available nucleotide or peptide sequences were found to be homologous to the *E. coli amtA* coding region. Hydropathy and hydrophobic moment analysis of the AmtA peptide do not indicate any candidate transmembrane domains. Furthermore, random insertion of Tn*phoA* produced numerous fusions[7] which map in the *cpdB* gene but none which map in *amtA*. Since prokaryotic signal sequences are also lacking and shock protein is not required for ^{14}C-methylammonium transport, it is very unlikely that the AmtA protein contains domains which are exported from the cytoplasm. The function of the AmtA peptide is unclear at present. One possibility is that the AmtA protein may play a role similar to that of TrkA in K⁺ transport (see Chapter IIC). Although consensus motifs for nucleotide binding are not evident in the peptide sequence, the hypothesis that the AmtA protein is a peripheral component of an NH_4^+ transport system could be considered as a guide for future experiments.

VIII. AmtB: A NEW CLASS OF NH_4^+ TRANSPORT MUTANT

Study of additional Amt⁻ isolates has revealed one new complementation group. As shown above, *E. coli* strain AJ1577 also meets the criteria for a structural gene mutant in NH_4^+ transport (Table 2). Plasmids bearing *amtA*

```
  1: MLDQVCQLAR NAGDAIMQVY DGTKPMDVVS KADNSPVTAA DIAAHTVIMD GLRRTLTPDVP

 61: VLSEEDPPGW EVRQHWQRYW LVDPLDGTKE FIKRNGEFTV NIALIDHGKP ILGVVYAPVM

121: NVMYSAAEGK AWKEECGVRK QIQVRDARPP LVVTSRSHAD AELKEYLQQL GEHQTTSIGS

181: SLKFCLVAEG QAHVYPRFGP TNIWDTAAGH AVAAAAGAHV HDWQGKPLDY TPRESFLNPG

241: FRVSIY
```

FIGURE 3. The peptide sequence of the *E. coli* AmtA protein. The sequence is oriented from amino to carboxyl terminus and was determined from the open reading frame of the *amtA* gene sequence.[17]

```
 841: CACACTGGCG GCAATCAGCG TGGCCAGGAG CGTTGCGCTA AACTTAATCAT CAGGGACAT

 901: CCTTTTATCA TCGGGAATAC GAAAGAAAAG GGAGAATAAA CGTCTTACTTA TAGAACAGT
                                                     CRP              -35
 961: GAAGAATGCC ACAATTTTAC GCTTTGAAAA TGATGACACT ATCACAGTTGG CGCATTCAT
                             -10                °°        °        °°°°°°°°°          SD
1021: TAACGATAGG GTATAAGTAA AACAATAAGT TAACACCGCT CACAGAGACGA GGTGGAGAA
             ORF→
1081: ATGTTTAGATC AAGTATGCCA GCTTGCACGG AATGCAGGCG ATGCCATTATG CAGGTCTAC

1141: GACGGGACGA AACCGATGGA CGTCGTCAGC AAAGCGGACA ATTCTCCGGTA ACGGCAGCG
```

FIGURE 4. Nucleotide sequence of the 5' region of the *E. coli amtA* gene. The sequence is oriented 5' (*Sal*I end) to 3' (*Hind*III end) with the same nucleotide numbers as the complete sequence.[17] The complete sequence is also available under GenBank accession number M55170. The initiating methionine codon (ORF), Shine-Dalgarno (SD) site, and the -10 and -35 promoter elements are overlined. The predicted site for cAMP receptor protein binding is bracketed, and nucleotides which are identical to the consensus sequence[19] are marked by open circles.

were not able to complement the mutation of AJ1577. However, these plasmids were fully capable of complementing the mutations of two independent AmtA⁻ isolates. We suggest that AJ1577 will serve to define AmtB, a new complementation group of ammonium transport mutants. Future studies should focus on molecular cloning of the putative *amtB* gene and, perhaps, additional genes which could encode transmembrane components of the *E. coli* high-affinity ammonium/methylammonium transport system or contain Ntr-regulated elements. Such steps seem necessary before the role of the *amtA* and other gene products in ammonium transport can be determined.

ACKNOWLEDGMENTS

This work was supported by grants DMB-8414973 and DMB-8715825 from the National Science Foundation (U.S.A.).

REFERENCES

1. **Stevenson, R. and Silver, S.,** Methylammonium uptake by *Escherichia coli:* evidence for a bacterial NH₄⁺ transport system, *Biochem. Biophys. Res. Commun.*, 75, 1133, 1977.
2. **Servin-Gonzalez, L. and Bastarrachea, F.,** Nitrogen regulation of synthesis of the high affinity methylammonium transport system of *Escherichia coli, J. Gen. Microbiol.*, 130, 3071, 1984.
3. **Jayakumar, A. and Barnes, E. M., Jr.,** A filtration method for measuring cellular uptake of [¹⁴C]methylamine and other highly permeant solutes, *Anal. Biochem.*, 135, 475, 1983.
4. **Cornell, N. W.,** Rapid fractionation of cell suspensions with the use of bromonated hydrocarbons, *Anal. Biochem.*, 102, 326, 1980.
5. **Jayakumar, A., Epstein, W., and Barnes, E. M., Jr.,** Characterization of ammonium (methylammonium)/potassium antiport in *Escherichia coli, J. Biol. Chem.*, 260, 7528, 1985.
6. **Woolfolk, C. A., Shapiro, B., and Stadtman, E. R.,** Regulation of glutamine synthetase. I. Purification and properties of glutamine synthetase from *Escherichia coli, Arch. Biochem. Biophys.*, 116, 177, 1966.
7. **Jayakumar, A., Hwang, S. J., Fabiny, J. M., Chinault, A. C., and Barnes, E. M., Jr.,** Isolation of an ammonium or methylammonium ion transport mutant of *Escherichia coli* and complementation by the cloned gene, *J. Bacteriol.*, 171, 996, 1989.
8. **Buurman, E. T., de Mattos, M. J. T., and Neijssel, O. M.,** Futile cycling of ammonium ions via the high affinity potassium uptake system (Kdp) of *Escherichia coli, Arch. Microbiol.*, 155, 391, 1991.
9. **Rhoads, D. B., Waters, F. B., and Epstein, W.,** Cation transport in *Escherichia coli.* VIII. Potassium transport mutants, *J. Gen. Physiol.*, 67, 325, 1976.
10. **Jayakumar, A., Schulman, I., MacNeil, D., and Barnes, E. M., Jr.,** Role of the *Escherichia coli glnALG* operon in regulation of ammonium transport, *J. Bacteriol.*, 166, 281, 1986.

11. **Jayakumar, A., Hong, J.-S., and Barnes, E. M., Jr.,** Feedback inhibition of ammonium (methylammonium) ion transport in *Escherichia coli* by glutamine and glutamine analogs, *J. Bacteriol.,* 169, 553, 1987.

12. **Magasanik, B.,** Reversible phosphorylation of an enhancer binding protein regulates the transcription of bacterial nitrogen utilization genes, *Trends Biochem. Sci.,* 13, 475, 1988.

13. **Reitzer, L. and Magasanik, B.,** Ammonia assimilation and the biosynthesis of glutamine, glutamate, aspartate, asparagine, L-alanine, and D-alamine, in *Escherichia coli and Salmonella typhimurium: Cellular and Molecular Biology,* Neidhardt, F. C., Ed., American Society for Microbiology, Washington, D.C., 1987, 1318.

14. **Wohlheuter, R. M., Schutt, H., and Holzer, H.,** Regulation of glutamine synthesis in vivo in *E. coli,* in *The Enzymes of Glutamine Metabolism,* Prusiner, S. and Stadtman, E. R., Academic Press, New York, 1973, 45.

15. **Stadtman, E. R., Mura, E., and Chock, P. B.,** The interconvertible enzyme cascade that regulates glutamine synthetase activity, in *Glutamine: Metabolism, Enzymology and Regulation,* Mora, J. and Palacios, R., Eds., Academic Press, New York, 1980, 41.

16. **Jayakumar, A., Rudd, K. E., Fabiny, J. M., and Barnes, E. M., Jr.,** Localization of the *Escherichia coli amtA* gene to 95.8 minutes, *J. Bacteriol.,* 173, 1572, 1991.

17. **Fabiny, J. M., Jayakumar, A., Chinault, A. C., and Barnes, E. M., Jr.,** Ammonium transport in *Escherichia coli:* localization and nucleotide sequence of the *amtA* gene, *J. Gen. Microbiol.,* 137, 983, 1991.

18. **Liu, J., Burns, D. M., and Beacham, I. R.,** Isolation and sequence analysis of the gene *(cpdB)* encoding periplasmic 2′,3′-cyclic phosphodiesterase, *J. Bacteriol.,* 165, 1002, 1986.

19. **Liu, J. and Beacham, I. R.,** Transcription and regulation of the *cpdB* gene in *Escherichia coli* K12 and *Salmonella typhimurium* LT2: evidence for modulation of constitutive promoters by cyclic AMP-CRP complex, *Mol. Gen. Genet.,* 222, 161, 1990.

Chapter IIIC

FUTILE CYCLING OF K$^+$ AND NH$_4^+$ IONS IN *ESCHERICHIA COLI*

Ed T. Buurman, M. Joost Teixeira de Mattos, and Oense M. Neijssel

TABLE OF CONTENTS

0-8493-6982-7/93/$0.00 + $.50

I. INTRODUCTION TO THE CHEMOSTAT CULTURE TECHNIQUE

In most of the experiments described in this contribution, chemostat cultures have been used. For a better understanding of the implications of the data obtained with such cultures a brief explanation of the principles of this culture technique will be given; more extensive reviews have been published elsewhere.[1,2]

Inoculation of cells in batch culture is usually followed by a period of adaptation (lag phase), after which the microbes grow with a constant doubling time, resulting in an exponential increase in biomass concentration until the stationary phase is attained. A variation of growth rate in such cultures can be achieved by changing the medium, e.g., using a "rich" or a "poor" carbon and energy source like glucose or acetate, respectively. However, even when a medium of constant composition is used the growth rate varies, and in a batch culture this can be observed, for a very short period of time, during the transition from the log to the stationary phase. This slowing down of the growth rate is caused by changes in the culture medium; most common is a rapid decrease of the concentration of one of the essential nutrients.

Monod[3] observed that in a medium in which all other nutrients were present in excess of the organisms' growth requirements, the specific growth rate of the organisms ("μ") was a function of the concentration of the carbon and energy source ("s") and could be described by the following formula:

$$\mu = \mu_{max} \cdot s/(K_s + s) \tag{1}$$

The specific growth rate is called "specific" because it is normalized with respect to biomass concentration and is defined as $(1/x) \cdot dx/dt$, x being the biomass concentration; μ_{max} and K_s are constants: μ_{max} is the maximal specific growth rate, and K_s is the nutrient concentration at which $\mu = 1/2\mu_{max}$ (in analogy with the Michaelis-Menten equation). Thus, if one were able to maintain a constant s at a value near or below K_s, one would be able to grow microbes at a constant but submaximal specific growth rate.

One of the culture techniques that allows the experimenter to grow microbes at such submaximal specific growth rates is the chemostat. The apparatus consists of a culture vessel in which culture pH value and temperature can be kept constant. To this vessel medium is added at a constant rate, and it is assumed that the fresh medium is instantaneously and perfectly mixed with the culture fluid. Since the vessel is equipped with a weir the volume of the fluid in the culture vessel remains constant and the (constant) influx of fresh medium equals the (constant) efflux of culture fluid. The flow rate divided by the culture volume equals the dilution rate ("D"), which usually is expressed in h^{-1}.

When microorganisms are inoculated into the culture vessel, they will start to grow after some time at a maximal rate (s $\gg$ K_s) like in batch culture, and biomass concentration will increase when $\mu_{max} > D$. During growth the microorganisms will consume nutrients and gradually s will become lower, causing a slowing down of the growth rate. Eventually a steady state is reached in which addition of fresh medium is essential for the continuation of growth, the increase in biomass concentration in the culture vessel due to growth is counterbalanced by washout through the weir, and the following relationship holds: $D = \mu$. It must be emphasized that, since the organisms in the culture vessel still grow exponentially, as soon as $\mu \neq D$ the biomass concentration in the culture vessel will change according to the equation $x_t = x_0 \cdot e^{(\mu - D)t}$. In this way the experimenter can grow a microbe at any particular μ by changing the medium flow rate, and as long as the D is set at a value below μ_{max} a steady state is possible.

A microorganism has a K_s value for every nutrient, but the medium of a chemostat culture generally should be composed in such a manner that when the culture is in a steady state the concentration of only *one* nutrient determines the growth rate; this nutrient is called the growth-limiting nutrient. The concentrations of all other nutrients in the culture fluids should be far above their respective K_s values. The concentration of the growth-limiting nutrient in the culture vessel is usually very low, and it is therefore difficult to measure K_s values accurately because rapid sampling is required.[4,5]

One of the most interesting features of a chemostat culture in steady state is that all culture parameters (concentrations of biomass, nutrients, and products) remain essentially constant. Since, at least in principle, this steady state can be maintained for an indefinite period of time, many measurements can be made of these culture parameters, and this allows very precise quantitative analyses of the process under study. Culture parameters are the specific rates of nutrient consumption (or product formation) and the growth yield. As with μ, the specific rate of nutrient consumption ("q", in millimoles per gram dry weight per hour) is called "specific" because it is normalized with respect to culture dry weight. This is an essential point because it allows a comparison between different culture conditions with different biomass concentrations. It can be determined easily: one measures the concentration of the nutrient in the medium (or gas) that is led into the culture vessel and in the culture fluids (or effluent gas). Clearly, the difference between those two concentrations is the amount consumed (or produced) by the microbial population in the vessel. When the culture dry weight and the dilution rate are known, one can calculate q. The growth yield ("Y" in grams of biomass per mole of nutrient consumed) indicates how much biomass is formed per mole of nutrient. The two parameters are interdependent according to the equation

$$\mu = Y \cdot q \tag{2}$$

Finally, in order to interpret data on specific oxygen consumption (q_{O_2}) and specific energy production (q_{ATP}) rates of growing microbial cultures, a closer examination of their determination is required. In cultures carrying out a fermentation, degradation of an energy source such as glucose occurs via glycolysis, which makes a calculation of the stoichiometries of ATP formed per product rather simple (e.g., one ATP per pyruvate, two ATPs per acetate, etc.). Thus, from the amounts of fermentation products found in the extracellular fluids one can easily derive the specific rate of ATP formation. In aerobic cultures (or cultures carrying out anaerobic respiration) the same method could be used, but there are some complications. The reason is that often most of the biologically available energy is generated via the respiratory chain in the form of a proton motive force and not of ATP. A crucial assumption is the efficiency of the respiratory chain (in particular, the number of proton-translocating loops — in classical terms, the P/O ratio). Therefore, a conversion of q_{O_2} into a q_{ATP} is critically dependent on the value of the P/O ratio that has been chosen. The fact that the values of the P/O ratios of bacterial respiratory chains are rather uncertain and that, in addition, these might very well depend on the growth conditions[6] could bedevil estimations of q_{ATP} under these conditions. Therefore, q_{O_2} data are often directly compared and interpreted like q_{ATP} data; it should be kept in mind, however, that a difference in q_{O_2} could stem from altered P/O ratios.

II. FUTILE CYCLING OF POTASSIUM IONS

In nutrient-limited chemostat cultures there is continuous competition among individual cells for the acquisition of the growth-limiting substrate, and only those cells which are maximally adapted (i.e., grow faster at a given s than the others, or maintain μ at a lower s) remain. Equation 2 predicts that in nutrient-limited environments there will be a selective advantage for those organisms that are able to increase their yield on the growth-limiting nutrient (Y) and/or increase q (the specific rate of uptake). A classic example of the first strategy is the adaptation of *Bacillus subtilis* to phosphate-limited growth conditions: the organisms respond to such environments by substituting teichoic acid (a phosphate-containing polymer in the cell wall) with teichuronic acid, an acidic polymer devoid of phosphate.[7]

In this contribution we will concentrate on the other strategy: an increased rate of nutrient uptake. In many cases the quantitatively most important rate-limiting step for the scavenging capacity seems to reside in transport of the compound across the cytoplasmic membrane or "trapping" inside the cell of substrates toward which the cytoplasmic membrane is very permeable (e.g., ammonia, glycerol). If one assumes (solely for reasons of simplicity) that the specific consumption rate follows Michaelis-Menten kinetics:

$$q = v = V_{max} \cdot s/(s + K_m) \tag{3}$$

it is clear that bacteria can increase q at a fixed s (or maintain q at a lower s) by increasing the conversion capacity or the affinity of either the transport system or the enzymatic "trapping" reaction for the substrate (i.e., increasing V_{max} or lowering K_m, respectively). Physiologically this means that a decrease in substrate concentration will result in increased synthesis of transport systems already present and/or synthesis of transport systems with an increased affinity. Examples of both strategies have been observed: in *Klebsiella pneumoniae* (formly *Aerobacter aerogenes* or *Klebsiella aerogenes*), for instance, the V_{max} of the uptake system for glucose (the phosphoenolpyruvate-dependent phosphotransferase system) increased with decreasing values of D in glucose-limited chemostat cultures (i.e., with decreasing extracellular glucose concentration).[8] Also, rather more appropriately, potassium-limited growth conditions cause *Escherichia coli* to synthesize the high-affinity potassium transport system (Kdp), which has a much lower K_m toward K^+ than the constitutive, low-affinity potassium uptake system (Trk)[9] (see Chapters IIA to IIC).

It is generally assumed that organisms growing under potassium-limited conditions must maintain a large concentration gradient of K^+ ions across the cytoplasmic membrane. The concentration of K^+ ions in the extracellular fluids of growing cultures has been found to be as low as 50 nM to 0.1 μM,[4,10] whereas the intracellular concentrations can be calculated to be in the range of 50 to 500 mM[11] (see Chapter IIA). One should be cautious, however, with a direct calculation of the magnitude of the K^+ gradient from these data. Both specific binding of K^+ ions to K^+-requiring enzymes and aspecific binding to polyanions will result in overestimation of the gradient. This is illustrated by the fact that only 70% of K^+ content of the cells can be released by cold osmotic shock with distilled water.[12] One should also take into account that it may not be realistic to assume that the activity of the K^+ ions in the cytoplasm is equivalent to the concentration of K^+ ions in the cytoplasm. Nevertheless, the fact that cells apparently can restore turgor pressure by means of accumulation of K^+ ions[11] implies that this ion will be present at a relatively high concentration and activity.

From chemostat culture studies it has become clear that, with all microorganisms tested thus far, potassium-limited conditions impose a heavy energy burden on the cell (Table 1). The reasons for this high energy demand are not immediately obvious. Assuming that the uptake of K^+ ions in those organisms is regulated in a way similar to that in *E. coli*, one could argue that uptake via a high-affinity transport system is energetically more demanding than K^+ transport via a low-affinity transport system and that this difference is the cause for the increased rate of energy consumption. The problem is that calculations, carried out with Equation 2, show that the specific rate of K^+ transport in organisms such as *K. pneumoniae* will range from about 0.02 mmol per gram dry weight per hour (at $\mu = 0.1$ h^{-1}, when the K^+ content of the cell is 0.8% [w/w][19]) to approximately 0.34 mmol per gram dry weight per hour (at $\mu = 0.8$ h^{-1} and a K^+ content of 1.68% [w/w][19]). Thus, even if one assumes that transport via a Kdp (or Kdp-like) system would cost twice as many ATP equivalents as transport via a Trk (or

TABLE 1
Specific Oxygen Consumption Rates and Specific ATP Production Rates of Comparable Glucose- and Potassium-Limited Chemostat Cultures of Various Organisms

Organism	Limitation	q_{O_2}	q_{ATP}	Ref.
Anaerobic Cultures				
Klebsiella pneumoniae	Glucose		32	13
	Potassium		51	
Clostridium butyricum	Glucose		18	14
	Potassium		35	
Aerobic Cultures				
Klebsiella pneumoniae	Glucose	10		15
	Potassium	28		
Bacillus stearothermophilus	Glucose	5.3		16
	Potassium	11.2		
Escherichia coli	Glucose	4.3		17
	Potassium	20		
Pseudomonas putida	Glucose	10		18
	Potassium	36		

Note: q_{O_2} and q_{ATP} are expressed in millimoles per gram dry weight per hour. Growth conditions of the separate strains differed from one another.

Trk-like) system, the observed differences in q_{ATP} or q_{O_2} shown in Table 1 cannot be explained.

Hueting et al.[15] found that potassium-limited cultures of *K. pneumoniae* expressed increased q_{O_2} values as compared to carbon-limited conditions (28 mmol per gram dry weight per hour vs. 10 mmol per gram dry weight per hour [Table 1]), but they also observed an immediate potassium efflux (approximately 0.3 mmol per gram dry weight per hour) from potassium-limited cells as soon as the energy source was exhausted. Mulder et al.[20] calculated that q_{ATP} values of aerobic, potassium-limited cultures of *E. coli* decreased from 120 to 80 mmol ATP per gram dry weight per hour (at D = 0.3 h^{-1}) when the Kdp operon was deleted. Furthermore, the presence of the Kdp system resulted in increased potassium exchange rates.[4]

A major step forward in solving this problem was made by Van Dam and co-workers.[20] Their explanation for these observations was futile cycling of K$^+$ ions. The high transmembrane gradient of K$^+$ ions supposedly caused leakage of these ions from the cytoplasm into the exterior. Of course, from a physiological point of view, one would expect that cells growing in steady state would prevent leakage of K$^+$ ions since leakage would lead to an increase of the extracellular concentration of K$^+$ ions and, therefore, decrease the scavenging efficiency of the organism, but it has been suggested that cycling of K$^+$ ions is essential for homeostasis of cytoplasmic pH.[21] In order to compensate for this loss, K$^+$ ions are taken up again via an energy-driven uptake system. As a consequence, energy must be consumed constantly in order to keep the intracellular K$^+$ concentration high. These additional energy demands will depend on the cycle rate on the one hand and the difference in energy produced by K$^+$ efflux and the energy consumed by subsequent K$^+$ uptake on the other hand. It can be safely assumed that under potassium-limited conditions uptake of potassium ions will occur mainly via Kdp, but the nature of the potassium leak(s) has been subject of a number of studies.

Potassium efflux by nonspecific leakage through the cytoplasmic membrane probably results in the energetically most expensive cycle since leakage will not yield biologically useful energy. However, K$^+$ ions are lost from *E. coli* cells lacking potassium uptake systems at a low rate of 10 μmol per gram dry weight per hour.[22,23] On the other hand, efflux and reuptake of K$^+$ ions via the same transport system would hardly cost energy when energy invested during uptake would be regained during efflux.[24] These exchange rates seem to be rather low in potassium-limited *E. coli*: via Kdp about 0.1 mmol per gram dry weight per hour (V$_{max}$ = 2.4 mmol per gram dry weight per hour, [K$^+$]$_{out}$ = 0.1 μM, K$_m$ = 2 μM) and via Trk 0.8 mmol gram dry weight per hour ([K$^+$]$_{out}$ = 60 μM).[24] To the best of our knowledge it is not known yet whether efflux systems (see Chapter IIE) are energy linked and whether they are active under steady-state potassium-limited conditions.

Mulder et al.[20] suggested that leakage of K$^+$ ions could occur via reversed action of potassium uptake systems that are not capable of maintaining the large gradient of K$^+$ ions provoked by the Kdp system. They have proposed that leakage of K$^+$ ions in wild-type *E. coli* growing in potassium-limited chemostat culture occurs mainly via the Trk system at a rate of 9 mmol K$^+$ per gram dry weight per hour (V$_{max}$ = 180 mmol per gram dry weight per hour, [K$^+$]$_{out}$ = 0.1 μM, K$_m$ of Kdp for K$^+$ = 2 μM).[4] However, when the high energy demands observed under potassium-limited conditions would lower the ATP pool and thereby inactivate Trk,[25] extensive leakage of K$^+$ ions via this system would be prevented.

In conclusion, the results described above strongly suggest that cycling of K$^+$ ions is a real phenomenon, but that the amount of energy spilled per K$^+$ ion cycled must be rather high (ranging from 4 to 5 mmol of ATP[20] to 50 mmol O$_2$[15] consumed per millimole K$^+$ cycled). Of course, this could be caused by incorrect assumptions in the calculation of q$_{ATP}$ values or underestimation of potassium cycling rates, but it also could indicate that potassium-limited conditions evoke energy-dissipating reactions in a different way.

III. FUNCTIONAL REPLACEMENT OF POTASSIUM IONS BY AMMONIUM IONS

By growing the organisms under a nutrient limitation, one can assess what the minimal requirement of the organisms for this nutrient is. The

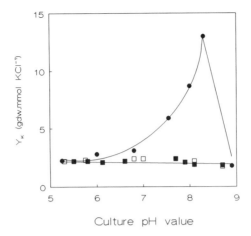

FIGURE 1. Yield on potassium (Y_K, in grams of dry weight per mole K^+) of potassium-limited chemostat cultures of *Klebsiella pneumoniae* (D = 0.28 h^{-1}, temp = 35°C) grown at various culture pH values with either ammonium chloride (●), sodium glutamate (□), or sodium nitrate (■) as the sole nitrogen source. (Data from Reference 30.)

potassium content of *Candida utilis* cells grown in potassium-limited medium was three- to tenfold lower then those grown under potassium-excess conditions,[26] which implies that the yield was highest under potassium-limited conditions. Similarly, since *B. subtilis* requires more potassium than *K. pneumoniae*,[19] the amount of biomass formed per mole potassium (Y_K) of the former organism is smaller than that of the latter when both are grown under potassium-limited conditions at the same growth rate.

The cellular potassium content of *K. pneumoniae* grown under various limitations hardly differed from one another when the standard medium was used, but increased both with cellular ribosomal RNA content and medium osmolarity.[27] This is in agreement with *in vitro* translation studies which proved that K^+ ions are essential for ribosomal activity[28] and batch culture experiments which showed that the potassium content of *E. coli* depends on medium osmolarity.[11] From this it was concluded that in Gram-negative organisms K^+ ions were required in high concentration in order both to maintain a sufficient turgor pressure across the cytoplasmic membrane[11] and to activate the ribosomal apparatus (see Chapter IIA).

In view of the above-mentioned physiological roles of K^+ ions, it was rather surprising to find that the steady-state dry weights (or Y_Ks) of potassium-limited chemostat cultures of *K. pneumoniae, B. stearothermophilus, E. coli, Pseudomonas* species, and even *C. utilis* increased at alkaline culture pH values, in spite of the fact that both dilution rate and osmolarity of the culture were kept constant[18,29-31] (Figure 1). Different hypotheses could be put forward to explain this phenomenon: (1) accumulation of storage polymers, such as glycogen; (2) less efficient potassium uptake at acidic culture pH values; and (3) substitution of K^+ ions by other cation(s) at alkaline pH values.

This decreased potassium requirement was investigated in more detail using a Gram-negative (*K. pneumoniae*),[30] a Gram-positive (*B. stearothermophilus*),[30] and a eukaryotic (*C. utilis*) organism,[31] and in all cases it was found that the presence of ammonium ions in the medium was responsible for this increase; a change in culture pH value did not influence the dry weight whenever sodium nitrate or sodium glutamate was used as the sole nitrogen source (Figure 1). Furthermore, neither a change in nitrogen source nor variation of the culture pH value changed the macromolecular cell composition;[30,31] therefore, it was concluded that ammonium ions could functionally replace a large part of the K^+ ions required. In this connection, it should be realized that the radius of the NH_4^+ ion (1.43 Å) is close to that of K^+ ions (1.33 Å), and it has already been shown that the larger Rb^+ ion (1.48 Å) can functionally replace K^+.[32,33]

The next question was why the observed effects were so dependent on the culture pH value. Kleiner[34] has calculated that the permeability of the cytoplasmic membrane of *K. pneumoniae* toward ammonia (NH_3) is very high (see Chapter IIIA); it is comparable to that toward H_2O. So, when the cells are growing in steady state the difference between the intracellular and extracellular NH_3 concentration will be vanishingly small. As a consequence, the intracellular NH_4^+ concentration will be determined by both the NH_3 concentration and the intracellular pH value, while the NH_3 concentration is a function of the culture pH value. Thus, at culture pH values higher than the cytoplasmic pH value, ammonium ions will be accumulated intracellularly to levels that are higher than those present in the culture medium. In this connection it must be pointed out that all media used in the cited chemostat studies contain 100 m*M* NH_4Cl. One can calculate that with an intracellular pH value of 7.5 and at a culture pH value of 8, the intracellular ammonium ion concentration will be about 300 m*M*. This value is comparable to the intracellular K^+ concentration in Gram-negative organisms such as *E. coli*.

In the mechanism described above, transport of ammonium ions does not play a role, although the presence of an ammonium ion transport system in *K. pneumoniae* and *E. coli* has been described. It is highly unlikely that it would have been present in the experiments described above because it has been observed that this transport system is not synthesized in the presence of high concentrations of ammonium ions in the medium. Nevertheless, it is important to emphasize that when the rate at which active ammonium transport occurs is small in comparison with the rate of diffusion of ammonia and the ammonia/ammonium equilibration rate, the presence of an ammonium ion carrier will not contribute to a net ammonium ion accumulation. Instead, energy-linked transport of ammonium ions will only result in influx of ammonium ions and efflux of ammonia molecules. The net effect will be a futile cycle and waste of energy due to both the energy requirement of ammonium ion uptake and extrusion of protons in order to maintain pH homeostasis in the cell.[20]

IV. FUTILE CYCLING OF AMMONIUM IONS

The high similarity between K^+ and NH_4^+ ions, as discussed in the previous section, led us to investigate whether potassium transport systems also might be able to transport ammonium ions. If this would occur *in vivo*, one would predict that this activity could generate the type of futile cycle described in the last paragraph above. This theory was investigated by growing *E. coli* FRAG-1 (a wild-type strain with regard to potassium transport systems) in chemostat culture both in the presence and in the absence of a high concentration of ammonium ions. The latter condition could be met by growing this strain with D,L-alanine as the sole nitrogen source, which led to extracellular ammonium ion concentrations of less than 1 mM.

Both glucose and (the carbon skeleton of) alanine were mainly converted into biomass, acetate, and carbon dioxide (Table 2). Since metabolism of alanine via the formation of pyruvate will yield less ATP than metabolism of glucose (due to the lack of substrate level phosphorylation and energy expenditure in transport of alanine across the cell membrane), one would predict the q_{O_2} value to be higher when alanine is used as a carbon and nitrogen source as compared to growth with ammonium chloride. However, no gross differences were found, with the exception of potassium-limited conditions. Here, the q_{O_2} value was found to be significantly higher when ammonium chloride served as the nitrogen source. Similar results were obtained with potassium-limited cultures of *B. stearothermophilus* or *K. pneumoniae*.[30]

In order to establish whether the high specific respiration rate was a result of the derepression of Kdp under potassium-limited growth conditions,[36] the above-mentioned experiments were repeated with *E. coli* FRAG-5, a derivative of *E. coli* FRAG-1 lacking the Kdp system.[36] Under potassium-excess conditions, both strains exhibited the same energy requirement when cultivated with D,L-alanine as the nitrogen source (Table 2). However, in contrast to *E. coli* FRAG-1, *E. coli* FRAG-5 showed no difference in q_{O_2} value between potassium-limited cultures grown with ammonium chloride and those grown with alanine as the sole nitrogen source. In theory, the ammonium ion transport system also could be involved in the experiments described above; however, as expected, in cells of *E. coli* FRAG-1 harvested from potassium-limited chemostat cultures grown with ammonium chloride no uptake of [^{14}C]-methylammonium could be detected.[37]

In analogy with ammonia, the unprotonated form of the methylammonium ion, methylamine, has also been reported to be able to traverse the cytoplasmic membrane rapidly.[34,38] It was therefore investigated whether methylammonium ions induced a futile cycle in cells that contained the Kdp system. This was found not to be the case: potassium-limited cultures grown with D,L-alanine as the nitrogen source for cell synthesis did not show significantly increased q_{O_2} values in the presence of methylammonium ions in the medium.[37]

The data show that the simultaneous presence of large concentrations of ammonium ions and the Kdp system is required for an increased energy

TABLE 2
Influence of the Nitrogen Source on the Specific Rates of Glucose, Alanine, and Oxygen Consumption and of Product Formation in *Escherichia coli* FRAG-1 and *E. coli* FRAG-5

Strain	Nitrogen source	Glucose	Alanine	Product Biomass	Acetate	Pyruvate	CO$_2$	O$_2$	n
FRAG-1	Alanine	5.0 ± 0.7	3.0 ± 0.2	3.0 ± 0.3	4.8 ± 0.4	0.8 ± 0.7	12.7 ± 1.5	13.3 ± 1.3	10
FRAG-5	Alanine	4.2 ± 0.4	3.1 ± 0.2	3.1 ± 0.2	4.4 ± 0.4	0	11.5 ± 1.5	11.7 ± 1.3	8
FRAG-1	NH$_4$Cl	8.0 ± 0.7	—	3.1 ± 0.1	5.1 ± 1.0	0.9 ± 1.1	16.2 ± 1.0	17.1 ± 1.4	18
FRAG-5	NH$_4$Cl	5.6 ± 0.7	—	3.1 ± 0.2	3.6 ± 0.6	0	11.6 ± 0.8	11.5 ± 0.8	11

Note: *E. coli* were grown in potassium-limited chemostat culture (D = 0.3 h^{-1}, pH = 7.0, temp = 35°C). Glucose and the carbon skeleton of alanine were carbon sources. The medium concentration of ammonium chloride was 100 m*M* and of D,L-alanine 50 m*M* (+75 m*M* NaCl). The molecular formula for biomass has been taken as C$_4$H$_7$O$_2$N^{35}. Rates have been expressed in millimoles per gram dry weight per hour. Carbon recoveries were between 86 and 95%; *n* = number of determinations. (Data from Buurman, E. T., Teixeira de Mattos, M. J., and Neijssel, O. M., *Arch. Microbiol.*, 155, 391, 1991. With permission.)

requirement of potassium-limited cultures of *E. coli* FRAG-1. Whenever this condition is not met (i.e., when alanine serves as the nitrogen source with *E. coli* FRAG-5, or when methylamine replaces ammonia), potassium-limited conditions show about the same q_{O_2} values as other carbon source excess conditions. The most plausible explanation for this observation is that futile cycling of ammonium ions occurs via the Kdp system. Of course, this does not exclude futile cycling of K$^+$ ions via the high- and low-affinity potassium transport systems as suggested by Mulder et al.[20] However, the data obtained with cultures grown with D,L-alanine as the sole nitrogen source indicate that the energy demands due to potassium cycling are less than those due to ammonium cycling. The same can be said for possible futile cycling of ammonium ions via potassium transport systems other than the Kdp system: under the culture conditions that were tested, the possible contributions of these cycles to the increased energy demand were too low to be detected.

Furthermore, the respiration rate of potassium-limited cultures in the absence of futile ammonium cycle did not differ from those of other carbon-excess cultures of *E. coli* FRAG-1. Of course, one could interpret this result by assuming that each limitation evokes dissipation of energy ("ATP") to the same extent, but in a way which depends on the nature of the limitation. Alternatively, generation of ATP from glucose also could occur in a less efficient way in terms of ATP generated per glucose converted as soon as the carbon and energy source is present in excessive amounts.

It is not clear how specific Kdp is toward monovalent cations. K$^+$ (radius 1.33 Å), Tl$^+$ (radius 1.47 Å),[39] and Rb$^+$ ions (radius 1.48 Å)[10] are transported *in vivo*. However, whereas Kdp-ATPase activity *in vitro* was stimulated by the presence of potassium ions, addition of rubidium, ammonium (radius 1.43 Å), or thallium ions was not successful.[40] To the best of our knowledge the ability of Kdp to transport methylammonium ions has never been tested, but the fact that derepression of the Kdp resulted neither in futile cycling of methylammonium ions *in vivo* nor in [^{14}C]-methylammonium uptake activity *in vitro*[37] suggests that methylammonium ions are not a substrate of Kdp. This is an important observation since it implies that not all ammonium ion transporting activities can be determined by experiments using [^{14}C]-methyl-ammonium ions.

Recently it has become clear that there are more potassium transport systems which in addition seem to be activated by ammonium ions. Golby et al.[41] suggested that the low-affinity potassium transport system of *Rhodobacter capsulatus* was capable of giving rise to futile cycling of ammonium ions, and Epstein and co-workers found that the presence of ammonium ions inhibited transport of K$^+$ ions via the low-affinity potassium transport systems of *E. coli* (but not via Kdp).[42] Finally, K$^+$-stimulated ATPase activity of the high-affinity potassium transport system of *B. acidocaldarius* was also stimulated by the presence of ammonium ions.[43]

V. CONCLUDING REMARKS

As pointed out above, it is our opinion that transport of ammonium ions into cells with a membrane highly permeable toward ammonia molecules[34] will not result in a physiologically relevant increase in the cytoplasmic content of this ion. From this it follows that we feel that it is rather unlikely that during evolution organisms have acquired both a cell membrane with this property and one or more transport systems for scavenging the cells' surroundings for remaining amounts of ammonium ions (which possibly originated from ammonia molecules *leaked out of* the cells, after being formed intracellularly by deamination of nitrogenous compounds, nitrogen fixation, or denitrification, leading to a process called "cyclic retention"[34]) in order to build up a high intracellular concentration of this ion. One wonders, therefore, why cells synthesize a (methyl)ammonium transport system under some growth conditions and what its physiological function is. This problem is further illustrated by the fact that *E. coli* and *K. pneumoniae* adapt to ammonia-limited environments by switching from an assimilatory pathway with a low affinity for ammonium ions or ammonia molecules (i.e., glutamate dehydrogenase) to one with a high affinity (i.e., glutamine synthetase plus glutamate synthase),[45] all of which are *cytosolic* enzymes. In other words, this adaptation suggests that even if an ammonium ions transport system were active, it could not accumulate ammonium ions intracellularly to levels high enough for the glutamate dehydrogenase to be functional.

It should be emphasized that the ability of a transport system to transport a certain substrate does not *a priori* define its physiological function (e.g., magnesium transport systems that are able to transport cobaltous ions[46]). Thus, it may be that the (methyl)ammonium ion transport activity is simply the activity of a system able to transport this ion in the absence of its (as yet unknown) physiological substrate. This hypothesis is strengthened by the fact that, in spite of a decrease of the ammonium ion concentration by the use of D,L-alanine as the nitrogen source both in chemostat culture (30 μM NH$_4^+$) and in batch culture, the (methyl)ammonium transporting activity remained low, whereas in the view of the "cyclic retention" hypothesis an ammonium ion uptake system still would be useful to take up ammonia released from the cells after deamination of alanine.

A very speculative hypothesis is that transport of ammonium ions into the cell is followed by subsequent metabolite transfer via enzyme-enzyme complexes, as has been proposed for glycolytic and the citric acid cycle enzymes.[47,48] Since the transported ammonium ions then would not be part of the free ammonium ion pool (i.e., would not interfere with the ammonium/ammonia equilibrium), activity of the transport system would effectively increase the supply of substrate for the ammonium assimilating machinery.

We hope that this contribution has shown that physiological studies with chemostat cultures have provided important information on the energetic consequences of the presence of multiple uptake systems for K$^+$ ions.

Moreover, they have uncovered an interesting relationship between the physiological functions of K^+ and ammonium ions. It is also clear that some quite fascinating problems are still unsolved. In this regard, a combined approach by molecular geneticists and microbial physiologists could be particularly rewarding.

REFERENCES

1. **Tempest, D. W.,** The continuous cultivation of micro-organisms. I. Theory of the chemostat, in *Methods in Microbiology,* Vol. 2, Norris, J. R. and Ribbons, D. W., Eds., Academic Press, London, 1970, 259.
2. **Evans, C. G. T., Herbert, D., and Tempest, D. W.,** The continuous cultivation of micro-organisms. II. Construction of a chemostat, in *Methods in Microbiology,* Vol. 2, Norris, J. R. and Ribbons, D. W., Eds., Academic Press, London, 1970, 277.
3. **Monod, J.,** *Recherches sur la Croissance des Cultures Bactériennes,* Ph.D. thesis, 2nd ed., Hermann, Paris, 1958, 58.
4. **Mulder, M. M.,** Energetic Aspects of Bacterial Growth: A Mosaic Non-Equilibrium Thermodynamic Approach, Ph.D. thesis, University of Amsterdam, Amsterdam, The Netherlands, 1988.
5. **Rutgers, M., Balk, P. A., and Van Dam, K.,** Effect of concentration of substrates and products on the growth of *Klebsiella pneumoniae* in chemostat cultures, *Biochim. Biophys. Acta,* 977, 142, 1989.
6. **Jones, C. W.,** Aerobic respiratory systems in bacteria, in *Microbial Energetics,* 27th Symp. Soc. General Microbiology, Haddock, B. A. and Hamilton, W. A., Eds., Cambridge University Press, London, 1977, 23.
7. **Ellwood, D. C. and Tempest, D. W.,** Control of teichoic acid and teichuronic acid biosyntheses in chemostat cultures of *Bacillus subtilis* var. *niger, Biochem. J.,* 111, 1, 1969.
8. **O'Brien, R. W., Neijssel, O. M., and Tempest, D. W.,** Glucose phospho*enol*pyruvate phosphotransferase activity and glucose uptake rate of *Klebsiella aerogenes* in chemostat culture, *J. Gen. Microbiol.,* 116, 305, 1980.
9. **Laimins, L. A., Rhoads, D. B., and Epstein, W.,** Osmotic control of kdp operon expression in *Escherichia coli, Proc. Natl. Acad. Sci. U.S.A.,* 78, 464, 1981.
10. **Walderhaug, M. O., Dosch, D. C., and Epstein, W.,** Potassium transport in bacteria, in *Ion Transport in Procaryotes,* Rosen, B. P. and Silver, S., Eds., Academic Press, New York, 1987, 84.
11. **Epstein, W.,** Osmoregulation by potassium transport in *Escherichia coli, FEMS Microbiol. Rev.,* 39, 73, 1986.
12. **McLaggan, D., Logan, T. M., Lynn, D. G., and Epstein, W.,** Involvement of τ-glutamyl peptides in osmoadaptation of *Escherichia coli, J. Bacteriol.,* 172, 3631, 1990.
13. **Teixeira de Mattos, M. J. and Tempest, D. W.,** Metabolic and energetic aspects of the growth of *Klebsiella aerogenes* NCTC 418 on glucose in anaerobic chemostat culture, *Arch. Microbiol.,* 134, 80, 1983.
14. **Crabbendam, P. M., Neijssel, O. M., and Tempest, D. W.,** Metabolic and energetic aspects of the growth of *Clostridium butyricum* on glucose in chemostat culture, *Arch. Microbiol.,* 142, 375, 1985.
15. **Hueting, S., De Lange, T., and Tempest, D. W.,** Energy requirement for maintenance of the transmembrane potassium gradient in *Klebsiella aerogenes:* a continuous culture study, *Arch. Microbiol.,* 123, 183, 1979.

16. **Pennock, J. and Tempest, D. W.**, Metabolic and energetic aspects of the growth of *Bacillus stearothermophilus* in glucose-limited and glucose-sufficient chemostat culture, *Arch. Microbiol.*, 150, 452, 1988.

17. **Neijssel, O. M.**, personal communication, 1991.

18. **Hardy, G. P. M. A.**, personal communication, 1988.

19. **Tempest, D. W.**, Quantitative relationships between inorganic cations and anionic polymers in growing bacteria, in *Microbial growth*, 19th Symp. Soc. General Microbiology, Meadow, P. M. and Pirt, S. J., Eds., Cambridge University Press, London, 1969, 111.

20. **Mulder, M. M., Teixeira de Mattos, M. J., Postma, P. W., and Van Dam, K.**, Energetic consequences of multiple K^+ uptake systems, *Biochim. Biophys. Acta*, 851, 223, 1986.

21. **Booth, I. R.**, Regulation of cytoplasmic pH in bacteria, *Microbiol. Rev.*, 49, 359, 1985.

22. **Elmore, M. J., Lamb, A. J., Ritchie, G. Y., Douglas, R. M., Munro, A., Gajewska, A., and Booth, I. R.**, Activation of potassium efflux from *Escherichia coli* by glutathione metabolites, *Mol. Microbiol.*, 4, 405, 1990.

23. **Bakker, E. P. and Mangerich, W. E.**, *N*-Ethylmaleimide induces K^+-H^+ antiport activity in *Escherichia coli* K-12, *FEBS Lett.*, 140, 177, 1982.

24. **Rhoads, D. B. and Epstein, W.**, Cation transport in *Escherichia coli*. X. Regulation of K^+ transport, *J. Gen. Physiol.*, 72, 283, 1978.

25. **Stewart, L. M. D., Bakker, E. P., and Booth, I. R.**, Energy coupling to K^+ uptake via the Trk system in *Escherichia coli*, *J. Gen. Microbiol.*, 131, 77, 1985.

26. **Aiking, H. and Tempest, D. W.**, Growth and physiology of *Candida utilis* NCYC 321 in potassium-limited chemostat culture, *Arch. Microbiol.*, 108, 117, 1976.

27. **Tempest, D. W. and Meers, J. L.**, The influence of NaCl concentration of the medium on the potassium content of *Aerobacter aerogenes* and on the inter-relationships between potassium, magnesium and ribonucleic acid in the growing bacteria, *J. Gen. Microbiol.*, 54, 319, 1969.

28. **Pestka, S.**, Peptidyl-puromycin synthesis on polyribosomes from *Escherichia coli*, *Proc. Natl. Acad. Sci. U.S.A.*, 69, 624, 1972.

29. **Hommes, R. W. J.**, personal communication, 1988.

30. **Buurman, E. T., Pennock, J., Tempest, D. W., Teixeira de Mattos, M. J., and Neijssel, O. M.**, Replacement of potassium ions by ammonium ions in different microorganisms grown in potassium-limited chemostat culture, *Arch. Microbiol.*, 152, 58, 1989.

31. **Buurman, E. T.**, The Effect of Cations on Microbial Metabolism and Growth Energetics, Ph.D. thesis, University of Amsterdam, Amsterdam, The Netherlands, 1991.

32. **Lester, G.**, Requirement of potassium by bacteria, *J. Bacteriol.*, 75, 426, 1958.

33. **Aiking, H. and Tempest, D. W.**, Rubidium as a probe for function and transport in the yeast *Candida utilis* NCYC 321, growing in chemostat culture, *Arch. Microbiol.*, 115, 215, 1977.

34. **Kleiner, D.**, Energy expenditure for cyclic retention of NH_3/NH_4^+ during N_2 fixation by *Klebsiella pneumoniae*, *FEBS Lett.*, 187, 237, 1985.

35. **Herbert, D.**, Stoicheiometric aspects of microbial growth, in *Continuous Culture*, Vol. 6, *Applications and New Fields*, Dean, A. C. R., Ellwood, D. C., and Evans, C. G. T., Eds., Ellis Horwood, Chichester, England, 1976, 1.

36. **Rhoads, D. B., Waters, F. B., and Epstein, W.**, Cation transport in *Escherichia coli*. VIII. Potassium transport mutants, *J. Gen. Physiol.*, 67, 325, 1976.

37. **Buurman, E. T., Teixeira de Mattos, M. J., and Neijssel, O. M.**, Futile cycling of ammonium ions via the high affinity potassium uptake system (Kdp) of *Escherichia coli*, *Arch. Microbiol.*, 155, 391, 1991.

38. **Zarlengo, M. and Abrams, A.**, Selective penetration of ammonia and alkylamines into *Streptococcus fecalis* and their effect on glycolysis, *Biochim. Biophys. Acta*, 71, 65, 1963.

39. **Epstein, W., Wieczorek, L., Siebers, A., and Altendorf, K.,** Potassium transport in *Escherichia coli:* genetic and biochemical characterization of the K⁺-transporting ATPase, *Biochem. Soc. Trans.,* 12, 235, 1984.

40. **Siebers, A. and Altendorf, K.,** The K⁺-translocating Kdp-ATPase from *Escherichia coli, Eur. J. Biochem.,* 178, 131, 1988.

41. **Golby, P., Carver, M., and Jackson, J. B.,** Membrane ionic currents in *Rhodobacter capsulatus.* Evidence for electrophoretic transport of K⁺, Rb⁺, and NH₄⁺, *Eur. J. Biochem.,* 187, 589, 1990.

42. **Epstein, W.,** personal communication, 1991.

43. **Bakker, E. P.,** personal communication, 1991.

44. **Kleiner, D.,** Bacterial ammonium transport, *FEMS Microbiol. Rev.,* 32, 87, 1985.

45. **Meers, J., Tempest, D. W., and Brown, C. M.,** "Glutamine(amide):2-oxoglutarate amino transferase oxido-reductase (NADP)", an enzyme involved in the synthesis of glutamate by some bacteria, *J. Gen. Microbiol.,* 64, 187, 1970.

46. **Hmiel, S. P., Snavely, M. D., Miller, C. G., and Maguire, M. E.,** Magnesium transport in *Salmonella typhimurium:* characterization of magnesium influx and cloning of a transport gene, *J. Bacteriol.,* 168, 1444, 1986.

47. **Srivastava, D. K. and Bernhard, S. A.,** Metabolite transfer via enzyme-enzyme complexes, *Science,* 234, 1081, 1986.

48. **Srere, P. A.,** Citric acid cycle redux, *Trends Biochem. Sci.,* 15, 411, 1990.

Index

INDEX

A

AAS, see Atomic absorption spectroscopy
Acetate, 81, 156, 180, 198
Acetate kinase, 79
Acetobacterium woodii, 172
Acetogenic bacteria, 172, 198
Acetogens vs. methanogens, 198
Acetyl-CoA, 79, 142, 172, 198
Acetyl phosphate, 79
Acholeplasma laidlawii, 241, 245
Acidaminococcus fermentans, 84
Acidophiles, 55, 218; see also specific types
ACMA, see 9-Amino-6-chloro-2-
    methoxyacridine
Acridine, 293
Adenosine triphosphatases (ATPases), see
    also specific types
  arsenate-resistance, 226
  classification of, 226
  F-type, 226, 246, 269, 278
  of fungi, 246
  ion-motive, 226–227
  membrane-bound, 269
  potassium-translocating, 214, 270, 339,
    340
  of protozoa, 246
  P-type, see P-type ATPases
  sodium-translocating, see Sodium-
    translocating ATPase
  vanadate-sensitive, 285–286
  V-type, 226, 246, 270, 283
  of yeasts, 245
Adenosine triphosphate (ATP)-binding
    proteins, 113; see also specific types
Aerobacter aerogenes, see *Klebsiella
    pneumoniae*
Aerobic oxidation, 156
Aerobic respiratory chains, 334
Affinity chromatography, 231, 244, 340
AF-Red chromatography, 340
AIB, see α-Aminoisobutyric acid
Alanine, 59, 91, 342, 420, 421
Alanine carriers, 57
Alanine-proline hinge, 91
Alcohols, 181–182; see also specific types
Algae, 26, 350, 381; see also specific types
Alkaline phosphatase, 89

Alkaliphiles, 10, 15, 26, 78; see also
    specific types
  ATP synthesis in, 148, 150
  extreme (extremophiles), 102, 117–119
  genetics of, 108
  in pH regulation, 103
  sodium cycles in, 102–105
Alkalophiles, 55; see also specific types
Amber-suppressor mutations, 312
Amiloride, 42, 106, 164, 194
Amino acids, 8, 55, 355, 387, 390; see also
    specific types
α-Aminobutyrate, 292
9-Amino-6-chloro-2-methoxyacridine
    (ACMA), 143, 149
Aminoglycosides, 218, 219; see also specific
    types
α-Aminoisobutyric acid (AIB), 116, 117,
    132, 342–344
Ammonia, 180, 380; see also Ammonium
  diffusion of, 283, 381–382
  transport of, 382
  uptake of, 382
Ammonia/ammonium cyclic retention, 389
Ammonia carriers, 382
Ammonia gradients, 382
Ammonium, 26, 268; see also Ammonia
  distribution of, 384–385
  in *Escherichia coli,* 215
  excess, 401
  external, 403
  functional replacement of potassium by,
    417–419
  futile cycling of, 259, 420–422
  intracellular, 382, 387, 402
  kinetics of uptake of, 390
  leakage of, 389
  properties of, 384–385
  rubidium and, 399
  transport of, see Ammonium transport
  uptake of, 383, 390
Ammonium chloride, 420
Ammonium gradients, 382–383
Ammonium/potassium antiporter, 215, 293,
    386, 399
Ammonium salts, 34
Ammonium transport, 379–391
  ammonia diffusion and, 381–382

Q

U

V

W

X

Y